CLINICAL MICROBIOLOGY
MADE RIDICULOUSLY SIMPLE

MARK GLADWIN, M.D.
WILLIAM TRATTLER, M.D.
C. SCOTT MAHAN, M.D.

MedMaster, Inc., Miami

Clinical Microbiology Made Ridiculously Simple aims at providing general principles of microbiology and is not intended as a working guide to patient drug administration. Please refer to the manufacturer's package insert for recommended drug dosage, undesirable effects, contraindications and drug interactions.

Cover by Richard March

Made in the United States of America

Published by
MedMaster, Inc.
P.O. Box 640028
Miami, FL 33164

Special Thanks to:

Contributing Chapter Authors

Prions
Hans Henrick Larsen, MD (Chapter 33)

One Step Toward the Post-Antibiotic Era?
Development and Spread of Antimicrobial Resistance
Earnest Alexander, Pharm.D.

The Agents of Bioterrorism
Luciana Borio, MD and Clarence Lam

Rest of the RNA Viruses
Amy Guillet Agrawal, MD and John Beigel, MD

Herpesviridae
John Beigel, MD

The Illustrators
Mark Gladwin, M.D.
Rin Carroll
Siobhan Arnold
Gail Gerlach
Steve Goldberg, M.D.
Jennifer Graeber
Elizabeth Mahan

The Book Reviewers
Stephen Goldberg, M.D.
Phyllis Goldenberg

Other Content Editors
James Hwe
Earnest Alexander, Pharm.D.
Henry Masur, MD
Amy Guillet Agrawal, MD
John Beigel, MD

Mnemonics and Other Contributions
David Flum, M.D.
Mike Stevens, M.D.
Gregory Schrank
Sundeep Grandhe

Family
Christina Gladwin, Ph.D.
Hugh Gladwin, Ph.D.
Marcia Trattler
Henry Trattler, M.D.
Meredith Trattler
Jill Trattler
Ali, Jeremy, and Joshua Trattler
And Flora . . .

PREFACE

A well-developed knowledge of clinical microbiology is critical for the practicing physician in any medical field. Bacteria, viruses, and protozoans have no respect for the distinction between ophthalmology, pediatrics, trauma surgery, or geriatric medicine. As a physician you will be faced daily with the concepts of microbial disease and antimicrobial therapy. Microbiology is one of the few courses where much of the "minutia" is regularly used by the practicing physician.

This book attempts to facilitate the learning of microbiology by presenting the information in a clear and entertaining manner brimming with memory aids.

Our approach has been to:

1) Write in a conversational style for rapid assimilation.

2) Include numerous figures serving as "visual memory tools" and summary charts at the end of each chapter. These can be used for "cram sessions" after the concepts have been studied in the text.

3) Concentrate more on clinical and infectious disease issues that are both interesting and vital to the actual practice of medicine.

4) Create a conceptual, organized approach to the organisms studied so the student relies less on memory and more on logical pathophysiology.

The text has been updated to include current information on rapidly developing topics, such as HIV and AIDS (vaccine efforts and all the new anti-HIV medications), Avian Influenza H5N1, SARS Coronavirus, Ebola virus, Hantavirus, *E. coli* outbreaks, Mad Cow Disease, brand-new antimicrobial antibiotics, and agents of bioterrorism

The mnemonics and cartoons in this book do not intend disrespect for any particular patient population or racial or ethnic group but are solely presented as memory devices to assist in the learning of a complex and important medical subject.

We welcome suggestions for future editions. Despite our best efforts to proofread and edit, mistakes often slip by. We welcome corrections and new mnemonic ideas from students who read this book (please email to *gladwinmt@upmc.edu*. If we publish your rockin' mnemonic (short rhymes, phrases, or other memory techniques that make the information easier to assimilate and memorize) we will recognize your name with the contribution!

MARK GLADWIN, MD
WILLIAM TRATTLER, MD
C. SCOTT MAHAN, MD

CONTENTS

BIOTERRORISM DEFENSE UPDATES:
http://www.medmaster.net/BioterrorismDefense.html

ATLAS OF MICROBIOLOGY
http://www.medmaster.net (free download)

PART 1. BACTERIA

CHAPTER 1. BACTERIAL TAXONOMY

All organisms have a name consisting of two parts: the genus followed by the species (i.e., *Homo sapiens*). Bacteria have been grouped and named primarily on their morphological and biochemical/metabolic differences. However, bacteria are now also being classified according to their immunologic and genetic characteristics. This chapter focuses on the Gram stain, bacterial morphology, and metabolic characteristics, all of which enable the clinician to rapidly determine the organism causing a patient's infection.

GRAM STAIN

Because bacteria are colorless and usually invisible to light microscopy, colorful stains have been developed to visualize them. The most useful is the Gram stain, which separates organisms into 2 groups: gram-positive bugs and gram-negative bugs. This stain also allows the clinician to determine whether the organism is round or rod-shaped.

For any stain you must first smear the substance to be stained (sputum, pus, etc.) onto a slide and then heat it to fix the bacteria on the slide.

There are 4 steps to the Gram stain:

1) Pour on crystal violet stain (a blue dye) and wait 60 seconds.

2) Wash off with water and flood with iodine solution. Wait 60 seconds.

3) Wash off with water and then "decolorize" with 95% alcohol.

4) Finally, counter-stain with safranin (a red dye). Wait 30 seconds and wash off with water.

When the slide is studied microscopically, cells that absorb the crystal violet and hold onto it will appear blue. These are called **gram-positive** organisms. However, if the crystal violet is washed off by the alcohol, these cells will absorb the safranin and appear red. These are called **gram-negative** organisms.

Gram-positive = BLUE
I'm positively BLUE over you!!

Gram-negative = RED
No (negative) RED commies!!

The different stains are the result of differences in the cell walls of gram-positive and gram-negative bacteria.

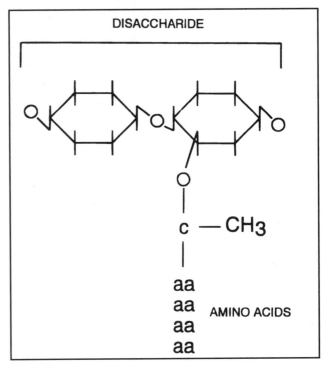

Figure 1-1

Both gram-positive and gram-negative organisms have more than 1 layer protecting their cytoplasm and nucleus from the extracellular environment, unlike animal cells, which have only a single cytoplasmic membrane composed of a phospholipid bilayer. The layer just outside the bacterial cytoplasmic membrane is the **peptidoglycan layer** or cell wall. It is present in both gram-positive and gram-negative organisms.

Fig. 1-1. The **peptidoglycan layer** or cell wall is composed of repeating disaccharides with 4 amino acids in a side chain extending from each disaccharide.

Fig. 1-2. The amino-acid chains of the peptidoglycan covalently bind to other amino acids from neighboring chains. This results in a stable cross-linked structure. The enzyme that catalyzes the formation of this linkage is called **transpeptidase** and is located in the inner cytoplasmic membrane. The antibiotic penicillin binds to and inhibits this enzyme. For this reason the enzyme is also called **penicillin binding protein** (see page 160).

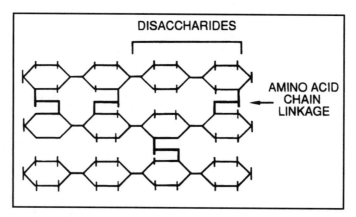

Figure 1-2

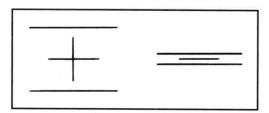

Figure 1-3

Fig. 1-3. The gram-positive cell wall is very thick and has extensive cross-linking of the amino-acid side chains. In contrast, the gram-negative cell wall is very thin with a fairly simple cross-linking pattern.

Fig. 1-4. The gram-positive cell envelope has an outer cell wall composed of complex cross-linked peptidoglycan, teichoic acid, polysaccharides, and other proteins. The inner surface of the cell wall touches the cytoplasmic membrane. The cytoplasmic membrane contains proteins that span the lipid bilayer. The bacterial cytoplasmic membrane (unlike that of animals) has no cholesterol or other sterols.

An important polysaccharide present in the gram-positive cell wall is teichoic acid. It acts as an antigenic determinant, so it is important for serologic identification of many gram-positive species.

Fig. 1-5. The gram-negative cell envelope has 3 layers, not including the periplasmic space. Like gram-positive bacteria, it has 1) a cytoplasmic membrane surrounded by 2) a peptidoglycan layer. 3) In addition, a gram-negative cell has a unique outer cell membrane.

The inner cytoplasmic membrane (as in gram-positive bacteria) contains a phospholipid bilayer with embedded proteins. Gram-negative bacteria have a periplasmic space between the cytoplasmic membrane and an extremely thin peptidoglycan layer. This periplasmic space is filled with a gel that contains proteins and enzymes. The thin peptidoglycan layer does

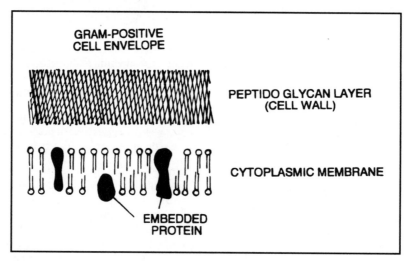

Figure 1-4

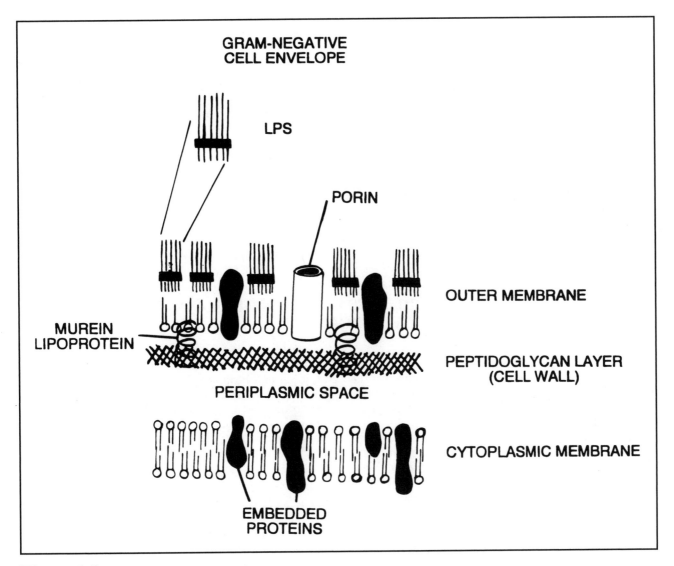

GRAM-NEGATIVE CELL ENVELOPE

LPS

PORIN

OUTER MEMBRANE

MUREIN LIPOPROTEIN

PEPTIDOGLYCAN LAYER (CELL WALL)

PERIPLASMIC SPACE

CYTOPLASMIC MEMBRANE

EMBEDDED PROTEINS

Figure 1-5

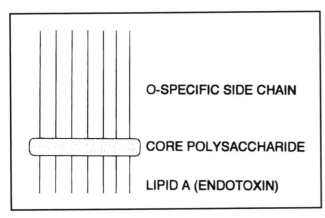

O-SPECIFIC SIDE CHAIN

CORE POLYSACCHARIDE

LIPID A (ENDOTOXIN)

Figure 1-6

not contain teichoic acid, although it does have a small helical lipoprotein called **murein lipoprotein**. This lipoprotein is important because it originates from the peptidoglycan layer and extends outward to bind the unique third outer membrane. This last membrane is similar to other cell membranes in that it is composed of two layers of phospholipid (bilayer) with hydrophobic tails in the center. What makes it unique is that the outermost portion of the bilayer contains lipopolysaccharide (LPS).

Fig. 1-6. Lipopolysaccharide (LPS) is composed of 3 covalently linked components:

1) Outer carbohydrate chains of 1–50 oligosaccharide units that extend into the surrounding media. These differ

GRAM-POSITIVE CELLS	GRAM-NEGATIVE CELLS
2 Layers: 1. Inner cytoplasmic membrane 2. Outer thick peptidoglycan layer (60–100% peptidoglycan)	3 Layers: 1. Inner cytoplasmic membrane 2. Thin peptidoglycan layer (5–10% peptidoglycan) 3. Outer membrane with lipopolysaccharide (LPS)
Low lipid content	High lipid content
NO endotoxin	Endotoxin (LPS) – lipid A
NO periplasmic space	Periplasmic space
NO porin channel	Porin channel
Vulnerable to lysozyme and penicillin attack	Resistant to lysozyme and penicillin attack

Figure 1-7 DIFFERENCES BETWEEN GRAM-POSITIVE AND GRAM-NEGATIVE ORGANISMS

from one organism to another and are antigenic determinants. This part is called the **O-specific side chain** or the **O-antigen**. Think of **O** for Outer to help remember this.

2) The center part is a water soluble **core polysaccharide**.

3) Interior to the core polysaccharide is the third component, **lipid A**, which is a disaccharide with multiple fatty acid tails reaching into the membrane. Lipid A is toxic to humans and is known as the gram-negative **endotoxin**. When bacterial cells are lysed by our efficiently working immune system, fragments of membrane containing lipid A are released into the circulation, causing fever, diarrhea, and possibly fatal endotoxic shock (also called septic shock).

Embedded in the gram-negative outer membrane are **porin** proteins, which allow passage of nutrients. These are also unique to gram-negative organisms.

What does this mean clinically?

The differences between gram-positive and gram-negative organisms result in varied interactions with the environment. The gram-positive thickly meshed peptidoglycan layer does not block diffusion of low molecular weight compounds, so substances that damage the cytoplasmic membrane (such as antibiotics, dyes, and detergents) can pass through. However, the gram-negative outer lipopolysaccharide-containing cell membrane blocks the passage of these substances to the peptidoglycan layer and sensitive inner cytoplasmic membrane. Therefore, antibiotics and chemicals that attempt to attack the peptidoglycan cell wall (such as penicillins and lysozyme) are unable to pass through.

Interestingly, the crystal violet stain used for Gram staining is a large dye complex that is trapped in the thick, cross-linked gram-positive cell wall, resulting in the gram-positive blue stain. The outer lipid-containing cell membrane of the gram-negative organisms is partially dissolved by alcohol, thus washing out the crystal violet and allowing the safranin counterstain to take.

Fig. 1-7. Summary of differences between gram-positive and gram-negative bacteria.

BACTERIAL MORPHOLOGY

Bacteria have 4 major shapes:

1) **Cocci**: spherical.
2) **Bacilli**: rods. Short bacilli are called **coccobacilli**.
3) **Spiral forms**: comma-shaped, S-shaped, or spiral-shaped.
4) **Pleomorphic**: lacking a distinct shape (like jello).

The different shaped creatures organize together into more complex patterns, such as pairs (diplococci), clusters, strips, and single bacteria with flagella.

Fig. 1-8. Bacterial morphology.

SO, WHAT ARE THE NAMES?!!!!

Gram-Positive

Start by remembering that there are 7 classic gram-positive bugs that cause disease in humans, and basically every other organism is gram-negative.

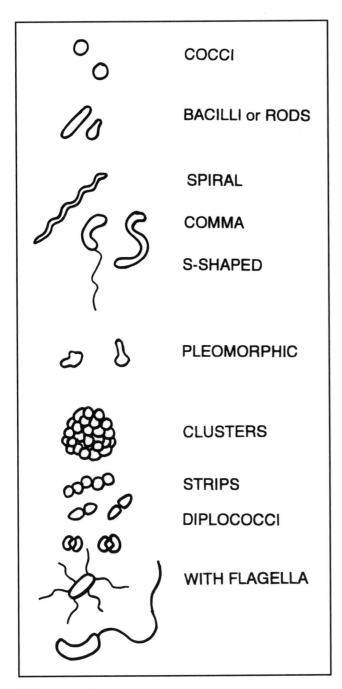

COCCI

BACILLI or RODS

SPIRAL

COMMA

S-SHAPED

PLEOMORPHIC

CLUSTERS

STRIPS

DIPLOCOCCI

WITH FLAGELLA

Figure 1-8

Of the gram-positives, 3 are cocci, and the other 4 are rod-shaped (bacilli).

The 3 gram-positive cocci both have the word **coccus** in their names:

1) *Streptococcus* and 2) *Enterococcus* form strips of cocci.

3) *Staphylococcus* forms clusters of cocci.

Two of the 4 gram-positive rods produce **spores** (spheres that protect a dormant bacterium from the harsh environment). They are:

4) *Bacillus*
5) *Clostridium*

The last 2 gram-positive rods do not form spores:

6) *Corynebacterium*
7) *Listeria*

Gram-Negative

Of the gram-negative organisms, there are only two groups of gram-negative cocci. Both are actually diplococci (look like 2 coffee beans kissing): *Neisseria* and *Moraxella*.

There is also just 1 group of spiral-shaped organisms: the Spirochetes. This group includes the bacterium *Treponema pallidum*, which causes syphilis.

The rest are gram-negative rods or pleomorphic.

Exceptions:

1) **Mycobacteria** are weakly gram-positive but stain better with a special stain called the **acid-fast stain** (See Chapter 15). This special group includes organisms that cause tuberculosis and leprosy.

2) **Spirochetes** have a gram-negative cell wall but are too small to be seen with the light microscope and so must be visualized with a special **darkfield microscope**. Spirochetes are all very slender and tightly coiled. From the inside out, they have a cytoplasm surrounded by an inner cytoplasmic membrane. Like all gram-negative bacteria they then have a thin peptidoglycan layer (cell wall) surrounded by the LPS containing outer lipoprotein membrane. However, spirochetes are surrounded by an additional phospholipid-rich outer membrane with few exposed proteins; this is thought to protect the spirochetes from immune recognition ("stealth" organisms). Axial flagella come out of the ends of the spirochete cell wall, but rather than protrude out of the outer membrane (like other bacteria shown in **Figure 2-1**), the flagella run sideways along the spirochete under the outer membrane sheath. These specialized flagella are called **periplasmic flagella**. Rotation of these periplasmic flagella spins the spirochete around and generates thrust, propelling them forward.

3) **Mycoplasma** do not have a cell wall. They only have a simple cell membrane, so they are neither gram-positive nor gram-negative.

Fig. 1-9. Summary of morphological differences among the bacteria.

MORPHOLOGY	GRAM-POSITIVE	GRAM-NEGATIVE
Circular (Coccus)	Streptococcus Enterococcus Staphylococcus	Neisseria Moraxella
Rod (Bacillus)	Corynebacterium Listeria Bacillus Clostridium Mycobacterium (acid-fast)	ENTERICS (live in the GI tract): • Escherichia coli • Shigella • Salmonella • Yersinia • Klebsiella • Proteus • Enterobacter • Serratia • Vibrio • Campylobacter • Helicobacter • Pseudomonas • Bacteroides (anaerobic) Haemophilus Bordetella Legionella Yersinia Francisella Brucella Pasteurella Gardnerella
Spiral		Spirochetes: • Treponema • Borrelia • Leptospira
Branching filamentous growth (like fungi)	Actinomyces (anaerobic) Nocardia (partially acid-fast)	
Pleomorphic		Chlamydia Rickettsiae
No cell wall	Mycoplasma	

Figure 1-9   MORPHOLOGICAL DIFFERENCES AMONG THE BACTERIA

CYTOPLASMIC STRUCTURES

Bacterial DNA usually consists of a single circle of double-stranded DNA. Smaller adjacent circles of double-stranded DNA are called plasmids; they often contain antibiotic resistance genes. **Ribosomes** are composed of protein and RNA and are involved in the translation process, during the synthesis of proteins. Bacteria, which are **procaryotes**, have smaller ribosomes (70S) than animals (80S), which are **eucaryotes**. Bacterial ribosomes consist of 2 subunits, a large subunit (50S) and a small subunit (30S). These numbers relate to the rate of sedimentation. Antibiotics, such as erythromycin and tetracycline, have been developed that attack like magic bullets. They inhibit protein synthesis preferentially at the bacterial ribosomal subunits while leaving the animal ribosomes alone. Erythromycin works at the 50S subunit, while tetracycline blocks protein synthesis at the 30S subunit.

METABOLIC CHARACTERISTICS

Bacteria can be divided into groups based on their metabolic properties. Two important properties include: 1) how the organism deals with oxygen, and 2) what the organism uses as a carbon and energy source. Other properties include the different metabolic end-products that bacteria produce such as acid and gas.

	OBLIGATE AEROBES	FACULTATIVE ANAEROBES	MICROAEROPHILIC	OBLIGATE ANAEROBES
Gram-positive	*Nocardia* (weakly acid-fast) *Bacillus cereus*	*Staphylococcus* *Bacillus anthracis* *Corynebacterium* *Listeria* *Actinomyces*	*Enterococcus* *Streptococcus* Some species of *streptococci* are facultative anaerobes	*Clostridium*
Gram-negative	*Neisseria* *Pseudomonas* *Bordetella* *Legionella* *Brucella*	Most other gram-negative rods	*Spirochetes* • *Treponema* • *Borrelia* • *Leptospira* *Campylobacter*	*Bacteroides*
Acid-fast	*Mycobacterium* *Nocardia*			
No cell wall		*Mycoplasma*		

* *Chlamydia* and *Rickettsia* do not have the metabolic machinery to utilize oxygen. They are energy parasites, and must steal their host's ATP.

Figure 1-10 OXYGEN SPECTRUM

Oxygen

How bacteria deal with oxygen is a major factor in their classification. Molecular oxygen is very reactive, and when it snatches up electrons, it can form hydrogen peroxide (H_2O_2), superoxide radicals (O_2^-), and a hydroxyl radical ($OH\bullet$). All of these are toxic unless broken down. In fact, our very own macrophages produce these oxygen radicals to pour over bacteria. There are 3 enzymes that some bacteria possess to break down these oxygen products:

1) **Catalase** breaks down hydrogen peroxide in the following reaction:

$$2H_2O_2 \rightarrow 2H_2O + O_2$$

2) **Peroxidase** also breaks down hydrogen peroxide.
3) **Superoxide dismutase** breaks down the superoxide radical in the following reaction:

$$O_2^- + O_2^- + 2H^+ \rightarrow H_2O_2 + O_2$$

Bacteria are classified on a continuum. At one end are those that love oxygen, have all the preceding protective enzymes, and cannot live without oxygen. On the opposite end are bacteria which have no enzymes and pretty much kick the bucket in the presence of oxygen:

1) **Obligate aerobes**: These critters are just like us in that they use glycolysis, the Krebs TCA cycle, and the electron transport chain with oxygen as the final electron acceptor. These guys have all the above enzymes.

2) **Facultative anaerobes**: Don't let this name fool you! These bacteria are aerobic. They use oxygen as an electron acceptor in their electron transfer chain and have catalase and superoxide dismutase. The only difference is that they **can** grow in the absence of oxygen by using fermentation for energy. Thus they have the **faculty to be anaerobic** but prefer aerobic conditions. This is similar to the switch to anaerobic glycolysis that human muscle cells undergo during sprinting.

3) **Microaerophilic bacteria (also called aerotolerant anaerobes)**: These bacteria use fermentation and have no electron transport system. They can tolerate low amounts of oxygen because they have superoxide dismutase (but they have no catalase).

4) **Obligate anaerobes**: These guys hate oxygen and have no enzymes to defend against it. When you are working on the hospital ward, you will often draw blood for culture. You will put the blood into 2 bottles for growth. One of these is an anaerobic growth media with no oxygen in it!

Fig. 1-10. The oxygen spectrum of the major bacterial groups.

Carbon and Energy Source

Some organisms use light as an energy source (phototrophs), and some use chemical compounds as an energy source (chemotrophs). Of the organisms that use chemical sources, those that use inorganic sources, such as ammonium and sulfide, are called **autotrophs**. Others use organic carbon sources and are called **heterotrophs**. All the medically important bacteria are **chemoheterotrophs** because they use chemical and organic compounds, such as glucose, for energy.

Fermentation (glycolysis) is used by many bacteria for oxygen metabolism. In fermentation, glucose is broken down to pyruvic acid, yielding ATP directly. There are different pathways for the breakdown of glucose to pyruvate, but the most common is the **Embden-Meyerhof pathway**. This is the pathway of glycolysis that we have all studied in biochemistry. Following fermentation the pyruvate must be broken down, and the different end products formed in this process can be used to classify bacteria. Lactic acid, ethanol, propionic acid, butyric acid, acetone, and other mixed acids can be formed.

Respiration is used with the aerobic and facultative anaerobic organisms. Respiration includes glycolysis, Krebs tricarboxylic-acid cycle, and the electron transport chain coupled with oxidative phosphorylation. These pathways combine to produce ATP.

Obligate intracellular organisms are not capable of the metabolic pathways for ATP synthesis and thus must steal ATP from their host. These bacteria live in their host cell and cannot survive without the host. Examples of obligate intracellur organisms are *Chlamydia* and *Rickettsia*. They are energy parasites because they need their host's ATP as an energy source. They possess a special cell membrane transport system to steal ATP.

Further metabolic differences (such as sugar sources used, end products formed, and the specific need for certain nutrients) figure in classifying bacteria and will be discussed in the chapters covering specific organisms.

CHAPTER 2.   CELL STRUCTURES,
VIRULENCE FACTORS AND TOXINS

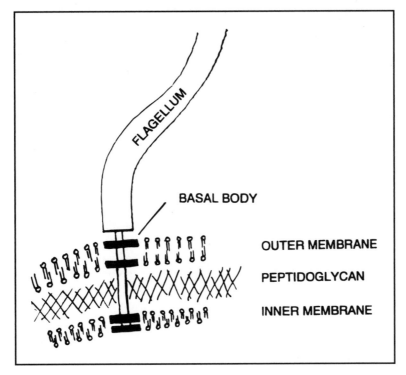

Figure 2-1

Virulent organisms are those that can cause disease. The **virulence** of an organism is the degree of organism pathogenicity. Virulence depends on the presence of certain cell structures and on bacterial exotoxins and endotoxins, all of which are virulence factors.

CELL STRUCTURES AS VIRULENCE FACTORS

Flagella

Fig. 2-1. Flagella are protein filaments that extend like long tails from the cell membranes of certain gram-positive and gram-negative bacteria. These tails, which are several times the length of the bacterial cell, move the bacteria around. The flagellum is affixed to the bacteria by a **basal body**. The basal body spans through the entire cell wall, binding to the inner and outer cell membrane in gram-negative bacteria and to the inner membrane in gram-positive bugs (the gram-positive bacteria don't have an outer membrane). The basal body spins around and spins the flagellum. This causes the bacterial flagella to undulate in a coordinated manner to move the bacteria toward a chemical concentration gradient or away from the gradient. This movement is called **chemotaxis**.

Fig. 2-2. Bacteria can have a single polar flagellum (**polar** means at one end of the cell) as is the case with *Vibrio cholera*, or many peritrichous flagella (all around the cell) as is the case with *Escherichia coli* and *Proteus mirabilis*. *Shigella* does not have flagella.

Spirochetes are all very slender and tightly coiled. Axial flagella come out of the ends of the spirochete cell wall, but rather than protrude out of the outer membrane (like other bacteria shown in **Figure 2-1**), the flagella run sideways along the spirochete under their unique outer membrane sheath. These specialized flagella are called **periplasmic flagella**. Rotation of

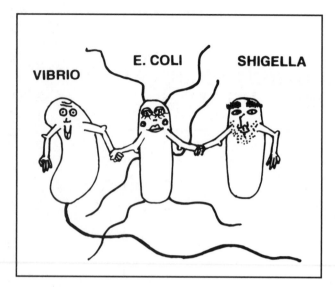

Figure 2-2

these periplasmic flagella spins the spirochete around and generates thrust, propelling them forward.

Pili

Pili (also called **fimbriae**) are straight filaments arising from the bacterial cell wall, making the bacterium look like a porcupine.

Fig. 2-3. Pili are much shorter than flagella and do not move. Pili can serve as adherence factors (in which case they are called **adhesins**). Many bacteria possess adhesins that are vital to their ability to cause disease. For example, *Neisseria gonorrhea* has pili that allow it to bind to cervical cells and buccal cells to cause gonorrhea. *Escherichia coli* and *Campylobacter jejuni* cannot cause diarrhea without their adhesins to bind to the intestinal epithelium, and *Bordetella pertussis* uses its adhesin to bind to ciliated respiratory cells and cause whooping cough. Bacteria that do not produce these pili cannot grab hold of their victim; they lose their virulence and thus cannot infect humans. There are also special pili, discussed in the next chapter, called **sex pili**.

Capsules

Capsules are protective walls that surround the cell membranes of gram-positive and gram-negative

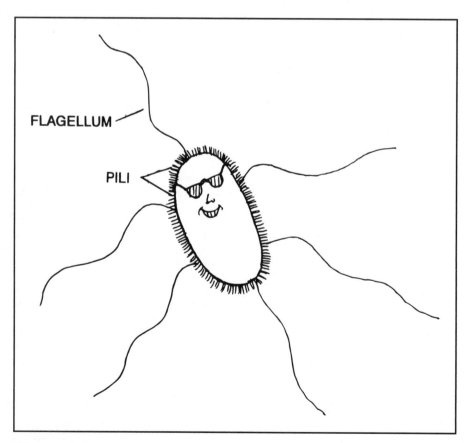

Figure 2-3

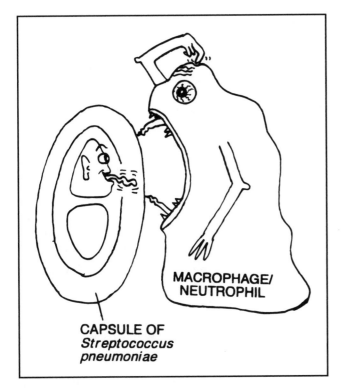

Figure 2-4

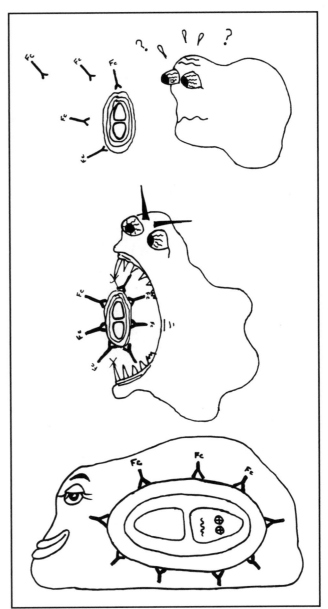

Figure 2-5

bacteria. They are usually composed of simple sugar residues. Bacteria secrete these sugar moieties, which then coat their outer wall. One bacterium, *Bacillus anthracis*, is unique in that its capsule is made up of amino acid residues.

Fig. 2-4. Capsules enable bacteria to be more virulent because macrophages and neutrophils are unable to phagocytize the encapsulated buggers. For example, *Streptococcus pneumoniae* has a capsule. When grown on media, these encapsulated bacteria appear as smooth (S) colonies that cause rapid death when injected into mice. Some *Streptococcus pneumoniae* do not have capsules and appear as rough (R) colonies on agar. These rough colonies have lost their virulence and when injected into mice do not cause death.

Two important tests enable doctors to visualize capsules under the microscope and aid in identifying bacteria:

1) **India ink stain**: Because this stain is not taken up by the capsule, the capsule appears as a transparent halo around the cell. This test is used primarily to identify the fungus *Cryptococcus*.

2) **Quellung reaction**: The bacteria are mixed with antibodies that bind to the capsule. When these antibodies bind, the capsule swells with water, and this can be visualized microscopically.

Antibodies directed against bacterial capsules protect us as they help our macrophages and neutrophils bind to and eat the encapsulated bacteria. The process of antibodies binding to the capsule is called **opsonization**.

Fig. 2-5. Once the antibodies have bound to the bacterial capsule (*opsonization*), the macrophage or neutrophil can then bind to the Fc portion of the antibody and gobble up the bacteria. A vaccine against *Streptococcus pneumoniae* contains antigens from the 23 most common types of capsules. Immunization with this vaccine elicits

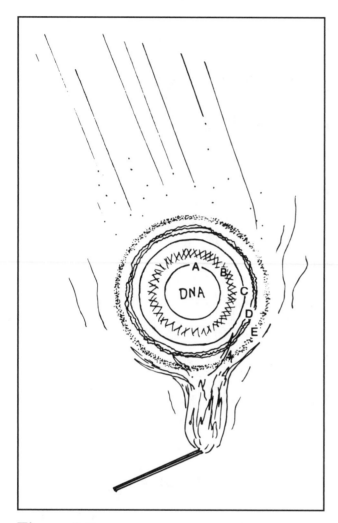

Figure 2-6

an immune response against the capsular antigens and the production of antibodies that protect the individual against future infections by this organism.

Endospores

Endospores are formed by only 2 genera of bacteria, both of which are gram-positive: the aerobic *Bacillus* and the anaerobic *Clostridium*.

Fig. 2-6. Endospores are metabolically dormant forms of bacteria that are resistant to heat (boiling), cold, drying and chemical agents. They have a multi-layered protective coat consisting of:

A) A cell membrane
B) A thick peptidoglycan mesh
C) Another cell membrane
D) A wall of keratin-like protein
E) An outer layer called the **exosporium**

Spores form when there is a shortage of needed nutrients and can lie dormant for years. Surgical instruments are heated in an autoclave, which uses steam under pressure, to 121 °C for 15 minutes, in order to ensure the destruction of *Clostridium* and *Bacillus* spores. When the spore is exposed to a favorable nutrient or environment, it becomes active again.

Biofilms

A biofilm is an extracellular polysaccharide network, similar to the capsule polysaccharides, that forms a mechanical scaffold around bacteria. The biofilm allows bacteria to bind to prosthetic devices, like intravenous catheters, and protects them from attack by antibiotics and the immune system. *Staphylococcus epidermidis* often forms biofilms on intravascular catheters and leaches out to cause bacteremia and catheter related sepsis. Imagine bacteria secreting their polysaccharide concrete around themselves to form a biological bunker. These biofilms are often very difficult for antibiotics to penetrate. The most effective way to cure an infection involving a prosthetic device is to remove the device.

Facultative Intracellular Organisms

Many bacteria are phagocytosed by the host's macrophages and neutrophils yet survive within these white blood cells unharmed!!! These bacteria inhibit phagosome-lysosome fusion, thus escaping the host's deadly hydrogen peroxide and superoxide radicals. Inside the cells these bacteria are safe from antibodies and other immune defenses.

FACULTATIVE INTRACELLULAR ORGANISMS
1. *Listeria monocytogenes*
2. *Salmonella typhi*
3. *Yersinia*
4. *Francisella tularensis*
5. *Brucella*
6. *Legionella*
7. *Mycobacterium*
8. *Nocardia*

Figure 2-7

Fig. 2-7. Facultative intracellular organisms. A helpful mnemonic to remember these eight organisms is: **List**en **Sal**ly **Yer** **Fr**iend **Bruce** **M**ust **Le**ave **N**ow (*Listeria, Salmonella, Yersinia, Francisella, Brucella, Legionella, Mycobacterium, Nocardia*).

TOXINS

Exotoxins

Exotoxins are proteins that are released by both gram-positive and gram-negative bacteria. They may cause many disease manifestations. Exotoxins are released by most of the major gram-positives. Gram-negative bacteria such as *Vibrio cholera, Escherichia coli*, and others can also excrete exotoxins. Severe diseases caused by bacterial exotoxins include anthrax (Saddam Hussein's threatened germ warfare agent), botulism, tetanus, and cholera.

Neurotoxins are exotoxins that act on the nerves or motor endplates to cause paralysis. Tetanus toxin and botulinum toxin are examples.

Enterotoxins are exotoxins that act on the gastrointestinal (GI) tract to cause diarrhea. Enterotoxins inhibit NaCl resorption, activate NaCl secretion, or kill intestinal epithelial cells. The common end result is the osmotic pull of fluid into the intestine, which causes diarrhea. The enterotoxins cause 2 disease manifestations:

1) **Infectious diarrhea**: Bacteria colonize and bind to the GI tract, continuously releasing their enterotoxins locally. The diarrhea will continue until the bacteria are destroyed by the immune system or antibiotics (or the patient dies secondary to dehydration). Examples: *Vibrio cholera, Escherichia coli, Campylobacter jejuni,* and *Shigella dysenteriae.*

2) **Food poisoning**: Bacteria grow in food and release enterotoxin in the food. The enterotoxin is ingested resulting in diarrhea and vomiting for less than 24 hours. Examples: *Bacillus cereus* and *Staphylococcus aureus.*

Pyrogenic exotoxins stimulate the release of cytokines and can cause rash, fever, and toxic shock syndrome (see page 42). Examples: *Staphylococcus aureus* and *Streptococcus pyogenes.*

Tissue invasive exotoxins allow bacteria to destroy and tunnel through tissues. These include enzymes that destroy DNA, collagen, fibrin, NAD, red blood cells, and white blood cells.

Miscellaneous exotoxins, which are the principle virulence factors for many bacteria, can cause disease unique to the individual bacterium. Often the exact role of the exotoxin is poorly understood.

Fig. 2-8. This chart gives many of the important exotoxins and compares their mechanisms of action. Glance over the chart now and return to it as you study individual bacteria.

Fig. 2-9. Exotoxin subunits in *Bacillus anthracis, Clostridium botulinum, Clostridium tetani, Corynebacterium diphtheriae,* and *Vibrio cholera.* Their exotoxins are all composed of 2 polypeptide subunits bound together by disulfide bridges. One of these subunits (called **B** for **binding** or **H** for **holding** on) binds to the target cell. The other subunit (called **A** for **action** or **L** for **laser**) then enters the cell and exerts the toxic effect. Picture these subunits as a key (B and H) and a gun (A and L) bound together by disulfide bonds. The key opens the cell, and then the gun does its damage.

Endotoxins

Remember from the last chapter that **endotoxin is lipid A**, which is a piece of the outer membrane lipopolysaccharide (LPS) of gram-negative bacteria (see **Fig. 1-6**). Lipid A/endotoxin is very toxic and is released when the bacterial cell undergoes lysis (destruction). Endotoxin is also shed in steady amounts from living bacteria. Sometimes, treating a patient who has a gram-negative infection with antibiotics can worsen the patient's condition because all the bacteria are lysed, releasing large quantities of endotoxin. **Endotoxin differs from exotoxin in that it is not a protein excreted from cells, but rather is a normal part of the outer membrane that sort of sheds off, especially during lysis.** Endotoxins pathogenic to humans have only been confirmed in gram-negative bacteria.

Septic Shock

Septic shock (endotoxic shock) is a common and deadly response to both gram-negative and gram-positive infection. In fact, septic shock is the number one cause of death in intensive care units and the 13th most common cause of death in the U.S. (Parrillo, 1990). To better understand septic shock, let us back up and review some terms.

Bacteremia: This is simply bacteria in the bloodstream. Bacteria can be detected by isolating the offending critters in blood cultures. Bacteremia can occur silently and without symptoms. Brushing your teeth results in transient bacteremia with few systemic consequences. Bacteremia can also trigger the immune system, resulting in sepsis and possibly death.

Sepsis: Sepsis refers to bacteremia that causes a systemic immune response to the infection. This response can include high or low temperature, elevation of the white blood cell count, and fast heart rate or breathing rate. Septic patients are described as "looking sick."

Septic shock: Sepsis that results in dangerous drops in blood pressure and organ dysfunction is called septic shock. It is also referred to as **endotoxic shock** because endotoxin often triggers the immune response that results in sepsis and shock. Since gram-positive bacteria

ORGANISM	TOXIN	MECHANISM
NEUROTOXINS		
Clostridium tetani	• Tetanospasmin (tetanus toxin)	1. H *(Heavy) subunit*: binds to neuronal gangliosides 2. L *(Light) subunit*: blocks release of inhibitory neurotransmitters (glycine, GABA) from Renshaw inhibitory interneurons
Clostridium botulinum	• Botulinum toxin	• Inhibits acetylcholine release from motor neuron endplates at neuromuscular junctions
ENTEROTOXINS		A. Infectious Diarrhea
Vibrio cholerae	• Choleragen	1. *Five B subunits*: binds to GM1 gangliosides on intestinal cell membranes 2. *Two A subunits*: carry out the ADP-ribosylation of the GTP-binding protein. This activates membrane associated adenylate cyclase, which converts ATP to cAMP. Elevated levels of cAMP induces the secretion of NaCl and inhibits reabsorption of NaCl
1. *E. coli* 2. *Campylobacter jejuni* 3. *Bacillus cereus*	• *E. coli* heat **labile** toxin (LT) Structurally similar to choleragen	
1. *E. coli* 2. *Y. enterocolitica*	• *E. coli* heat **stable** toxin (ST)	• No effect on concentration of cAMP. Rather, it binds to a receptor on the intestinal brush border and activates guanylate cyclase to produce **G**MP. This results in inhibition of resorption of NaCl
1. *Shigella dysenteriae* 2. Enterohemorrhagic *E. coli* 3. Enteroinvasive *E. coli*	1. Shiga toxin 2. Shiga-like toxin (When "shiga-toxin" is released by bacteria other than *Shigella*)	1. *Five B subunits*: bind to intestinal epithelial cells 2. *A subunit*: inhibits protein synthesis by inactivating the 60S ribosomal subunit. This kills intestinal epithelial cells
ENTEROTOXINS		B. Food Poisoning
Staphylococcus aureus	• Staphylococcal heat stable toxin	
Bacillus cereus	• Heat stable toxin	

Figure 2-8   EXOTOXINS

RESULTS	NOTES

• **Tetanus**: continuous motor neuron activity. Uncontrolled muscle contractions with lockjaw and tetanic paralysis of respiratory muscles	1. Vaccine: formalin inactivated tetanus toxin (Part of **DTaP** vaccine):
	  1. **D**iphtheria
	  2. **T**etanus
	  3. **a**cellular **P**ertussis
	2. Toxin gene carried on plasmid

• **Botulism**: Flaccid paralysis with respiratory muscle paralysis	1. Most potent exotoxin 2. Toxin obtained by lysogenic conversion
• **Cholera**: Increasing **cyclic AMP** levels result in increased intraluminal NaCl, which osmotically pulls fluid and electrolytes into the intestinal tract. This causes diarrhea and dehydration	• **Death by dehydration**
• Increasing **cyclic GMP** levels inhibit NaCl resorption by intestinal epithelial cells. This results in increased osmotic pull of fluid and electrolytes into the intestinal tract, causing diarrhea	
• Shiga toxin kills absorptive intestinal epithelial cells, resulting in sloughing off of dead cells and poor absorption of fluid and electrolytes from the intestinal tract	1. Bloody diarrhea 2. May be responsible for hemolytic uremic syndrome 3. Inhibits protein synthesis in a manner analogous to the antiribosomal antibiotics (erythromycin, tetracycline, etc.)
• Diarrhea and vomiting that lasts for less than 24 hours	• Toxins are deposited on food colonized with toxin-producing *Staphylococcus*
1. Vomiting that lasts for less than 24 hours. 2. Limited diarrhea	1. *B. cereus* endospores survive low temperature cooking. Then, this bacterium grows and deposits this toxin on food 2. *B. cereus* can also produce food poisoning by secretion of a heat labile enterotoxin (similar to that of *E. coli*)

ORGANISM	TOXIN	MECHANISM
	PYROGENIC TOXINS	
Streptococcus pyogenes Group A streptococci	• Streptococcus pyrogenic toxin	• Activates the endogenous mediators of sepsis, such as the cytokine interleukin-1
Staphylococcus aureus	• Toxic shock syndrome toxin (TSST-1)	• Activates the endogenous mediators of sepsis, such as the cytokine interleukin-1
	TISSUE INVASIVE TOXINS	
Streptococcus pyogenes	1. Hemolysins/Streptolysin O and S 2. Streptokinase 3. DNAases 4. Hyaluronidase 5. NADase	1. Lyses red blood cells 2. Activates plasminogen to lyse fibrin clots 3. Hydrolyzes DNA 4. Breaks down proteoglycans 5. Hydrolyzes NAD
Staphylococcus aureus	Many of the above, and: 1. Lipases 2. Penicillinase 3. Staphylokinase 4. Leukocidin 5. Exfoliatin 6. Factors that bind complement	1. Hydrolyzes lipids 2. Destroys penicillins 3. Activates plasminogen to lyse fibrin clots 4. Lyses white blood cells 5. Epithelial cell lyses 6. Cripples host complement defense
Clostridium perfringens	• More than 12 lethal toxins, named by Greek letters: Alpha toxin (lecithinase) is the most important (and most lethal)	• Alpha toxin: lecithinase hydrolyzes lecithin in cell membranes, resulting in cell death
	MISCELLANEOUS EXOTOXINS	
Bacillus anthracis	• Anthrax toxin (three components): 1. Edema Factor (EF): 2. Lethal Factor (LF) 3. Protective Antigen (PA)	1. *Protective Antigen* (PA): binding (B)subunit, which allows entry of EF into the target cell 2. *Edema Factor* (EF): (A subunit) Calmodulin-dependent adenylate cyclase, increases cAMP, which impairs neutrophil function & causes massive edema (disrupts water hemostasis) 3. *Lethal Factor* (LF): is a zinc metalloprotease that inactivates protein kinase. This toxin stimulates the macrophage to release tumor necrosis factor alpha and interleukin-1Beta, which contributes to death in anthrax.
Corynebacterium diphtheriae	• Diphtheria toxin	1. *B subunit*: binds to heart and neural tissue 2. *A subunit*: ADP ribosylates elongation factor (EF$_2$), thereby inhibiting translation of human mRNA into proteins

Figure 2-8 (continued)

CHAPTER 2. CELL STRUCTURES, VIRULENCE FACTORS AND TOXINS

RESULTS	NOTES
• **Scarlet fever**	• Obtains exotoxin from a temperate bacteriophage by lysogenic conversion
• **Toxic shock syndrome**: 1. Fever 2. Rash 3. Desquamation 4. Diarrhea 5. Hypotension (shock)	• *Streptococcus pyogenes* can also cause toxic shock syndrome
• Tissue destruction: 1. Abscesses 2. Skin infections 3. Systemic infection	
1. Tissue destruction: A. Abscesses B. Skin infections C. Systemic infection 2. Exfoliatin is responsible for scalded skin syndrome in infants	
• Tissue destruction and gas gangrene	
• **Anthrax** • Edema factor: an extracellular adenylate cyclase which gets internalized by "defensive" phagocytic cells. Adenylate cyclase is activated by calmodulin, increasing the concentration of cAMP within neutrophils and macrophages. This inhibits their ability to phagocytose bacteria	1. All 3 components are required for activity of this toxin 2. PA is the B (binding) subunit, while both EF and LF are A (action) subunits of the anthrax toxin.
• **Diphtheria**: 1. Myocarditis (heart) 2. Peripheral nerve palsies 3. Central nervous system effects	1. This exotoxin can be considered a human antibiotic, as it inhibits protein synthesis, just as tetracycline and erythromycin inhibit protein synthesis in bacteria

17

ORGANISM	TOXIN	MECHANISM
Corynebacterium diphtheriae (continued)		

Bordetella pertussis	• Four toxins:
	  1. Pertussis toxin
	    B Subunit: binds to target cells
	    A Subunit: inhibits phagocytosis
	  2. Extracytoplasmic adenylate cyclase
	  3. Filamentous hemagglutinin: allows binding to ciliated epithelial cells
	  4. Tracheal cytotoxin
		1. Pertussis toxin
		  B Subunit: binds to target cells
		  A Subunit: activates membrane G proteins to activate membrane bound adenylate cyclase (thus increasing cAMP levels). This inhibits macrophage & neutrophil phagocytosis
		2. *Extracytoplasmic adenylate cyclase*: Similar to *Bacillus anthracis* edema factor, which impairs chemotaxis and phagocytosis
		3. *Filamentous hemagglutinin*: allows binding to ciliated epithelial cells
		4. *Tracheal cytotoxin*: damages respiratory epithelial cells

Clostridium difficile	1. Toxin A 2. Toxin B	1. *Toxin A*: causes fluid secretion and mucosal inflammation, leading to diarrhea 2. *Toxin B*: cytotoxic to colonic epithelial cells
Pseudomonas aeruginosa	• Pseudomonas exotoxin A	• Inhibits protein synthesis by inhibiting Elongation Factor 2(EF$_2$): same mechanism as the diphtheria toxin

Figure 2-8   (continued)

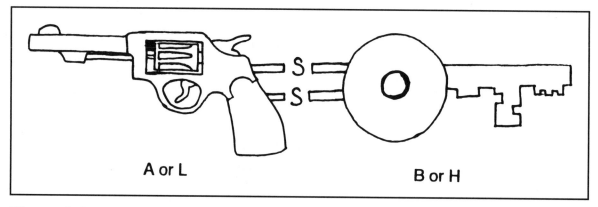

Figure 2-9

RESULTS	NOTES
	2. Vaccine: formalin inactivated tetanus toxin (Part of **DTaP** vaccine): A. **Diphtheria** B. **Tetanus** C. **acellular Pertussis** 3. Obtains exotoxin from a temperate bacteriophage by lysogenic conversion
• **Whooping cough**	• Vaccine: formalin inactivated tetanus toxin (Part of **DTaP** vaccine): 1. **Diphtheria** 2. **Tetanus** 3. **acellular Pertussis**
• **Pseudomembranous enterocolitis**: colonic inflammation, with pseudo-membrane formation. • Clinically: diarrhea (often bloody), fever, & abdominal pain	1. Antibiotic-associated diarrhea 2. Hypervirulent strain NAP1/BI/027 produces additional "binary toxin"
	• Note that diphtheria toxin has the same action as *Pseudomonas* exotoxin A, but they have different targets 1. Exotoxin A: liver 2. Diphtheria toxin: heart

M. Gladwin, W. Trattler, and S. Mahan, *Clinical Microbiology Made Ridiculously Simple* ©MedMaster

and fungi can also trigger this adverse immune response, the term **septic shock** is more appropriate and inclusive.

The chain of events that lead to sepsis and often death begins with a localized site of infection of gram-negative or gram-positive bacteria or fungi. From this site or from the blood (bacteremia), the organisms release structural components (such as endotoxin and/or exotoxin) that circulate in the bloodstream and stimulate immune cells such as macrophages and neutrophils. These cells, in response to the stimulus, release a host of proteins that are referred to as **endogenous mediators** of sepsis.

The most famous endogenous mediator of sepsis is **tumor necrosis factor (TNF)**. TNF is also called **cachectin** because it is released from tumors, producing a wasting (weight loss) syndrome, called cachexia, in cancer patients. Injecting TNF into experimental animals produces hypotension and death (shock). In sepsis, TNF triggers the release of the **cytokine interleukin-1** from macrophages and endothelial cells, which in turn triggers the release of other cytokines and prostaglandins. This churning maelstrom of mediators at first defends the body against the offending microorganisms, but ultimately turns against the body. The mediators act on the blood vessels and organs to produce vasodilatation, hypotension, and organ system dysfunction.

The mortality rate for septic shock is high: up to 40% of patients will die, even with intensive care and antibiotic therapy. For every organ system that fails the mortality rises. Usually two organs are involved (vascular system with hypotension and lungs with hypoxia) and the mortality rate is about 40%. For each additional organ failure (renal failure, etc.) add 15–20% mortality!

ORGAN SYSTEM	EFFECT ON ORGAN SYSTEM	DAMAGING EFFECT ON BODY
Vascular system	Vasodilation	1. Decreased blood pressure 2. Organ hypoperfusion
Heart	Myocardial depression	1. Decreased cardiac output 2. Decreased blood pressure 3. Organ hypoperfusion
Kidneys	Acute renal failure	1. Decreased urine output 2. Volume overload 3. Accumulation of toxins
Lungs	Adult respiratory distress syndrome	• Hypoxia
Liver	Hepatic failure	1. Accumulation of metabolic toxins 2. Hepatic encephalopathy
Brain	Encephalopathy	• Alteration in mental status
Coagulation system	Disseminated intravascular coagulation	1. Clotting 2. Bleeding

Figure 2-10   EFFECTS OF SEPTIC SHOCK

Treatment

The most important principle of treatment is to find the site of infection and the bug responsible and eradicate it! The lung is the most common site (pneumonia) followed by the abdomen and urinary tract. In one-third of cases a site of infection is not identified. Antibiotic therapy is critical with a 10 to 15 fold increased mortality when antibiotics are delayed. Even while working up the site of infection you should start broad coverage antibiotics (called empiric therapy). In other words, as soon as the patient looks sick, start blasting your shotgun at all potential targets. Fire early and hit everything.

Blood pressure must be supported with fluids and drugs (dopamine and norepinephrine are commonly used) and oxygenation maintained (intubation and mechanical ventilation is often required).

In the last decades, efforts to block the inflammatory cascade with monoclonal antibodies against endotoxin, tumor necrosis factor, and interleukin-1, anti-inflammatory agents such as ibuprofen and steroids, and a host of other investigational agents (tumor necrosis factor soluble receptor, nitric oxide antagonists, activated protein C and antioxidant compounds), have met with disappointing results. Most of these treatments have failed to reduce mortality in clinical trials. Similarly, hydrocortisone has long been felt to be beneficial for patients with septic shock, but the largest study to date, published in 2008, failed to show any benefit. Despite this negative study corticosteroid use still has its proponents for use in refractory septic shock.

Fig. 2-10.   The end organ effects of septic shock.

Mnemonic—4 bacteria that produce exotoxins that increase levels of **cAMP**:

c = cholera (*Vibrio cholera*)

A = anthrax (*Bacillus anthracis*)

M = Montezuma's revenge (popular name for enterotoxigenic *E. coli*)

P = pertussis (*Bordetella pertussis*)

(mnemonic courtesy of Gregory Schrank)

References

Parrillo JE. Pathogenic mechanisms of septic shock. N Engl J Med 1993;328:1471–1477.

Parrillo JE, moderator. Septic shock in humans: advances in the understanding of pathogenesis, cardiovascular dysfunction, and therapy. Ann Intern Med 1990; 113: 227–42.

Recommended Review Articles:

Annane D, Bellissant E, Cavaillon JM. Septic Shock. Lancet 2005;365:63–78.

Hotchkiss RS, Karl IE. The pathophysiology and treatment of sepsis. N Engl J Med 2003;348:138–50.

Sprung C, Annane D, Keh D, et al. Hydrocortisone therapy for patients with septic shock. NEJM 2008;358(2):111–124.

CHAPTER 3.   BACTERIAL SEX GENETICS

The bacterial chromosome is a double-stranded DNA molecule that is closed in a giant loop. Because there is only one copy of this molecule per cell, bacteria exist in a **haploid** state. Bacteria do not have nuclear membranes surrounding their DNA.

This chapter does not attempt to cover all the details of bacterial genetics, such as replication, transcription, and translation. These topics are covered extensively in genetics courses. Instead, this chapter covers the mechanisms of bacterial exchange of genetic information. You see, procaryotes have it rough as they do not engage in sexual union with other bacteria. They undergo gene replication, forming an exact copy of their genome, and then split in two, taking a copy with each half (binary fission). The cells of higher organisms (eucaryotes) contribute a set of gametes from each parent and thus ensure genetic diversity. So how do the sexless creatures undergo the genetic change so necessary for survival?

One mechanism is simple mutation. However, it is rare for a single point mutation to change an organism in a helpful manner. Point mutations usually result in nonsense or missense (does this make sense?). There are 4 ways in which bacteria are able to exchange genetic fragments: 1) transformation, 2) transduction, 3) conjugation (so much for celibacy), and 4) transposon insertions.

CHANGE = SURVIVAL

The exchange of genetic material allows for the sharing of genes that code for proteins, such as those that provide antibiotic resistance, exotoxins, enzymes, and other virulence factors (pili, flagella, and capsules). Scientists can take advantage of these exchange mechanisms for genetic engineering and chromosomal mapping. Read on . . . but only if you are over 21 years old.

TRANSFORMATION

Naked DNA fragments from one bacterium, released during cell lysis, bind to the cell wall of another bacterium. The recipient bacterium must be **competent**, which means that it has structures on its cell wall that can bind the DNA and take it up intracellularly. Recipient competent bacteria are usually of the same species as the donor. The DNA that has been brought in can then incorporate itself into the recipient's genome if there is enough homology between strands (another

reason why this transfer can only occur between closely related bacteria).

The famous example of this type of exchange is the experiment conducted by Frederick Griffith in 1928. He used the *Streptococcus pneumoniae* bacteria, which are classified into many different types based on differences in their cellular capsule. Griffith used smooth encapsulated pneumococci, which cause violent infection and death in mice, and rough nonencapsulated pneumococci, which do not kill mice. You can think of the encapsulated pneumococci as smooth hit men who kill mice, and the rough nonencapsulated pneumococci as only acting rough (they are pushovers and can't kill a flea). Griffith heat-killed the smooth encapsulated bad guys and injected them, along with the live rough nonencapsulated pushovers, into mice. Lo and behold, the mice died, and when he cultured out bacteria from the blood, he could only find **live smooth** encapsulated pneumococci. The gene encoding the capsule had been released from the heat-killed bacteria and became incorporated into the living rough nonencapsulated bacteria. The rough bacteria were thus **transformed** into virulent encapsulated smooth bacteria.

Scientists now use this method extensively for inserting recombinant DNA and for mapping genes on chromosomes. It can be used in mapping because the frequency of transformation leading to two traits being transferred is relative to their distance apart on the genome. The closer they are to each other, the more likely that they will be transferred together.

TRANSDUCTION

Transduction occurs when a virus that infects bacteria, called a **bacteriophage**, carries a piece of bacterial DNA from one bacterium to another. To understand this topic, let us digress for a moment and talk about bacteriophages.

Fig. 3-1. Bacteriophages resemble most viruses in having a protein coat called a **capsid** that surrounds a molecule of DNA or RNA. They look almost like spiders with long skinny necks.

The phage will bind by its tail fibers to specific receptors on the bacterial cell surface. This is called **adsorption**. The phage then undergoes **penetration**. Much like a spider squatting down and sinking in its stinger, the phage pushes the long hollow tube under its neck sheath through the bacterial cell wall and cytoplasmic

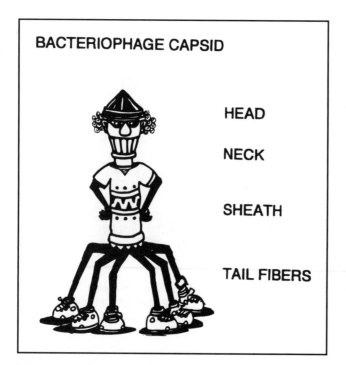

Figure 3-1

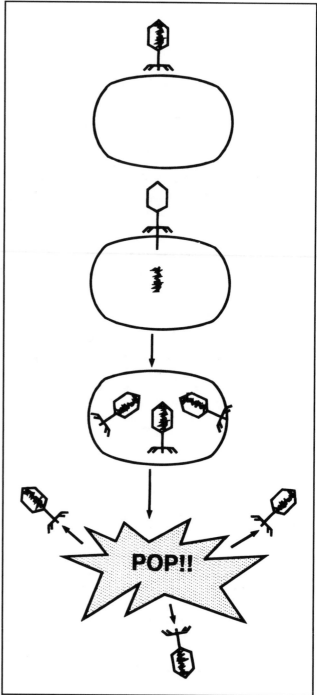

Figure 3-2

membrane. DNA in the head is injected through the tube into the bacterium.

Fig. 3-2. Following adsorption and penetration, the injected DNA takes over the host bacteria's RNA polymerase for the transcription of phage DNA to messenger RNA (mRNA). New capsids, DNA, and enzymes are formed, and the bacterial cell fills with new phages. At some point the cell can hold no more particles and lyses, releasing the phages.

To make things more complicated, there are two types of phages, **virulent phages** and **temperate phages**. Virulent phages behave as shown in **Fig. 3-2**, infecting the bacteria, reproducing, and then lysing and killing the bacteria. On the other hand, temperate phages have a good temperament and do not immediately lyse the bacteria they infect. The temperate phage undergoes adsorption and penetration like the virulent phage but then, rather than undergoing transcription, its DNA becomes incorporated into the bacterial chromosome. The DNA then waits for a command to activate.

Fig. 3-3. The integrated temperate phage genome is called a **prophage**. Bacteria that have a prophage integrated into their chromosome are called **lysogenic** because at some time the repressed prophage can become activated. Once activated, the prophage initiates the production of new phages, beginning a cycle that ends with bacterial cell lysis. So temperate phages, although of good temperament, are like little genetic time bombs.

Lysogenic immunity is the term used to describe the ability of an integrated bacteriophage (prophage) to block a subsequent infection by a similar phage. The first temperate phage to infect a bacteria produces a repressor protein. This "survival of the fittest" adaptation

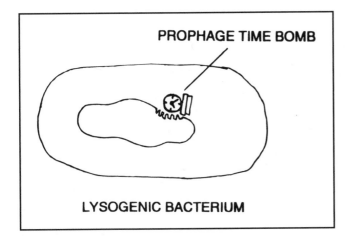

Figure 3-3

ensures that the first temperate phage is the bacteria's sole occupant.

Now that we understand bacteriophages, let's discuss how these phages can carry bacterial DNA from one bacterium to another. This process is called **transduction**. Just as there are two types of phages, there are two types of transduction. Virulent phages are involved in **generalized** transduction and temperate phages in **specialized** transduction.

Generalized Transduction

Generalized transduction occurs as follows. After phage penetration into a host bacterium, the phage DNA is transcribed, replicated, and translated into capsids and enzymes. At this same time the bacterial DNA is repressed and eventually destroyed. Sometimes pieces of the bacterial DNA are left intact. If these pieces are the same size as the phage DNA, they can accidentally be packed into the phage capsid head. Following lysis of the cell and release of the phages, the one phage with bacterial DNA in its head can then infect another bacterium. It will inject the piece of bacterial DNA that it is "accidentally" carrying. If there is some homology between the newly injected strand and the recipient bacterial genome, the piece may become incorporated. The gene on that piece could encode a protein that the recipient did not originally have, such as a protein that inactivates an antibiotic. In generalized transduction, the bacteriophage is only carrying bacterial DNA, so the recipient cell will survive (since no viral genes that encode for replication and lysis are present). This type of genetic transfer is more effective than transformation because the transferred DNA piece is protected from destruction during transfer by the phage capsid that holds it.

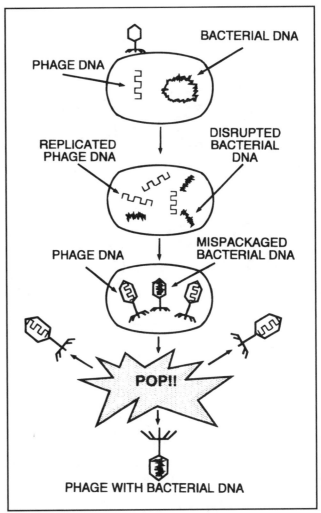

Figure 3-4

Fig. 3-4. Generalized transduction

A) Adsorption and penetration occur. The viral DNA is drawn as a thin line, and the bacterial circular DNA is drawn as a thick circle.

B) Destruction of the bacterial DNA leaves some intact (thick) pieces. The phage DNA has undergone replication.

C) Capsids are translated and packed. The middle one has been packed with a bacterial DNA fragment.

D) Cell lysis occurs, liberating phages including the phage with bacterial DNA.

Specialized Transduction

Specialized transduction occurs with temperate phages. Remember that the temperate phage penetrates, and then its DNA becomes incorporated into the

23

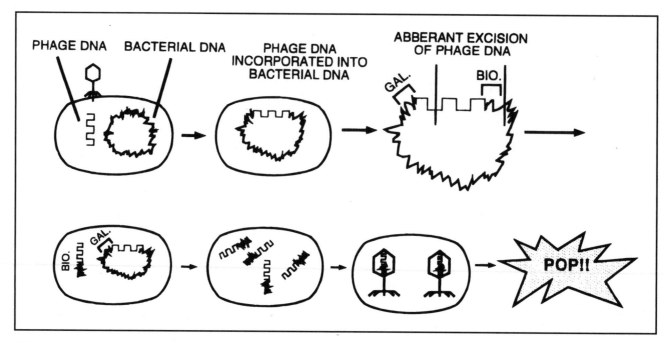

Figure 3-5

bacterial chromosome. It is then called a prophage, and the bacterium is now lysogenic (**Fig. 3-3**). Normally the prophage just waits doing nothing, but it can eventually become active. If it becomes active, the prophage DNA is spliced out of the bacterial chromosome and is then replicated, translated, and packaged into a capsid. Sometimes there is an error in splicing, and a piece of bacterial DNA that lies at one side of the prophage will be cut, replicated, and packaged with the phage DNA. This may result in a transfer of that piece of bacterial DNA to another bacteria.

Fig. 3-5. Specialized transduction occurs with phage lambda in *Escherichia coli*. The site of insertion of the lambda prophage lies between the *Escherichia coli* gene for biotin synthesis and galactose synthesis. If a splicing error occurs, the biotin (BIO) gene or the galactose (GAL) gene (but not both, as the piece of DNA spliced is of a set length) will be carried with the phage DNA and packaged. Thus the gene for biotin synthesis can now be transferred to another bacteria that does not have that capability. You will frequently hear about this form of gene acquisition; it is called **lysogenic conversion**. For example, the gene for *Corynebacterium diphtheria's* exotoxin is obtained by lysogenic conversion.

CONJUGATION

Conjugation is bacterial sex at its best: hot and heavy! In conjugation DNA is transferred directly by cell-to-cell contact, resulting in an extremely efficient exchange of

genetic information. The exchange can occur between unrelated bacteria and is the major mechanism for transfer of antibiotic resistance.

For conjugation to occur, one bacterium must have a **self-transmissible plasmid**, also called an **F plasmid** (for fertility, not the other word!). Plasmids are circular double-stranded DNA molecules that lie outside the chromosome and can carry many genes, including those for drug resistance. F plasmids encode the enzymes and proteins necessary to carry out the process of conjugation. Bacteria that carry F plasmids are called F(+) cells. In conjugation, an F(+) donor cell will pass its F plasmid to an F(−) recipient cell, thus making the recipient F(+).

Fig. 3-6. The self-transmissible plasmid (F plasmid) has a gene that encodes enzymes and proteins that form the sex penis, that is, **sex pilus**.

This long protein structure protrudes from the cell surface of the donor F(+) bacterium and binds to and penetrates the cell membrane of the recipient bacterium (this is finally getting juicy!). Now that a conjugal bridge has formed, a nuclease breaks off one strand of the F plasmid DNA, and this single strand of DNA passes through the sex pilus (conjugal bridge) to the recipient bacterium.

Fig. 3-7. As one DNA strand is passed through the conjugal bridge, the remaining strand is paired with new nucleotide bases (dotted line). The same thing happens to the strand that passes to the other cell. At the

Figure 3-6

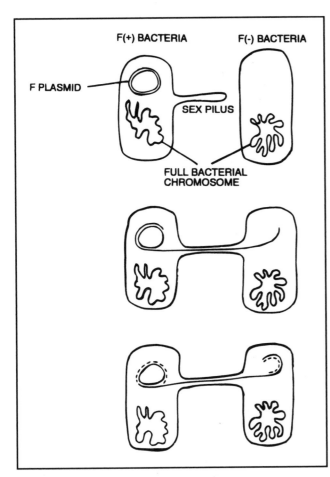

Figure 3-7

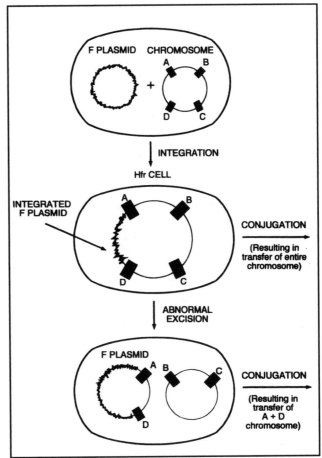

Figure 3-8

end of the sexual union, the conjugal bridge breaks down and both bacteria have double-stranded circular F plasmids. The recipient F(−) cell is now F(+).

Fig. 3-8. Rarely, the extra-chromosomal F plasmid becomes integrated into the neighboring bacterial chromosome much in the same way as a temperate bacteriophage does. The bacterial cell is then called a **Hfr cell** (High frequency of chromosomal recombinants). This integration can result in two unique mechanisms of DNA transfer:

1) The F plasmid that is now together with the entire bacterial circular DNA undergoes normal conjugation with an F(−) cell. The entire bacterial chromosome (including the integrated F plasmid) will transfer from the Hfr cell to the recipient cell.

2) The integrated F plasmid in the Hfr cell may be excised at a different site from that of integration. This can result in an F plasmid that now also contains a segment of chromosomal DNA. These plasmids are called **F′** (**F prime**) plasmids. This F′ conjugation is analogous to

specialized transduction because in both situations a nearby segment of chromosomal DNA is picked up "accidentally" and can be transferred to other bacterial cells.

Some plasmids are non-self-transmissible plasmids. These plasmids do not have the genes necessary for directing conjugation. They do replicate within their host bacterium, however, and continue to be passed on as the bacteria divide in binary fission.

Plasmids are tremendously important medically. Certain plasmids encode enzymes that degrade antibiotics (penicillinase), or generate virulence factors (such as fimbriae and exotoxins).

TRANSPOSONS

Fig. 3-9. Transposons are mobile genetic elements. You can visualize them as DNA pieces with legs. These pieces of DNA can insert themselves into a donor chromosome **without having DNA homology**. They can carry genes for antibiotic resistance and virulence factors.

Transposons insert into the DNA of phages, plasmids, and bacterial chromosomes. They do not replicate

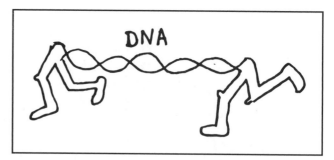

Figure 3-9

independently but are copied during their host's DNA transcription. When transposons leave the DNA they are incorporated in, there is frequently aberrant excision and the transposon can carry new DNA away to another site. The importance of transposons clinically is that a transposon gene that confers a particular drug resistance can move to the plasmids of different bacterial genera, resulting in the rapid spread of resistant strains.

GRAM-POSITIVE BACTERIA

CHAPTER 4.   STREPTOCOCCI

Tests for Strep and Staph

Streptococci and staphylococci are both gram-positive spheres (cocci) and are responsible for a wide variety of clinical diseases. It is often necessary to differentiate between these two organisms to prescribe the appropriate antibiotic. The first way to differentiate them is to examine their appearance on a Gram stain. Streptococci line up one after the other like a **strip** of button candy, while staphylococci appear as a cluster that can be visualized as a cluster of hospital **staff** members posing for a group shot (**Fig. 4-1**).

Fig. 4-1. A second method to differentiate streptococci from staphylococci involves the enzyme **catalase**. A quick look at our **staff** (Staph) picture reveals that a CAT has joined them, so the **staff** picture is CAT(alase) positive. That is, staphylococci possess the enzyme catalase, whereas streptococci do not. Staphylococci are thus referred to as catalase positive while streptococci are catalase negative. Catalase converts H_2O_2 (hydrogen peroxide, which is used by macrophages and neutrophils) into H_2O and O_2. To test for catalase, a wire loop is rubbed across a colony of gram-positive cocci and mixed on a slide with H_2O_2. If bubbles appear, the enzyme catalase must be present, and so staphylococci are present. (See Fig. 5-2).

Figure 4-1

Streptoccal Classification

Certain species of streptococci can either completely or partially hemolyze red blood cells (RBCs). The streptococci are divided into three groups based on their specific hemolytic ability. The streptococci are incubated overnight on a blood agar plate. **Beta-hemolytic** streptococci completely lyse the RBCs, leaving a clear zone of hemolysis around the colony. **Alpha-hemolytic** streptococci only partially lyse the RBCs, leaving a greenish discoloration of the culture medium surrounding the colony. This discolored area contains unlysed RBCs and a green-colored metabolite of hemoglobin. **Gamma-hemolytic** streptococci are unable to hemolyze the RBCs, and therefore we should really not use the word "hemolytic" in this situation (the term non-hemolytic streptococci is often used to avoid confusion).

The streptococci can also be classified based on the antigenic characteristics of the C carbohydrate (a carbohydrate found on the cell wall). These antigens are called **Lancefield antigens** and are given letter names (from A, B, C, D, E, through S). Historically, the Lancefield antigens have been used as a major way of differentiating the many streptococci. However, there are so many different types of streptococci that we now rely less on the Lancefield antigens and more on a combination of tests such as the above mentioned patterns of hemolysis, antigenic composition (including Lancefield), biochemical reactions, growth characteristics, and genetic studies. Although there are more than 30 species of streptococci, only 5 are significant human pathogens. Three of these pathogens have Lancefield antigens: Lancefield group A, B and D. The other two pathogenic species of the streptococcal genus do not have Lancefield antigens, and are therefore just called by their species names: One is *Streptococcus pneumoniae* and the other is actually a big group of streptococci collectively called the Viridans group streptococci.

GROUP A BETA-HEMOLYTIC STREPTOCOCCI
(also called *Streptococcus pyogenes*)

These organisms are so-named because they possess the Lancefield group A antigen and are beta-hemolytic on blood agar. They are also called *Streptococcus pyogenes* (which means pus-producing) and cause the diseases "strep throat," scarlet fever, rheumatic fever, and post-streptococcal glomerulonephritis.

The components of the streptococcal cell wall that are antigenic include:

1) **C carbohydrate**: The C carbohydrate was used by Rebecca Lancefield to divide streptococci into groups. *Streptococcus pyogenes* has the "Lancefield Group A" type of C carbohydrate.

2) **M protein** (80 types): This is a major virulence factor for the group A streptococcus. It inhibits the activation of complement and protects the organism from phagocytosis. However, it is also the weakest point in the organism's defense, because plasma (B) cells generate antibodies against the M protein. These antibodies bind to the M protein (opsonization), aiding in the destruction of the organism by macrophages and neutrophils.

Beta-hemolytic group A streptococci also have many enzymes that contribute to their pathogenicity:

1) **Streptolysin O**: The *O* stands for oxygen labile as it is inactivated by oxygen. This enzyme destroys red and white blood cells and is the reason for the beta-hemolytic group A streptococci's beta-hemolytic ability. This enzyme is also antigenic. Following pharyngeal or systemic beta-hemolytic group A streptococcal infection, anti-streptolysin O (ASO) antibodies develop. On the wards you may order ASO titers on a patient's blood to confirm recent infection.

2) **Streptolysin S**: The **S** stands for oxygen *S*table. This is also responsible for beta-hemolysis but is not antigenic.

3) **Pyrogenic exotoxin** (also called **erythrogenic toxin**): This is found in only a few strains of beta-hemolytic group A streptococci, but when these strains invade they can cause scarlet fever.

Some strains produce pyrogenic exotoxins that are superantigens. The exotoxins directly superstimulate T cells to pour out inflammatory cytokines. This causes a streptococcal toxic shock syndrome (Holm, 1996). More on scarlet fever and toxic shock syndrome later . . .

4) Other enzymes include **streptokinase** (activates the proteolytic enzyme plasmin, which breaks up fibrin blood clots), **hyaluronidase, DNAases, anti-C5a peptidase**, and others (see **Fig. 2-8**).

Staphylococcus aureus has many enzymes that are similar to those of streptococci. You will learn about these in the next chapter.

Beta-hemolytic group A streptococci cause 4 types of disease by **local invasion and/or exotoxin release**. These include:

1) Streptococcal pharyngitis
2) Streptococcal skin infections
3) Scarlet fever
4) Streptococcal toxic shock syndrome

Beta-hemolytic group A streptococci can also cause 2 **delayed antibody mediated diseases**:

1) Rheumatic fever
2) Glomerulonephritis

Local Invasion/Exotoxin Release

1) **Streptococcal pharyngitis**: This is the classic strep throat with red swollen tonsils and pharynx, a purulent exudate on the tonsils, high temperature, and swollen lymph nodes. It usually lasts 5 days (penicillin therapy speeds recovery).

Because exudative pharyngitis (pus on tonsils) can be caused by non-streptococcal organisms (like viruses) a throat swab should be sent for a **rapid antigen detection test (RADT)**. These tests can be completed in minutes and are highly specific for *Streptococcus pyogenes* and immunologically detect group A carbohydrate antigen. In children negative RADTs should be backed up by a throat culture due to the high incidence of strep throat in this population and only moderate sensitivity of the RADT.

"Mom, my throat hurts!!!"

2) **Skin infections**: Skin infections can range from folliculitis (infections of the hair follicles), pyoderma, erysipelas, cellulitis (a deep infection of the skin cells, producing red, swollen skin which is hot to the touch), and impetigo (a vesicular, blistered, eruption, most common in children, that becomes crusty and flaky and is frequently found around the mouth). These skin infections can also be caused by *Staphylococcus aureus*. Therefore, treatment for these infections consists of a penicillinase resistant penicillin like dicloxacillin, which covers both group A beta-hemolytic streptococci and *Staphylococcus aureus*.

Erysipelas is a streptococcal infection of the superficial skin, the dermis only. It has a specific appearance: a raised, bright red rash with a sharp border that advances from the initial site of infection. Unlike cellulitis, erysipelas is only rarely caused by *Staphylococcus aureus*.

Pyoderma is a pustule, usually on the extremity or face, that breaks down after 4–6 days to form a thick crust. It heals slowly and leaves a depigmented area.

"Mom, my throat hurts and my skin is disintegrating!!!!"

Necrotizing Fasciitis ("Flesh-eating Streptococcus"): This type of group A beta-hemolytic streptococcal infection has actually been around for years but may indeed be on the rise (news coverage certainly is). Certain strains have **M** proteins that block phagocytosis, allowing the bacteria to move rapidly through tissue. Streptococci enter through a break in the skin caused by trauma and then follow a path along the fascia which lies

Figure 4-2

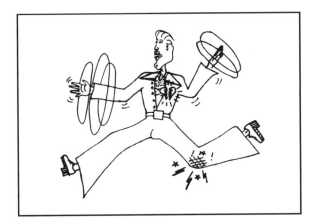

Figure 4-3

between the subcutaneous tissue and muscle. Within a day the patient develops swelling, heat, and redness that moves rapidly from the initial skin infection site. A day later the skin color changes from red to purple to blue, and large blisters (bullae) form. Later the skin dies and muscle may also become infected (myositis).

This infection must be recognized early and the fascia surgically removed. Rapid antibiotic therapy is crucial. Group A beta-hemolytic streptococci are still exquisitely sensitive to penicillin **G**. It may be wise to add clindamycin, as this drug rapidly shuts down streptococcal metabolism and will block toxin production (Holm, 1996; Stevens, 1988). Even with antibiotics and surgery the mortality rate is high (>50%).

Necrotizing fasciitis can also be caused by *Staphylococcus, Clostridium* species, gram-negative enterics, or mixed infection with more than one of these bacteria (Stevens, 1992).

Fournier's gangrene is a form of necrotizing fasciitis involving the male genital area and perineum; it is often caused by mixed organisms but can be caused by *Streptococcus pyogenes.*

3) **Scarlet fever**: Certain beta-hemolytic group A streptococci not only cause a sore throat, but also produce an exotoxin called either **pyrogenic toxin** or **erythrogenic toxin**. This exotoxin is acquired by lysogenic conversion (see Chapter 3). The exotoxin produces fever (so it is pyrogenic) and causes a scarlet-red rash. The rash begins on the trunk and neck, and then spreads to the extremities, sparing the face. The skin may peel off in fine scales during healing.

"Mom, my body is turning scarlet!!!!"

Fig. 4-2. "MOM, help!!!" Pharyngitis, impetigo, and scarlet fever. Note that scarlet fever actually spares the face.

4) **Streptococcal toxic shock syndrome**: It is now clear that beta-hemolytic group A streptococci can cause toxic shock syndrome like that caused by *Staphylococcus aureus*. Similar to scarlet fever, streptococcal toxic shock syndrome is also mediated by the release of pyrogenic toxin. See Chapter 5 and **Fig. 5-9** for more details. Treat severe *Streptococcus pyogenes* infections (severe skin infections, necrotizing fasciitis, streptococcal toxic shock syndrome) with high dose penicillin and with clindamycin. This is because *Streptococcus pyogenes* remains very sensitive to penicillin, with minimal resistance, and the clindamycin inhibits the bacterial ribosome and thus shuts down protein synthesis of pyrogenic toxin and the M protein.

Delayed Antibody-Mediated Disease

1) **Rheumatic fever**:

With the advent of penicillin, rheumatic fever is now uncommon. It usually strikes children 5–15 years of age. When it occurs, it has been shown to follow untreated beta-hemolytic group A streptococcal **pharyngitis** (but NOT after a skin infection). The 6 major manifestations of rheumatic fever are:

a) Fever.
b) Myocarditis (heart inflammation).
c) Joint swelling (arthritis).
d) Chorea (uncontrolled dance-like movements of the extremities) which usually begins 2–3 weeks after the pharyngitis. This is also called Sydenham's chorea or St. Vitus dance.
e) Subcutaneous nodules (rubbery nodules just under the skin).
f) Rash, called **erythema marginatum** because it has a red margin that spreads out from its center.

Fig. 4-3. Picture John Travolta in the movie **Rheumatic Fever**, the upcoming sequel to *Saturday*

Night Fever. His **heart** is damaged from the stress of the hours of disco dancing, his **joints** are aching from dropping to his knees, and his arms are moving rhythmically in a disco **choreiform** jam.

Rheumatic fever is antibody-mediated. There are antigens in the heart that are similar to the antigens of the beta-hemolytic group A streptococci. Therefore, the antibodies that form to eradicate this particular streptococcus also cross-react with antigens in the heart. This immunologic attack on the heart tissue causes heart inflammation, called myocarditis. Patients may complain of chest pain and may develop arrhythmias or heart failure.

Over years, likely after recurrent infections with streptococci, the heart becomes permanently damaged. The most frequently damaged site of the heart is the mitral valve, followed by the aortic valve. These damaged valves may become apparent **many** years (10–20) after the initial myocarditis, and can be picked up on physical exam because they produce heart murmurs. So, **there is an initial myocarditis, and many years later rheumatic valvular heart disease develops**.

These patients are susceptible to recurrent bouts of rheumatic fever and further heart damage. To prevent further damage to the heart (which is permanent and irreversible), prophylactic penicillin therapy is required for much of the patient's life. This will prevent future beta-hemolytic group A streptococcal infections, which if they occur will elicit more of the cross-reacting antibodies.

The joint pain of rheumatic fever is classified as an acute migratory polyarthritis, which is to say that joint pains arise at various sites throughout the day and night. Fortunately, there is no permanent injury to the joints.

2) **Acute post-streptococcal glomerulonephritis**:

This is an antibody-mediated inflammatory disease of the glomeruli of the kidney. It occurs about one week after infection of either the **pharynx OR skin** by **nephritogenic** (having the ability to cause glomerulonephritis) strains of beta-hemolytic group A streptococci. Fortunately, only a few strains of beta-hemolytic group A streptococci are nephritogenic. Certain antigens from these nephritogenic streptococci induce an antibody response. The resulting antigen-antibody complexes travel to and are deposited in the glomerular basement membrane, where they activate the complement cascade. This leads to local glomerular destruction in the kidney.

Clinically, a child will show up in your office, and his mother will complain that his face is puffy. This is caused by the retention of fluid from his damaged kidney. His urine is darker than normal (tea or coca-cola

Figure 4-4

colored) due to hematuria (blood in the urine). The child may also have hypervolemia secondary to fluid retention, which can cause high blood pressure. Upon further questioning you may be able to elicit the fact that he had a sore throat or skin infection a week or so ago. This type of glomerular disease usually has a good prognosis (especially in the pediatric population).

"Mom, my urine is tea colored!!!!"

Fig. 4-4. Acute post-streptococcal glomerulonephritis causes tea colored urine (hematuria).

GROUP B STREPTOCOCCI
(also called *Streptococcus agalactiae*)

These streptococci are also beta-hemolytic. When thinking of group B streptococci, think of group **B** for **BABY**.

About 25% of women carry these bugs vaginally, and a baby can acquire these bacteria during delivery. These organisms cause neonatal (< 3 months of age) meningitis, pneumonia, and sepsis.

Neonates with meningitis do not present with a stiff neck, which is the classic sign seen in adults. Instead, they display nonspecific signs, such as fever, vomiting, poor feeding, and irritability. If you even suspect meningitis, you must act rapidly because every minute counts. Diagnosis of meningitis is made by a lumbar puncture. Antibiotics are often started prior to the results of the lumbar puncture if meningitis is suspected. The organisms that must be covered by the antibiotics include *Escherichia coli*, *Listeria monocytogenes*, and group B streptococcus. These are the 3 most common pathogens associated with meningitis in infants younger than 3 months of age.

Clinical Pearl: Three bacteria are responsible for most meningitis acquired by the baby coming out of the birth canal (within the first 3 months of age): *Listeria monocytogenes, Escherichia coli*, and Group B Streptococcus. Two bacteria cause meningitis later in life after the maternal antibodies passively given to fetus wane and before new antibodies develop: *Neisseria meningitides* and *Haemophilus influenzae*.

Group B Streptococcus can also infect pregnant women, causing bacteremia and sepsis. Secondary infection of the fetus results in stillbirth or spontaneous abortion in about 30% of cases.

There has been an increase in the incidence of Group B streptococcal infections in non-pregnant adults, causing pneumonia or sepsis in people with co-existent medical complications like diabetes, malignancy, renal or liver failure and neurological disease (stroke) and in the elderly (>65 years of age; especially in nursing homes).

Viridans Group Streptococci

The members of this huge group include the Mitis group (*S. mitis, S. sanguis, S. parasanguis, S. gordonii, S. crista, S. infantis, S. oralis, S. peroris*), Salivarius group (*S. salivarius, S. vestibularis, S. thermophilus*), the Mutans group (*S. mutans, S. sobrinus, S. criceti, S. rattus, S. downei, S. macacae*), and the Anginosus group (*S. anginosus, S. constellatus, and S. intermedius*). Note that many of these names refer to the mouth and saliva since the viridans group streptococci are indigenous to the GI tract. These critters represent more than 30% of the culturable bacteria from dental plaque, gingival crevices, the tongue, and saliva!

This is a big, heterogeneous group of streptococci that are not identified based on one Lancefield group. **Viridis** is the Latin word for **green**, and most of the viridans streptococci are alpha-hemolytic, producing greenish discoloration on blood agar. They are normal human gastro-intestinal (G.I.) tract flora that are frequently found in the nasopharynx and gingival crevices.

The viridans streptococci cause 3 main types of infection: dental infections, endocarditis, and abscesses.

1) **Dental infections**: Some of the viridans streptococci, especially *S. mutans*, can bind to teeth and ferment sugar, which produces acid and dental caries (cavities!!).

2) **Endocarditis**: Dental manipulations send showers of these organisms into the bloodstream. Subsequently, they can implant on the endocardial surface of the heart, most commonly on a previously damaged heart valve (such as from old rheumatic fever, a congenital heart defect, or mitral valve prolapse). These bacteria produce an extracellular dextran that allows them to cling to car-

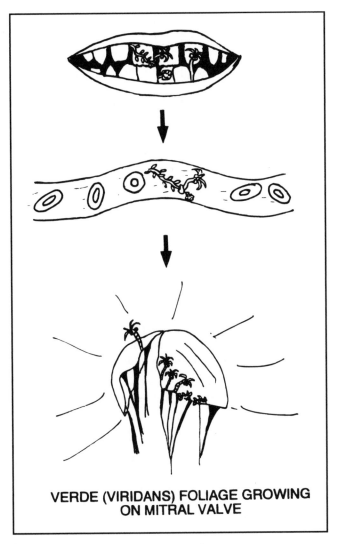

VERDE (VIRIDANS) FOLIAGE GROWING ON MITRAL VALVE

Figure 4-5

diac valves. This results in subacute bacterial endocarditis (SBE), characterized by a slow (hence "subacute") growth and piling up of bacteria on the heart valve (like a pile of bacteria on a petri dish). Clinically, a patient with subacute bacterial endocarditis slowly develops low-grade fevers, fatigue, anemia, and heart murmurs secondary to valve destruction. In contrast, acute infective endocarditis is caused by a staphylococcal infection, often secondary to IV drug abuse, and is characterized by an abrupt onset of shaking chills, high spiking fevers, and rapid valve destruction.

Fig. 4-5. When you think of **virid**ans streptococci, think of *VERDE*, which is the word for "green" in Spanish. Now picture the Verde (green) foliage between some incisors—you know, palm trees, vines, the works. When these teeth are pulled by dentists, the Verde foliage

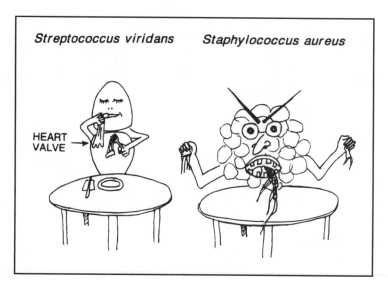

Figure 4-6

enters the blood stream and settles on leaflets of the heart valves, especially valves which have been previously damaged (such as valves damaged by rheumatic fever).

Fig. 4-6. Viridans Streptococcus is eating heart valves slowly, while *Staphylococcus aureus* is eating fast (Notice that these organisms appear as a strip and cluster respectively!). Viridans Streptococcus, slowly eats away at the valve just as a plant slowly grows into soil. This is in sharp contrast to *Staphylococcus aureus*, who received his Olympic gold (aureus) medals for his ability to rapidly bind to and destroy the heart valves. Therefore, subacute bacterial endocarditis (SBE) is caused by viridans Streptococcus, while acute bacterial endocarditis is the disease associated with *Staphylococcus aureus*. Note that group D streptococci (discussed below) can also cause subacute bacterial endocarditis.

Interestingly, the streptococci work together as a team to establish SBE. Initially, *Streptococcus pyogenes* causes rheumatic fever, which damages the heart valves. Now, viridans Streptococcus or the group D streptococci can more easily adhere to the heart valves and cause SBE!!!

3) **Abscesses**: There is a subgroup of the viridans streptococci formerly referred to as the *Streptococcus* Milleri or Intermedius group and now called the ***Anginosus species* group** (comprised of *Streptococcus intermedius, S. Constellatus*, and *S. Anginosus*) which are microaerophilic and are part of the normal G.I. tract flora. These oxygen hating critters are often found in abscesses in the brain or abdominal organs. They are found alone in pure cultures or in mixed cultures with anaerobes (like *Bacteroides fragilis*).

A clinical pearl is that if a *Streptococcus intermedius* group bacteria grows in the blood you should suspect that there is an abscess hiding in an organ and you should consider investigating with a CAT scan with contrast.

Streptococcus InterMeDius and AnginoSus
IMe**D**iately **AS**sess for **ABSCESS**

GROUP D STREPTOCOCCI
(Enterococci and Non-enterococci)

These bacteria, which can be alpha or gamma-hemolytic, traditionally have been divided into two subgroups: the **enterococci** (comprised of *Enterococcus faecalis* and *Enterococcus faecium*) and the **non-enterococci** (comprised of many organisms including *Streptococcus bovis* and *Streptococcus equinus*). Recently the enterococci have been shown to be sufficiently different from the streptococci to be given their own genus enterococcus. *S. bovis* and *S. equinus* are still classified as streptococci.

Enterococcus (faecalis and faecium)

The enterococci take up residence in the human intestines and are considered normal bowel flora. They are variably hemolytic and unique in that they all grow well in 40% bile or 6.5% NaCl. Clinically, the enterococci are commonly the infecting agents in urinary tract infections, biliary tract infections (as they grow well in bile), bacteremia, and subacute bacterial endocarditis (SBE). While these bugs are not as virulent as *Streptococcus pyogenes*, they are always around in the G.I. tract and prey on weak hospitalized patients. In fact, the enterococci are currently the second to third most common cause of hospital acquired (nosocomial) infection! In hospitalized patients the enterococci frequently cause urinary tract infections, wound infections, native and prosthetic valve endocarditis (like Viridans group streptococci), and bacteremia and sepsis after infecting intravenous catheters.

NEWS FLASH!!!! Read all about it! Enterococcus now resistant to ampicillin and vancomycin!

The enterococci are resistant to most of the drugs we use to kill gram positive bacteria. We usually treat enterococcal infections with ampicillin plus an aminoglycoside. However, many enterococcal strains are now resistant to both of these agents; in these cases we treat with vancomycin (see **Fig. 17-17**). Now our worst nightmare has been realized: vancomycin resistant enterococci (VRE) have developed and have been spreading in the U.S. The resistance property is carried on a gene that is transferable. These VRE isolates have acquired a chromosomal transposon DNA element called **vanA** that encodes a series of proteins that modify the D-alanine-D-alanine terminus of the peptidoglycan cell wall, changing it to D-alanine-D-**lactate**, which has a low affinity for vancomycin. Even more scary, this vanA gene can be transferred to the really nasty *Staphylococcus aureus* (See **Chapter 34** to learn more about antibiotic resistance)!

The treatment of multiple resistant enterococci (VRE) is very difficult and requires complicated antibiotic susceptibility testing, infectious disease specialty consultation and the use of some new classes of antibiotics. Agents used include dalfopristine/quinupristine, which only covers *E. faecium*, and daptomycin or linezolid which have activity against both *E. faecium* and *E. faecalis*.

Non-Enterococci (*Streptococcus bovis and equinus*)

Like the enterococci, *Streptococcus bovis* is hardy, growing in 40% bile (but not in 6.5% NaCl). It lives in the G.I. tract, and it causes similar diseases.

An important unique property is that there is a remarkable association between *S. bovis* infection and colon cancer!!! In some series 50% of people with *S. bovis* bacteremia have a colonic malignancy. We do not know if *S. bovis* is a cause of colon cancer or just a marker of the disease.

BOVIS in the BLOOD: Better Beware, CANCER in the BOWEL

Streptococcus pneumoniae
(Alias the pneumococcus; No Lancefield antigen)

The pneumococcus is a very important organism because it is a major cause of bacterial pneumonia and meningitis in adults, and otitis media in children. **P**neumococcus is to **P**arents what group **B** streptococcus is to **B**abies.

The pneumococcus does not have Lancefield antigens! Under the microscope, they appear as lancet-shaped gram-positive cocci arranged in pairs (diplococci).

The major virulence factor of the pneumococcus is its polysaccharide capsule, which protects the organism from phagocytosis. Fortunately, the capsule is antigenic, and antibodies specific for the capsule can neutralize the pneumococcus. The only problem is that there are 84 different capsule serotypes, so surviving an infection with this organism only provides immunity to 1 out of the 83 possible capsule types.

There are 2 important lab tests to identify the pneumococcus:

1) **Quellung reaction**: When pneumococci on a slide smear are mixed with a small amount of antiserum (serum with antibodies to the capsular antigens) and methylene blue, the capsule will appear to swell. This technique allows for rapid identification of this organism.

2) **Optochin sensitivity**. *Streptococcus pneumoniae* is alpha-hemolytic (partial hemolysis-greenish color) but *Streptococcus viridans* is also alpha-hemolytic! To differentiate the two, a disc impregnated with optochin (you don't want to know the real name) is placed on the agar dish. The growth of *Streptococcus pneumoniae* will be inhibited, while *Streptococcus viridans* will continue to grow.

Streptococcus pneumoniae is the most common cause of pneumonia in adults. Pneumococcal pneumonia occurs suddenly, with shaking chills (rigors), high fevers, chest pain with respirations, and shortness of breath. The alveoli of one or more lung lobes fill up with white blood cells (pus), bacteria, and exudate. This is seen on the chest X-ray as a white consolidated lobe. The patient will cough up yellow-green phlegm that on Gram stain reveals gram-positive lancet-shaped diplococci.

Fig. 4-7. The "pneumococcal warrior." He is a mighty foe, with "capsule" armor, a lung emblem on his shield, and a lancet-shaped diplococcus lance. The lung emblem on his shield shows the severe lobar pneumonia caused by this organism. Note the consolidation of the middle right lobe and the lower left lobe, which accompany fever and shaking chills.

Streptococcus pneumoniae is also the most common cause of otitis media (middle ear infection) in children and the most common cause of bacterial meningitis in adults. The classic sign of meningitis, nuchal rigidity

Figure 4-7

(a stiff neck) is usually present in an adult with meningitis.

Fig. 4-8. Otitis media (in children mostly): The pneumococcal warrior's lance zips through the ears of an enemy soldier!!

Otitis media is caused by three main bacteria, *Streptococcus pneumoniae* (≈30% of cases), *Haemophilus influenzae* (≈25%) and *Moraxella catarrhalis* (≈15–20%).

Fig. 4-9. Meningitis: Our warrior is smashing his enemy's head with a hammer!!

The first pneumococcal vaccine (the pneumovax) has 23 of the most common capsular polysaccharide antigens. It is given to people for whom pneumococcal pneumonia would be exceptionally deadly, such as immunocompromised or elderly folk. Individuals without spleens (asplenic) or with HIV disease are unable to defend themselves against encapsulated bacteria and should also be vaccinated. Unfortunately, the polysaccharide vaccine has low immunogenicity and

Figure 4-8

efficacy in children. Initially, a heptavalent conjugate vaccine containing 7 (thus heptavalent) capsular polysaccharide antigens from serotypes 4, 6B, 9V, 14, 18C, 19F, and 23F was recommended for all children. When first introduced, this vaccine had almost 100% efficacy in the prevention of invasive pneumococcal infections in children, but as the years have passed the pneumococci have adapted and now new invasive subtypes are emerging. In 2009, a new 13 valent conjugate vaccine was introduced to keep up with these new pneumococcal serotypes that had emerged. This new vaccine still covers the prior serotypes, but adds coverage for 6 additional ones. Because serotypes **3, 6B, 9V, 14, 19F**, and **23F** are the most common causes of otitis media (bold serotypes are covered by vaccine), it also has been shown to reduce cases of otitis media in children (Eskola, et. al. N Engl J Med 2001; 344:403–9). Due to its improved immunogenicity, as of September 2014, the 13 valent conjugate vaccine is now recommended in all adults ≥65 years of age (in addition to the 23 valent polysaccharide vaccine).

Fig. 4-10. We see the doctor shooting a hole through our warrior (the pneumococci) with the antibody tipped pneumovax (pneumococcal pneumonia vaccine).

NEWS FLASH!!!! Read all about it! *Streptococcus pneumoniae* now resistant to penicillins!

Certain strains of *Streptococcus pneumoniae* are now showing intermediate level resistance to penicillin

Figure 4-9

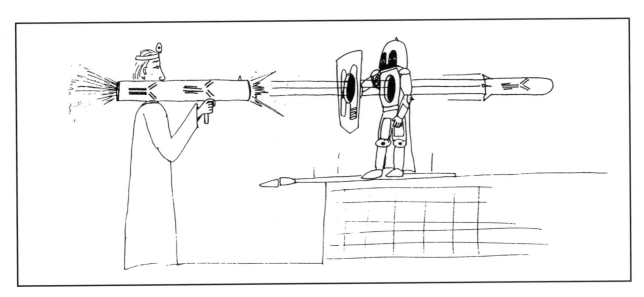

Figure 4-10

(minimal inhibitory concentrations (MIC) of 0.1–1.0 micrograms penicillin per ml blood) and even high level resistance (MIC > 2.0 micrograms/ml blood). In some European countries 2/3 of strains have intermediate or high level resistance! In the U.S. about 25% of strains have intermediate to high level resistance; the percentage is much higher in day care settings where children are frequently given antibiotics. Worse yet, the pneumococcus is also acquiring resistance to erythromycin, trimethoprim/sulfamethoxazole, and chloramphenicol.

The good news is that high dose penicillin (1 million units every 4 hours) and the cephalosporins are effective against bugs with intermediate level resistance. In areas where high level resistant strains are common, vancomycin will have to be added.

Unfortunately, we are witnessing dramatic changes in the way we treat this common and dangerous critter.

Fig. 4-11. Summary of streptococcal groups.

References

Factor SH, Levine OS, et al. Invasive group A streptococcal disease: risk factors for adults. Emerg Infect Dis. 2003; 9(8):970–977.

Holm SE, Invasive group A streptococcal infections. N. Eng. J. Med. 1996; 335:590–591.

Mandell GL, Bennett JE, Dolin R, eds. Principles and Practice of Infectious Diseases; 6th edition. New York: Churchill Livingstone, 2005; Streptococci chapters 2360–2457.

Shulman ST, Bisno AL, et al. Clinical Practice Guideline for the Diagnosis and Management of Group A Streptococcal Pharyngitis: 2012 Update by the Infectious Diseases Society of America. Clin Infect Dis 2012; 55:1279–1282.

Stevens DL. Invasive group A streptococcus infections. Clinical Infectious Diseases 1992; 14:2–13.

Stevens DL, Gibbons AE, et al. The eagle effect revisited: efficacy of clindamycin, erythromycin, and penicillin in the treatment of streptococcal myositis. J. Infect. Dis. 1988; 158:23–8.

Tuomanen EI, Austrian R, Masure HR. Mechanisms of disease: pathogenesis of pneumococcal infection. N Engl J Med 1995; 332(19):1280–1284.

Recommended Review Articles:

Johansson L, Thulin P, Low DE, Norrby-Teglund A. Getting under the skin: the immunopathogenesis of Streptococcus pyogenes deep tissue infections. Clin Infect Dis 2010;51(1): 58–65.

Lappin E, Ferguson AJ. Gram-positive toxic shock syndromes. Lancet Infect Dis. 2009; 9(5):281–90.

GRAM-POSITIVE COCCI	METABOLISM	VIRULENCE	TOXINS

Lancefield group A: *Streptococcus pyogenes*	1. Catalase-negative
	2. Microaerophilic
	3. Beta-hemolytic, due to enzymes that destroy red and white blood cells
	  A. Streptolysin O:
	    a. Oxygen labile
	    b. Antigenic
	  B. Streptolysin-S
	    a. Oxygen stable
	    b. Non-antigenic
		1. M-protein (70 types)
		  a. Adherence factor
		  b. Anti-phagocytic
		  c. Antigenic: Induces antibodies which can lead to phagocytosis
		2. Lipoteichoic acid: adherence factor
		3. Streptokinase
		4. Hyaluronidase
		5. DNAase
		6. Anti-C5a peptidase
			• Erythrogenic or Pyrogenic Toxin (produced only by lysogenized Group A Streptococci): responsible for scarlet fever
			2. Toxic shock syndrome toxin (similar to, but different from the staph exotoxin TSST-1)

Lancefield group B: *Streptococcus agalactiae*	1. Catalase-negative 2. Facultative anaerobe 3. Beta-hemolytic		

Figure 4-11   STREPTOCOCCI

PATHOLOGY	TREATMENT	DIAGNOSTICS	MISCELLANEOUS

DIRECT INVASION/TOXIN
1. Pharyngitis:
  A. Red, swollen
    tonsils and
    pharynx
  B. Purulent exudate
    on tonsils
  C. Fever
  D. Swollen lymph nodes
2. Skin Infections:
  A. Folliculitis
  B. Cellulitis
  C. Impetigo
  D. Necrotizing fasciitis
3. Scarlet fever:
  fever and scarlet
  red rash on body
4. Toxic shock
  syndrome
 	
ANTIBODY MEDIATED
1. Rheumatic fever
  (may follow
  streptococcal pharyngitis):
  A. Fever
  B. Myocarditis: heart
    inflammation
  C. Arthritis: migratory
    polyarthritis
  D. Chorea
  E. Rash: erythema
    marginatum
  F. Subcutaneous
    nodules
  • 10–20 years after
    infection, may develop
    permanent heart valve
    damage
2. Acute post-streptococcal
  glomerulonephritis:
  tea-colored urine,
  following streptococcal
  skin or pharynx infection
	1. Penicillin G
	2. Penicillin V
	3. Erythromycin
	4. Penicillinase-resistant
	  penicillin: in skin
	  infections, where
	  staphylococci could be
	  the responsible organism
	• Following rheumatic
	  fever:
	  A. Patients are placed
	    on continuous
	    prophylactic
	    antibiotics to
	    prevent repeat strep
	    throat infections
	    that could potentially
	    lead to a repeat
	    case of rheumatic
	    fever
	• For invasive streptococcus
	  pyogenes infections, such
	  as necrotizing fasciitis or
	  streptococcal toxic shock
	  syndrome, consider
	  adding clindamycin
		1. Gram stain: gram-
		  positive cocci
		  in chains
		2. Culture on standard
		  laboratory media.
		  Growth is inhibited
		  by bacitracin
		  (*S. Pyogenes* is
		  the only beta-hemolytic
		  streptococcus which
		  is sensitive to bacitracin)
		3. Pharyngitis: Throat swab
		  rapid antigen detection
		  test (RADT) is specific for
		  Streptococcus pyogenes
		  and immunologically
		  detects group A carbo-
		  hydrate antigen.
			1. Dick Test: once
			  commonly used to
			  confirm Scarlet Fever
			  diagnosis
			2. C-Carbohydrate:
			  used for Lancefield
			  groupings

1. Neonatal meningitis
2. Neonatal pneumonia
3. Neonatal sepsis
4. Sepsis in pregnant women
  (with secondary infection
  of fetus)
5. Increasing incidence of
  infections in elderly >65
  years of age and patients
  with diabetes or neuro-
  logical disease: causes
  sepsis and pneumonia.
	• Penicillin G	1. Gram stain of
		  cerebrospinal fluid (CSF)
		  or urine
		2. Culture of CSF,
		  urine or blood
			• Part of normal flora
			  (25% of pregnant
			  women carry Group B
			  streptococci in their
			  vagina)

GRAM-POSITIVE COCCI	METABOLISM	VIRULENCE	TOXINS
Lancefield group D: 2 sub-types: 1. Enterococci: • *Streptococcus faecalis* • *Streptococcus faecium* 2. Non-enterococci • *Streptococcus bovis* • *Streptococcus equinus*	1. Catalase-negative 2. Facultative anaerobe 3. Usually gamma-hemolytic, but may be alpha-hemolytic	• Extracellular dextran helps them bind to heart valves	
Streptococcus viridans	1. Catalase-negative 2. Facultative anaerobe 3. Alpha-hemolytic	• Extracellular dextran helps them bind to heart valves	
Streptococcus pneumoniae (pneumococci)	1. Catalase-negative 2. Facultative anaerobe 3. Alpha-hemolytic	• Capsule (83 serotypes)	• Pneumolysin: binds to cholesterol in host-cell membranes (but its actual effect is unknown)

Figure 4-11   (continued)

PATHOLOGY	TREATMENT	DIAGNOSTICS	MISCELLANEOUS

1. Subacute bacterial endocarditis
2. Biliary tract Infections
3. Urinary tract Infections (especially the enterococci)
	1. Ampicillin, sometimes combined with an aminoglycoside
	2. Resistant to penicillin G
	3. Emerging resistance to vancomycin
	4. For vancomycin resistant organisms (VRE) consider linezolid, daptomycin, and nitrofurantoin.
		1. Gram stain
		2. Culture:
		  A. Enterococci can be cultured in:
		    1. 40% bile
		    2. 6.5% sodium chloride
		  B. Nonenterococci can only grow in bile
			• *S. Bovis* associated with colonic malignancies

1. Subacute bacterial endocarditis 2. Dental caries (cavities): caused by *Streptococcus mutans* 3. Brain or liver abscesses: caused by *Streptococcus intermedius group*	• Penicillin G	1. Gram stain 2. Culture 3. Resistant to optochin	• Part of the normal oral flora (found in the nasopharynx and gingival crevices) and GI tract

1. Pneumonia
2. Meningitis
3. Sepsis
4. Otitis media (in children)
	1. Penicillin G (IM)
	2. Erythromycin
	3. Ceftriaxone
	4. Vaccine: made against the 23 most common capsular antigens. Vaccinate individuals who are susceptible, such as elderly folk or asplenic individuals (including being functionally asplenic due to sickle cell anemia)
	5. Heptavalent and the newer 13 valent conjugated vaccines are effective at preventing otitis media and pneumonia.
		A. Gram stain: reveals gram-positive diplococci
		B. Culture: does not grow in presence of:
		  1. Optochin
		  2. Bile
		C. Positive Quellung test: swelling when tested against antiserum containing anti-capsular antibodies
			• Quellung reaction: technique used to detect encapsulated bacteria (such as *S. pneumoniae* and *H. influenzae*)

M. Gladwin, W. Trattler, and S. Mahan, *Clinical Microbiology Made Ridiculously Simple* ©MedMaster

CHAPTER 5. STAPHYLOCOCCI

Staphylococci are forever underfoot, crawling all over hospitals and living in the nasopharynx and skin of up to 50% of people. While at times they cause no symptoms, they can become mean and nasty. They will be one of your future enemies, so know them well.

The 3 major pathogenic species are *Staphylococcus aureus*, *Staphylococcus epidermidis*, and *Staphylococcus saprophyticus*.

It is extremely important to know how to differentiate staphylococci from streptococci because most staphylococci are penicillin G resistant! You can do 3 things to differentiate them—Gram stain, catalase test, and culture.

1) **Gram stain**:

Fig. 5-1. Staphylococci lie in grape-like clusters as seen on Gram stain. Visualize this cluster of hospital staff posing for a group photo. *Staphylococcus aureus* is catalase-positive, thus explaining the cats in the group photo. *Staphylococcus aureus* (**aureus** means "gold") can be differentiated from the other beta-hemolytic cocci by their elaboration of a golden pigment when cultured on sheep blood agar. Notice that our hospital **Staff** (**Staph**) all proudly wear gold medals around their necks.

2) **Catalase test**: All staphylococci have the enzyme **catalase** (streptococci do not!).

Fig. 5-2. Catalase testing, showing a cluster of staphylococci (catalase-positive) blowing oxygen bubbles. To test, rub a wire loop across a colony of gram-positive cocci and mix on a slide with H_2O_2. If bubbles appear, this indicates that H_2O_2 is being broken down into oxygen bubbles and water; catalase-positive staphylococci are present.

3) **Culture**: *Staphylococcus aureus* and certain streptococci are beta-hemolytic (completely hemolyze red blood cells on an agar plate), but *Staphylococcus aureus* can be differentiated from the other beta-hemolytic cocci by their elaboration of a golden pigment on sheep blood agar.

Now that we can differentiate staphylococci from streptococci, it is important to know which species of staphylococcus is the actual pathogen. The key point: Of the 3 pathogenic staphylococcal species, only *Staphylococcus aureus* is **coagulase positive**!!! It elaborates the enzyme, coagulase, which activates prothrombin, causing blood to clot. In **Fig. 5-1**, note how all the Gold-Medalists (*Staphylococcus **aureus***) hang out together to show each other their gold medals. You can think of them as **coagulating** together. So when a gram-positive coccus in clusters is isolated in culture, the microbiology laboratory will do a coagulase test. If they report to you that the test demonstrates coagulase *positive*

Figure 5-1

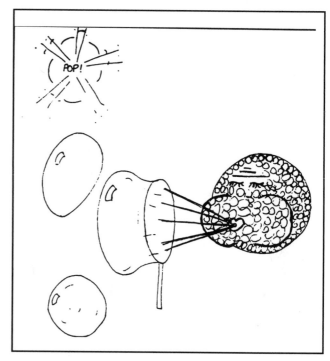

Figure 5-2

gram-positive cocci in clusters you know you have *Staphylococcus aureus*. If they report coagulase *negative* gram-positive cocci in clusters, think of *Staphylococcus epidermidis* or *Staphylococcus saprophyticus*.

Staphylococcus aureus

This critter has a microcapsule surrounding its huge peptidoglycan cell wall, which in turn surrounds a cell membrane containing penicillin binding protein (also called transpeptidase—see page 160). Numerous powerful defensive and offensive protein weapons stick out of the microcapsule or can be excreted from the cytoplasm to wreak havoc on our bodies:

Proteins That Disable Our Immune Defenses

1) **Protein A**: This protein has sites that bind the Fc portion of IgG. This may protect the organism from **opsonization** and phagocytosis.

2) **Coagulase**: This enzyme can lead to fibrin formation around the bacteria, protecting it from phagocytosis.

3) **Hemolysins** (4 types): Alpha, beta, gamma, and delta. They destroy red blood cells, neutrophils, macrophages, and platelets.

4) **Leukocidins**: They destroy leukocytes (white blood cells). Community acquired methicillin resistant *Staphylococcus aureus* (CA-MRSA) produces a particular leukocidin called **Panton-Valentine Leukocidin (PVL)**, which is associated with a propensity to form abscesses (more on this later).

5) **Penicillinase**: This is a secreted form of beta-lactamase. It disrupts the beta-lactam portion of the penicillin molecule, thereby inactivating the antibiotic (see Chapter 16).

6) **Novel penicillin binding protein**: This protein, also called transpeptidase, is necessary for cell

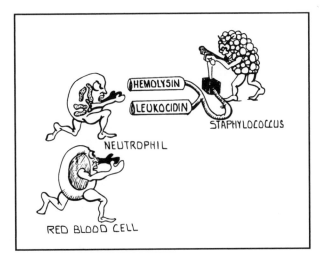

Figure 5-4

wall peptidoglycan formation and is inhibited by penicillin. Some strains of *Staphylococcus aureus* have new penicillin binding proteins that are resistant to penicillinase-resistant penicillins and cephalosporins.

Fig. 5-3. *Staphylococcus aureus* wielding protein A and coagulase shields, defending itself from attacking antibodies and phagocytosis.

Fig. 5-4. Hapless red blood cell following a neutrophil, running to destruction at the hands of *Staphylococcus aureus* and its hemolysin and leukocidin dynamite.

Proteins to Tunnel Through Tissue

1) **Hyaluronidase** ("Spreading Factor"): This protein breaks down proteoglycans in connective tissue.

2) **Staphylokinase**: This protein lyses formed fibrin clots (like streptokinase).

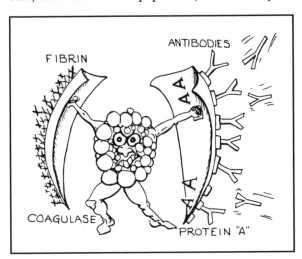

Figure 5-3

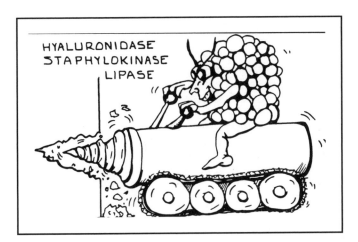

Figure 5-5

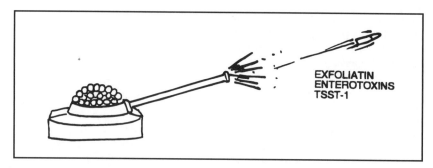

Figure 5-6

3) **Lipase**: This enzyme degrades fats and oils, which often accumulate on the surface of our body. This degradation facilitates *Staphylococcus aureus*' colonization of sebaceous glands.

4) **Protease**: destroys tissue proteins.

Fig. 5-5. *Staphylococcus aureus* produces proteins that allow the bacteria to tunnel through tissue.

Exotoxin Assault Weaponry

1) **Exfoliatin**: A diffusible exotoxin that causes the skin to slough off (**scalded skin syndrome**).

2) **Enterotoxins** (heat stable): Exotoxins which cause food poisoning, resulting in vomiting and diarrhea.

3) **Toxic Shock Syndrome toxin** (TSST-1): This exotoxin is analogous to the pyrogenic toxin produced by Lancefield group A beta-hemolytic streptococci, but is far more deadly. This exotoxin causes toxic shock syndrome and is found in 20% of *Staphylococcus aureus* isolates. These pyrogenic toxins are called superantigens and bind to the MHC class II molecules on antigen presenting cells (such as macrophages). The toxin-MHC II complex causes a massive T cell response and outpouring of cytokines, resulting in the toxic shock syndrome described below. (see **Fig. 5-9**).

Fig. 5-6. *Staphylococcus aureus* produces exotoxin assault weaponry.

Staphylococcus aureus causes a broad range of human disease, and can infect almost any organ system. The diseases can be separated into 2 groups:

Disease caused by **exotoxin release**:

1) Gastroenteritis (food poisoning).
2) Toxic shock syndrome.
3) Scalded skin syndrome.

Disease resulting from **direct organ invasion** by the bacteria:

1) Pneumonia
2) Meningitis
3) Osteomyelitis
4) Acute bacterial endocarditis

5) Septic arthritis
6) Skin infections
7) Bacteremia/sepsis
8) Urinary tract infection

Diseases Caused by Exotoxin Release

1) **Gastroenteritis**: Staphylococci can grow in food and produce an exotoxin. The victim will then eat the food containing the pre-formed toxin, which then stimulates peristalsis of the intestine with ensuing nausea, vomiting, diarrhea, abdominal pain, and occasionally fever. The episode lasts 12 to 24 hours.

Fig. 5-7. *Staphylococcus aureus* gastroenteritis. "I told you not to eat the mayonnaise, sweetheart!"

2) **Toxic Shock Syndrome**: You may have heard about toxic shock syndrome and super-absorbent tampons. It now appears that these tampons, when left in place for a long time, in some way stimulate *Staphylococcus aureus* to release the exotoxin TSST-1. This exotoxin penetrates the vaginal mucosa and is a potent stimulator of both tumor necrosis factor (TNF) and interleukin-1. TSST-1 also dramatically enhances susceptibility to endotoxin. Tampon use is not the only cause of this syndrome, since men and non-menstruating women can also be affected.

Fig. 5-8. Infected sutures in surgical wounds, cutaneous and subcutaneous infections, and infections following childbirth or abortion can all be foci from which *Staphylococcus aureus* can release its TSST-1 exotoxin.

Fig. 5-9. Toxic shock syndrome is caused by *Staphylococcus aureus* releasing TSST-1. This toxin creates symptoms that you can think of as a hybrid between food poisoning (enterotoxins) and the streptococcal pyrogenic toxin that produces scarlet fever. The syndrome involves the sudden onset of high fever, nausea, vomiting, and watery diarrhea (enterotoxin-like syndrome), followed in a few days by a diffuse erythematous (red) rash (like scarlet fever). The palms and soles undergo desquamation (fine peeling of the skin) late in the course of the illness. The toxic shock syndrome is also associated with septic shock as described in Chapter 2: blood pressure

Figure 5-7

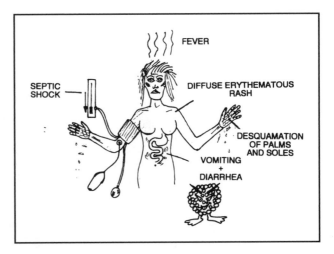

Figure 5-9

may bottom out (frank shock) and the patient may suffer severe organ system damage (such as acute respiratory distress syndrome or acute renal failure).

Figure 5-8

Treatment includes cleaning the infected foci, removal of the tampon or drainage of an infected wound, along with supportive care. Antibiotics can help by killing the bacteria and preventing more exotoxin from being produced. However, antibiotics are not curative because it is the exotoxin, not the bacteria, which causes the clinical manifestations.

3) **Staphylococcal Scalded Skin Syndrome**: This disease is similar in pathogenesis to toxic shock syndrome. A *Staphylococcus aureus* strain, which produces **exfoliative toxin A and B**, establishes a localized infection and releases a diffusible toxin that exerts distant effects. Unlike toxic shock syndrome, it usually affects neonates with local infection of the recently severed umbilicus or older children with skin infections. Clinically, it causes cleavage of the middle epidermis, with fine sheets of skin peeling off to reveal moist red skin beneath. Healing is rapid and mortality low. The doctor must rule out a drug allergy, since this can present similarly and may result in death if the use of the offending drug is not halted.

Disease Resulting from Direct Organ Invasion

Fig. 5-10. Diseases caused by direct organ invasion by *Staphylococcus aureus*. Visualize the Staph-wielding wizard. (Note the cluster of staphylococci at the head of his staff.) The pathology includes:

1) **Pneumonia**: *Staphylococcus aureus* is a rare but severe cause of community-acquired bacterial pneumonia. Pneumonia is more common in hospitalized patients. It usually follows a viral influenza (flu) upper respiratory illness, with abrupt onset of fever, chills, and lobar consolidation of the lung, with rapid destruction of the lung parenchyma, resulting in cavitations (holes in the lung). This violent destructive pneumonia frequently causes effusions and empyema (pus in the pleural space).

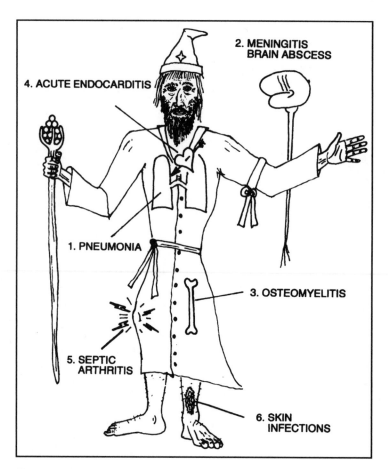

Figure 5-10

2) **Meningitis, Cerebritis, Brain Abscess**: These patients can present with high fever, stiff neck, headache, obtundation, coma, and focal neurologic signs.

3) **Osteomyelitis**: This is a bone infection that usually occurs in boys under 12 years of age. The infection spreads to the bone hematogenously, presenting locally with warm, swollen tissue over the bone and with systemic fever and shakes.

4) **Acute Endocarditis**: This is a violent destructive infection of the heart valves with the sudden onset of high fever (103–105 F°), chills, and myalgias (like a bad flu). The patient with staphylococcal endocarditis may have no history of valvular disease and may not have a murmur. Vegetations grow rapidly on the valve, causing valvular destruction and embolism of vegetations to the brain (left heart valve involvement) or lung (right heart valve infected). Intravenous drug users develop a right-sided tricuspid valve endocarditis and may present with pneumonia caused by bacterial embolization from this infected valve. Endocarditis caused by *Streptococcus viridans* and Group D Streptococci has a more gradual onset (see **Fig. 4-6**).

5) **Septic Arthritis**: Invasion of the synovial membrane by *Staphylococcus aureus* results in a closed infection of the joint cavity. Patients complain of an acutely painful red swollen joint with decreased range of motion. *Staphylococcus aureus* is the most common pathogen causing this disease in the pediatric age group and in adults over the age of 50. Without prompt treatment, many patients will permanently lose the function of the involved joint. Diagnosis requires examination of the synovial fluid, which will characteristically appear yellowish and turbid, with a huge number of neutrophils (>100,000), as well as a positive Gram stain or culture. Therapy requires drainage of the joint and antimicrobial therapy.

6) **Skin Infections**: Minor skin infections are almost exclusively caused by either *Streptococcus pyogenes* (Group A beta-hemolytic) or by *Staphylococcus aureus*. It is clinically impossible to differentiate the two and, in fact, both may be involved. Streptococci can be treated with penicillin G, but staphylococci are often resistant. The trick is to find an antibiotic that covers both streptococci and staphylococci, which is becoming increasingly difficult with the emergence of CA-MRSA (more to come). In an outpatient setting, clindamycin is often a good choice.

Skin infections caused by staphylococci or streptococci usually follow a major or minor break in the skin, with scratching of the site spreading the infection:

44

a) **Impetigo**: This contagious infection usually occurs on the face, especially around the mouth. Small vesicles lead to pustules, which crust over to become honey-colored, wet, and flaky.

b) **Cellulitis**: This is a deeper infection of the cells. The tissue becomes hot, red, shiny and swollen.

c) **Local Abscesses, Furuncles, and Carbuncles**: An **abscess** is a collection of pus. Infection of a hair follicle produces a single pus-filled crater with a red rim. This infection can penetrate deep into the subcutaneous tissue to become a **furuncle**. These may bore through to produce multiple contiguous, painful lesions communicating under the skin called **carbuncles**. Significant abscesses must be surgically drained.

d) **Wound infections**: Any skin wound can be infected with Staphylococcus aureus, resulting in an abscess, cellulitis, or both. When a sutured post-surgical wound becomes infected, it must be reopened and often left open to heal by secondary intention (from the bottom of the wound outward).

7) **Blood and catheter infections: Staphylococcus aureus** can migrate from the skin and colonize central venous catheters resulting in bacteremia, sepsis, and septic shock, as well as endocarditis.

Methicillin-Resistant *Staphylococcus aureus* (MRSA)

Most staphylococci are penicillin-resistant because they secrete penicillinase. Methicillin, Nafcillin, and other penicillinase-resistant penicillins are not broken down by penicillinase, thus enabling them to kill most strains of Staphylococcus aureus. MRSA is a strain of Staphylococcus aureus that has acquired multi-drug resistance, even against methicillin and nafcillin. This resistance is mediated by an acquired chromosomal DNA segment (mecA) encoding a new **penicillin binding protein 2A** that can take over the job of peptidoglycan cell wall assembly when the normal penicillin binding protein (transpeptidase) is inhibited. Until recently, most strains of MRSA developed in the hospital under the influence of heavy antibiotic pressure. These strains had been classified as health care or hospital acquired MRSA or **HA-MRSA** (this distinction is starting to go away). Generally these hospital acquired MRSA infections have exhibited extensive antibiotic resistance. In these cases vancomycin remains one of the most useful antibiotics. Unfortunately, strains of Staphylococcus aureus resistant to vancomycin are now being reported. These **VRSA** isolates have acquired a chromosomal transposon DNA element called **vanA** from vancomycin resistant enterococcus (VRE) that encodes a series of proteins that modify the D-alanine-D-alanine terminus of the peptidoglycan cell wall, changing it to D-alanine-D-**lactate**, which has a low affinity for vancomycin.

Community Acquired Methicillin-Resistant *Staphylococcus aureus* (CA-MRSA)

In recent years we have seen the emergence of multiple clones of MRSA that have been dubbed "community acquired" MRSA due to its emergence outside the hospital. There have been numerous highly publicized outbreaks of **CA-MRSA** infections among sports teams. Humans are frequently colonized in their nasopharynx and skin folds with Staphylococcus aureus, and this leads to a propensity to develop skin and soft tissue infections with the same organism. Exposure to CA-MRSA in close contact settings may lead to the contacts becoming carriers or developing skin and soft tissue infections. CA-MRSA carries a gene which encodes for the Panton-Valentine Leukocidin (**PVL**) toxin which is associated with a propensity to form skin abscesses. Both **HA-MRSA** and **CA-MRSA** pass their resistance genes via mobile genetic elements that can be passed between bacteria. The genes encoding methicillin resistance are carried on a genomic strand called **SCC*mec*** (staphylococcal cassette chromosome conferring resistance to **methicillin**). It just so happens that the old HA-MRSA strains have an SCC*mec* element that is large and bulky because it also carries multiple resistance elements to antibiotics in addition to methicillin. As a result, transfer of this mobile element is a bulky time-consuming process. In contrast, the new kid on the block **CA-MRSA** has a much smaller SCC*mec* transferable element that is easily transferred among staph bacteria, but carries few resistant mutations with it. As a result CA-MRSA is much more efficient at spreading its seed (or SCC*mec* in this case) and it is now the predominant methicillin resistant staphylococcus bacterium acquired both in and outside the hospital. Fortunately, CA-MRSA still tends to be susceptible to several oral antibiotics such as clindamycin and trimethoprim-sulfamethoxazole.

For more information about the myriad mechanisms of Staphylococcus aureus resistance to antibiotics, please refer to **Chapter 34**.

Staphylococcus epidermidis

This organism is part of our normal bacterial flora and is widely found on the body. Unlike Staphylococcus aureus, it is **coagulase-negative**.

This organism normally lives peacefully on our skin without causing disease. However, compromised hospital patients with Foley urine catheters or intravenous lines can become infected when this organism migrates from the skin along the tubing.

Staphylococcus epidermidis is a frequent skin contaminant of blood cultures. Contamination occurs when the needle used to draw the blood passes through skin covered with Staphylococcus epidermidis. Drawing blood from 2 sites will help determine if growth of Staphylococcus epidermidis represents a real bacteremic

GRAM-POSITIVE COCCI	METABOLISM	VIRULENCE	TOXINS
Staphylococcus aureus	1. Catalase-positive 2. **Coagulase-positive** 3. Facultative anaerobe	*Protective Proteins*: 1. Protein A: binds IgG, preventing opsonization and phagocytosis 2. Coagulase: allows fibrin formation around organism 3. Hemolysins 4. Leukocidins 5. Penicillinase *Tissue-Destroying Proteins*: 1. Hyaluronidase: breaks down connective tissue 2. Staphylokinase: lyses formed clots 3. Lipase	*Assault Weaponry* 1. Exfoliatin: scalded skin syndrome 2. Enterotoxin: food poisoning 3. Toxic shock syndrome toxin (TSST-1)
Staphylococcus epidermidis	1. Catalase-positive 2. **Coagulase-negative** 3. Facultative anaerobe	1. Polysaccharide capsule: adheres to a variety of prosthetic devices. Forms a **biofilm.** 2. Highly resistant to antibiotics!	
Staphylococcus saprophyticus	1. Catalase-positive 2. **Coagulase-negative** 3. Facultative anaerobe		

Figure 5-11   STAPHYLOCOCCI

infection or is merely a contamination. If only one of the samples grows *Staphylococcus epidermidis*, you can suspect that this is merely a skin contaminant. However, if 2 cultures are positive, the likelihood of bacteremia with *Staphylococcus epidermidis* is high.

Staphylococcus epidermidis also causes infections of prosthetic devices in the body, such as prosthetic joints, prosthetic heart valves, and peritoneal dialysis catheters. In fact, *Staphylococcus epidermidis* is the most frequent organism isolated from infected indwelling prosthetic devices. The organisms have a polysaccharide capsule that allows adherence to these prosthetic materials.

Staphylococcus epidermidis often forms **biofilms** on intravascular catheters and leaches out to cause bacteremia and catheter related sepsis. A biofilm is an extracellular polysaccharide network, similar to the capsule polysaccharides, that forms a mechanical scaffold around bacteria. The biofilm allows bacteria to bind to prosthetic devices, like intravenous catheters, and protects them from attack by antibiotics and the immune system. Imagine bacteria secreting their polysaccharide concrete around themselves to form a biological bunker.

Staphylococcus saprophyticus

This organism is a leading cause (second only to *E. coli*) of urinary tract infections in sexually active young women. It is most commonly acquired by females (95%) in the community (NOT in the hospital). This organism is coagulase-negative.

CLINICAL	TREATMENT	DIAGNOSTICS
A. *Exotoxin Dependent* 　1. Gastroenteritis (food poisoning): Rapid onset of vomiting & diarrhea, with rapid recovery 　2. Toxic shock syndrome: 　　A. High fever 　　B. Nausea and vomiting 　　C. Watery diarrhea 　　D. Erythematous rash 　　E. Hypotension 　　F. Desquamation of palms and soles 　3. Scalded skin syndrome B. *Direct Invasion* 　1. Pneumonia 　2. Meningitis 　3. Osteomyelitis (in children) 　4. Acute bacterial endocarditis 　5. Septic arthritis 　6. Skin infection 　7. Bacteremia/sepsis 　8. Urinary tract infection	1. Penicillinase-resistant penicillins: nafcillin (IV) and dicloxacillin (oral). 2. 1st generation cephalosporins: cefazolin (IV), cephalexin (oral) treat with intravenous 3. Clindamycin (IV and oral) If MRSA: 1. Vancomycin (IV) 2. Daptomycin (IV) 3. Clindamycin (IV and oral) 4. Trimethaprim-sulfamethoxazole (IV and oral) 5. Linezolid (IV and oral) 6. Ceftaroline (IV) 7. Telavancin (IV) 8. Dalbavancin (IV) 9. Oritavancin (IV) 10. Tedizolid (IV and po) 11. Tigecycline	1. Gram stain: reveals gram-positive cocci in clusters 2. Culture: 　A. Beta-hemolytic 　B. Produces a golden yellow pigment. 3. Metabolic 　A. Catalase-positive 　B. Coagulase-positive 4. Polymerase chain reaction (PCR) detection of ribosomal RNA.
A. Nosocomial infections: 　1. Prosthetic joints 　2. Prosthetic heart valves 　3. Sepsis from intravenous lines 　4. Urinary tract infections B. Frequent skin contaminant in blood cultures!	• Vancomycin (since resistant to multiple antibiotics)	1. Gram stain; reveals gram-positive cocci in clusters 2. Culture 3. Metabolic 　A. Catalase-positive 　B. Coagulase-negative
• Urinary tract infections in sexually active women	• Penicillin	1. Gram stain: reveals gram-positive cocci in clusters 2. Culture: gamma-hemolytic 3. Metabolic 　A. Catalase-positive 　B. Coagulase-negative

M. Gladwin, W. Trattler, and S. Mahan, *Clinical Microbiology Made Ridiculously Simple* ©MedMaster

Fig. 5-11.　Summary chart of staphylococci.

Recommended Review Articles:

Bamberger DM and Boyd SE. Management of Staphylococcus aureus infections. Am Fam Physician 2005;72:2474–81.

Boucher H, Miller LG, Razonable RR. Serious infections caused by methicillin-resistant Staphylococcus aureus. Clin Infect Dis. 2010;51 Suppl 2:S183–97.

Davis SL, et al. Epidemiology and outcomes of community-associated methicillin-resistant Staphylococcus aureus infection. J Clin Microb 2007;45(6):1705–1711.

Eckmann C, Dryden M. Treatment of complicated skin and soft-tissue infections caused by resistant bacteria: value of linezolid, tigecycline, daptomycin and vancomycin. Eur J Med Res. 2010;15(12):554–63.

Henderson DK. Managing methicillin-resistant staphylococci: a paradigm for preventing nosocomial transmission of resistant organisms. Am J Med 2006;119:S45–52.

Jeyaratnam D, Reid C, Kearns A and Klein J. Community acquired MRSA: an alert to paediatricians. Arch Dis Child 2006;91:511–2.

Liu C, Bayer A, et al. Clinical practice guidelines by the Infectious Diseases Society of America for the treatment of methicillin-resistant Staphylococcus Aureus infections in adults and children. Clin Infect Dis 2011;52:1–38.

Mehnert-Kay SA. Diagnosis and management of uncomplicated urinary tract infections. Am Fam Physician 2005;72:451–6.

Moran GJ, et al. Methicillin-resistant S. aureus infections among patients in the emergency department. NEJM 2006;355(7):666–674.

CHAPTER 6. *BACILLUS* AND *CLOSTRIDIUM* (SPORE-FORMING RODS)

There are 6 medically important gram-positive bacteria: 2 are cocci, and 4 are rods (bacilli). Two of the rods are spore-formers and 2 are not. We have already discussed the 2 gram-positive cocci (streptococci and staphylococci). In this chapter we will examine the 2 gram-positive **spore-forming** rods, *Bacillus* and *Clostridium*.

Bacillus and *Clostridium* cause disease by the release of potent exotoxins (see **Fig. 2-8**). They differ biochemically by their like or dislike of oxygen. *Bacillus* enjoys oxygen (so is **aerobic**), while *Clos*tridium multiply in an **anaerobic** environment. In an air tight **Clos**et, if you will!

BACILLUS

There are 2 pathogenic species of gram-positive, **aerobic**, spore-forming rods: *Bacillus anthracis* and *Bacillus cereus*. *Bacillus anthracis* causes the disease anthrax while *Bacillus cereus* causes gastroenteritis (food poisoning).

Bacillus anthracis
(Anthrax)

Bacillus anthracis is unique in that it is the only bacterium with a capsule composed of protein (poly-D-glutamic acid). This capsule prevents phagocytosis. *Bacillus anthracis* causes **anthrax**, a disease that primarily affects herbivores (cows and sheep). Humans are exposed to the spores of *Bacillus anthracis* during direct contact with infected animals or soil, or when handling infected animal products, such as hides or wool. In the U.S., cases have followed contact with goat hair products from Haiti, such as drums or rugs. Human-to-human transmission has never been reported.

Bacillus anthracis forms a spore which is very stable, resistant to drying, heat, ultraviolet light, and disinfectants, and can survive dormant in the soil for decades. Once it is introduced into the lungs, intestine, or a skin wound, it germinates and makes toxins. The germination and expression of plasmid encoded virulence factors (on plasmids **pXO1** and **pXO2** is regulated by an increase in temperature to 37 °C, carbon dioxide concentration, and serum proteins. So the spore actually activates only when introduced into the host! Because of its small size (1–2 μm, ideal for inhalation into alveoli), stability and very high mortality associated with pulmonary anthrax (after spore inhalation), it is felt to be an ideal candidate

for biological terrorism and warfare. This fear came to fruition in the United States in October 2001 when 22 cases of confirmed or suspected anthrax were identified. The initial cases were discovered in postal workers in New Jersey and Washington DC. Three letters were found which contained anthrax spores and based on molecular typing were indistinguishable from the infecting agent. Eleven affected individuals developed cutaneous anthrax and recovered with appropriate therapy. Of nine confirmed cases of pulmonary anthrax four died. In addition, there were two suspected cases of inhalation anthrax and both patients died.

Fig. 6-1. Anthony has Anthrax. This figure demonstrates how the *Bacillus anthracis* spores are contracted from contaminated products made of hides and goat hair. The spores can germinate on skin abrasions (cutaneous anthrax), be inspired into the lungs (respiratory anthrax), or ingested into the gastrointestinal tract (GI anthrax). The spores are often phagocytosed by macrophages in the skin, intestine, or lung and then germinate, becoming active (vegetative) gram-positive rods. The bacteria are released from the macrophage, reproduce in the lymphatic system, and then invade the bloodstream (up to 10–100 million bugs per milliliter of blood!!!).

With a cutaneous anthrax infection (the most common route of entry), *Bacillus anthracis* rapidly multiplies and releases a potent exotoxin. This exotoxin causes localized tissue necrosis, evidenced by a painless round black lesion with a rim of edema. This lesion is called a "malignant pustule" because without antibiotic therapy (penicillin), *Bacillus anthracis* can continue to proliferate and disseminate through the bloodstream, which can cause death. The skin lesion usually resolves spontaneously in 80–90% of cases, but sometimes severe skin edema and shock occur.

Pulmonary anthrax, called woolsorter's disease, is not actually pneumonia. The spores are taken up by macrophages in the lungs and transported to the hilar and mediastinal lymph nodes where they germinate. Mediastinal hemorrhage occurs resulting in mediastinal widening (enlarged area around and above the heart seen on chest radiograph and CT scan) and pleural effusions.

Gastrointestinal anthrax frequently results in death and fortunately is rare. Outbreaks have followed the ingestion of spores (often from contaminated meat). *Bacillus anthracis* matures and replicates within the intestine, where it releases its exotoxin. The exotoxin causes a

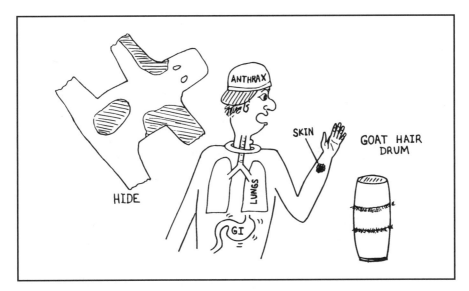

Figure 6-1

necrotic lesion within the intestine. Patients present with vomiting, abdominal pain, and bloody diarrhea.

The release of exotoxin is the major reason why anthrax carries such a high mortality rate. These toxins are encoded on a plasmid called **pXO1**. The exotoxin contains 3 separate proteins, which by themselves are nontoxic but together produce the systemic effects of anthrax:

1) **Edema factor (EF)** This is the active A subunit of this exotoxin and is a calmodulin-dependent adenylate cyclase. It increases cAMP, which impairs neutrophil function and causes massive edema (disrupts water homeostasis).

2) **Protective antigen (PA)** promotes entry of EF into phagocytic cells (similar to a B subunit of the other A–B toxins, see discussion of exotoxins in Chapter 2).

3) **Lethal factor (LF)** is a zinc metalloprotease that inactivates protein kinase. This toxin stimulates the macrophage to release tumor necrosis factor α and interleukin-1β, which contribute to death in anthrax.

A second plasmid, **pXO2**, encodes three genes necessary for the synthesis of a **poly-glutamyl** capsule. This capsule inhibits phagocytosis of the vegetative bacteria. Both plasmids are critical for bacterial virulence.

Rapid identification and the prompt use of ciprofloxacin or doxycycline are critical in preventing the high mortality associated with systemic infection by *Bacillus anthracis*. Since the episode of bioterrorism in 2001 more sensitive and rapid methods of diagnosis have been developed, including obtaining a nasal swab for polymerase chain reaction which has the ability to detect as few as 3 spores, and improved serologic techniques are now available. Individuals taking part in high-risk

activities (petting goats or cows in countries where this disease is rampant) and military personnel should be given a vaccine composed of the protective antigen (PA). Animals are vaccinated with living cultures attenuated by the loss of their antiphagocytic protein capsule. These living vaccines are considered too dangerous for human use.

Bacillus cereus
("Be serious")

Bacillus cereus is different from *Bacillus anthracis* in that it is motile, non-encapsulated, and resistant to penicillin. *Bacillus cereus* causes food poisoning (nausea, vomiting, and diarrhea). Food poisoning occurs when *Bacillus cereus* deposits its spores in food, which then survive the initial cooking process. The bacteria then germinate in the food and begin releasing their enterotoxin. To inactivate the spores, the cooked food must be exposed to high temperatures and/or refrigeration.

Bacillus cereus can secrete 2 types of enterotoxins, which cause different kinds of food poisoning:

1) A **heat-labile toxin** similar to the enterotoxin of cholera and the LT from *Escherichia coli* (see **Fig. 2-8**) causes nausea, abdominal pain and diarrhea, lasting 12–24 hours.

2) A **heat-stable toxin** produces a clinical syndrome similar to that of *Staphylococcus aureus* food poisoning, with a short incubation period followed by severe nausea and vomiting, with limited diarrhea.

When a patient is rushed to the hospital with food poisoning, and examination of the food reveals *B. cereus*, the best way to respond when your attending orders you to treat the patient with antibiotics is "**Be serious**, Dr. Goofball." Since the food poisoning is caused by the pre-formed enterotoxin of *Bacillus cereus*, antibiotic therapy will not alter the course of this patient's symptoms.

CLOSTRIDIUM

Clostridium are also gram-positive spore-forming rods. However, they are **anaerobic**, and can therefore be separated from the aerobic spore-forming rods (*Bacillus*) by anaerobic culture. This group of bacteria is responsible for the famous diseases botulism, tetanus, gas gangrene, and pseudomembranous colitis.

Clostridium harm their human hosts by secreting extremely powerful exotoxins and enzymes. Rapid diagnosis of a clostridial infection is crucial, or your patient will die!!!!

Clostridium botulinum
(Botulism)

Clostridium botulinum produces an extremely lethal neurotoxin that causes a rapidly fatal food poisoning. The neurotoxin blocks the release of acetylcholine (ACh) from presynaptic nerve terminals in the autonomic nervous system and motor endplates, causing **flaccid** muscle paralysis.

Adult Botulism

Eating smoked fish or home-canned vegetables is associated with the transmission of botulism. *Clostridium botulinum* spores float in the air and can land on food. If the food is cooked thoroughly, the spores will die. However, if the food with the spores is not cooked sufficiently, and is then placed into an anaerobic environment (like a glass jar, can, or zip-lock freezer bag), *Clostridium botulinum* matures and synthesizes its neurotoxin. Those who consume the contents of the jar when it is opened weeks later **will be ingesting the potent neurotoxin**. These afebrile patients initially develop bilateral cranial nerve palsies causing double vision (diplopia) and difficulty swallowing (dysphagia). This is followed by general muscle weakness, which rapidly leads to sudden respiratory paralysis and death. Patients must immediately be treated with an antitoxin, which can neutralize only the unbound free neurotoxin in the bloodstream. Intubation and ventilatory support is critical until the respiratory muscles resume activity.

Infant Botulism

Infant botulism occurs when infants **ingest food contaminated with *Clostridium botulinum* spores** (cases have followed ingestion of fresh honey contaminated with spores). The spores germinate and

Clostridium botulinum colonizes the infant's intestinal tract. From this location, botulism toxin is released.

Initially, the infant will be constipated for two to three days. This is followed by difficulty swallowing and muscle weakness. These "floppy" babies must be hospitalized and given supportive therapy. A recent well performed study confirmed the benefit of human botulism immunoglobulin intravenous (BIG-IV) for infant botulism. BIG-IV differs from the standard equine botulism antitoxin used for adult patients in that it is human derived rather than horse derived and therefore does not predispose to side effects such as serum sickness or anaphylaxis. (McDonald LC, et al. NEJM, 2006).

Wound Botulism

Wound botulism, the least common presentation of botulism, is associated with puncture wounds or deep space infections, which provide an ideal environment for spores from the soil or environment to germinate and produce toxin. The presentation is similar to adult botulism except for absence of prodromal gastrointestinal symptoms and a longer incubation period. In addition, patients are more likely to have a fever and an associated elevation of their white count. Treatment is surgical debridement of devitalized tissue, anti-toxin, and usually antibiotics to cover clostridia and other co-existent pathogens.

Fig. 6-2. Botulism. The adult is eating home-canned beans with neurotoxin while the infant is eating honey with spores. The adult often requires intubation and ventilatory support while the baby is merely "floppy."

Clostridium tetani
(Tetanus)

Clostridium tetani causes **tetanus**, a disease that classically follows a puncture wound by a rusty nail but can follow skin trauma by any object contaminated with spores. *Clostridium tetani* spores, which are commonly found in soil and animal feces, are deposited in the wound and can germinate as long as there is a localized anaerobic environment (necrotic tissue). From this location, *Clostridium tetani* releases its exotoxin, called **tetanospasmin**.

The tetanus toxin ultimately causes a sustained contraction of skeletal muscles called **tetany**.

Fig. 6-3. Tetany occurs after the tetanus toxin is taken up at the neuromuscular junction (end plate) and is transported to the central nervous system. There the toxin acts on the inhibitory Renshaw cell interneurons, preventing the release of GABA and glycine, which are inhibitory neurotransmitters. This **inhibition** of **inhibitory** interneurons allows motor neurons to send

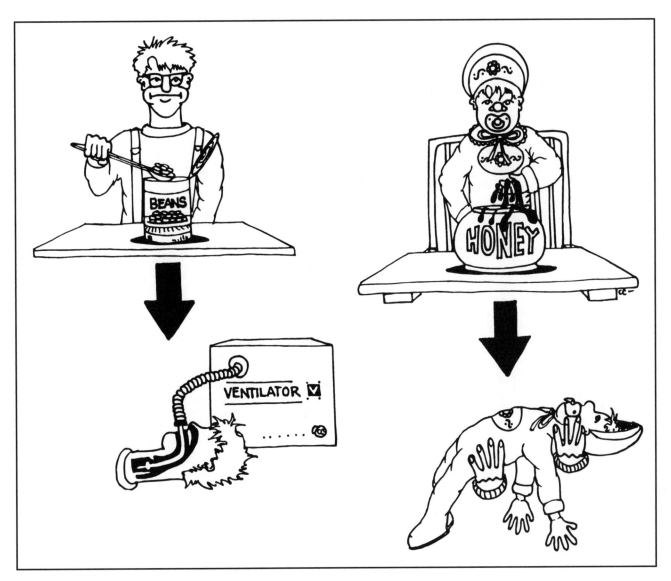

Figure 6-2

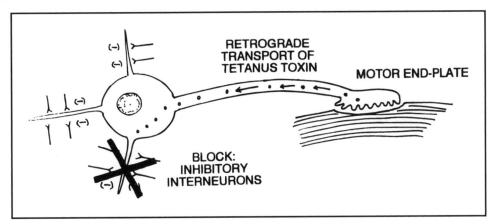

Figure 6-3

Figure 6-4

a high frequency of impulses to muscle cells, which results in a sustained tetanic contraction.

Fig. 6-4.   Clinically, the patient with tetanus presents with severe muscle spasms, especially in the muscles of the jaw (this is called **trismus**, or **lockjaw**). The affected patient exhibits a grotesque grinning expression, called **risus sardonicus**, which is due to spasm of the facial muscles. Mortality is high once the stage of lockjaw has been reached.

Because of the high mortality of tetanus, prophylactic immunization with formalin-inactivated toxin (tetanus toxoid) is performed once every ten years in the U. S. This booster serves to regenerate the circulating antibodies against tetanus toxin, that were first generated via childhood immunizations. You may not remember your first shot (you were probably just 2 months old at the time), but all children in the U. S. are immunized with a series of DPT (diphtheria-pertussis-tetanus) shots at ages 2, 4,

6, and 18 months, followed by a booster before entry into school (4–6 years). This regimen provides protection from tetanus (along with diphtheria and pertussis). However, the protection from tetanus only lasts about 10 years so booster shots of tetanus are given every 10 years.

In the emergency room you will encounter 3 types of patients with skin wounds:

1) Patients who were immunized as a child and received periodic boosters but the last shot was **more than 10 years ago**. These patients are given another booster.

2) Patients who have **never been immunized**. Not only do these patients need a booster, but they should also receive preformed antibodies to the tetanus toxin called human tetanus immune globulins.

3) Patients who come to the hospital having already developed tetanus. The big picture is to clear the toxin and the toxin-producing bacteria and to keep the patient alive until the toxin has cleared. This is accomplished in the following 5 steps of therapy:

a) Neutralize circulating toxin with human tetanus immune globulins.
b) Give an immunization booster to stimulate the patient's own immune system to develop antitetanus toxin antibodies.
c) Clean the wound, excising any devitalized tissue, to remove any remaining source of *Clostridium tetani*.
d) Antibiotics (metronidazole or penicillin) may help to clear the remaining toxin-producing bacteria.
e) Provide intensive supportive therapy until the toxin is cleared. Muscle relaxants may have to be administered, and the patient may have to be placed on a ventilator.

Fig. 6-5.   Both botulism and tetanus can lead to respiratory failure requiring mechanical ventilation. In botulism, there is flaccid muscle paralysis as the

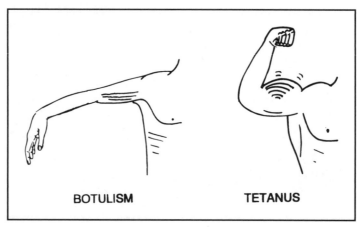

Figure 6-5

acetylcholine at the motor end plate is blocked. In tetanus there is constant muscle contraction, as the inhibitory signals are blocked. The tetanic contraction of the respiratory muscles also results in respiratory failure.

Clostridium perfringens
(Gas Gangrene)

Everyone has heard of gas gangrene. Prior to antibiotics, *Clostridium perfringens* devastated soldiers wounded in battle. This bacterium, whose spores can be found in the soil, matures in anaerobic conditions and produces gas. The spores can contaminate wounds from battle or other trauma. Deep wounds with lots of dead tissue create an anaerobic environment that offers an excellent home for *Clostridium perfringens*. As this anaerobic organism grows, it releases its battery of exotoxin enzymes (see **Fig. 2-8**), causing further tissue destruction.

Clinically, there are 3 classes of infection with *Clostridium perfringens*:

1) **Cellulitis/wound infection**: Necrotic skin is exposed to *Clostridium perfringens*, which grows and damages local tissue. Palpation reveals a moist, spongy, crackling consistency to the skin due to pockets of gas; this is called **crepitus**.

2) **Clostridial myonecrosis**: *Clostridium perfringens*, inoculated with trauma into muscle, secretes exotoxins that destroy adjacent muscle. These anaerobic bacteria release other enzymes that ferment carbohydrates, resulting in gas formation. A computerized tomogram (CT) scan reveals pockets of gas within the muscles and subcutaneous tissue. As the enzymes degrade the muscles, a thin, blackish fluid exudes from the skin.

Clostridial myonecrosis is fatal unless identified and treated very early. Hyperbaric oxygen, antibiotics (such as penicillin), and removal of necrotic tissue can be life-saving.

3) **Diarrheal illness**: *Clostridium perfringens* spores can germinate in foods such as meats, poultry and gravy. Ingestion of large amounts of bacteria can lead to toxin production in the gut and subsequent watery diarrhea. A more severe diarrheal illness caused by one subtype of *Clostridium perfringens* can lead to hemorrhagic necrosis of the jejunum (this severe form is uncommon in the U.S.).

Clostridium difficile
(Pseudomembranous Enterocolitis)

While you may never see a case of anthrax, tetanus, or botulism in your career, this will certainly NOT be the case with *Clostridium difficile*. You will tangle with this critter frequently. *Clostridium difficile* is the pathogen responsible for antibiotic-associated pseudo-membranous colitis (diarrhea), which can follow the use of broad spectrum antibiotics. These antibiotics can wipe out part of the normal intestinal flora, allowing the pathogenic *Clostridium difficile* that is sometimes present to superinfect the colon. Once *Clostridium difficile* grows in abundance, it then releases its exotoxins. Toxin A causes diarrhea, and Toxin B is cytotoxic to the colonic cells. This disease is characterized by severe diarrhea, abdominal cramping, and fever.

Since 2002, a newer potentially more pathogenic strain of *Clostridium difficile*, NAP1/BI/027, has been circulating in Canada and the United States and is associated with selected outbreaks. This new strain produces 15 to 20 times more toxin than prior strains, as well as a newly defined **binary toxin CDT**. Patients infected with this new strain have been sicker and more resistant to initial therapy.

Because of *Clostridium difficile* it becomes very *difficile* (difficult) to give patients antibiotics.

Examination by colonoscopy can reveal red inflamed mucosa and areas of white exudate called **pseudomembranes** on the surface of the large intestine. Necrosis of the mucosal surface occurs underneath the pseudomembranes.

When a patient develops diarrhea while on antibiotics (or within a month of being on antibiotics), *Clostridium difficile* must be considered as a possible cause. There are several diagnostic tests available for *C. difficile*. PCR testing for toxin A and B genes is the preferred test due to its excellent sensitivity and specificity and quick turn-around time (results can be available within an hour). Enzyme immunoassay (EIA) for the toxins A and B is still commonly done but it only has a sensitivity of about 75%. Other tests such as EIA for glutamate dehydrogenase antigen and cell culture are variably performed.

Treatment includes discontinuing the initial antibiotic and then beginning appropriate therapy. As *C. difficile* has become a worsening problem, potential treatments have blossomed. Metronidazole is recommended for mild cases of *C. difficile* related diarrhea and vancomycin for severe disease or relapse. Both antibiotics effectively kill *C. difficile*. Vancomycin is unique in that it is not absorbed when taken orally and therefore remains in the gastrointestinal tract at the site of infection. Think of the Vancomycin "Van" cruising down the gastrointestinal tract to run over hapless *Clostridium difficile* bacteria. Note that IV vancomycin does not get into the gastrointestinal tract and is worthless against colitis due to *C. difficile*! Metronidazole is well absorbed into the bloodstream and is carried to the site of infection via the colonic blood vessels, so that it can be given orally or IV. see Chapter 18, **Fig. 18-6**).

Unfortunately relapse due to *C. difficile* is increasingly common (10-25%) and with each relapse further recurrences may occur. A new agent, fidaxomicin, was approved in 2011 and has been shown to be superior to vancomycin in treating recurrences of *C. difficile* infection. For those with an adventurous spirit... fecal transplantation has been shown to be highly effective at preventing recurrent disease in persons who have had multiple relapses of *C. difficile* infection.

Fig. 6-6. Summary of the Gram-positive spore-forming rods

References

Arnon SS, Schechter R, Maslanka SE, Jewell, NP, Hatheway CL. Human botulism immune globulin for the treatment of infant botulism. New Eng J Med 2006;354:462–471.

Bartlett JG, Gerding DN. Clinical recognition and diagnosis of *Clostridium difficile* infection. Clin Infect Dis 2008;46: S12–8.

Gerding DN, Johnson S. Management of *Clostridium difficile* infection: Thinking inside and outside the Box. Clin Infect Dis 2010;51:1306–13.

Jernigan DB, et al. Investigation of bioterrorism-related anthrax, United States, 2001: epidemiologic findings. Emerg Infect Dis 2002;10:1019–1028.

Louie TJ, Miller MA, et al. Fidaxomicin versus Vancomycin for *Clostridium difficile* infection. NEJM 2011;364:422–31.

ORGANISM	RESERVOIR	TRANSMISSION	METABOLISM	VIRULENCE
Bacillus anthracis	• Herbivores (zoonotic) A. Sheep B. Goats C. Cattle	• Endospores 1. Cutaneous 2. Inhalation 3. Ingestion	• Aerobic (but since it can grow without oxygen, it is classified as a facultative anaerobe)	1. Unique **protein capsule** (polymer of gamma-D-glutamic acid): antiphagocytic 2. **Non**-motile 3. Virulence depends on acquiring 2 plasmids. One carries the gene for the protein capsule; the other carries the gene for its exotoxin
Bacillus cereus		• Endospores	• Aerobic	1. **No** Capsule 2. Motile
Clostridium botulinum	1. Soil 2. Stored vegetables: • Home-canned • Zip-lock storage bags 3. Smoked fish 4. Fresh honey: associated with infant botulism	• Endospores (heat resistant)	• **Anaerobic**	• Motile: flagella (so H-antigen positive)

Figure 6-6   GRAM-POSITIVE SPORE-FORMING RODS

McDonald LC, et al. An epidemic, toxin gene-variant stain of *Clostridium difficile*. New Eng J Med 2005;353:2433–2441.

Machem CC, Walter FG. Wound botulism. Vet Hum Toxicol 1994;36:233–237.

van Nood E, Vrienze A, et al. Duodenal infusion of donor feces for recurrent *Clostridium difficile*. NEJM 2013;368:407–15.

Update: Investigation of bioterrorism-related anthrax and interim guidelines for the exposure management and antimicrobial therapy, October 2001. MMWR Weekly 2001;50:909–919.

Warny M, et al. Toxin production by an emerging strain of *Clostridium difficile* associated with outbreaks of severe disease in North America and Europe. Lancet 2005;366:1079–1084.

Recommended Review Articles:

Inglesby TV, O'Toole T, Henderson DA, et al. Anthrax as a biological weapon, 2002: updated recommendations for management. JAMA. 2002;287(17):2236–52

Kelly CP, LaMont JT. Clostridium difficile—more difficult than ever. N Engl J Med. 2008;359(18):1932–40.

Pellizzari R, et al. Tetanus and Botulism Neurotoxins: mechanisms of action and therapeutic uses. Philosophical Transactions of the Royal Society of London 1999;354:259–268.

Salkind AR Clostridium difficile: an update for the primary care clinician. South Med J. 2010;103(9):896–902.

Swartz MN. Recognition and management of anthrax—an update. N Engl J Med. 2001;345(22):1621–6. Epub 2001 Nov 6.

TOXINS	CLINICAL	TREATMENT	DIAGNOSTICS

Exotoxin: 3 proteins
a. Protective antigen (PA)
b. Edema factor (EF)
c. Lethal factor (LF)
	Anthrax
	1. Cutaneous (95%): painless black vesicles; Can be fatal if untreated
	2. Pulmonary (woolsorter's disease)
	3. GI: abdominal pain, vomiting and bloody diarrhea
	  • Infection results in permanent immunity (if the patient survives)
		1. Ciprofloxacin
		2. Doxycycline
		3. Raxibacumab (monoclonal antibody for use in inhalational anthrax)
		4. Vaccine: for high-risk individuals
		  A. Vaccine is composed of the protective antigen (PA)
		  B. Animal vaccine is composed of a live strain, attenuated by loss of its protein capsule
			1. Gram stain
			2. Culture
			3. Serology
			4. PCR of nasal swab

Enterotoxins A. Heat labile: similar to enterotoxin of cholera and *E. coli*. B. Heat stable: produces syndrome similar to that of *Staphylococcus aureus* food poisoning, but with limited diarrhea	• **Food poisoning**: nausea, vomiting and diarrhea	1. Vancomycin 2. Clindamycin 3. Resistant to beta-lactam antibiotics 4. No treatment for food poisoning ("Be serious, Dr. Goofball: food poisoning is caused by the pre-formed enterotoxin)	• Culture specimen from suspected food source
1. Neurotoxin: inhibits release of acetylcholine from peripheral nerves 2. Toxin is not secreted. Rather it is released upon the death of the bacterium	***Food-borne botulism***: 1. Cranial nerve palsies 2. Muscle weakness 3. Respiratory paralysis ***Infant botulism***: 1. Constipation 2. Flaccid paralysis ***Wound botulism***: 1. Similar to Food-borne except absence of GI prodromal symptoms	1. Antitoxin (for food-borne and wound botulism) 2. Human botulism immunoglobulin (for infant botulism) 3. Penicillin 4. Hyperbaric oxygen 5. Supportive therapy: including incubation and ventilatory assistance	1. Gram stain 2. Culture: requires anaerobic conditions 3. Patient's serum injected into mice results in death

ORGANISM	RESERVOIR	TRANSMISSION	METABOLISM	VIRULENCE
Clostridium tetani	• Soil	• Endospores: introduced through wound	• **Anaerobic**	• Motile: flagella (so H-antigen positive)
Clostridium perfringens	• Ubiquitous: 1. Soil 2. GI tract of humans & mammals	• Endospores	• **Anaerobic**	• **NON**-motile
Clostridium difficile	1. Intestinal tract 2. Endospores found in hospitals and nursing homes	• Fecal-oral: ingestion of endospores	• **Anaerobic**	• Motile: flagella (so H-antigen positive)

Figure 6-6   (continued)

TOXINS	CLINICAL	TREATMENT	DIAGNOSTICS

• **Tetanospasmin**: inhibits release of GABA and glycine (both inhibitory neurotransmitters) from nerve cells, resulting in sustained muscle contraction	***Tetanus***
	1. Muscle spasms
	2. Lockjaw (trismus)
	3. Risus sardonicus
	4. Respiratory muscle paralysis
		1. Tetanus toxoid: vaccination with formalin-inactivated toxin (toxoid). Part of the DPT vaccine
		2. Antitoxin: human tetanus immune globulin (preformed anti-tetanus antibodies)
		3. Clean the wound
		4. Metronidazole or penicillin
		5. Supportive therapy: may require ventilatory assistance
		  • Vaccine: **DTap**
		    1. **D**iphtheria
		    2. **T**etanus
		    3. **a**cellular **P**ertussis
			1. Gram stain: gram-positive rods, often with an endospore at one end, giving them the appearance of a **drumstick**
			2. Culture: requires anaerobic conditions

1. Alpha toxin: lecithinase (splits lecithin into phosphocholine and diglyceride) 2. 11 other tissue destructive enzymes	***Gaseous Gangrene*** A. Cellulitis/wound infection B. Clostridial myonecrosis: fatal if untreated C. Watery diarrhea: associated with food-borne ingestion	1. Radical surgery (may require amputation) 2. Penicillin 3. Hyperbaric oxygen	1. Gram stain 2. Culture: requires anaerobic conditions
1. Toxin A: diarrhea 2. Toxin B: cytotoxic to colonic epithelial cells	• **Pseudomembranous enterocolitis**: antibiotic-associated diarrhea	1. Metronidazole 2. Oral vancomycin 3. Fidaxomicin 4. Fecal transplant 5. Discontinue unnecessary antibiotics	1. Immunoassay for *C. difficile* toxin 2. PCR for toxin A and B genes

M. Gladwin, W. Trattler, and S. Mahan, *Clinical Microbiology Made Ridiculously Simple* ©MedMaster

CHAPTER 7. *CORYNEBACTERIUM AND LISTERIA* (NON-SPORE-FORMING RODS)

We have examined 3 gram-positive cocci (*Streptococcus, Staphylococcus,* and *Enterococcus*) and 2 gram-positive spore-producing rods (*Bacillus* and *Clostridium*). Now we will discuss the other 2 gram-positive rods (both non-spore-formers): *Corynebacterium diphtheriae* and *Listeria monocytogenes.* Both of these gram-positive rods infect patients in the **pediatric** age group.

CORYNEBACTERIUM DIPHTHERIAE

Corynebacterium diphtheriae is the pathogen responsible for diphtheria. It colonizes the pharynx, forming a grayish **pseudomembrane** composed of fibrin, leukocytes, necrotic epithelial cells, and *Corynebacterium diphtheriae* cells. From this site, the bacteria release a powerful exotoxin into the bloodstream, which specifically damages heart and neural cells by interfering with protein synthesis.

Fig. 7-1. Visualize the invading *Corynebacterium diphtheriae* organisms as a tiny invading army overrunning the throat and building a launching platform on the pharynx. The army quickly constructs exotoxin rockets. From the safety of their pharynx base, they fire off deadly rockets to the heart and central nervous system.

While working in the pediatric emergency room, you see a child with a sore throat and fever. There is a dark inflammatory exudate on the child's pharynx, which appears darker and thicker than that of strep throat. Although you may feel the urge to scrape off this tightly adherent pseudomembrane, you must resist this temptation, because bleeding will occur and the systemic absorption of the lethal exotoxin will be enhanced.

Being the brightest medical student in the pediatric emergency room, you immediately recognize that this child probably has diphtheria. Realizing that you are dealing with an extremely powerful exotoxin, you quickly **TELL** yo**UR** In**TE**rn not to "**loaf** around" (Loeffler's). Immediately send the throat and nasopharynx swabs for culture on **potassium tellurite agar** and **Loeffler's coagulated blood serum** media. However, these culture results will not be ready for days!!! You may try a Gram stain of a specimen from the pseudomembrane, but gram-positive rods are not always seen. Since there is no time to loaf with diphtheria, it is often best to proceed rapidly to treatment via the following 3-step method.

1) **Antitoxin**: The diphtheria antitoxin only inactivates circulating toxin, which has not yet reached its target tissue, so this must be administered quickly to prevent damage to the heart and nervous system.

2) **Penicillin** or **erythromycin**: Either antibiotic will kill the bacteria, preventing further exotoxin release and rendering the patient non-contagious.

3) **DPT vaccine**: The child must receive the DPT vaccine, as infection by *Corynebacterium diphtheriae* does not always result in immunity to future infection by this organism. The DPT vaccine stands for: **D** = Diphtheria; **P** = Pertussis (Whooping Cough); and **T** = Tetanus. The diphtheria portion contains formalin inactivated diphtheria toxin (see Chapter 6, Tetanus section for more on DPT).

Now that therapy has been administered, we can sit back, relax, and confirm our clinical suspicion of diphtheria. On the potassium-tellurite plate, colonies of *Corynebacterium diphtheriae* become gray to black within 24 hours. With Loeffler's coagulated blood serum, incubation for 12 hours followed by staining with methylene blue will reveal rod-shaped pleomorphic bacteria.

Fortunately for nonimmunized children, not all *Corynebacterium diphtheriae* secrete this exotoxin. Just as Group A beta-hemolytic streptococci must first be lysogenized by a temperate bacteriophage to produce the erythrogenic toxin that causes scarlet fever, *Corynebacterium diphtheriae* first must be lysogenized by a temperate bacteriophage which codes for the diphtheria exotoxin.

This powerful exotoxin contains two subunits. The B subunit binds to target cells and allows the A subunit to enter the cell. Once inside the cell, the A subunit blocks protein synthesis by inactivating elongation factor (EF_2), which is involved in translation of eucaryotic mRNA into proteins (See **Fig. 2-8**). Notice an interesting comparison: Anti-ribosomal antibiotics are specifically designed to inhibit protein synthesis in bacterial (procaryotic) cells. Similarly, this exotoxin specifically inhibits protein synthesis in humans (eucaryotes). Thus this exotoxin can be considered a "human antibiotic," because its damage to heart and neural cells can be lethal.

Other *Coryneform bacteria* (also called the diphtheroids)

The Greek word "koryne" means "club" and the word "bacterion" means "little rod"; other bacteria that share these morphological features with *Corynebacterium diphtheriae* have been called either the Coryneform bacteria or

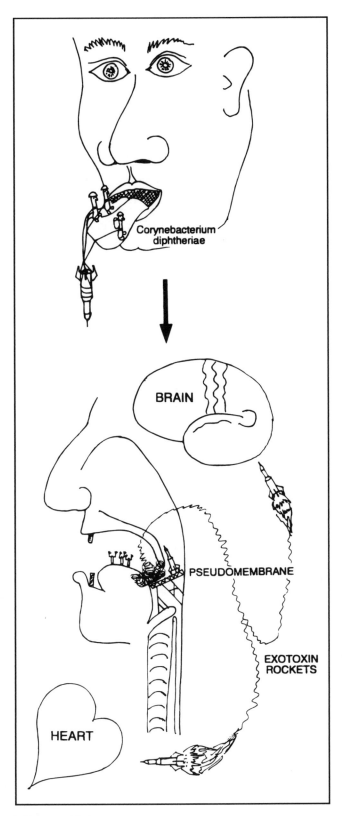

Figure 7-1

the diphtheroids. These bacteria include the genera *Corynebacterium, Arcanobacterium, Brevibacterium, Microbacterium* and others. These Coryneform bacteria are normal inhabitants of water, soil and the human skin and mucous membranes. They are frequently isolated as "contaminants" in cultures in hospitalized patients and you will often receive a report from the micro lab on one of your patients stating that the early culture is growing a diphtheroid or gram positive rod. While this usually does not indicate an active infection, one must be careful because these bacteria are increasingly causing both community acquired infections (native valve endocarditis, urinary tract infections, prostatitis, and periodontal infections) and nosocomial (hospital acquired) infections. In hospitalized patients, these bacteria can cause surgical wound infections, catheter and prosthetic device related infections, and native and prosthetic valve endocarditis. Common nosocomial pathogens include *Corynebacterium jeikeium, Corynebacterium urealyticum, Corynebacterium amycolatum,* and *Corynebacterium striatum.*

Another thing that sets the Coryneform bacteria apart from *Corynebacterium diptheriae* is that they are usually quite resistant to antibiotics and require treatment with intravenous vancomycin. Consultation with an infectious diseases physician is advised to help determine if an actual infection exists and to guide antibiotic therapy.

Rhodococcus equi (formerly *Corynebacterium equi*)

This gram positive, aerobic nonmotile, bacillary bacteria (rod-like that can grow long, curved, clubbed and even form branching short filaments) typically infects animals like cattle, sheep, deer, bears, dogs and cats. It has been isolated in manure and soil and if inhaled by an immunocompromised person (organ transplant recipient or HIV infected) it can form a necrotizing pneumonia that looks like that caused by *Mycobacterium tuberculosis* or *Nocardia*: the infection forms infiltrates, single or multiple nodules that cavitate and pleural effusions. **Clinical Pearl**: Upper lobe lung nodules and cavities that form air-fluid levels are characteristic of *Rhodococcus* infection, while the upper lung cavities caused by *Mycobacterium tuberculosis* rarely form air-fluid levels. Also, *Rhodococcus* may stain **partially acid-fast**, leading to even more diagnostic confusion when trying to differentiate from tuberculosis.

LISTERIA MONOCYTOGENES

Listeria monocytogenes is a small facultatively anaerobic, non-spore forming gram-positive rod that when isolated in blood is often first identified by the micro lab as a diphtheroid (be careful because you might think this is just a contaminant!). It has 1–5 flagella and when grown at 25 °C exhibits a tumbling motility. Because it can grow at low temperatures, it is often cultured at 4–10 °C, so called cold-enrichment, to differentiate it from other bacteria. It has a major virulence factor, **listeriolysin O**, that allows it to escape the phagolysosomes of macrophages and avoid intracellular killing.

To remember why this bug *List*eriosis is bad, think of this **List**: Pregnant women, neonates, and meningitis in the elderly and immunocompromised:

1) The first group at high risk of infection is **pregnant women**. Infection usually occurs in the third trimester, when the cell-mediated immunity decreases. Interestingly, meningitis is unusual in pregnant women who usually suffer from bacteremia and sepsis. The bacteria infects the fetus and 22% of these infections result in neonatal death; surviving babies are often born prematurely with active infection. Because *Listeria* is acquired from ingestion of contaminated foods (infected coleslaw, milk, soft cheeses, butter and deli meats) pregnant women are often told to avoid soft cheese and cold cuts.

2) The second group at risk is the **fetus and neonate**. Infection is acquired in utero as described above and can also be contracted from an asymptomatic mother, with vaginal colonization with *Listeria*, during vaginal birth. This mode of infection results in neonatal meningitis presenting about 2 weeks post-partum. Since the advent of *Haemophilus influenzae* vaccination (HiB vaccine-see **Chapter 11**), *Listeria monocytogenes* now causes about 20% of all neonatal meningitis.

Clinical Pearl: Three bacteria are responsible for most meningitis acquired by the baby coming out of the birth canal (within first 3 months of age): *Listeria monocytogenes*, *Escherichia coli*, and *Group B Streptococcus*. Two bacteria cause meningitis later in life after the maternal antibodies passively given to fetus wane and before new antibodies develop: *Neisseria meningitides* and *Haemophilus influenzae*.

3) The third group of patients at risk for *Listeria* meningitis is the **elderly** and the **immunocompromised**. In fact, *Listeria* is the second most common cause of meningitis, after *Pneumococcus*, in people >50 years of age and is the most common cause of meningitis in patients with lymphoma, on corticosteroids or receiving organ transplantation. It also causes meningitis frequently in patients with AIDS.

You may wonder why this organism invades neonates and certain immunosuppressed patients but not an immune **competent** host. The main reason is that *Listeria monocytogenes* is a resistant fellow, able to hide out and survive within certain immune cells, such as macrophages and neutrophils that can phagocytose, or engulf, foreign objects such as bacteria. Since they can survive either outside or within cells, *Listeria monocytogenes* is called a **facultative intracellular organism** (see **Fig. 2-7**). However, in immune competent hosts, the immune system can release factors that activate the macrophage, so that these cells can now destroy the "vagrant" bacteria within them. Immunologists refer to this immune system-mediated method of destroying *Listeria* as **cell-mediated immunity**. However, neonates (up to 3 months of age) and immunosuppressed patients are unable to activate their phagocytic cells, thus allowing *Listeria monocytogenes* to flourish and infect the meninges. Since pregnancy may also depress cell-mediated immunity, *Listeria monocytogenes* can infect pregnant women as well, who may develop meningitis or remain asymptomatic carriers.

Fig. 7-2. **A)** The macrophage of a neonate or an immunosuppressed patient. **B)** The macrophage of an immune competent person.

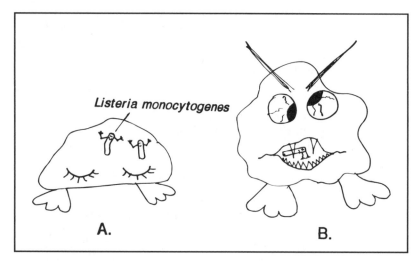

Figure 7-2

When meningitis develops in a patient who is at high risk for *Listeria monocytogenes*, it is important to treat it empirically with antibiotics that will cover this bacterium. After a lumbar puncture confirms that this is a bacterial meningitis (cerebrospinal fluid analysis reveals a high number of neutrophils, a high protein level, a low glucose, and the Gram stain of the cerebrospinal fluid may demonstrate gram-positive rods), we must add either **ampicillin** or **trimethoprim-sulfamethoxazole** to the antibiotic regimen. These are 2 antibiotics that cover *Listeria monocytogenes*.

Clinical Pearl: In all adults over 50 years of age and in immunocompromised patients who develop an acute meningitis, add ampicillin or trimethoprim sulfamethoxazole to empirically cover *Listeria monocytogenes*!

Fig. 7-3. Summary of the non-spore-forming gram-positive rods.

Recommended Review Articles:

Allerberger F, Wagner M. Listeriosis: a resurgent foodborne infection. Clin Microbiol Infect. 2010;16(1):16–23.

Gidengil CA, Sandora TJ, Lee GM. Tetanus-diphtheria-acellular pertussis vaccination of adults in the USA. Expert Rev Vaccines. 2008;7(5):621–34.

Nohynek H, Madhi S, Grijalva CG. Childhood bacterial respiratory diseases: past, present, and future. Pediatr Infect Dis J. 2009;28(10 Suppl):S127–32.

NAME	MORPHOLOGY	TRANSMISSION	METABOLISM	VIRULENCE
Corynebacterium diphtheriae	1. Gram-positive rods (very pleomorphic and club-shaped) 2. Non-spore-forming 3. Non-motile	• Respiratory droplets from a carrier	1. Facultative anaerobe 2. Catalase-positive	• Pseudomembrane forms in the pharynx, which serves as a base from where it secretes its toxin
Listeria monocytogenes	1. Gram-positive rods 2. Non-spore-forming 3. Motile: tumbling motility is seen when grown at 25 °C.	1. Ingestion of contaminated raw milk or cheese from infected cows 2. Vaginally (during birth) 3. Transplacental infection of fetus from bacteremic mother	1. Facultative anaerobe 2. Catalase-positive 3. Beta-hemolytic on blood agar	1. Motile (via flagella): so has H-antigen 2. Hemolysin: (like streptolysin O) a. Heat labile b. Antigenic

Figure 7-3   NON SPORE-FORMING GRAM-POSITIVE RODS

TOXINS	CLINICAL	TREATMENT	DIAGNOSTICS	MISCELLANEOUS

• Exotoxin (coded by a bacteriophage): *A subunit*: blocks protein synthesis by inactivating EF_2 *B subunit*: provides entry into cardiac and neural tissue
 	
 • This exotoxin is like an anti-human antibiotic, as it inhibits eucaryotic protein synthesis, just as tetracycline inhibits protein synthesis in bacteria
	Diphtheria
	1. Mild sore throat with fever initially
	2. Pseudomembrane forms on pharynx
	3. Myocarditis causing A-V conduction block and dysrhythmia
	4. Neural involvement
	  a. Peripheral nerve palsies
	  b. Guillain Barre-like syndrome
	  c. Palatal paralysis and cranial neuropathies
		1. Antitoxin
		2. Penicillin or erythromycin
		3. Vaccine: DPT
		  • Diphtheria: formalin Inactivated exotoxin, as antibodies to the B-subunit are protective
		  • Pertussis
		  • Tetanus
			1. Gram stain: gram-positive pleomorphic rods (sometimes described as looking like Chinese letters)
			2. Culture:
			  A. Potassium tellurite: get dark black colonies
			  B. Loeffler's medium: after 12 hours of growth, stain with methylene blue. Reddish (Babes-Ernst) granules can be seen
				1. Obtains exotoxin from a temperate bacteriophage by lysogenic conversion
				2. Schick test: injection of diphtheria exotoxin into the skin, to determine whether a person is susceptible to infection by *C. diphtheriae*

• **Listeriolysin O and phospholipases:** allows escape from the phagolysosomes of macrophages	1. Neonatal meningitis 2. Meningitis in immunosuppressed patients and the elderly (>50) 3. Septicemia in pregnant women	1. Ampicillin 2. Trimethoprim/ sulfamethoxazole	1. Gram stain: gram-positive rods 2. Culture: can grow at temperatures as low as 0 °C. So use cold enrichment technique to isolate from mixed flora	• Facultative intracellular parasite • Cell-mediated immunity is protective

M. Gladwin, W. Trattler, and S. Mahan, *Clinical Microbiology Made Ridiculously Simple* ©MedMaster

GRAM-NEGATIVE BACTERIA

CHAPTER 8. *NEISSERIA*

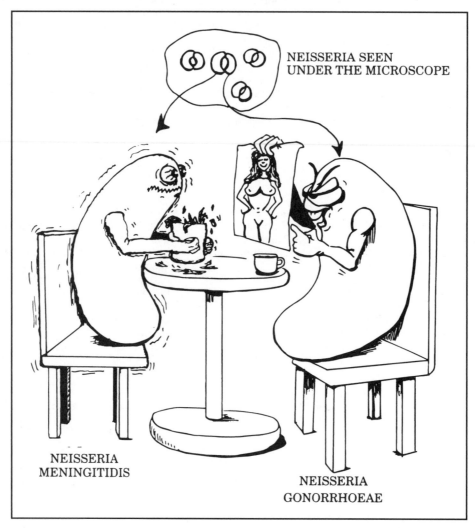

NEISSERIA SEEN UNDER THE MICROSCOPE

NEISSERIA MENINGITIDIS

NEISSERIA GONORRHOEAE

Figure 8-1

It's time to examine the only pathogenic gram-negative cocci, *Neisseria*. These guys hang out in pairs and are thus called diplococci. Each coccus is shaped like a kidney bean, and a pair of cocci sticks together with their concave sides facing each other, almost making the diplococcus look like a small doughnut.

Two species cause disease in humans: ***Neisseria meningitidis*** and ***Neisseria gonorrhoeae***.

Fig. 8-1. Meet the 2 pathogenic kidney beans, which have been removed from the microscope slide. They are sitting together at the breakfast table. Notice that they sit facing each other, forming a gram-negative doughnut-shaped diplococcus. The bean on the left, *Neisseria meningitidis*, drinks a pot of coffee and becomes very **nervous** and **irritable** (**central nervous system irritation—meningitis**). The other pathogenic kidney bean is *Neisseria gonorrhoeae*, who is a **pervert** (notice how he is displaying the latest center-fold pin-up). He enjoys hanging out on sexual organs and swimming in "sexual fluids." He causes the **sexually transmitted disease (STD) gonorrhea**.

NEISSERIA MENINGITIDIS

Besides causing meningitis, *Neisseria meningitidis* (also called the **meningococcus**) causes life-threatening sepsis (meningococcemia).

Virulence factors of the meningococcus include:

1) **Capsule**: A polysaccharide capsule surrounds the bacterium and is antiphagocytic, as long as there are no specific antibodies to coat (opsonize) the bacterium. *Neisseria meningitidis* is classified into serogroups based on different capsular polysaccharides, which are antigenic (stimulate a human antibody response). There are at least 13 serogroups of meningococcus designated A, B, C, D, E, H, I, K, L, X, Y, Z, and W-135 (I would give the people who named these an F-grade for leaving the F, G, and J letters out of the order!). Meningitis is usually caused by serogroups A, B, and C.

2) **Endotoxin (LPS)**: The meningococci can release blebs of endotoxin, which causes blood vessel destruction (hemorrhage) and sepsis (Chapter 2, page 13). The blood vessel hemorrhage is seen on the skin as tiny, round, red dots of hemorrhage called **petechiae** (a petechial rash). This same hemorrhaging process can damage the adrenal glands.

3) **IgA1 protease**: This is only found in pathogenic species of *Neisseria*. This enzyme cleaves IgA (a type of antibody) in half.

4) *Neisseria meningitidis* can extract iron from human transferrin via a non-energy requiring mechanism.

5) **Pili**: *Neisseria meningitidis* possess pili that allow attachment to human nasopharyngeal cells and undergo antigenic variation to avoid attack by the immune system.

Although *Neisseria meningitidis* has all of the above virulence factors, it usually blends in and becomes part of the normal flora of the nasopharynx. These individuals (about 5% of the population) are called carriers. Carriers are lucky, as this asymptomatic nasopharyngeal infection allows them to develop anti-meningococcal antibodies (this is called natural immunization).

High-risk groups are:

1) **Infants aged 6 months to 2 years**
2) **Army recruits**
2) **College freshmen**

During pregnancy, maternal protective antibodies cross the placenta and provide protection to the newborn, but only for the first few months of life. Infants will not manufacture their own protective antibodies for

a few years. Therefore, there is a window period (from age 6 months to 2 years) when infants are extremely susceptible to a meningococcal infection (meningitis or septicemia). Note that *Haemophilus influenzae* has a similar antibody-free window.

A second scenario for an invasive meningococcal infection occurs when army recruits from all over the United States are placed together in close quarters and must survive "boot camp." In this close-knit group, carrier rates are greater than 40%. Each army recruit may be a carrier of a particular strain of meningococcus that the other army recruits' immune systems have never been exposed to, increasing susceptibility to invasive disease. Further, due to the mentally and physically exhausting training, the immune system's ability to defend itself is weakened.

Freshmen entering college and living in dormitories are also at high risk for reasons similar to army recruits. You might remember "Fresh**men**ingitis."

Meningococcal Disease

Neisseria meningitidis spreads via respiratory secretions and usually lives asymptomatically in the nasopharynx. Rarely, the bacteria will invade the bloodstream (bacteremia) from the nasopharynx, resulting in meningitis and/or deadly sepsis (called meningococcemia). The classic "clue" to an invasive meningococcal infection is the appearance of a **petechial rash**. This rash is due to the release of endotoxin from the meningococcus, causing vascular necrosis, an inflammatory reaction, and hemorrhage into the surrounding skin. Note that the diplococci can be seen (Gram stain) or cultured from biopsies of the petechiae.

1) **Meningococcemia**: The intravascular multiplication of *Neisseria meningitidis* results in an abrupt onset of spiking fevers, chills, arthralgia (joint pains), and muscle pains, as well as the petechial rash. These patients usually look acutely ill. Once in the bloodstream, the meningococci rapidly disseminate throughout the body. This can lead to meningitis and/or fulminant meningococcemia.

2) **Fulminant meningococcemia (Waterhouse-Friderichsen syndrome)**: This is septic shock (see Chapter 2, page 13). Bilateral hemorrhage into the adrenal glands occurs, which causes adrenal insufficiency. Abrupt onset of hypotension and tachycardia occurs, along with rapidly enlarging petechial skin lesions. Disseminated intravascular coagulation (DIC) and coma may develop. Death can occur rapidly (6–8 hours).

3) **Meningitis**: This is the most common form of meningococcal disease, usually striking infants < 1 year of age. Infants usually display nonspecific findings of an infection, including fever, vomiting, irritability, and/or lethargy. A bulging open anterior fontanelle may

be a sign of meningitis in neonates, while slightly older infants may display a stiff neck, as well as positive Kernig's and Brudzinski's signs.

The classic petechial skin rash may occur when meningococcemia occurs in conjunction with meningitis. This allows the physician to make a presumptive diagnosis of meningococcal meningitis even before performing a diagnostic spinal tap.

Clinical Pearl: Three bacteria are responsible for most meningitis acquired by the baby coming out of the birth canal (within first 3 months of age): *Listeria monocytogenes*, *Escherichia coli* and *Group B Streptococcus*. Two bacteria cause meningitis later in life after the maternal antibodies passively given to fetus wane and before new antibodies develop: *Neisseria meningitidis* and *Haemophilus influenzae*.

H. influenzae used to be the most common cause of bacterial meningitis in children and adults with *N. meningitidis* close behind, but now with the success of the *H. influenzae* type B capsular conjugate vaccine, *N. meningitidis* is number one in the United States and associated with a 13% mortality.

Diagnosis involves Gram stain and culture of the meningococcus from blood, cerebrospinal fluid, or petechial scrapings. *Neisseria* grow best on blood agar that has been heated so that the agar turns brown (called chocolate agar). The classic medium for culturing *Neisseria* is called the **Thayer-Martin VCN** media. This is chocolate agar with antibiotics, which are included to kill competing bacteria.

V stands for **vancomycin**, which kills gram-positive organisms.

C stands for **colistin (polymyxin)** which kills all gram-negative organisms (except *Neisseria*).

N stands for **nystatin**, which eliminates fungi.

Therefore, only *Neisseria* (both *Neisseria meningitidis* and *Neisseria gonorrhoeae*) are able to grow on this culture medium. The addition of a high concentration of CO_2 further promotes the growth of *Neisseria*.

In the laboratory, the differentiation between the *Neisseria* species is based on *Neisseria meningitidis'* ability to produce acid from maltose metabolism, while *Neisseria gonorrhoeae* cannot!

Prompt treatment with **penicillin G** or **ceftriaxone** is required at the first indication of disseminated meningococcemia. Close contacts of an infected patient are treated with **rifampin** or **ciprofloxacin**. Immunization with purified capsular polysaccharides from certain strains (groups A, C, Y, and W135) is currently available and used for epidemics and in high-risk groups. In late 2014 and early 2015 two meningococcal serogroup B vaccines were approved by the FDA for use in individuals age 10- 25 years of age. Previously, there were no licensed vaccines for serogroup B in the United

States, despite serogroup B being responsible for one-third of meningococcal disease in the US.

The problem with the current vaccines is that they do not induce sufficient immunity in children less than 2 years of age, and the duration of the immunity in adults is only about 2–4 years. Just like the case for *H. influenzae* type B capsular conjugate vaccine, new vaccines are being developed against *N. meningitidis* that are conjugated with mutant diphtheria toxin or tetanus toxoid proteins to elicit a stronger, more lasting, immune response earlier in life (at 4 months of age).

NEISSERIA GONORRHOEAE

Neisseria gonorrhoeae, often called the **gonococcus**, causes the second most common sexually transmitted disease, gonorrhea (chlamydial infections are slightly more common).

Virulence factors of the gonococcus include:

1) **Pili**: *Neisseria gonorrhoeae* has complex genes coding for their pili. These genes undergo multiple recombinations, resulting in the production of pili with hyper-variable amino acid sequences. These changing antigens in the pili protect the bacteria from our antibodies, as well as from vaccines aimed at producing antibodies directed against the pili.

The pili adhere to host cells, allowing the gonococcus to cause disease. They also serve to prevent phagocytosis, probably by holding the bacteria so close to host cells that macrophages or neutrophils are unable to attack.

2) **Outer membrane protein porins** (PorA and PorB, formerly called protein I) appear to promote invasion into epithelial cells

3) **Opa proteins** are another class of outer membrane proteins that promote adherence and invasion into epithelial cells; these are so named as their expression results in opaque colonies.

Overall, the pili, porins and Opa proteins allow the gonococci to bind to a fallopian tube non-ciliated epithelial cell. The gonococcal endotoxin (LPS) then destroys the cilia on neighboring cells. The gonococcus is then taken up by endocytosis, transported in the endocytotic vacuole (where it multiplies) and is released into the subepithelial space, where it can cause more systemic infection.

Gonococcal Disease in Men

A man who has unprotected sex with an infected person can acquire a *Neisseria gonorrhoeae* infection. This organism penetrates the mucous membranes of the urethra, causing inflammation of the urethra (urethritis). Although some men will remain asymptomatic, most will complain of painful urination along with a purulent urethral discharge (pus can be expressed from the tip of the penis). Both asymptomatic and symptomatic men can pass this infection to another sexual partner.

Possible complications of this infection include epididymitis, prostatitis, and urethral strictures. Fortunately, this disease is easily cured by a small dose of ceftriaxone.

Men having sex with men (referred to as **MSM**) results in **rectal gonococcal infection** and this is often the only site of infection in this group. This infection is usually asymptomatic but can cause anal pruritis, tenesmus, and/or rectal bleeding and purulent discharge.

Gonococcal Disease in Women

Like men, women can also develop a gonococcal urethritis, with painful burning on urination and purulent discharge from the urethra. However, urethritis in women is more likely to be asymptomatic with minimal urethral discharge.

Neisseria gonorrhoeae also infects the columnar epithelium of the cervix, which becomes reddened and friable, with a purulent exudate. A large percentage of women are asymptomatic. If symptoms do develop, the woman may complain of lower abdominal discomfort, pain with sexual intercourse (dyspareunia), and a purulent vaginal discharge. Both asymptomatic and symptomatic women can transmit this infection.

A gonococcal infection of the cervix can progress to **pelvic inflammatory disease** (PID, or "pus in dere"). PID is an infection of the uterus (**endometritis**), fallopian tubes (**salpingitis**), and/or ovaries (**oophoritis**). Clinically, patients can present with fever, lower abdominal pain, abnormal menstrual bleeding, and cervical motion tenderness (pain when the cervix is moved by the doctor's examining finger). Menstruation allows the bacteria to spread from the cervix to the upper genital tract. It is therefore not surprising that over 50% of cases of PID occur within one week of the onset of menstruation. The presence of an intrauterine device (IUD) increases the risk of a cervical gonococcal infection progressing to PID. *Chlamydia trachomatis* is the other major cause of PID (see Chapter 13, page 111).

Complications of PID include:

1) **Sterility**: The risk of sterility appears to increase with each gonorrhea infection. Sterility is most commonly caused by scarring of the fallopian tubes, which occludes the lumen and prevents sperm from reaching the ovulated egg.

2) **Ectopic pregnancy**: The risk of a fetus developing at a site other than the uterus is significantly increased with previous fallopian tube inflammation (salpingitis). The fallopian tubes are the most common site for an ectopic pregnancy. Again, with scarring down of the fallopian tubes, there is resistance to normal egg transit down the tubes.

3) **Abscesses** may develop in the fallopian tubes, ovaries, or peritoneum.

4) **Peritonitis**: Bacteria may spread from ovaries and fallopian tubes to infect the peritoneal fluid.

5) **Peri-hepatitis (Fitz-Hugh-Curtis syndrome)**: This is an infection by *Neisseria gonorrhoeae* of the capsule that surrounds the liver. A patient will complain of right upper quadrant pain and tenderness. This syndrome may also follow chlamydial pelvic inflammatory disease.

Gonococcal Disease in Both Men and Women

1) **Gonococcal bacteremia**: Rarely, *Neisseria gonorrhoeae* can invade the bloodstream. Manifestations include fever, joint pains, and skin lesions (which usually erupt on the extremities). Pericarditis, endocarditis, and meningitis are rare but serious complications of a disseminated infection.

2) **Septic arthritis**: Acute onset of fever occurs along with pain and swelling of 1 or 2 joints. Without prompt antibiotic therapy, progressive destruction of the joint will occur. Examination of synovial fluid usually reveals increased white blood cells. Gram stain and culture of the synovial fluid confirms the diagnosis, revealing gram-negative diplococci *within* the white blood cells. Gonococcal arthritis is the most common kind of septic arthritis in young, sexually active individuals.

Gonorrhea may also be transmitted anally or orally, wherever there is a transmission of body fluids.

Gonococcal Disease in Infants

Neisseria gonorrhoeae can be transmitted from a pregnant woman to her child during delivery, resulting in **ophthalmia neonatorum**. This eye infection usually occurs on the first or second day of life and can damage the cornea, causing blindness. Because neonatal *Chlamydia* eye infections are also a threat, **erythromycin** eye drops, which are effective against both *Neisseria gonorrhoeae* and *Chlamydia*, are given to all newborns. Gonococcal conjunctivitis can also occur in adults.

Diagnosis and Treatment

Diagnosis of *Neisseria gonorrhoeae* infection is best made by Gram stain and culture on Thayer-Martin VCN medium. Pus can be removed from the urethra by inserting a thin sterile swab. When this is Gram stained and examined under the microscope, the tiny doughnut-shaped diplococci can be seen **within** the white blood cells.

In the past, simple use of sulfonamides and penicillin G used to whip this bug, but now it has developed multiple mechanisms of antibiotic resistance (see **Chapter 34**):

1) **Plasmids** that can be conjugally transferred between gonococci and other bacteria. There is a TEM-1

type beta-lactamase (penicillinase) encoding plasmid (called the Pc^r determinant). There is a plasmid with the *tetM* gene sequence that encodes a protein that protects ribosomes from the effects of tetracycline.

2) **Chromosomally mediated antibiotic resistance** to beta-lactams, tetracycline and now the fluoroquinolones is a big problem. The *mtr* gene locus encodes an efflux pump that prevents accumulation of antibiotics in cells. The *penA* locus represents a mutation that alters penicillin binding protein 2 (the transpeptidase required to synthesize peptidoglycan) to reduce its affinity for penicillin. Multiple mutations in the chromosomal *gyrA* and *gyrB* genes that encode the DNA gyrases confer resistance to ciprofloxacin.

We may be fast approaching the day of untreatable gonorrhea! The recommended treatment of choice is currently ceftriaxone, a third generation cephalosporin (see page 181) combined with a 1 gram single dose of azithromycin. Until 2012, ceftriaxone alone was considered sufficient, but reduced susceptibility to cephalosporins as a result of several gene mutations has led to increased resistance.

The azithromycin will also cover *Chlamydia trachomatis*, because up to 50% of patients will be concurrently infected with this beta-lactam-resistant (ceftriaxone included) bacteria.

MORAXELLA (BRANHAMELLA) CATARRHALIS

The greater family of Neisseriaceae is composed of five genera: *Neisseria, Moraxella* (subgenera *Branhamella*), *Kingella, Acinetobacter,* and *Oligella*. We will only discuss *Moraxella* and *Kingella* as they are important human pathogens you will need to know about.

Moraxella (Branhamella) catarrhalis causes two major diseases: otitis media and upper respiratory infection in patients with chronic obstructive pulmonary disease (COPD or emphysema) or in the elderly:

Otitis media: this middle ear infection occurs in about 80% of all children by 3 years of age. It is caused by three main bacteria, *Streptococcus pneumoniae* ($\approx$30% of cases), *Haemophilus influenzae* ($\approx$25%) and *Moraxella catarrhalis* ($\approx$15–20%).

GRAM-NEGATIVE DIPLOCOCCI	RESERVOIR	MORPHOLOGY & METABOLISM	VIRULENCE	TOXINS
Neisseria meningitidis	1. Nasopharynx of humans **only**. • Immunity can develop to particular strains • Strict human parasite 2. Spread by respiratory transmission	1. Kidney bean shaped with concave sides facing each other, forming the appearance of a doughnut 2. Gram-negative diplococci 3. Facultative-anaerobe 4. Grows best in high CO_2 environment 5. Ferments **m**altose and **g**lucose – easy to remember, since there is both an "**m**" and "**g**" in **m**enin**g**itidis	1. Capsule: a. 13 serotypes based on antigenicity of capsule polysaccharides b. Serotypes A, B, & C are associated with epidemics of meningitis (usually type B) 2. IgA$_1$ protease 3. Have unique proteins that can extract iron from transferrin, lactoferrin and hemoglobin 4. Pili: for adherence	1. Endotoxin: Lipopolysaccharide (LPS) 2. **No** exotoxins

Figure 8-2   NEISSERIA

COPD exacerbations: worsening of wheezing, shortness of breath and cough. These exacerbations are often associated with an acquisition of a new strain of nontypeable *Haemophilus influenzae* or infection with *Moraxella catarrhalis* (30% of cases). *Moraxella catarrhalis* also causes pneumonia in the elderly.

Kingella kingae is the most frequent human pathogen of the *Kingella* genera. It frequently colonizes the throats of young children and can cause septic arthritis and osteomyelitis in children. In children and adults it can cause endocarditis of native and prosthetic valves.

Clinical Pearl: Kingella kingae is commonly grouped with several slow growing gram negative pathogens known to cause endocarditis called the **HACEK** group of bacteria:

Haemophilus species
Actinobacillus species
Cardiobacterium species
Eikenella species
Kingella species

Fig. 8-2. Summary of *Neisseria*.

Recommended Review Articles:

Bolan GA, Sparling, et al. The emerging threat of untreatable Gonococcal infection. NEJM 2012;366:485–7.
Cook RL, Hutchison SL, Ostergaard L, et al. Systematic review: noninvasive testing for Chlamydia trachomatis and Neisseria gonorrhoeae. Ann Intern Med 2005;142:914–25.
Crossman SH. The challenge of pelvic inflammatory disease. Am Fam Physician 2006;73:859–64.
Deeks ED. Meningococcal quadrivalent (serogroups A, C, w135, and y) conjugate vaccine (Menveo): in adolescents and adults. BioDrugs. 2010;24(5):287–97
Gottlieb SL, Berman SM, Low N. Screening and treatment to prevent sequelae in women with Chlamydia trachomatis genital infection: how much do we know? J Infect Dis. 2010;201 Suppl 2:S156–67.
Pathan N, Faust SN, and Levin M. Pathophysiology of meningococcal meningitis and septicaemia. Arch Dis Child 2003;88:601–7.

CLINICAL	TREATMENT	DIAGNOSTICS	MISCELLANEOUS
1. Asymptomatic carriage in the nasopharynx 2. Meningitis: A. Fever B. Stiff neck (nuchal rigidity) C. Vomiting D. Lethargy or altered mental status E. Petechial rash 3. Septicemia (meningococcemia): A. Fever B. Petechial rash C. Hypotension D. Fulminant meningococcemia (**Waterhouse-Friderichsen Syndrome**): hemorrhage of the adrenal glands along with hypotension and the petechial rash	1. There are two separate vaccines — one against capsular antigens: A, C, Y & W-135; the second is against capsular antigen B. 2. Antibiotics: A. Penicillin G B. Ceftriaxone (or other third generation cephalosporins) C. Rifampin and ciprofloxacin are used for prophylaxis of close contacts of infected persons.	1. Gram-stain 2. Culture A. Culture specimen on blood agar that has been heated to 80 °C. for 15 minutes (called **chocolate agar**) B. Selective media: prevents growth of bacteria • Thayer Martin VCN V = Vancomycin C = Colistin N = Nystatin C. Cell wall contains cytochrome oxidase which oxidizes dye tetramethylphenylene diamine from colorless to deep pink. Used to ID colonies D. PCR of bacteria DNA in clinical specimens	1. Neonates are very susceptible from 6 to 24 months, when protective antimeningococcal IgG is low 2. Army recruits are also at high risk (with carriage rates of greater than 40%)

GRAM-NEGATIVE DIPLOCOCCI	RESERVOIR	MORPHOLOGY & METABOLISM	VIRULENCE	TOXINS
Neisseria gonorrhoeae	1. Humans only (no immunity to repeated infections) 2. Sexually transmitted	1. Kidney bean shaped with concave sides facing each other, forming the appearance of a doughnut 2. Gram-negative diplococci 3. Facultative-anaerobe 4. Grows best in high CO_2 environment 5. Ferments only **g**lucose (not maltose) - easy to remember, since there is only a "**g**" (no "m") in **g**onorrhoeae	1. Pili: A. Adherence to epithelial cells B. Antigenic variation C. Antiphagocytic: binds bacteria tightly to host cell, protecting it from phagocytosis 2. IgA$_1$ protease 3. Outer membrane proteins: Protein I: porin Protein II (opacity protein): presence associated with dark, opaque colonies • for adherence 4. Have unique proteins that can extract iron from transferrin, lactoferrin and hemoglobin	1. Endotoxin: Lipopoly-saccharide (LPS) 2. **No** exotoxins
Moraxella (Branhamella) catarrhalis	• Part of the normal respiratory flora			

Figure 8-2   (continued)

70

CLINICAL	TREATMENT	DIAGNOSTICS	MISCELLANEOUS
1. Asymptomatic (but still infectious) 2. Men: urethritis 3. Women: cervical gonorrhea, which can progress to pelvic inflammatory disease (PID) *Complications of PID* A. Sterility B. Ectopic pregnancy C. Abscess D. Peritonitis E. Perihepatitis 4. Both men and women: A. Gonococcal bacteremia B. Septic arthritis: gonococcal arthritis is the most common cause of septic arthritis in sexually active individuals 5. Neonates: **Ophthalmia neonatorum** conjunctivitis in newborns. *N. gonorrhoeae* is acquired during passage through an infected birth canal. Conjunctivitis usually erupts within the first 5 days	1. **Antibiotic of choice**: Ceftriaxone 250 mg IM and Azithromycin 1 gm orally x 1 2. **Second line**: A. Cefixime + azithromycin or doxycycline B. Spectinomycin (not available in the U.S.) 3. For **ophthalmia neonatorum**: • Erythromycin eye drops should be given immediately following birth, for prophylaxis against both *N. gonorrhoeae* and *Chlamydia trachomatis* conjunctivitis • Infants with ophthalmia neonatorum require systemic treatment with ceftriaxone. Erythromycin syrup should also be provided to cover for possible concurrent Chlamydial disease (This is important, as failure to treat neonatal *Chlamydia* conjunctivitis can lead to Chlamydial pneumonia)	1. Gram-stain of urethral pus reveals the tiny gram-negative doughnut-shaped diplococci within white blood cells 2. Culture A. Culture specimen on **chocolate agar** B. Selective media: prevents growth of other bacteria • Thayer Martin VCN V = Vancomycin C = Colistin N = Nystatin C. Cell wall contains cytochrome oxidase which oxidizes dye tetramethylphenylene diamine from colorless to deep pink. Used to ID colonies D. PCR of bacterial DNA in clinical specimens	• No immunity following infection: a person can be reinfected numerous times
1. Otitis media in children 2. Can cause other respiratory tract infections, such as sinusitis, bronchitis, & pneumonia 3. COPD exacerbations	1. Azithromycin or clarithromycin 2. Amoxicillin with clavulanate 3. Oral second or third generation cephalosporin 4. Trimethoprim/ sulfamethoxazole		• Resistant to penicillin

M. Gladwin, W. Trattler, and S. Mahan, *Clinical Microbiology Made Ridiculously Simple* ©MedMaster

CHAPTER 9.   THE ENTERICS

The **enterics** are gram-negative bacteria that are part of the normal intestinal flora or cause gastrointestinal disease. The family, genus, and species of all the enterics are organized in the chart at the end of this chapter so that you will not be confused with the different names. Many of these bacteria are referred to simply by their genus name because there are so many different species in some groups. The main groups are **Enterobacteriaceae, Vibrionaceae, Pseudomonadaceae** and **Bacteroidaceae**.

These organisms are also divided into groups based upon **biochemical** and **antigenic** properties.

Biochemical Classification

Some of the important biochemical properties of the organisms, which can be measured in the lab, are:

1) The **ability to ferment lactose** and convert it into gas and acid (which can be visualized by including a dye that changes color with changes in pH). *Escherichia coli* and most of the enterobacteriaceae ferment lactose while *Salmonella, Shigella* and *Pseudomonas aeruginosa* do not.

2) The **production of H_2S**, ability to hydrolyze urea, liquefy gelatin, and decarboxylate specific amino acids.

Some growth media do 2 things at once: 1) They contain chemicals that inhibit the growth of gram-positive bacteria that may be contaminating the sample. 2) They have indicators that change color in the presence of lactose fermentation. The 2 that you should know are:

1) **EMB agar** (Eosine Methylene Blue): Methylene blue inhibits gram-positive bacteria, and colonies of lactose fermenters become deep purple to black in this medium. *Escherichia coli* colonies take on a metallic green sheen in this medium.

2) **MacConkey agar**: Bile salts in the medium inhibit gram-positive bacteria, and lactose fermenters develop a pink-purple coloration.

In today's modern laboratories there are plastic trays with up to 30 different media that measure many biochemical reactions, including those just described. A colony of unknown bacteria is inoculated onto each medium and incubated. A computer then interprets the results and identifies the bacteria.

Fecal Contamination of Water

A classic method for determining whether water has been contaminated with feces demonstrates some of the practical uses of biochemical reactions and some important

properties of *Escherichia coli*. Follow this discussion for the overall big picture.

You are traveling through Uruguay and wind up in a village whose people are suffering from a terrible diarrhea. After giving intravenous fluids to scores of babies, you start to wonder whether the cause of this infection can be eradicated. When questioned, the villagers tell you that they obtain their water from a common river. You know that the enterics are transmitted by the fecal-oral route, and you begin to wonder if there is fecal contamination of the river water. How will you prove that the water is fecally contaminated?

Escherichia coli to the rescue! You see, *Escherichia coli* is a coliform, which means that it is a normal inhabitant of the intestinal tract. Think of *E. **coli*** = **coli**-form = **colo**n. *Escherichia coli* is normally not found outside the intestine. So if you find *Escherichia coli* in the village stream water, it does not necessarily mean that *Escherichia coli* is causing the diarrhea, but it **does** tell you that there is fecal matter in the river and that some enteric may be responsible. You pull out your tattered copy of *Clinical Micro Made Ridiculously Simple* and begin the test.

1) **Presumptive Test**: You add the river water samples to test tubes containing nutrient broth (like agar) that contains **lactose**. These tubes contain an inverted vial that can trap gas and a dye indicator that changes color if acid is produced. You let the sample grow for a day. If lactose is fermented, gas is produced and the dye is visualized. You now know that either *Escherichia coli* or a nonenteric bacteria that ferments lactose is in the water. To find out which you continue . . .

2) **Confirmed Test**: Streak EMB agar plates with the water samples, and the *Escherichia coli* should form colonies with a metallic green sheen. Also *Escherichia coli* can grow at 45.5C° but most nonenterics cannot, so you can grow 2 plates at 45.5C° and 37C° and compare the colonies on both.

3) **Completed Test**: Colonies that were metallic green are placed in the broth again. If they produce acid and gas, then you know the river water contains *Escherichia coli*.

You travel upstream and find an outhouse that has been built in a tree hanging over the water. You inform the villagers about the need to defecate in areas that do not have river runoff and teach them to build latrines. Within a few weeks the epidemic has ended!

Antigenic Classification

The enterics form many groups, based on cell surface structures that bind specific antibodies (antigenic

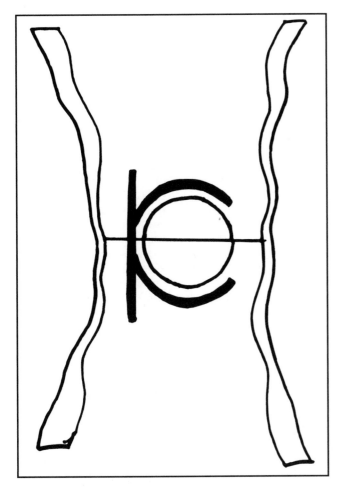

Figure 9-1

determinants). The enterics have 3 major surface antigens, which differ slightly from bug to bug.

1) **O antigen**: This is the most external component of the lipopolysaccharide (LPS) of gram-negative bacteria. The O antigen differs from organism to organism, depending on different sugars and different side-chain substitutions. Remember **O** for **Outer** (see **Fig. 1-6**, for more information on LPS).

2) **K antigen**: This is a capsule (Kapsule) that covers the O antigen.

3) **H antigen**: This antigenic determinant makes up the subunits of the bacterial flagella, so only bacteria that are motile will possess this antigen. *Shigella* does not have an H antigen. *Salmonella* has H antigens that change periodically, protecting it from our antibodies.

Fig. 9-1. The O antigen forms the outer part of the cell membrane, the K antigen wraps around the cell like a capsule, and the arms of the H antigen become wavy flagella.

Pathogenesis

The organisms in this chapter produce 2 types of disease:

1) **Diarrhea** with or without systemic invasion.

2) **Various other infections** including urinary tract infections, pneumonia, bacteremia, and sepsis, especially in debilitated **hospitalized** patients.

Diarrhea

A useful concept in understanding diarrhea produced by these organisms is that there are different clinical manifestations depending on the "depth" of intestinal invasion:

1) **No cell invasion**: The bacteria bind to the intestinal epithelial cells but do not enter the cell. Diarrhea is caused by the release of exotoxins (called enterotoxins in the GI tract), which causes electrolyte and fluid loss from intestinal epithelial cells or epithelial cell death. Watery diarrhea **without** systemic symptoms (such as fever) is the usual picture. Enterotoxigenic *Escherichia coli* and *Vibrio cholera* are examples.

2) **Invasion of the intestinal epithelial cells**: The bacteria have virulence factors that allow binding and invasion into cells. Toxins are then released that destroy the cells. The cell penetration results in a systemic immune response with local white blood cell infiltration (leukocytes in the stool) as well as fever. The cell death results in red blood cell leakage into the stool. Examples: Enteroinvasive *Escherichia coli*, *Shigella*, and *Salmonella enteritidis*.

3) **Invasion of the lymph nodes and bloodstream**: Along with abdominal pain and diarrhea containing white and red cells, this deeper invasion results in **systemic symptoms** of fever, headache, and white blood cell count elevation. The deeper invasion can also result in mesenteric lymph node enlargement, bacteremia, and sepsis. Examples: *Salmonella typhi*, *Yersinia enterocolitica*, and *Campylobacter jejuni*.

Various Other Infections

The enterics are normal intestinal inhabitants and usually live with us in peaceful harmony. In the hospital and nursing homes, however, some bad things happen. They acquire antibiotic resistance and can cause disease in debilitated patients. They can invade the debilitated patients when Foley catheters are in the urethra or when a patient aspirates vomitus that has been colonized by the enterics. Because of this hospital acquisition, you will often hear them described as **the hospital-acquired gram-negatives** or **nosocomial gram-negatives**. Examples: *Escherichia coli*, *Klebsiella* *pneumoniae, Proteus mirabilis, Enterobacter, Serratia,* and *Pseudomonas aeruginosa.*

FAMILY ENTEROBACTERIACEAE

Escherichia coli

Escherichia coli normally resides in the colon without causing disease. However, there is an amazing amount of DNA being swapped about among the enterics by conjugation with plasmid exchange, lysogenic conversion by temperate bacteriophages, and direct transposon mediated DNA insertion (see Chapter 3). When *Escherichia coli* acquires virulence in this manner, it can cause disease:

> **Nonpathogenic *Escherichia coli* (normal flora)** + Virulence factors = DISEASE.

Virulence factors include the following:

1) Mucosal interaction:
   a) Mucosal adherence with pili (colonization factor).
   b) Ability to invade intestinal epithelial cells.
2) Exotoxin production:
   a) Heat-labile and stable toxin (LT and ST).
   b) Shiga-like toxin.
3) Endotoxin: Lipid A portion of lipopolysaccharide (LPS).
4) Iron-binding siderophore: obtains iron from human transferrin or lactoferrin.

Diseases caused by *Escherichia coli* in the presence of virulence factors include the following:

1) Diarrhea.
2) Urinary tract infection.
3) Neonatal meningitis.
4) Gram-negative sepsis, occurring commonly in debilitated hospitalized patients.

Escherichia coli Diarrhea

Escherichia coli diarrhea may affect infants or adults. Infants worldwide are especially susceptible to *Escherichia coli* diarrhea, since they usually have not yet developed immunity. Since water lost in the stool is often not adequately replaced, death from *Escherichia coli* diarrhea is usually due to dehydration. About 5 million children die yearly from this infection.

Adults (and children) from developed countries, traveling to underdeveloped countries, are also susceptible to *Escherichia coli* diarrhea, since they have not developed immunity during their childhood. This **travelers' diarrhea** is the so-called **Montezuma's revenge** named after the Aztec chief killed at the hands of the Spanish explorer, Cortez.

The severity of *Escherichia coli* diarrhea depends on which virulence factors the strain of *Escherichia coli* possesses. We will discuss 3 groups of diarrhea-producing *Escherichia coli*. These have been named based on their virulence factors and the different diarrheal diseases they cause.

1) **Enterotoxigenic *Escherichia coli* (ETEC)**: This *Escherichia coli* causes traveler's diarrhea. It has pili (colonization factor) that help it bind to intestinal epithelial cells, where it releases exotoxins that are similar to the cholera exotoxins discussed on page 80. The toxins are the **heat *labile* *toxin* (LT)**, which is just like the cholera toxin, and the **heat stable *toxin* (ST)**. These exotoxins inhibit the reabsorption of Na^+ and Cl^- and stimulate the secretion of Cl^- and HCO_3^- into the intestinal lumen. Water follows the osmotic pull of these ions, resulting in water and electrolyte loss. This produces a severe watery diarrhea with up to 20 liters being lost a day!!! **The stool looks like rice water—just like cholera!**

2) **Enterohemorrhagic *Escherichia coli* (EHEC)**: These *Escherichia coli* also have a pili colonization factor like the ETEC but differ in that they secrete the powerful **Shiga-like toxin** (also called **verotoxin**) that has the same mechanism of action as the *Shigella* toxin (see page 76). They both inhibit protein synthesis by inhibiting the 60S ribosome, which results in intestinal epithelial cell death. So these bacteria hold onto the intestinal epithelial cells and shoot away with the Shiga-like toxin (see **Fig. 9-3**). The diarrhea is **bloody** (hemorrhagic), accompanied by severe abdominal cramps, and is called **hemorrhagic colitis**.

Hemolytic uremic syndrome (HUS) with anemia, thrombocytopenia (decrease in platelets), and renal failure (thus uremia), is associated with infection by a strain of EHEC, called ***Escherichia coli* 0157:H7**. Numerous outbreaks have occurred secondary to infected hamburger meat served at fast food chains, suggesting that cattle may be a reservoir for EHEC.

3) **Enteroinvasive *Escherichia coli* (EIEC)**: This disease is the same as that caused by Shigella (page 76). In fact, the main virulence factor is encoded in a plasmid shared by *Shigella* and *Escherichia coli*. This plasmid gives the bacteria the ability to actually **invade** the epithelial cells. EIEC also produces small amounts of Shiga-like toxin. The host tries to get rid of the invading bacteria, and this results in an immune-mediated inflammatory reaction with **fever**. White blood cells invade the intestinal wall, and the diarrhea is bloody with white blood cells. **Like shigellosis!**

Fig. 9-2. *Vibrio cholera, Escherichia coli,* and *Shigella dysenteriae* all holding hands. *Escherichia coli* can

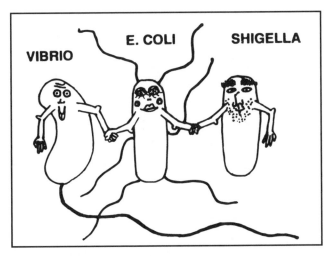

Figure 9-2

cause diarrhea indistinguishable from shigellosis and cholera. The big picture here is that the different types of diarrhea produced by *Escherichia coli* and the other enterics are dependent on virulence acquisition from plasmids, and there is active sharing of these factors. So *Escherichia coli* diarrhea can look just like cholera (rice-water stools) or just like shigellosis (diarrhea with blood and white cells).

Escherichia coli Urinary Tract Infections (UTIs)

The acquisition of a pili virulence factor allows *Escherichia coli* to travel up the urethra and infect the bladder (**cystitis**) and sometimes move further up to infect the kidney itself (**pyelonephritis**). *Escherichia coli* is the most common cause of urinary tract infections, which usually occur in women and hospitalized patients with catheters in the urethra. Symptoms include burning on urination (**dysuria**), having to pee frequently (**frequency**), and a feeling of fullness over the bladder. Culture of greater than 100,000 colonies of bacteria from the urine establishes the diagnosis of a urinary tract infection.

Escherichia coli Meningitis

Escherichia coli is a common cause of neonatal meningitis (group B streptococcus is first). During the first month of life, the neonate is especially susceptible.

Escherichia coli Sepsis

Escherichia coli is also the most common cause of gram-negative sepsis. This usually occurs in debilitated

hospitalized patients. Septic shock (see Chapter 2, page 13) due to the lipid A component of the LPS is usually the cause of death.

Escherichia coli Pneumonia

Escherichia coli is a common cause of hospital-acquired pneumonia.

Klebsiella pneumoniae

This enteric is encapsulated (O antigen) but is non-motile (no H antigen). *Klebsiella pneumoniae* prowls hospitals, causing sepsis (second most common after *Escherichia coli*). It also causes urinary tract infections in hospitalized patients with Foley catheters. Hospitalized patients and alcoholics (debilitated patients) are prone to a *Klebsiella pneumoniae* pneumonia, which is characterized by a bloody sputum in about 50% of cases. This pneumonia is violent and frequently destroys lung tissue, producing cavities. Thick sputum coughed up with *Klebsiella pneumoniae* classically looks like red currant jelly, which is the color of the O antigen capsule. The mortality rate is high despite antibiotic therapy.

Proteus mirabilis

This organism is very motile. In fact, when you smear the bacteria on a plate it will grow not as distinct round colonies, but rather as a confluence of colonies as the bacteria rapidly move and cover the plate. This organism is able to break down urea and is thus often referred to as the urea-splitting *Proteus*.

There are 3 strains of *Proteus* that have cross-reacting antigens with some *Rickettsia* (Chapter 13, **Fig. 13-11**). They are OX-19, OX-2, and OX-K. This is purely coincidental but serves as a useful clinical tool to determine if a person has been infected with *Rickettsia*. Serum is mixed with these *Proteus* strains to determine whether there are antibodies in the serum that react with the *Proteus* antigens. If these antibodies are present, this suggests that the patient has been infected with *Rickettsia*.

Proteus is another common cause of urinary tract infections and hospital-acquired (nosocomial) infections. Examination of the urine will reveal an alkaline pH, which is due to *Proteus*' ability to split urea into **NH₃** and CO_2.

Enterobacter

This highly motile gram-negative rod is part of the normal flora of the intestinal tract. It is occasionally responsible for hospital-acquired infections.

Serratia

Serratia is notable for its production of a **bright red pigment**. It can cause urinary tract infections, wound infections, or pneumonia.

Shigella

There are four species of *Shigella* (*dysenteriae, flexneri, boydii,* and *sonnei*) and all are **non**-motile. If you look back at the picture of *Escherichia coli* and *Shigella* holding hands (**Fig. 9-2**), you will see that *Shigella* has no flagella. *Shigella* does not ferment lactose and does not produce H₂S. These properties can be used to distinguish *Shigella* from *Escherichia coli* (lactose fermenter) and *Salmonella* (non-lactose fermenter, produces H₂S).

Humans are the only hosts for *Shigella*, and the dysentery that it causes usually strikes preschool age children and populations in nursing homes. Transmission by the fecal-to-oral route occurs via fecally contaminated water and hand-to-hand contact (Employees please wash hands!). *Shigella* is **never** considered part of the normal intestinal flora! It is always a pathogen.

Shigella is similar to enteroinvasive *Escherichia coli* (EIEC) in that they both invade intestinal epithelial cells and release **Shiga toxin**, which causes cell destruction. White cells arrive in an inflammatory reaction. The colon, when viewed via colonoscopy, has shallow ulcers where cells have sloughed off. The illness begins with fever (unlike ETEC and cholera, which do not invade epithelial cells and therefore do **not** induce a fever), abdominal pain, and diarrhea. The diarrhea may contain flecks of bright-red blood and pus (white cells). Patients develop diarrhea because the inflamed colon, damaged by the Shiga toxin, is unable to reabsorb fluids and electrolytes.

Fig. 9-3. Visualize Shazam Shigella with his Shiga blaster laser, entering the intestinal epithelial cells and blasting away at the 60S ribosome, causing epithelial cell death.

Shiga Toxin

This is the same toxin as in EHEC and EIEC, and its mechanism is the same. There is an A subunit bound to 5 B subunits. The B subunits (B for Binding) bind to the microvillus membrane in the colon, allowing the entry of the deadly A subunit (A for Action). The A subunits inactivate the 60S ribosome, inhibiting protein synthesis and killing the intestinal epithelial cell.

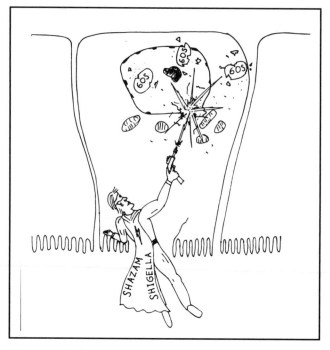

Figure 9-3

Salmonella
("The Salmon")

Salmonella is a non-lactose fermenter, is motile (like a salmon), and produces H₂S.

You will hear of *Salmonella*'s **Vi antigen**. This is a polysaccharide capsule that surrounds the O antigen, thus protecting the bacteria from antibody attack on the O antigen. This is just like the K antigen (just to confuse you!), but with *Salmonella* they named it **Vi** (for **virulence**).

While there are over 2000 Salmonella serotypes, recently all the clinically important Salmonella subtypes have been classified as a single species, *Salmonella cholerasuis*. Despite this attempt at simplification for clinical purposes Salmonella serotypes are often still divided into three groups: *Salmonella typhi, Salmonella cholerae-suis,* and *Salmonella enteritidis*. This will not be that difficult to remember because they are named according to the diseases they cause.

Salmonella differs from the other enterics because it lives in the gastrointestinal tracts of **animals** and infects humans when there is contamination of food or water with **animal feces**.

Fig. 9-4. Many animals can carry *Salmonella*. (Picture a salmon). In the U.S. there was even an epidemic of salmonellosis from pet turtles. Today in the U.S., *Salmonella* is most commonly acquired from eating chickens and uncooked eggs. *Salmonella typhi* is an

Figure 9-4

exception as it is **not** zoonotic (an infectious disease of animals that can be transmitted to man). *Salmonella typhi* is carried only by humans.

Salmonella (like *Shigella*) is **never** considered part of the normal intestinal flora! It is always pathogenic and can cause 4 disease states in humans: 1) the famous **typhoid fever**, 2) a **carrier state**, 3) **sepsis**, and 4) **gastroenteritis** (diarrhea).

Typhoid Fever

This illness caused by *Salmonella typhi* is also called **enteric fever**. *Salmonella typhi* moves one step beyond EIEC and *Shigella*. After invading the intestinal epithelial cells, it invades the regional lymph nodes, finally seeding multiple organ systems. During this invasion the bacteria are phagocytosed by monocytes and can survive intracellularly. So *Salmonella typhi* is a **facultative intracellular parasite** (see **Fig. 2-7**).

Fig. 9-5. Typhoid fever, caused by *Salmonella typhi*, depicted by a Salmon with fever (thermometer) and rose spots on its belly. Salmonellosis starts 1–3 weeks after

exposure and includes fever, headache, and abdominal pain that is either diffuse or localized to the right lower quadrant (over the terminal ileum), often mimicking appendicitis. As inflammation of the involved organs occurs, the spleen may enlarge and the patient may develop diarrhea and rose spots on the abdomen—a transient rash consisting of small pink marks seen only on light-skinned people.

Diagnose this infection by culturing the blood, urine, or stool. **Ciprofloxacin** or **ceftriaxone** are considered appropriate therapy.

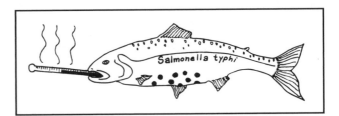

Figure 9-5

Carrier State

Fig. 9-6. Some people recovering from typhoid fever become chronic carriers, harboring *Salmonella typhi* in their gallbladders and excreting the bacteria constantly. These people are not actively infected and do not have any symptoms. A famous example occurred in 1906 when Typhoid Mary (Mary Mallon), an Irish immigrant who worked as a cook, spread the disease to dozens in New York City. (Again—employees please wash hands after using the toilet!) Some carriers actually require surgical removal of their gallbladders to cure them.

Sepsis

Fig. 9-7. Salmon cruising in the bloodstream to infect lungs, brain, or bone. This systemic dissemination is usually caused by *Salmonella choleraesuis* and does not involve the GI tract.

A pearl of wisdom:

Remember that *Salmonella* is encapsulated with the Vi capsule. Our immune system clears encapsulated bacteria by opsonizing them with antibodies (**see Fig. 2-5**), and then the macrophages and neutrophils in the spleen (the reticulo-endothelial system) phagocytose the opsonized bacteria. So, patients who have lost their spleens (asplenic), either from trauma or from sickle-cell disease, have difficulty clearing encapsulated bacteria and are more susceptible to *Salmonella* infections. **Patients with sickle-cell anemia are particularly prone to *Salmonella* osteomyelitis (bone infection).**

Vigorous and prolonged antibiotic therapy is required to treat *Salmonella* osteomyelitis.

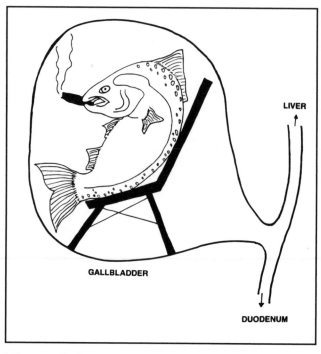

Figure 9-6

Diarrhea (Gastroenteritis)

Fig. 9-8. *Salmonella* diarrhea is the most common type of *Salmonella* infection and can be caused by any of hundreds of serotypes of nontyphoidal Salmonella. The presentation includes nausea, abdominal pain, and diarrhea that is either watery or, less commonly, contains mucous and trace blood. Fever occurs in about

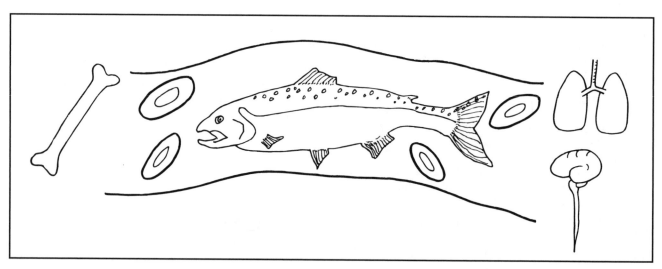

Figure 9-7

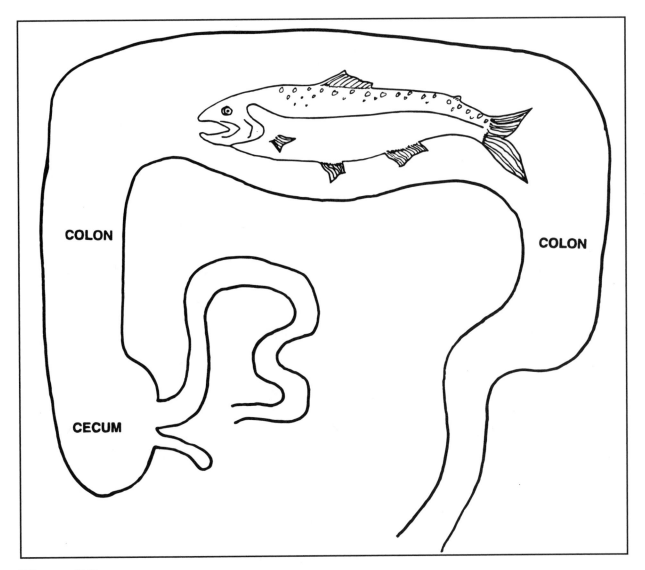

Figure 9-8

half the patients. This diarrhea is caused by a yet-uncharacterized cholera-like toxin (watery diarrhea) and sometimes also by ileal inflammation (mucous diarrhea).

Treatment usually involves only fluid and electrolyte replacement, as antibiotics do not shorten the course of the disease and do cause prolonged bacterial shedding in the stool. The diarrhea only lasts a week or less.

Yersinia enterocolitica

This motile gram-negative rod is another cause of acute gastroenteritis. Since **entero** is part of *Yersinia enterocolitica*'s name, it is not surprising that this organism is a cause of acute gastroenteritis. It is not really an enteric bacterium but is included here because it causes diarrhea. This organism is closely related to *Yersinia pestis*, which is the cause of the bubonic plague. Like *Yersinia pestis*, animals are a major source of *Yersinia enterocolitica; Yersinia enterocolitica* differs in that it is transferred by the fecal-oral route rather than the bite of a flea.

Following ingestion of contaminated foods, such as milk from domestic farm animals or fecally contaminated water, patients will develop fever, diarrhea, and abdominal pain. This pain is often most severe in the right lower quadrant of the abdomen, and therefore patients may appear to have appendicitis. Examination of the terminal ileum (located in the right lower quadrant) will reveal mucosal ulceration.

The pathogenesis of this organism is twofold:

1) **Invasion**: Like *Salmonella typhi*, this organism possesses virulence factors that allow binding to the intestinal wall and systemic invasion into regional lymph nodes and the bloodstream. Mesenteric lymph nodes swell, and sepsis can develop.

2) **Enterotoxin**: This organism can secrete an enterotoxin, very similar to the heat-stable enterotoxin of *Escherichia coli*, that causes diarrhea.

Diagnosis can be made by isolation of this organism from feces or blood. Treatment does not appear to alter the course of the gastroenteritis, but patients who have sepsis should be treated with antibiotics. Although refrigeration of food can wipe out many types of bacterial pathogens, *Yersinia enterocolitica* can survive and grow in the cold.

Other members of the Enterobacteriaceae family that you will hear of on the wards include *Edwardsiella*, *Citrobacter*, *Hafnia*, and *Providencia*.

FAMILY VIBRIONACEAE

Vibrio cholera

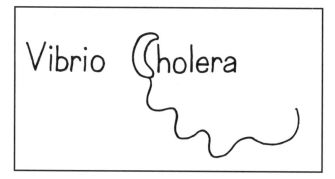

Figure 9-9

Fig. 9-9. As you can see, *Vibrio cholera* is a curved gram-negative rod with a single polar flagellum.

Cholera is the diarrheal disease caused by *Vibrio cholera*. The bacteria are transmitted by the fecal-oral route, and fecally contaminated water is usually the culprit. Adults in the U.S., especially travelers, and children in endemic areas are the groups primarily infected (immunity develops in adults in endemic areas). Recent epidemics have arisen secondary to poor disposal of sewage in many South American countries (400,000 cases in Latin America in 1991), and 1993 monsoon floods that mixed feces with potable water in Bangladesh.

The bacteria multiply in the intestine and cause the same disease as ETEC, but more severe. As with ETEC, there is no epithelial cell invasion. The bacteria attach to the epithelial cells and release the cholera toxin, which is called **choleragen**. The disease presents with the abrupt onset of a watery diarrhea (classically described as looking like **rice water**) with the loss of up to 1 liter of fluid per hour in severe cases. Shock from isotonic fluid loss will occur if the patient is not rehydrated. Like ETEC:

Cholera causes death by dehydration.

Physical findings such as diminished pulses, sunken eyes, and poor skin turgor will develop with severe dehydration.

Choleragen

This toxin has the same mechanism of action as *Escherichia coli's* LT toxin (although choleragen is coded on the chromosome, while LT is transmitted via a plasmid). There is one A subunit (Action) attached to five B subunits (Binding). The B subunit binds to the GM1 ganglioside on the intestinal epithelial cell surface, allowing entry of the A subunit. In the cell, the A subunit activates G-protein, which in turn stimulates the activity of a membrane-bound adenylate cyclase, resulting in the production of cAMP. Intracellular cAMP results in active secretion of Na and Cl as well as the inhibition of Na and Cl reabsorption. Fluid, bicarbonate, and potassium are lost with the osmotic pull of the NaCl as it travels down the intestine.

Microscopic exam of the stool should **not** reveal leukocytes (white cells) but may reveal numerous curved rods with fast darting movements. Treatment with fluid and electrolytes is lifesaving, and doxycycline will shorten the duration of the illness.

Vibrio parahaemolyticus

This organism is a marine bacterium that causes gastroenteritis after ingestion of uncooked seafood (sushi). This organism is the leading cause of diarrhea in Japan.

Campylobacter jejuni

(**Camping bacteria** in the **jejunum** with nothing better to do than cause diarrhea!)

This critter is **important**!!! This gram-negative rod that looks like *Vibrio cholera* (curved with a single polar flagellum) is often lost deep in textbooks. Don't let this happen. *Campylobacter jejuni*, ETEC, and the Rotavirus are the three most common causes of diarrhea in the world. Estimates are that *Campylobacter jejuni* causes up to 2 million cases of diarrhea a year in the U.S. alone.

This is a zoonotic disease, like most *Salmonella* (except *Salmonella typhi*), with reservoirs of *Campylobacter jejuni* in wild and domestic animals and in poultry. The fecal-oral route via contaminated water is often the mode of transmission. This organism can also be acquired by drinking unpasteurized milk. As with most diarrheal illness, children are the most commonly affected worldwide.

The illness begins with a prodrome of fever and headache, followed after half a day by abdominal cramps and a **bloody**, loose diarrhea. This organism invades the lining of the small intestine and spreads systemically as do *Salmonella typhi* and *Yersinia enterocolitica*. *Campylobacter jejuni* also secretes an LT toxin similar to that of *Escherichia coli* and an unknown cytotoxin that destroys mucosal cells. The exact role of these toxins in the pathogenesis of *campylobactor* diarrhea is still unknown.

Helicobacter pylori
(formerly called *Campylobacter pylori*)

This organism is the **most common cause of duodenal ulcers** and chronic gastritis (inflamed stomach). (Aspirin products rank second.) It is the second leading cause of gastric (stomach) ulcers, behind aspirin products. The evidence for this is as follows:

1) *Helicobacter pylori* can be cultured from ulcer craters.

2) Feeding human volunteers *Helicobacter pylori* causes ulcer formation and gastritis.

3) Pepto-Bismol, used for years for gastritis, has bismuth salts, which inhibit the growth of *Helicobacter pylori*.

4) Antibiotics help treat duodenal and gastric ulcer disease: Multiple studies have shown that treatment with combinations of bismuth salts, Metronidazole, ampicillin, and/or tetracycline, clears *Helicobacter pylori* and results in a dramatic decrease in both duodenal and gastric ulcer recurrence (Veldhuyzen van Zanten, 1994; Ransohoff, 1994; Sung, 1995).

Fig. 9-10. *Helicobacter pylori* causes duodenal and gastric ulcers and gastritis. Visualize a **Helicopter-bacteria** lifting the cap off a duodenal and gastric ulcer crater. If you have a more violent disposition, visualize an Apache **helicopter-bacteria** shooting hellfire missiles at the stomach.

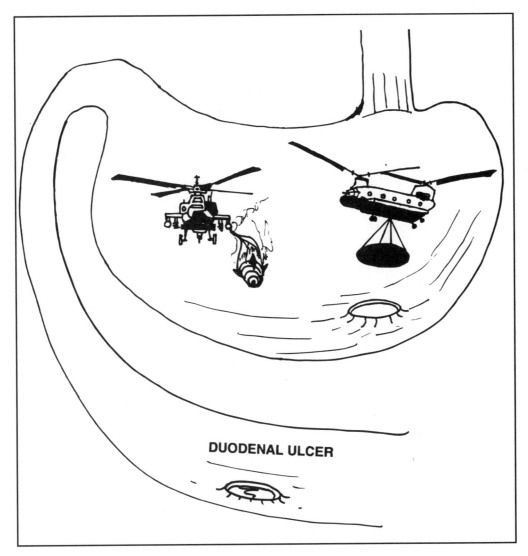

DUODENAL ULCER

Figure 9-10

GRAM-NEGATIVE RODS	RESERVOIR	TRANSMISSION	METABOLISM	VIRULENCE
ENTEROBACTERIACEAE				
Enterobacteriaceae generalities	• All are Gram-negative rods	1. Fecal-oral 2. Migration up the urethra 3. Colonization of catheters in hospitalized patients (Foley catheters, central lines, etc.)	1. Catalase-positive 2. Oxidase-negative 3. Ferments glucose 4. Facultative anaerobic	• Many of these organisms can acquire antibiotic resistance
Escherichia coli	• Humans: GI and urinary tract	1. Fecal-oral 2. Migration up the urethra 3. Colonization of catheters in hospitalized patients (Foley catheters, central lines, etc.) 4. Aspiration of oral *E. coli*	1. Indole-positive (makes indole from tryptophan) 2. Beta-hemolytic 3. Ferments lactose	1. Fimbriae (pili): colonization factor 2. Siderophore 3. Adhesins 4. Capsule (K-antigen) 5. Flagella (H-antigen)
Klebsiella pneumoniae			• Indole-negative • Ferments lactose	1. Capsule 2. Non-motile

Figure 9-11   ENTERIC BACTERIA

CHAPTER 9. THE ENTERICS

CHAPTER 9.   THE ENTERICS

TOXINS	CLINICAL	TREATMENT	DIAGNOSTICS	MISCELLANEOUS
1. Many have enterotoxins 2. All have endotoxin: lipopolysaccharide (LPS)	1. Many organisms cause diarrhea 2. Various other infections including urinary tract infections, pneumonia and sepsis especially (in debilitated hospitalized patients)		1. Eosine methylene blue agar (EMB): inhibitory to gram-positive bacteria 2. MacConkey agar: Contains bile salts in the media that inhibit gram-positive bacteria	*Antigenic Classification* 1. O-antigen: Outer portion of LPS. 2. K-antigen: Kapsule 3. H-antigen: Flagella

Enterotoxins
1. LT (Heat labile): increases cAMP (same as cholera toxin)
2. ST (heat stable): increases cGMP
3. Shiga-like toxin (verotoxin): inhibits protein synthesis by inactivating the 60S ribosomal subunit
	1. Newborn meningitis
	2. Urinary tract infection
	3. Hospital acquired sepsis
	4. Hospital acquired pneumonia
	5. Diarrhea
	  A. Noninvasive strain (Enterotoxigenic): releases LT and ST toxins, causing traveler's diarrhea
	  B. Enterohemorrhagic: bloody diarrhea; no fever, no pus in stool; secretes Shiga-like toxin: causes hemorrhagic colitis and hemolytic uremic syndrome (*E. coli* strain O157:H7)
	  C. Enteroinvasive: bloody diarrhea (with pus in stool) and fever. Also secretes small amounts of Shiga-like toxin.
		1. Cephalosporins
		2. Aminoglycosides
		3. Trimethoprim & sulfamethoxazole
		4. Fluoroquinolones
			1. Gram stain
			2. Culture (specimen may be urine, sputum, CSF or blood). Can grow at 45.5 °C.
			3. Pathogenic strains may be isolated from stool
			• *E. coli* ferments lactose. Its colonies produce a green metallic sheen on EMB agar and are pink-purple on MacConkey agar.
				• Index organism for fecal contamination of water

	1. Pneumonia, with significant lung necrosis and bloody sputum, commonly in alcoholics, or those with underlying lung disease 2. Hospital acquired urinary tract infections and sepsis	1. Third generation cephalosporin 2. Ciprofloxacin		

GRAM-NEGATIVE RODS	RESERVOIR	TRANSMISSION	METABOLISM	VIRULENCE
Proteus mirabilis			1. Urease: splits urea into NH_3 & CO_2 2. Indole-negative 3. Does not ferment lactose.	• Motile (swarming)
Shigella dysenteriae	• Humans	• Fecal-oral transmission	1. No H_2S production 2. Does not ferment lactose.	1. Invades submucosa of intestinal tract, but not the lamina propria 2. NON-motile: No H-antigen (since no flagella)
Salmonella typhi Non-typhi groups of *Salmonella*	• *S. typhi* is found only in humans • Zoonotic: 1. Pet turtles 2. Chickens 3. Uncooked eggs	• *S. typhi* is transmitted via fecal-oral route	1. Produces H_2S 2. Does not ferment lactose.	1. Motile (H-antigen) 2. Capsule (called the VI antigen): protects from intracellular killing 3. Siderophore
Yersinia enterocolitica	• Zoonotic: can be found in pigs	1. Ingestion of contaminated food or water 2. Unpasteurized milk	1. Non-lactose fermenter 2. Virulence factors are temperature sensitive; expressed at 37 °C	1. V and W antigens 2. Motile

Figure 9-11 (continued)

TOXINS	CLINICAL	TREATMENT	DIAGNOSTICS	MISCELLANEOUS
No toxin	1. Urinary tract infection: urine has a high pH due to urease. May get stones in the bladder 2. Sepsis	1. Ampicillin 2. Trimethoprim & sulfamethoxazole	1. Culture: colonies swarm over entire culture plate 2. Examination of urine shows a high pH (from splitting urine into NH_3 and CO_2)	• Weil-Felix test: a test that uses antibodies against certain strains of proteus to diagnose rickettsial diseases (as certain rickettsiae share similar antigens)
• Shiga toxin: Inactivates the 60S ribosome, inhibiting protein synthesis and killing intestinal epithelial cells	BLOODY diarrhea with mucus and pus (similar to enteroinvasive *E. coli*)	1. Fluoroquinolones 2. Azithromycin 3. Trimethoprim & sulfamethoxazole	• Stool culture: never part of the normal intestinal flora	• IgA is best for immunity

	1. Enteric Fever
	  A. Typhoid fever:
	    1. Fever
	    2. Abdominal pain
	    3. Liver or spleen enlargement
	    4. Rose spots on abdomen
	  B. Paratyphoid fever (similar to typhoid fever, but caused by non-typhoid *Salmonella*)
	2. Chronic carrier state
	3. Gastroenteritis
	4. Sepsis
	5. Osteomyelitis: especially in sickle cell patients
		1. Ciprofloxacin
		2. Ceftriaxone
		3. Trimethoprim & sulfamethoxazole
		4. Azithromycin
		*Salmonella gastroenteritis: there is little benefit from antibiotic treatment—can prolong carrier state
			• Culture: blood, stool or urine may contain *S. typhi*.
			• never part of the normal intestinal flora
				A. Facultative intracellular parasite:
				  1. Lives within macrophages in lymph nodes
				  2. Can live in gallbladder for years (carriers secrete *S. typhi* in stools)
				B. Persons who are asplenic or have nonfunctioning spleens (sickle cell anemia) are at increased risk of infection by this organism

• Enterotoxin similar to the heat stable toxin of *E. coli*: increases cGMP levels	• Acute enterocolitis, with fever, diarrhea and abdominal pain.	• Antibiotics do not alter the course of the diarrhea. However, patients with positive blood culture should be treated with antibiotics	1. Stool or blood cultures may be positive 2. Examination of the terminal ilium with colonoscopy will reveal mucosal ulceration	1. Can survive refrigeration. 2. Closely related to *Yersinia pestis*, which is the cause of bubonic plague

GRAM-NEGATIVE RODS	RESERVOIR	TRANSMISSION	METABOLISM	VIRULENCE
VIBRIONACEAE				
Vibrio cholerae		1. Fecal-oral transmission 2. Morphology: short comma shaped, gram-negative rod, with a single polar flagellum	1. Oxidase-positive 2. Ferments sugars (except lactose)	1. Motile (H-antigen) 2. Mucinase: digests mucous layer so *V. cholerae* can attach to cells. 3. Fimbriae: helps with attachment to cells. 4. Non-invasive!!
Vibrio parahaemolyticus	• Fish	1. Consumption of raw fish 2. Morphology: short comma shaped, with a single polar flagellum	• Halophilic (likes salt)	1. Motile (H-antigen) 2. Capsule
Campylobacter jejuni	• Zoonotic: wild and domestic animals, and poultry	1. Uncooked meat (especially poultry) 2. Unpasteurized milk 3. Fecal-oral 4. Morphology: curved gram-negative rods, with a single polar flagellum	1. Microaerophilic 2. Oxidase positive 3. Optimum temperature is 42 °C	1. Motile (H-antigen) 2. Invasive
Helicobacter pylori		Morphology: curved gram-negative rods, with a tuft of polar flagellum	1. Microaerophilic 2. Urease-positive	
BACTEROIDACEAE				
Bacteroides fragilis		• Part of the normal flora of the intestine	1. Anaerobic 2. Gram-negative rod	

Figure 9-11 (continued)

TOXINS	CLINICAL	TREATMENT	DIAGNOSTICS	MISCELLANEOUS
• Choleragen (enterotoxin): like LT of *E. coli*; increases levels of cAMP, causing secretion of electrolytes from the intestinal epithelium. This results in secretion of fluid into the intestinal tract.	• Cholera: severe diarrhea with rice water stools. No pus in stools	1. Replace fluids 2. Doxycycline 3. Fluoroquinolone	1. Dark field microscopy of stool reveals motile organisms that are immobilized with antiserum 2. Grows as flat yellow colonies on selective media: thiosulfate-citrate-bile salts-sucrose (TCBS) agar	Death by dehydration; children affected in endemic areas 1991: Latin America epidemic 1993: Epidemic in Bangladesh and India
• Hemolytic cytotoxin	• Cause of 25% of food poisoning in Japan (diarrhea for 3 days)	1. Doxycycline 2. Fluoroquinolone *(unclear if antibiotics change clinical course)	• Requires thiosulfate & bile salts	
1. Enterotoxin: similar to cholera toxin and the LT of *E. coli* 2. Cytotoxin: destroys mucosal cells	• Secretory or bloody diarrhea	1. Fluoroquinolone 2. Erythromycin	1. Microscopic exam of stool reveals motile, curved gram-negative rods 2. Selective media with antibiotics at 42 °C.	• One of the three most common causes of diarrhea in the world
• No toxin	1. Duodenal ulcers 2. Chronic gastritis	1. Bismuth, ampicillin, metronidazole and tetracycline 2. Clarithromycin and omeprazole • Both regimens reduce duodenal ulcer relapse		
• Does not contain lipid A (so NO	• Abscesses in the gastrointestinal tract, pelvis and	1. Metronidazole 2. Clindamycin 3. Chloramphenicol	1. Gram-stain 2. Anaerobic culture	• Infection occurs when the organism

GRAM-NEGATIVE RODS	RESERVOIR	TRANSMISSION	METABOLISM	VIRULENCE
			3. Non-spore former 4. Polysaccharide capsule	
Bacteroides melaninogenicus		• Part of the normal flora of the intestine	1. Anaerobic 2. Gram-negative rod 3. Non-spore former 4. Polysaccharide capsule	
Fusobacterium			1. Anaerobic 2. Gram-negative rod 3. Non-spore former	

Figure 9-11 (continued)

TOXINS	CLINICAL	TREATMENT	DIAGNOSTICS	MISCELLANEOUS
Endotoxin)	lungs	4. Surgically drain abscesses		enters into the peritoneal cavity
• Does not contain lipid A (so NO Endotoxin)	1. Necrotizing anaerobic pneumonia 2. Periodontal disease	1. Metronidazole 2. Clindamycin		• Produce a black pigment when grown on blood agar
	1. Necrotizing anaerobic pneumonia 2. Periodontal disease 3. Abdominal and pelvic abscess 4. Otitis media	• Penicillin G	1. Gram-stain 2. Anaerobic culture	

M. Gladwin, W. Trattler, and S. Mahan, *Clinical Microbiology Made Ridiculously Simple* ©MedMaster

FAMILY BACTEROIDACEAE

We have spent so much time studying all the preceding enteric bacteria that you may be surprised to find out that 99% of the flora of our intestinal tract is made up of obligate anaerobic gram-negative rods comprising the family Bacteroidaceae. The mouth and vagina are also home to these critters.

Bacteroides fragilis

This bacterium is notable for being one of the few gram-negative bacteria that does not contain lipid A in its outer cell membrane (NO endotoxin!). However, it does possess a capsule.

You will become very familiar with *Bacteroides fragilis* while studying surgery. This bacterium has low virulence and normally lives in peace in the intestine. However, when a bullet tears into the intestine, when a seat belt lacerates the intestine in a car wreck, when abdominal surgery is performed with bowel penetration, or when the intestine ruptures secondary to infection (appendicitis) or ischemia, THEN the bacteria go wild in the peritoneal cavity, forming **abscesses**. An abscess is a contained collection of bacteria, white cells, and dead tissue. Fever and sometimes systemic spread accompany the infection.

This abscess formation is also seen in obstetric and gynecologic patients. Abscesses may arise in a patient with a septic abortion, pelvic inflammatory disease (tubo-ovarian abscess), or an intrauterine device (IUD) for birth control.

Bacteroides fragilis is rarely present in the mouth, so it is rarely involved in aspiration pneumonias.

Following abdominal surgery, antibiotics that cover anaerobes are given as prophylaxis against *Bacteroides fragilis*. These include clindamycin, metronidazole (Flagyl), chloramphenicol, and others (see Chapter 17, **Fig. 17-16**). If an abscess forms, it must be surgically drained.

Bacteroides melaninogenicus

This organism produces a black pigment when grown on blood agar. Hence, the name ***melaninogenicus***. It lives in the mouth, vagina, and intestine, and is usually involved in necrotizing anaerobic pneumonias caused by aspiration of lots of sputum from the mouth (during a seizure or drunken state). It also causes periodontal disease.

Fusobacterium

This bacterium is just like *Bacteroides melaninogenicus* in that it also causes periodontal disease and aspiration pneumonias. *Fusobacterium* can also cause abdominal and pelvic abscesses and otitis media.

ANAEROBIC GRAM-POSITIVE COCCI

Peptostreptococcus (strip or chain of cocci) and *Peptococcus* (cluster of cocci) are **gram-positive anaerobes** that are part of the normal flora of the mouth, vagina, and intestine. They are mixed with the preceding organisms in abscesses and aspiration pneumonias.

Members of the *Streptococcus viridans* group, discussed in Chapter 4, are mentioned here because they are gram-positive, microaerophilic, and are frequently isolated from abscesses (usually mixed with other anaerobic bacteria). These oxygen-hating critters have many names (such as *Streptococcus anginosus* and *Streptococcus milleri*) and are a part of the normal GI flora.

Fig. 9-11. Summary of enteric bacteria.

References

Bhan MK, Bahl R, Bhatnagar S. Typhoid and paratyphoid fever. Lancet 2005;366:749–762.

Costa F, D'Elios MM. Management of Helicobacter pylori infection. Expert Rev Anti Infect Ther. 2010;8(8):887–92.

Fischer Walker CL, Sack D, Black RE. Etiology of diarrhea in older children, adolescents and adults: a systematic review. PLoS Negl Trop Dis. 2010;4(8):e768.

Hill DR, Beeching NJ. Travelers' diarrhea. Curr Opin Infect Dis. 2010;23(5):481–7.

Meltzer E, Schwartz E. Enteric fever: a travel medicine oriented view. Curr Opin Infect Dis. 2010;(5):432–7.

CHAPTER 10.   HOSPITAL-ACQUIRED GRAM-NEGATIVES
By C. Scott Mahan, MD

These bugs often arise in hospitalized patients whose natural defenses have been compromised by intravenous catheters, endo-tracheal tubes, foley catheters and surgical incisions. Additionally, hospitalized persons are often given antibiotics that alter their microbiologic flora, and they are exposed to bacteria carried on the hands of healthcare personnel. In this setting, these crafty, often highly antibiotic-resistant organisms thrive!

In 2002 there were an estimated 1.7 million hospital-acquired infections and 99,000 associated deaths. Hospital-acquired infections were the sixth leading cause of death in the United States. Infections due to gram-negative organisms are of particular concern, as these organisms are masters of up-regulating and acquiring genes that code for antibiotic resistance. Hospital-acquired infections include pneumonia (often associated with endo-tracheal intubation and mechanical ventilation), urinary tract infections (associated with foley catheters), wound infections (associated with recent surgery and implanted devices), and bloodstream infections (associated with intravenous and intra-arterial lines). Just remember the **4 W's: wind** (pneumonia), **water** (urinary tract infection), **wound**, and **wires** (intravenous lines). See **Fig. 10-1**.

Fig. 10-1.   Common hospital acquired infections (the **4 W's**). Think of a patient that is on a ventilator thus

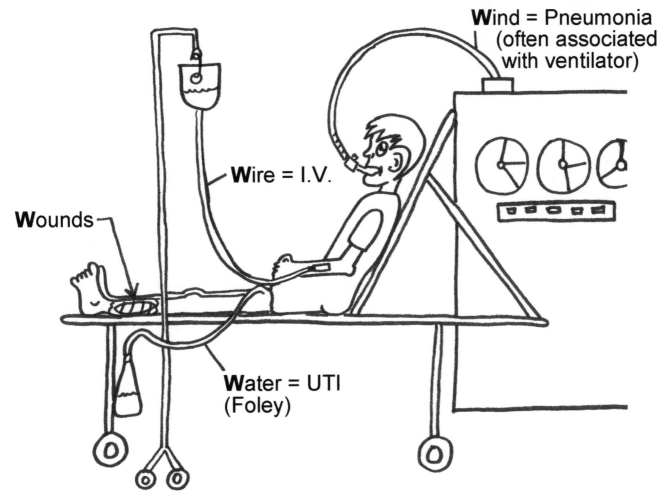

Figure 10-1

predisposing to pneumonia (**Wind**); he has a foley catheter making him at risk for a urinary tract infection (**Water**); he has had recent surgery on his leg (**Wound**); and he has multiple intravenous lines (**Wires**).

In the United States, gram-negative bacteria account for about 30% of hospital-acquired infections, and in the intensive care units, where lines and tubes are even more common, this number approaches 70%. The *Enterobacteriacae* family of gram-negative organisms such as *E.coli, Klebsiella*, and *Enterobacter* (see **Chapter 9**) is the most common group overall. Unfortunately, multi-drug resistant (MDR) gram-negative bacteria such as Pseudomonads (*Pseudomonas aeruginosa, Stenotrophomonas maltophilia*, and *Burkholderia cepacia*) and Acinetobacter are being increasingly reported. See **Fig. 10-2**.

Fig. 10-2. To remember these 4 often highly resistant gram-negative bacteria, envision the generals-*Pseudomonas, Acinetobacter*, and the troops, *Stenotrophomonas* and *Burkholderia*, sitting around a map of the world (really the hospitalized patient) and strategizing their attack.

Pseudomonas aeruginosa

You will hear so much about this bug while working in the hospital that you'll wish the Lord had never conjured it up. There are two reasons why it is so important:

1) It colonizes and infects sick, immunocompromised hospitalized patients.
2) The rascal is resistant to almost every antibiotic, so it has become an art to think up "anti-pseudomonal coverage." On discussions on rounds you will always need to consider the antibiotic coverage for *Pseudomonas*.

Pseudomonas aeruginosa is an obligate aerobic (non-lactose fermenter), gram-negative rod. It produces a green fluorescent pigment (**pyoverdin**) and a blue pigment (**pyocyanin**), which gives colonies and infected wound dressings a greenish-blue coloration. It

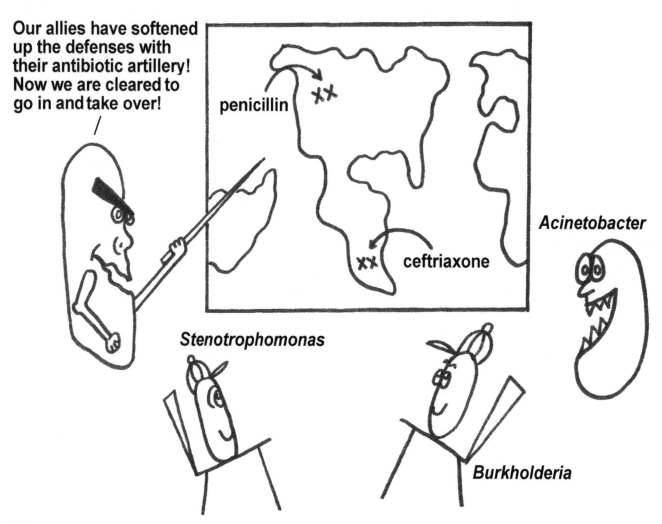

Figure 10-2

also produces a sweet grape-like scent, so wound dressings and agar plates are often sniffed for organism identification. *Pseudomonas aeruginosa* has weak invasive ability. Healthy people just don't get infections with this guy! However, once inside a weakened patient, the story changes. It elaborates numerous exotoxins including **exotoxin A**, which has the same mechanism of action as diphtheria toxin (stops protein synthesis) but is not antigenically identical. Some strains also possess a capsule that is antiphagocytic and aids in adhesion to target cells (in the lungs for example).

Important *Pseudomonas aeruginosa* Infections

1) Pneumonia (Wind)
a) Most cystic fibrosis patients have their lungs colonized with *Pseudomonas aeruginosa*. These patients develop a chronic pneumonia, which progressively destroys their lungs.

b) Immunocompromised patients (cancer patients and intensive care unit patients) are highly susceptible to pneumonia caused by *Pseudomonas aeruginosa*.

2) Osteomyelitis (related to Wounds)
a) Diabetic patients have an increased risk of developing foot ulcers infected with *Pseudomonas aeruginosa*. The infection can penetrate into the bone, resulting in osteomyelitis.

b) Intravenous (IV) drug abusers have an increased risk of osteomyelitis of the vertebrae or clavicle.

c) Children develop osteomyelitis secondary to puncture wounds to the foot. The classic injury that leads to pseudomonal osteomyelitis is stepping on a nail while wearing tennis shoes. Tennis shoes are often sweaty and create a nice moist environment for the water-loving pseudomonas.

3) Burn-wound infections (Wound):
This organism sets up significant infections of burn wounds, which can eventually lead to a fatal sepsis.

4) Sepsis (Wire):
Pseudomonas sepsis carries an extremely high mortality rate. This can come from infected "wires" or catheters, or from other sites of infection (Wind, Wound or Water).

5) Urinary tract infections, pyelonephritis (Water):
This occurs in debilitated patients in nursing homes and in hospitals. They often have urethral Foley catheters, which serve as a source of infection.

6) Endocarditis:
Staphylococcus aureus and *Pseudomonas aeruginosa* are frequent causes of right heart valve endocarditis in IV drug abusers.

7) Malignant external otitis:
A *Pseudomonas* external ear canal infection burrows into the mastoid bone, primarily in elderly diabetic patients.

8) Corneal infections:
This can occur in contact lens wearers.

MNEMONIC (courtesy of Eric Davies): *Pseudomonas aeruginosa*

BE PSEUDO
Burns
Endocarditis
Pneumonia
Sepsis
External malignant otitis media
UTI
Diabetic osteomyelitis

Treatment of *Pseudomonas* is complicated due to resistance to many antibiotics. When a pseudomonal infection is suspected, two antibiotics with anti-pseudomonal activity are often given until the organism is identified and drug susceptibilities are available (anti-pseudomonal antibiotics are reviewed in **Chapter 17**/Penicillin-Family Antibiotics). At this point, therapy can be continued with a single agent to which the organism is susceptible. Despite a long running debate on the value of two agents for "synergy" in treating pseudomonal infections, there is limited evidence that combining two anti-pseudomonal agents leads to improved outcomes.

Burkholderia cepacia

This oxidase-positive, aerobic gram-negative bacillus is one of several species that together form the *Burkholderia cepacia* complex. These are environmental organisms that are able to grow in water, soil, plants, and animals. In the hospital, *B. cepacia* has been isolated from several water and environmental sources and can be transmitted to patients. *B. cepacia* may also cause infections in burn and ventilated patients. Persons with cystic fibrosis (CF) are at greatest risk for disease due to this pathogen. CF patients may have a range of disease related to *B. cepacia* that includes asymptomatic carriage, bronchiectasis (dilated infected airways) or alternatively a rapidly progressive pneumonia with bacteremia. These organisms are often highly drug-resistant and require susceptibility testing to guide therapy.

Stenotrophomonas maltophilia

S. maltophilia is an increasingly common pathogen in hospitalized patients. While it is often just a colonizer and part of the normal respiratory flora, it can cause disease in hospitalized and immunocompromised persons. Infections include pneumonia (isolated in 3% of hospitalized persons with pneumonia) and line-related bacteremia. Due to its very narrow range of antibiotic susceptibility, it is often selected for and has a chance to thrive in persons placed on broad antibiotic coverage for other pathogens. Trimethoprim-sulfamethoxazole is the treatment of choice for this pathogen.

Acinetobacter

Acinetobacter species are very similar to Pseudomonas. They are aerobic gram-negative bacteria found in the soil and water and cause a wide range of infections in the hospital environment. *Acinetobacter baumannii* is the species most commonly isolated. *A. baumannii* can survive for extended periods on environmental surfaces, increasing its transmission

in healthcare settings. Much like *Pseudomonas, Acinetobacter* is a frequent cause of hospital-acquired pneumonia, line related bacteremias, burn infections, and foley catheter-associated urinary tract infections.

These guys can fool the lab technicians. At times they may appear gram-positive, and at other times they may even be misidentified as *Neisseria* species. This is because they can be coccobacillary (short rods) or coccal in appearance, and on solid media they often form diplococci similar to *Neisseria*.

A. baumannii can be a real challenge to treat. It has acquired multiple mechanisms of antibiotic resistance, making the choice of appropriate antibiotics difficult. Acinetobacter strains may be susceptible to aminoglycosides (such as gentamicin, tobramycin, and amikacin), carbapenems, polymixins (colistin, polymyxin E, and polymyxin B), tigecyline, and sulbactam (the B-lactamase

GRAM-NEGATIVE RODS	RESERVOIR	TRANSMISSION	METABOLISM	VIRULENCE
Pseudomonas aeruginosa	– Soil – Water – Plants – Animals – Intestinal flora – Skin	– Medical devices – Hands of healthcare workers	1. Obligate aerobe 2. Non-lactose fermenter 3. Oxidase-positive	1. Motile (polar flagella) 2. Hemolysin 3. Collagenase 4. Elastase 5. Fibrinolysin 6. Phospholipase C 7. DNAse 8. Antiphagocytic capsule (some strains)
Burkholderia cepacia	– Soil – Water – Plants – Animals – Lungs of cystic fibrosis patients	– Medical devices – Hands of healthcare workers – Between cystic fibrosis patients?	1. Oxidase-positive 2. Non-lactose fermenter	Extremely antibiotic and disinfectant resistant
Stenotrophomonas maltophilia	– Soil – Water – Plants – Animals – Normal respiratory flora	– Medical devices – Hands of healthcare workers	1. Oxidase-negative 2. Non-lactose fermenter	Extremely antibiotic and disinfectant resistant
Acinetobacter baumannii	– Soil – Water – Skin – Secretions	– Medical devices – Hands of healthcare workers	1. Obligate aerobe 2. Oxidase-negative 3. Non-lactose fermenter	Multiple acquired mechanisms of antibiotic resistance

Figure 10-3

inhibitor). Sometimes *Acinetobacter* strains are resistant or only intermediately sensitive to all available antibiotics. It is in these cases that creativity comes into play and one must try unusual antibiotic combinations for synergy as well as prolonged infusions. This difficult task often falls on the Infectious Disease specialist.

Prevention

The best bet to limit morbidity and mortality from hospital-acquired infections is to prevent them from occurring in the first place. Probably the three most important factors in preventing these infections are good hand-hygiene by all healthcare practioners, limiting the use of invasive devices (ventilators, foley catheters, and intravenous lines), and judicious use of antibiotics.

Fig. 10-3. Summary of hospital-acquired gram-negative bacteria.

References

Fishbain, J, Peleg, AY. Treatment of *Acinetobacter* Infections. Clin Infect Dis 2010;51(1):79–84.

Garnacho-Montero J, Amaya-Villar R. Multiresistant Acinetobacter baumannii infections: epidemiology and management. Curr Opin Infect Dis. 2010;23(4):332–9.

Guidelines for the Management of Adults with Hospital-acquired, Ventilator-associated, and Healthcare-associated Pneumonia. Am J Respir Crit Care Med 2005;171:388–416.

Jones RN. Microbial Etiologies of Hospital-acquiredBacterial Pneumonia and Ventilator-associated Bacterial Pneumonia. Clin Infect Dis 2010;51(S1):S81–S87.

Peleg AY, Hopper DC. Hospital-Acquired Infections Due to Gram-negative Bacteria. N Engl J Med 2010;362:1804–13.

TOXINS	CLINICAL	TREATMENT	DIAGNOSTICS	MISCELLANEOUS
1. Exotoxin A (similar to diphtheria toxin): inhibits protein synthesis by blocking EF$_2$	1. **Burns** 2. **Endocarditis** 3. **Pneumonia** 4. **Sepsis** 5. **External malignant** otitis media 6. **UTI** 7. **Diabetic** osteomyelitis	1. Timentin 2. Piperacillin 3. Imipenem 4. Doripenem 5. Aminoglycosides 6. Aztreonam 7. Ciprofloxacin 8. Ceftazidime 9. Cefepime 10. Polymixins	1. Culture: greenish metallic appearing colonies on blood agar, which have fruity (grape) smell	1. Common etiology for infection in neutropenic patients 2. Produces pigments when cultured: a. pyocyanin (blue pigment) b. pyoverdin (green pigment)
	1. Pneumonia in Cystic fibrosis patients. 2. Infections in patients with chronic granulomatous disease	1. Trimethoprim-sulfamethoxazole 2. Timentin 3. Ciprofloxacin 4. Ceftazidime 5. Carbapenems	1. May use selective media with colistin to select for growth.	
	1. Pneumonia in ventilated patients on broad antibiotics 2. Line-related bacteremia	1. Trimethoprim-sulfamethoxazole		1. Often a non-pathogenic colonizer of respiratory flora
1. Bacteriocin production 2. Protective capsule (inhibits phagocytosis)	1. Pneumonia 2. Line-related bacteremia 3. UTI 4. Burn/wound infections 5. Eye infections	1. Aminoglycosides 2. Carbapenems 3. Polymixins 4. Tigecyline 5. Sulbactam		1. May be mistaken for *Neisseria* 2. Guide therapy with antibiotic susceptibilities

CHAPTER 11. *HAEMOPHILUS, BORDETELLA, AND LEGIONELLA*

The gram-negative rods *Haemophilus influenzae, Bordetella pertussis* and *Legionella pneumophila* are grouped together because they are all acquired through the respiratory tract. This makes sense if you consider the species names: *influenzae* (the flu—an upper respiratory illness), *pertussis* (cough), and *pneumophila* (lung loving).

Haemophilus influenzae

The name *Haemophilus influenzae* describes some of its properties:

Haemophilus means "blood loving." This organism requires a blood-containing medium for growth. Hematin found in blood is necessary for the bacterium's cytochrome system. Blood also contains NAD+, needed for metabolic activity.

influenzae: This bacterium often attacks the lungs of persons debilitated by a viral influenza infection. During the 1890 and 1918 influenza pandemics, scientists cultured *Haemophilus influenzae* from the upper respiratory tracts of "flu" patients, leading them to incorrectly conclude that *Haemophilus influenzae* was the etiologic agent of the flu.

Haemophilus influenzae is an obligate human parasite that is transmitted via the respiratory route. Two important concepts help us understand how this critter causes disease:

1) A polysaccharide capsule composed of polyribitol ribose phosphate confers virulence: There are 6 types of capsules, designated a, b, c, d, e, and f. Of these, type b is commonly associated with invasive *Haemophilus influenzae* disease in children, such as meningitis, epiglottitis, and septic arthritis.

Capsule *b* = *b*ad

Nonencapsulated strains (also called **nontypeable**) of *Haemophilus influenzae* can colonize the upper respiratory tract of children and adults. They lack the virulent invasiveness of their encapsulated cousins and can only cause local infection. They frequently cause otitis media in children as well as respiratory disease in adults weakened by preexisting lung disease, such as chronic bronchitis from smoking or recent viral influenza infection.

Clinical Pearl: Patients with chronic obstructive pulmonary disease (COPD) get frequent infections with nontypeable *H. Influenzae* which causes a worsening of wheezing, shortness of breath and cough (COPD exacerbation). These exacerbations are often associated with an acquisition of a new strain of nontypeable H. influenzae.

2) Antibodies to the capsule are lacking in infants and children between 6 months and 3 years of age. The mother possesses antibodies against the b capsule which she has acquired in her lifetime. She passes these antibodies to the fetus transplacentally and in her breast milk. These "passively" acquired antibodies last for about 6 months. It takes 3–5 years of *Haemophilus influenzae* colonization and infection for children to develop their own antibodies. So there is a window during which children are sitting ducks for the invasive *Haemophilus influenzae*.

Haemophilus influenzae type b

1) **Meningitis**: This is the most serious infection caused by encapsulated *Haemophilus influenzae* type b. Prior to the introduction of vaccination of U.S. children in 1991, it was the main cause of meningitis in young children between the age of 6 months to 3 years (more than 10,000 cases per year). Following inhalation, this organism invades the local lymph nodes and blood-stream, and then penetrates into the meninges. Since infants usually do not display the classic stiff neck, non-specific signs such as fever, vomiting, and altered mental status are the clues to this potentially fatal infection.

Although mortality with appropriate antibiotics is less than 5%, up to half of infected children will still have permanent residual neurologic deficits, such as mental retardation, seizures, language delay, or deafness. When a bacterial meningitis is treated with antibiotics, the killed bacteria lyse and release cellular antigens, such as LPS lipid A (endotoxin), resulting in a violent immune response that destroys neurons as well as bacteria. Recent studies show that treatment with steroids 15–20 minutes before giving IV antibiotics will decrease this risk of developing neurologic deficits. It is theorized that the steroids limit the inflammatory response to the dead bacteria's antigens while allowing bacterial killing.

Clinical Pearl: Three bacteria are responsible for most meningitis acquired by the baby coming out of the birth canal (within first 3 months of age): *Listeria monocytogenes, Escherichia coli,* and *Group B Streptococcus*. Two bacteria cause meningitis later in life (6 months to 3 years) after the maternal antibodies passively given to fetus wane and before new antibodies develop: *Neisseria meningitides* and *Haemophilus influenzae*.

2) **Acute epiglottitis**: *Haemophilus influenzae* type b can also cause rapid swelling of the epiglottis, obstructing the respiratory tract and esophagus. Following a sore throat and fever, the child develops severe upper airway wheezing (stridor) and is unable to swallow. Excessive saliva will drool out of the child's mouth as it is unable to pass the swollen epiglottis. The large, red epiglottis looks like a red cherry at the base of the tongue. If you suspect this infection, do not examine the larynx unless you are ready to insert an endotracheal breathing tube because manipulation can cause laryngeal spasm. This may cause complete airway obstruction that can only be bypassed with a tracheotomy.

3) **Septic arthritis**: *Haemophilus influenzae* type b is the most common cause of septic arthritis in infants. Most commonly, a single joint is infected, resulting in fever, pain, swelling and decreased mobility of the joint. Examination of the synovial fluid (joint fluid) by Gram stain reveals the pleomorphic gram-negative rods.

4) **Sepsis**: Children between 6 months to 3 years present with fever, lethargy, loss of appetite, and no evidence of localized disease (otitis media, meningitis, or epiglottitis). Presumably the bacteria invade the bloodstream via the upper respiratory tract. Since the spleen is the most important organ in fighting off infection by encapsulated bacteria, it is not surprising that children with absent or non-functioning spleens (either by surgery or with sickle-cell disease) are at highest risk. Prompt identification and treatment will prevent *Haemophilus influenzae* type b from invading the meninges, epiglottis, or a joint.

Meningitis, epiglottitis, and bacterial sepsis are rapidly fatal without antibiotic therapy. Ampicillin used to be the drug of choice prior to the development of resistance. Ampicillin resistance is transmitted by a plasmid from strain to strain of *Haemophilus influenzae*. Currently, a third generation cephalosporin, such as **cefotaxime** or **ceftriaxone**, is the drug of choice for serious infections. **Ampicillin** or **amoxicillin** can be used for less serious infections, such as otitis media.

Vaccination
(Hib capsule vaccine)

The key to controlling this organism is to stimulate the early generation of protective antibodies in young children. However, it is difficult to stimulate antibody formation in the very young.

The first vaccine, consisting of purified type b capsule, was effective only in generating antibodies in children older than 18 months. Currently used vaccines are composed of *H. influenzae* capsule polyribitol ribose phosphate (PRP) conjugated with mutant diphtheria toxin protein, a *Neisseria meningitides* outer membrane protein, or tetanus toxoid to activate the T-lymphocytes and antibodies against the b capsule antigens. Vaccination with the Hib capsule of children in the U.S. at ages 2, 4, 6, and 15 months (along with the DPT and polio vaccines) has dramatically reduced the incidence of *Haemophilus influenzae* infection. Acute *Haemophilus influenzae* epiglottitis is now rarely seen in U.S. emergency rooms.

Hib, Hib, Hurray!

Other efforts involve immunizing women in the eighth month of pregnancy, resulting in increased antibody secretion in breast milk (passive immunization).

Haemophilus ducreyi

This species is responsible for the sexually transmitted disease **chancroid**. Clinically, patients present with a **painful** genital ulcer. Unilateral **painful** swollen inguinal lymph nodes rapidly develop in half of infected persons. The lymph nodes become matted and will rupture, releasing pus.

The **differential diagnosis** includes:

1) **Syphilis** (*Treponema pallidum*): It is extremely important to exclude syphilis as the cause of the ulcer. Remember that the ulcer of syphilis is painless and the associated adenopathy is bilateral, **painless**, and non-suppurative (no pus).

2) **Herpes** (Herpes simplex virus 1 and 2): Herpetic lesions start as vesicles (blisters), yet once they break they can be misdiagnosed as chancroid, especially because they are **painful**. Herpes is usually accompanied by systemic symptoms such as myalgias and fevers. Chancroid does not usually produce systemic symptoms.

3) **Lymphogranuloma venereum** (*Chlamydia trachomatis*): LGV has **painless** matted suppurative inguinal lymph nodes that develop much more slowly than chancroid. The primary ulcer of LGV disappears before the nodes enlarge, whereas with chancroid they coexist.

Gram stain of the ulcer may reveal gram-negative coccobacilli. Isolation of *H. ducreyi* from an ulcer or from a lymph node aspirate is considered diagnostic. A very promising approach to help with these cases of genital ulcers is a **polymerase chain reaction (PCR) multiplex** that amplifies and detects the bacterial DNA of *Haemophilus ducreyi*, *Treponema pallidum*, and herpes simplex 1 and 2, all at the same time!

Single dose treatment with 1-gram of oral azithromycin or 250 mg of intramuscular ceftriaxone is currently recommended. Effective treatment of genital ulcers is crucial, because these open lesions create a break in the skin barrier, increasing the risk of HIV transmission.

Other *Haemophilus* species

A few rare *Haemophilus* species deserve mention because they have fastidious growth requirements (variable need for X and V factor, and/or CO_2), grow slowly and cause up to 5% of cases of endocarditis. The species include *H. parainfluenzae, H. aphrophilus,* and *H. paraphrophilus.*

Clinical Pearl: These *Haemophilus* species are included in the **HACEK** group of bacteria:

HACEK are slow growing bacteria known to cause endocarditis:
Haemophilus species
Actinobacillus species
Cardiobacterium species
Eikenella species
Kingella species

Gardnerella vaginalis
(formerly *Haemophilus vaginalis*)

This organism causes bacterial vaginitis in conjunction with anaerobic vaginal bacteria. Women with vaginitis develop burning or pruritis (itching) of the labia, burning on urination (dysuria), and a copious, foul-smelling vaginal discharge that has a fishy odor. It can be differentiated from other causes of vaginitis (such as *Candida* or *Trichomonas*) by examining a slide of the vaginal discharge (collected from the vagina during speculum exam) for the presence of **clue cells**. Clue cells are vaginal epithelial cells that contain tiny pleomorphic bacilli within the cytoplasm.

Treat this infection with **metronidazole**, which covers *Gardnerella* as well as co-infecting anaerobes. As a note, this species was separated from the genus *Haemophilus* because it does not require X-factor or V-factor for growth in culture.

Bordetella pertussis

This bacterium is named:
Bordetella because it was discovered in the early 1900's by two scientists named Bordet and Gengou. It seems that Bordet got the better end of the deal!
Pertussis means "violent cough." *Bordetella pertussis* causes whooping cough.

Exotoxin Weapons

Bordetella pertussis is a violently militant critter with a (gram) negative attitude. He is a gram-negative rod armed to the hilt with 4 major weapons (virulence factors). These virulence factors allow him to attach to the ciliated epithelial cells of the trachea and bronchi. He evades the host's defenses and destroys the ciliated cells, causing whooping cough.

1) **Pertussis toxin**: Like many bacterial exotoxins this toxin has a **B** subunit that **Binds** to target cell receptors, "unlocks" the cell, allowing entry of the A subunit. The **A** subunit (**A** for **Action**) activates cell-membrane-bound G regulatory proteins, which in turn activate adenylate cyclase. This results in an outpouring of cAMP, which activates protein kinase and other intracellular messengers. The exact role of this toxin in whooping cough is not entirely clear, but it has 3 observed effects: a) histamine sensitization, b) increase in insulin synthesis, and c) promotion of lymphocyte production and inhibition of phagocytosis.

2) **Extra cytoplasmic adenylate cyclase**: When attacking the bronchi, *Bordetella pertussis* throws its adenylate cyclase grenades. They are swallowed by host neutrophils, lymphocytes, and monocytes. The internalized adenylate cyclase then synthesizes the messenger cAMP, resulting in impaired chemotaxis and impaired generation of H_2O_2 and superoxide. This weakens the host defense cells' ability to phagocytose and clear the bacteria.

3) **Filamentous hemagglutinin (FHA)**: *Bordetella pertussis* does not actually invade the body. It attaches to ciliated epithelial cells of the bronchi and then releases its damaging exotoxins. The FHA, a pili rod extending from its surface, is involved in this binding. Antibodies directed against the FHA prevent binding and disease, and thus they are protective.

4) **Tracheal cytotoxin**: This toxin destroys the ciliated epithelial cells, resulting in impaired clearance of bacteria, mucus, and inflammatory exudate. This toxin is probably responsible for the violent cough.

Whooping Cough

The number of cases of whooping cough has decreased dramatically since vaccination programs began. In the prevaccination era in the United States, there were approximately 100–300 thousand cases a year which occurred primarily in children between the ages of 1–5. With the introduction of vaccination in the 1940's case rates plummeted, with a nadir of only 1010 cases reported in 1976. Subsequently, there has been a miniresurgence of disease with over 25,000 cases in 2005. Most cases diagnosed today are in adolescents and adults (with waning vaccine-associated immunity). In response to the rise in pertussis cases, the U.S. now recommends a booster vaccination for adults ages 19–64.

Whooping cough is a highly contagious disease with transmission occurring via respiratory secretions on the hands or in an aerosolized form. A week-long incubation period is followed by 3 stages of the disease:

1) **Catarrhal stage**: This stage lasts from 1–2 weeks and is similar to an upper respiratory tract infection, with low-grade fevers, runny nose, sneezing, and mild

cough. It is during this period that the disease is most contagious.

2) **Paroxysmal stage**: The fever subsides and the infected individual develops characteristic bursts of nonproductive cough. There may be 15–25 of these attacks per day, and the person may appear normal between events. The attacks consist of 5–20 forceful coughs followed by an inspiratory gasp through the narrowed glottis. This inspiration sounds like a whoop. During these paroxysms of coughing the patient can become hypoxemic and cyanotic (blue from low oxygen), the tongue may protrude, eyes bulge, and neck veins engorge. Vomiting often follows an attack. The paroxysmal stage can last a month or longer. The illness is more severe in the young, with up to 75% of infants less than 6 months of age and 40% of infants and young children more than 6 months requiring hospitalization.

Infants and partially immunized (wearing off) children and adults may not have the typical whoop. Infants can have cough and apnea spells (no breathing). Adults may present with a persistent cough. In fact, as many as 20–30% of adolescents and adults with a chronic cough of greater than 1 week duration may have serological evidence of *B. pertussis*.

Clinical Pearl: *B. pertussis* is now a frequent cause of chronic unexplained cough in adolescents and adults. Other causes to exclude include asthma, post-nasal drip (allergic rhinitis), acid-reflux and use of ACE-inhibitors for blood pressure control.

Examination of the white blood cells will surprisingly reveal an increase in the lymphocyte count with just a modest increase in the neutrophils (more like a viral picture). The increased number of lymphocytes seems to be one of the manifestations of the pertussis toxin.

3) **Convalescent stage**: The attacks become less frequent over a month, and the patient is no longer contagious.

Since this organism will **not grow on cotton**, specimens for culture are collected from the posterior pharynx with a **calcium alginate** swab. This swab is inserted into the posterior nares and the patient is then instructed to cough. The swab is then wiped on a special culture medium with potato, blood, and glycerol agar, called the **Bordet-Gengou medium**. At most hospitals, identification of this bacterium can be made with rapid serological tests (ELISA) and polymerase chain reaction (PCR) assays.

Treatment is primarily supportive. Infants are hospitalized to provide oxygen, suctioning of respiratory secretions, respiratory isolation, and observation. Treatment of infected individuals with **erythromycin** in the prodromal or catarrhal stage may ameliorate symptoms and shorten the duration of infectiousness. Later therapy during the paroxysmal stage does not alter the course of illness but may decrease bacterial shedding. Household contacts should also receive erythromycin.

Vaccination

The vaccine currently used in the U.S. is an acellular vaccine with antigens for pertussis toxin, FHA, and one or two other antigens, depending on the vaccine manufacturer. It is combined with the formalin-inactivated tetanus and diphtheria toxoids to form the DTaP (**D**iptheria-**T**etanus- **a**cellular **P**ertussis) vaccine, and it is given at 2, 4, 6, 15–18 months, and 4–6 years of age. The heat-killed whole cell vaccine is no longer available in the U.S. but remains in use in the developing world. The whole cell vaccine has markedly more side-effects, which led to its eventual replacement in the U.S. Additionally, a **Tdap** vaccination is recommended for all adults in the U.S. ages 19–64 years old to boost waning immunity. Note the **d**, **a**, and **p** which are in small letters because lesser amounts of the diphtheria toxoid and acellular pertussis antigens are given as compared to that given in the childhood vaccine.

Legionella pneumophila
(Legionnaires' Pneumonia)

Legionella pneumophila is an aerobic gram-negative rod that is famous for causing an outbreak of pneumonia at an American Legion convention in Philadelphia in 1976 (thus its name).

This organism is ubiquitous in natural and man-made water environments. Aerosolized contaminated water is inhaled, resulting in infection. Sources that have been identified during outbreaks have included air conditioning systems, cooling towers, and whirlpools. Outbreaks have even been associated with organism growth in shower heads and produce mist machines in supermarkets!!! Person-to-person transmission has not been demonstrated.

Like *Mycobacterium tuberculosis*, this organism is a facultative intracellular parasite that settles in the lower respiratory tract and is gobbled up by macrophages. This means that once it has been phagocytosed, it inhibits phagosome-lysosome fusion, surviving and replicating intracellularly.

Two interesting and unusual things about *Legionella* deserve mention: The first is that *Legionella pneumophila* is a facultative intracellular parasite for free living amoebas (like *Naegleria* and *Acanthamoeba*; see **Chapter 31**). *Legionella* multiplies a thousand fold within the amoebas and when the amoeba encysts in the face of environmental hardship, the *Legionella* is protected. The second interesting fact is that Legionella can enter a low metabolic state and can survive in a **biofilm**. Disruption of this biofilm can result in massive release of Legionella into the water.

Legionella is responsible for diseases ranging from asymptomatic infection and a flulike illness called Pontiac fever to a severe pneumonia called Legionnaires' disease:

1) **Pontiac fever**: Like influenza, this disease involves headache, muscle aches, and fatigue, followed by fever and chills. Pontiac fever strikes suddenly and completely resolves in less than one week. Pontiac fever was so-named for the illness that struck 95% of the employees of the Pontiac, Michigan, County Health Department. The causative agent was identified as *Legionella pneumophila* carried by the air conditioning system.

2) **Legionnaires' disease**: Patients develop very high fevers and a severe pneumonia.

Legionella pneumophila is a common cause of community acquired pneumonia, accounting for an estimated 0.5–10% of all admitted pneumonia cases (2% is likely, the most accurate estimate). While it causes a classic lobar consolidative pneumonia that can be impossible to distinguish from pneumococcal pneumonia there are a few unusual clinical elements, such as a fever with pulse-temperature dissociation (high fever, low heart rate), severe headache, confusion, myalgia (muscle aches) sometimes associated with rhabdomyolysis (muscle breakdown with increased levels of serum CPK and myoglobinuria), cough (only half of the time productive of purulent sputum), hyponatremia, hypophosphatemia and elevated liver enzymes (AST, ALT, alkaline phosphatase, LDH). Diarrhea and abdominal pain also occur. Sometimes the systemic symptoms like fever, myalgias, confusion, abdominal pain and diarrhea precede the lung symptoms, leading to misdiagnosis of influenza or acute abdomen.

GRAM-NEGATIVE RODS	RESERVOIR	VIRULENCE	TOXINS
Haemophilus influenzae	• Man only (obligate human parasite) • Transmitted via respiratory route	1. Capsule: 6 types, a–f (b is most virulent) 2. Attachment pili 3. IgA_1 protease	• Cytolethal distending toxin (CDT) • Hemolysin
Haemophilus ducreyi	• Sexually transmitted disease		• No exotoxins

Figure 11-1 HAEMOPHILUS, BORDETELLA AND LEGIONELLA

To kill this bug the antibiotic has to be concentrated inside a cell, the macrophage, where the *Legionella* is hiding. Beta-lactams and aminoglycosides do not do this well so the mainstays of treatment for *Legionella pneumophila* are the **macrolides** (erythromycin, azithromycin, clarithromycin), **tetracyclines** (doxycycline) and **quinolones** (ciprofloxacin, levofloxacin, moxifloxacin). We call these drugs **"atypical coverage"** since they cover the atypical bacteria *Mycoplasma, Legionella,* and *Chlamydia*, which—in addition to viral pneumonia—all cause atypical pneumonia (atypical pneumonia was so named because the penicillins did not work for these pneumonias). Then attempt to determine the source of *Legionella*. Is the air conditioning system contaminated?

Fig. 11-1. Summary of *Haemophilus, Bordetella* and *Legionella*.

Recommended Review articles:

Carratalà J, Garcia-Vidal C. An update on Legionella. Curr Opin Infect Dis. 2010;23(2):152–7.

Cornia PB, Hersh AL, et al. Does this coughing adolescent or adult patient have pertussis? JAMA. 2010;304(8):890–6.

Mandell LA, Wunderink RG, et al.; Infectious Diseases Society of America; American Thoracic Society. Infectious Diseases Society of America/American Thoracic Society consensus guidelines on the management of community-acquired pneumonia in adults. Clin Infect Dis. 2007;44 Suppl 2:S27–72.

CLINICAL	TREATMENT	DIAGNOSTICS	MISCELLANEOUS
ENCAPSULATED *H. influenzae* (usually type B capsule) 1. **Meningitis**: *Haemophilus influenzae* type b is the primary cause of meningitis in infants from 3 to 36 months of age. Complications include mental retardation, seizures, deafness, and death 2. **Acute epiglottitis** 3. **Septic arthritis** in infants 4. **Sepsis**: especially in patients without functioning spleens 5. **Pneumonia** *NONENCAPSULATED* *H. influenzae* 1. **Otitis media** 2. **Sinusitis** 3. COPD exacerbation and pneumonia	1. Second or third generation cephalosporins (since *H. influenzae* can acquire ampicillin resistance by plasmids) 2. Hib vaccine: *H. influenzae* polysaccharide capsule of type b strain (Hib) is conjugated to diphtheria toxoid and given to children at 2, 4, 6, and 15 months (DTaP and oral polio are given at the same time). This has resulted in solid immunity during the critical 3 month to 3 year age, and has dramatically reduced the incidence of Hib infection (acute epiglottitis, meningitis, etc.) in the U.S. 3. Passive immunization: mother is immunized during 8th month of pregnancy to increase passive antibody transfer in breast milk	1. Gram stain 2. Culture specimen on blood agar that has been heated to 80 °C. for 15 minutes (now called **chocolate agar**). This high temperature lyses the red blood cells, releasing both hematin (called X factor) and NAD^+ (called V factor). Like the *Neisseria*, *H. influenzae* organisms grow best when the chocolate agar is placed in a high CO_2 environment at 37 °C. 3. Fluorescently labeled antibodies (ELISA and latex particle agglutination) 4. Positive Quellung test: due to its capsule (just like *Streptococcus pneumoniae*)	1. *Haemophilus influenzae* requires two factors for growth (both found in blood): • X factor: Hematin • V factor: NAD^+ 2. Note: *Haemophilus* stands for "blood loving"
• **Chancroid**: painful genital ulcer, often associated with unilateral swollen lymph nodes that can rupture, releasing pus	1. Azithromycin or erythromycin 2. Ceftriaxone (IM) 3. Ciprofloxacin	• Gram stain and culture of ulcer exudate and pus released from swollen lymph node	1. A sexually transmitted disease 2. Requires X factor (hematin) only.

GRAM-NEGATIVE RODS	RESERVOIR	VIRULENCE	TOXINS
Gardnerella vaginalis	• Sexually transmitted disease	• No capsule	• No exotoxins
Bordetella pertussis	• Man: highly contagious • Transmitted via respiratory route	1. Capsule 2. Beta-lactamase 3. Filamentous hemagglutinin (FHA): A pili rod that extends from the surface of *B. pertussis*, enabling the bacteria to bind to ciliated epithelial cells of the bronchi	1. **Pertussis toxin**: activates G proteins that increase cAMP, resulting in: A. Increased sensitivity to histamine B. Increased insulin release C. Increased number of lymphocytes in blood 2. Extracytoplasmic adenylate cyclase: "weakens" neutrophils, lymphocytes and monocytes 3. Filamentous hemagglutinin: allows binding to ciliated epithelial cells 4. Tracheal cytotoxin: kills ciliated epithelial cells
Legionella pneumophila	• Ubiquitous in man and natural water environments 1. Air conditioning systems 2. Cooling towers	1. Facultative intracellular parasite: Dot/Icm type IV secretion system inhibits macrophage phagosome/ endo/lysosome fusions 2. Cu-Zn superoxide dismutase and catalase-peroxidase protects bacteria from macrophage superoxide and hydrogen peroxide oxidative burst 3. Pili and flagella promote attachment and invasion 4. Secretion of protein toxins like RNAase, phospholipase A and phospholipase C.	• Cytotoxin: kills hamster ovary cells

Figure 11-1 (continued)

CLINICAL	TREATMENT	DIAGNOSTICS	MISCELLANEOUS
• **Bacterial vaginitis**: foul smelling vaginal discharge (with fishy odor), vaginal pruritus, and often dysuria	• Metronidazole	• **Clue cells**: vaginal epithelial cells that contain tiny pleomorphic gram-negative bacilli within the cytoplasm	• Does not require X factor or V factor for growth
Whooping Cough 1. Catarrhal phase: patient is highly contagious (1–2 weeks) A. Low grade fever, runny nose & mild cough B. Antibiotic susceptible during this stage 2. Paroxysmal phase (2–10 weeks) A. Whoop (bursts of non-productive coughs) B. Increased number of lymphocytes in blood smear C. Antibiotics ineffective during this stage 3. Convalescent stage	1. Erythromycin (most effective if given in catarrhal stage) 2. Vaccine: **DaPT** 1. **D**iptheria 2. **a**cellular **P**ertussis 3. **T**etanus (Given routinely at ages 2, 4, 6, 15 months and between 4–6 years.) 3. Treat household contacts with erythromycin.	1. Bordet-Gengou media: potatoes, blood and glycerol agar, with penicillin added 2. Rapid serologic tests (ELISA) • Collect specimen from posterior pharynx on a calcium alginate swab since *B. pertussis* will not grow on cotton 3. Direct fluorescein-labeled antibodies applied to nasopharyngeal specimens for rapid diagnosis. 4. PCR detection of bacterial DNA in respiratory secretions	• High risk groups: 1. Infants less than one year old 2. Adults (as immunity acquired from vaccine wears off)
1. **Pontiac fever**: headache, fever, muscle aches and fatigue, Self-limiting; recovery in a week is common. 2. **Legionnaires' Disease**: pneumonia: fever and non-productive cough	1. Azithromycin 2. Levofloxacin 3. Doxycycline	1. Culture on buffered charcoal yeast extract agar (L-cysteine is a critical ingredient) 2. Serology (IFA and ELISA) 3. Urinary antigen can be detected by radioim-munoassay with high sensitivity and specificity and will remain positive for months after infection. Urine antigen test only detects *L. pneumophila* **serogroup 1**, but this accounts for 90% of cases.	1. Facultative intracellular parasite: inside alveolar macrophages 2. Persons with compromised immune systems are especially susceptible

M. Gladwin, W. Trattler, and S. Mahan, *Clinical Microbiology Made Ridiculously Simple* ©MedMaster

CHAPTER 12. *YERSINIA, FRANCISELLA, BRUCELLA,* AND *PASTEURELLA*

These organisms have been included in the same chapter because they share many characteristics (*Pasteurella* only shares the first 2):

1) They are all gram-negative rods (bacilli).

2) All of these are zoonotic diseases (i.e., they are primarily diseases of animals).

3) These bacteria are very virulent and are able to penetrate any body area they touch. This can occur on the skin following an insect bite, animal bite, or direct contact with an animal. This can also occur in the lungs after inhalation of infected aerosolized matter.

4) From the site of contact (usually the skin) the bacteria are phagocytosed by macrophages. They can survive inside the macrophages and so are **facultative intracellular organisms**. They migrate to the regional lymph nodes, set up infection there, and then move to the bloodstream and other organs, such as the liver, spleen, and lungs.

Like other facultative intracellular organisms (see **Fig. 2-7**) immunity is cell-mediated, and intradermal injections of bacterial extracts will elicit a delayed-type-hypersensitivity (DTH) reaction. This reaction results in skin swelling and induration (hardening) at the injection site 1–2 days later. The presence of swelling indicates previous exposure to the bacteria and can be used as a diagnostic test (see discussion of DTH and intradermal skin testing in Chapter 15, page 144)

5) The common treatment is an **aminoglycoside** (gentamicin or streptomycin) and/or **doxycycline**, which must be given for a prolonged period so as to reach the hidden intracellular bacteria.

Yersinia pestis
(Bubonic Plague)

You have all heard of **bubonic plague** and that rats were somehow involved. Rats are the *PESTS* (*Yersinia pestis*) that harbor this disease, while fleas serve as vectors, carrying *Yersinia pestis* to humans. Bubonic plague destroyed one fourth of the population of Europe in the 14th century. Later outbreaks moved from China to India (where the disease killed 10 million) and in the 1900's to San Francisco. The organism now resides in squirrels and prairie dogs of the southwestern U.S.

The **F1, V**, and **W** virulence factors enable this organism to resist destruction after phagocytosis (facultative intracellular organism):

1) **Fraction 1 (F1)**: This capsular antigen has antiphagocytic properties.

2) **V and W antigens**: These antigens, which are a protein and lipoprotein respectively, are unique to the *Yersinia* genus. Their actions are unknown.

Fig. 12-1. Visualize a rat riding in a Fuel Injected (**F1**), **VW** bug, being pursued by a macrophage. These three virulence factors are involved in *Yersinia pestis'* resistance to destruction after phagocytosis.

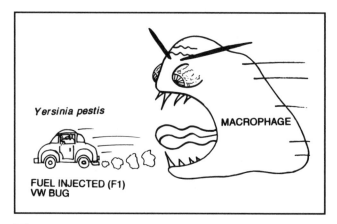

Figure 12-1

Fig. 12-2. *Yersinia pestis* is a gram-negative bacterium with a bipolar staining pattern. The ends of the rod-shaped bacterium take up more stain than the center. Three mammals fall prey to *Yersinia pestis*: wild rodents, domestic city rodents, and humans. The bacteria reside in the wild rodent population between epidemics and are carried from rodent to rodent by the flea. When wild rodents come into contact with domestic city rats (during droughts when wild rodents forage for food), fleas can then carry the bacteria to domestic rats. As the domestic rat population dies, the fleas become hungry and search out humans.

During interepidemic periods (we are in one now), bubonic plague may be contracted during camping,

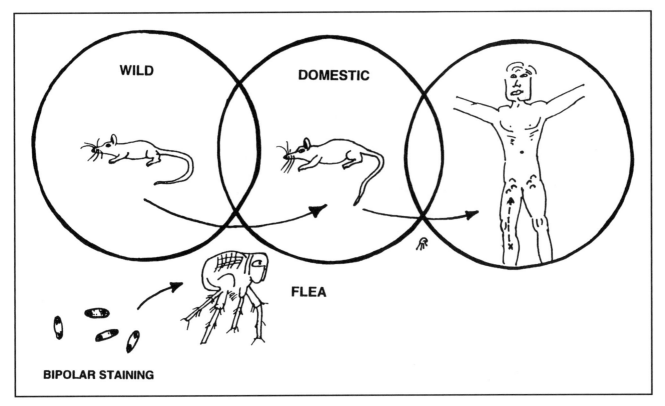

Figure 12-2

hunting or hiking. The human victim either touches a dead infected rodent or is bitten by an infected flea.

The bacteria invade the skin and are gobbled up by macrophages. They continue to reproduce intracellularly and within a week move to the nearest lymph nodes, usually the inguinal nodes (*boubon* is the Greek word for "groin"). The nodes swell like eggs and become hot, red, and painful. Fever and headache set in. The bacilli invade the bloodstream, liver, lungs, and other organs. Hemorrhages under the skin cause a blackish discoloration, leading people to call bubonic plague the "Black Death." Without treatment, death can occur in a few days. During epidemics, the disease can also be seen as pneumonic plague with pneumonia and human-to-human transmission by aerosolized bacteria.

If you see a patient who has been camping in Arizona or New Mexico and has developed fever, have a high index of suspicion. You may want to start gentamicin right away. You can't depend on the presence of swollen lymph nodes: Between 1980 and 1984, 25% of the cases in New Mexico did not have lymph node involvement.

This disease is deadly if untreated! About 75% of untreated people die!

Control of epidemics involves DDT for the fleas and destruction of the rats. If you only kill the rats, the starving fleas will feed on humans instead!

Another species of *Yersinia* called *Yersinia enterocolitica* infects the colon and is closely related to *Escherichia coli* (see Chapter 9 page 79).

Francisella tularensis
(Tularemia)

Tularemia is a disease that resembles bubonic plague so closely that it is always included in the differential diagnosis when considering bubonic plague. This disease is most commonly acquired from handling infected **rabbits** and from the bites of **ticks** and **deerflies**. More than a hundred creatures carry this bacterium, including rabbits, other mammals, and even reptiles and fish. Tularemia is distributed all over the U.S.

Fig. 12-3. *Francis* (*Francis*ella) the rabbit (rabbit vector) is playing in the *Tul*ips (*Tul*arensis). One ear has a tick, the other a deerfly.

Like *Yersinia pestis*, this organism is extremely virulent and can invade any area of contact, resulting in more than one disease presentation. The most important diseases caused by *Francisella tularensis* are the ulceroglandular and pneumonic diseases:

1) **Ulceroglandular tularemia**: Following the bite of a tick or deerfly, or contact with a wild rabbit,

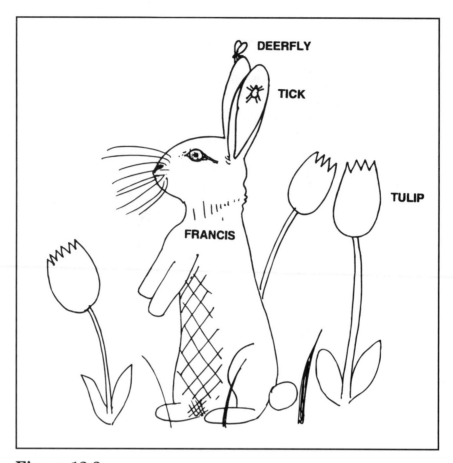

DEERFLY

TICK

TULIP

FRANCIS

Figure 12-3

a well-demarcated hole in the skin with a black base develops. Fever and systemic symptoms develop, and the local lymph nodes become swollen, red, and painful (sometimes draining pus). The bacteria can then spread to the blood and other organs. Note that these symptoms are almost identical to bubonic plague, but the skin ulcer is usually absent in the plague and the mortality rate is not nearly as high as in bubonic plague, reaching 5% for ulceroglandular tularemia.

2) **Pneumonic tularemia**: Aerosolization of bacteria during skinning and evisceration of an infected rabbit or hematogenous spread from the skin (ulceroglandular tularemia) to the lungs can lead to a lung infection (pneumonia).

Francisella tularensis can also invade other areas of contact such as the eyes (**oculoglandular tularemia**) and the gastrointestinal tract (**typhoidal tularemia**).

Because this bacterium is so virulent (just 10 organisms can cause disease), most labs will not culture it from blood or pus. For the same reason it is not advisable to drain the infected lymph nodes. Diagnosis rests on the clinical picture, a skin test similar to the PPD for tuberculosis, and the measurement of the titers of antibodies to *Francisella tularensis*.

Brucella
(Brucellosis)

All the names of *Brucella* species are based on the animal they infect:

- *Brucella melitensis* (goats)
- *Brucella abortus* (causes abortions in cows)
- *Brucella suis* (pigs)
- *Brucella canis* (dogs)

Humans acquire *Brucella* from direct contact with infected animal meat or aborted placentas, or ingestion of infected milk products. The incidence of this disease worldwide is greater than that of both bubonic plague and tularemia. In the U.S., however, it is not very common because cattle are immunized and milk is pasteurized.

Fig. 12-4. If you do see a patient with brucellosis, he will most likely be a worker in the meat-packing

Figure 12-4

Figure 12-5

industry (beef), a veterinarian, a farmer, or a traveler who consumes dairy (cow or goat) products in Mexico or elsewhere.

Like the other bacteria in this chapter, *Brucella* penetrates the skin, conjunctiva, lungs, or GI tract. However, neither buboes nor a primary skin ulcer appear. Penetration is followed by lymphatic spread, facultative intracellular growth in macrophages, and blood and organ invasion. The symptoms are systemic with fever, chills, sweats, loss of appetite, backache, headache, and sometimes lymphadenopathy. The fever usually peaks in the evening and slowly returns to normal by morning. The slow rise in temperature during the day, declining at night, has led to its other name, **undulant fever**. These symptoms can last from months to years, but fortunately the disease is rarely fatal.

Diagnosis of active disease is best made by culture of the organism from the blood, bone marrow, liver, or lymph nodes. Serologic examination that demonstrates elevated anti-*Brucella* antibodies suggests active disease. A skin test (with brucellergin) similar to that for tularemia is available, but a positive result only indicates exposure to the organism and does not prove that there is active brucellosis.

Pasteurella multocida

This organism is a gram-negative zoonotic organism. However, it is NOT a facultative intracellular organism!!!! This bacterium colonizes the mouths of cats much in the same way that *Streptococcus viridans* colonizes the human nasopharynx. It also causes disease in other mammals and birds.

Fig. 12-5. **Cat** chasing a **bird** in a **"Pasteur."**
This bacterium causes the most frequent wound infection following a cat or dog bite. When a patient comes in with a cat or dog bite (or scratch), it is important not to close the wound with sutures. A closed wound creates a pleasant environment for *Pasteurella multocida* growth, and the resulting infection can invade local joints and bones. Treat infected patients with penicillin or doxycycline.

Fig. 12-6. Summary of zoonotic gram-negative rods.

References

Chaanteau S, et al. Development and testing of a rapid diagnostic test for bubonic and pneumonic plague. Lancet 2003;361:211–216.

Gill V, Cunha B. Tularemia Pneumonia; Seminars in Respiratory Infections, Vol 12, No. 1; 1997; 61–67.

Pappas G, Akritidis N, Bosilkovski M, Tsianos E. Brucellosis. New Eng J Med 2005;352:2325–2336.

Titball R, Leary S. Plague. British Medical Bulletin 1998; 54: 625–633.

Recommended Review Articles:

Bitam I, Dittmar K, et al. Fleas and flea-borne diseases. Int J Infect Dis. 2010;14(8):e667–76.

Butler T. Plague into the 21st century. Clin Infect Dis. 2009; 49(5):736–42.

Lieberman JM. North American zoonoses. Pediatr Ann. 2009; 38(4):193–8.

ORGANISM	RESERVOIR	TRANSMISSION	METABOLISM	VIRULENCE	TOXINS
Yersinia pestis	1. Wild rodents 2. City rats 3. Squirrels and prairie dogs in the SW U.S.	1. Flea bite 2. Contact with infected animal tissue 3. Inhaled aerosolized organisms: human to human transmission occurs during epidemics	1. Facultative anaerobe 2. Virulence factors are temperature sensitive: only expressed at 37 °C. (temperature inside macrophages) 3. Virulence is plasmid mediated	1. Fraction 1 (F1): this capsular antigen is antiphagocytic 2. V and W proteins 3. Non-motile 4. Requires calcium at 37 °C. If insufficient calcium, *Y. pestis* alters its metabolism and protein production. This trait assists with its intracellular state	1. Pesticin: kills other bacteria (including *E. coli*) 2. Intracellular murine toxin: lethal to mice
Yersinia enterocolitica	• Wild & domestic animals	• Unpasteurized milk	1. Facultative anaerobe 2. Virulence factors are temperature sensitive: only expressed at 37 °C.	1. Invasive 2. V & W proteins 3. Motile at 25 °C	• Enterotoxin (like ST toxin of *E. coli*): increase cGMP levels
Francisella tularensis	1. Rabbits and squirrels 2. Ticks can serve as a reservoir	1. Bite of tick, deerfly or infected animals 2. Direct contact with infected animal tissue (usually rabbit) 3. Inhaled aerosolized organisms 4. Ingestion of contaminated meat or water 5. Easily transmitted to lab personnel	1. Obligate aerobe 2. Requires cysteine	1. Capsule: antiphagocytic 2. Non-motile	
Brucella *Brucella melitensis*: *Brucella abortus*: *Brucella suis*: *Brucella canis*: Goats Cattle Pigs Dogs		1. Direct contact with contaminated livestock or aborted placentas 2. Ingestion of infected milk products 3. Aerosolization in laboratory or possibly due to bioterrorism	• Obligate aerobe	1. Capsule 2. Non-motile 3. Tropism for erythritol, a sugar found in animal placentas	
Pasteurella multocida	• Part of the normal flora of domestic & wild animals	• Bite from dog or cat	• Facultative anaerobe	1. Capsule 2. Non-motile	

Figure 12-6 ZOONOTIC GRAM-NEGATIVE RODS

CLINICAL	TREATMENT	DIAGNOSTICS	MISCELLANEOUS

1. **Bubonic plague:**
   A. regional lymph nodes (usually groin) swell, and become red, hot and tender (called a bubo)
   B. High fever
   C. Conjunctivitis
2. **Sepsis**: bacteria survive in macrophages, and spread to blood and organs. Death occurs in 75% if untreated
3. **Pneumonic plague**: during epidemics, pneumonia occurs, as bacteria are spread from person to person by aerosolized respiratory secretions: 100% fatal if untreated
	1. Streptomycin or gentamicin
	2. Doxycycline
	3. Killed vaccine is effective only for a few months (attenuated vaccine is more effective but also has more side effects)
		1. Gram stain will reveal gram-negative rods with bipolar staining: the ends of these rod shaped bacteria take up stain more than the center
		2. Blood culture
		3. Culture of bubo aspirate
		4. Serology
		5. Rapid diagnostic test: antibody against F1 (capsular antigen)
			1. Facultative intracellular parasite
			2. *Yersinia* can accept plasmids from *E. coli*, and shares many antigens with enteric bacteria
			3. Subcutaneous hemorrhage results in a blackish skin discoloration, giving the name "Black Death"

1. **Enterocolitis**: focal ulcerations in ileum & mesenteric lymph nodes 2. Arthritis 3. Rash	1. Fluoroquinolone 2. Trimethoprim / sulfamethoxazole • Cephalosporin resistant!	• Cold enrichment of stool with saline selects for *Yersinia*	1. Facultative intracellular parasite 2. Bipolar staining
Tularemia 1. Ulceroglandular: at the site of tick bite or direct contact with contaminated rabbit, an ulcer develops, with swelling of local lymph nodes 2. Pneumonia: inhalation, or through the blood 3. Oculoglandular: direct inoculation into eyes 4. Typhoidal: ingestion results in gastrointestinal symptoms (abdominal pain) and fever	1. Gentamicin or streptomycin 2. Doxycycline 3. Attenuated vaccine: only for high risk individuals	1. Culture (but very dangerous due to its high infectivity): requires addition of cysteine to blood agar media 2. Skin test 3. Measure rise in IgG antibody titer (IgM is not very good)	• Facultative intracellular parasite

1. **Brucellosis:**
   • Undulating fever (fever peaks in the evening, and returns to normal by morning)
   • Weakness
   • Loss of appetite
2. Induces abortions in animals.
	1. Pasteurization of milk
	2. Treat with combination of doxycycline and one other drug (gentamicin, streptomycin, or rifampin).
	3. All cattle are immunized with a living attenuated strain of *Brucella abortus*
		1. Culture of blood, bone marrow (best yield), liver, or lymph nodes
		2. Serologic tests
		3. Skin test: indicates exposure only
			• Facultative intracellular parasite

• Wound infections (following dog or cat bites): may progress to infection of nearby bones and joints	1. Penicillin G 2. Doxycycline 3. Third generation cephalosporin	• Culture specimen on standard laboratory media	*Not* a facultative intracellular organism!

CHAPTER 13. *CHLAMYDIA, RICKETTSIA,* AND FRIENDS

Chlamydia and *Rickettsia* are 2 groups of gram-negative bacteria that are obligate intracellular parasites. This means they can survive only by establishing "residence" inside animal cells. They need their host's ATP as an energy source for their own cellular activity. They are **energy parasites**, using a cell membrane transport system that steals an ATP from the host cell and spits out an ADP. Both *Chlamydia* and *Rickettsia* have this **ATP/ADP translocator**. They differ in that *Rickettsia* can oxidize certain molecules and create ATP (via oxidative phosphorylation) while *Chlamydia* does not appear to have this cytochrome system and in fact has no mechanism for ATP production. The obligate intracellular existence brings up 2 questions:

Q: How do we grow and isolate these creatures when nonliving media do not contain ATP???

A: Indeed, the obligate intracellular existence makes it impossible to culture these organisms on nonliving artificial media. However, we can inoculate *Chlamydia* or *Rickettsia* into living cells (most commonly chick embryo yolk sac or cell culture).

Q: Are these bacteria really viruses, since they are very tiny and use the host's cell for their own reproduction????

A: Although *Chlamydia* and *Rickettsia* share a few characteristics with viruses (such as their small size and being obligate intracellular parasites), they have both RNA **and** DNA (while viruses have **either** DNA **or** RNA). Also, unlike viruses they both synthesize their own proteins and are sensitive to antibiotics.

Fig. 13-1. Comparison of *Chlamydia* and *Rickettsia* with bacteria and viruses.

Chlamydia and *Rickettsia* cause many distinct human diseases. *Chlamydia* spreads by person-to-person contact, while *Rickettsia* spreads by an arthropod vector.

CHLAMYDIA

Chlamydia is extremely tiny. It is classified as gram-negative because it stains red with Gram stain technique and has an inner and outer membrane. Unlike other gram-negative bacteria, it does not have a peptidoglycan layer and has no muramic acid.

Fig. 13-2. *Chlamydia* wearing his **CLAM** necklace next to a herpes virus demonstrating that *Chlamydia* is about the same size as some of the large viruses.

Chlamydia is especially fond of columnar epithelial cells that line mucous membranes. This correlates well with the types of infection that *Chlamydia* causes, including conjunctivitis, cervicitis, and pneumonia.

	BACTERIA	CHLAMYDIAE AND RICKETTSIAE	VIRUSES
Size(nm.)	300–3000	350	15–350
Obligatory intracellular parasites	No	YES	YES
Nucleic acids	RNA & DNA	RNA & DNA	RNA *OR* DNA
Reproduction	Fission	Complex cycle with fission	Synthesis and assembly
Antibiotic sensitivity	YES	YES	NO
Ribosomes	YES	YES	NO
Metabolic enzymes	YES	YES	NO
Energy production	YES	NO	NO

Figure 13-1   COMPARISON OF *CHLAMYDIA* AND *RICKETTSIA* WITH BACTERIA AND VIRUSES

Figure 13-2

The *Chlamydia* life cycle is complex as the bacteria exist in 2 forms:

1) **Elementary body (EB)**: This is a metabolically inert (does not divide), dense, round, small (300 nm.), infectious particle. The outer membrane has extensive disulfide bond cross-linkages that confer stability for extracellular existence.

Fig. 13-3. Think of the elementary body as an elementary weapon like the cannon ball, fired from host cell to host cell, spreading the infection.

2) **Initial body** (also called **reticulate body**): Once inside a host cell the elementary body inhibits phagosome-lysosome fusion, and grows in size to 1000 nm. Its RNA content increases, and binary fission occurs, forming the initial body (IB). Although the IB synthesizes its own DNA, RNA, and proteins, it requires ATP from the host. Therefore, *Chlamydia* is considered an energy parasite as well as an intracellular parasite.

Fig. 13-4. The *Chlamydia* life cycle:

A) The infectious particle is the elementary body (EB). The EB attaches to and enters (via endocytosis) columnar epithelial cells that line mucous membranes.

B) Once within an endosome, the EB inhibits phagosome-lysosome fusion and is not destroyed. It transforms into an initial body (IB).

C) Once enough IBs have formed, some transform back into EB.

D) The life cycle is completed when the host cell liberates the elementary body (EB), which can now infect more cells.

There are 3 species of *Chlamydia* that are known to cause disease in humans. The taxonomy is undergoing constant revision. At last glance the names were *Chlamydia trachomatis* which primarily infects the eyes, genitals, and lungs; *Chlamydophila psittaci* and *Chlamydophila pneumonia*, both of which primarily affect the lungs. All are treated with doxycycline, a macrolide, or a fluoroquinolone.

Fig. 13-5. Chlamydial diseases.

Chlamydia trachomatis

Fig. 13-6. *Chlamydia trachomatis* primarily infects the **eyes** and **genitals**. Picture a flower child with groovy **clam** eyeglasses and a **clam** bikini.

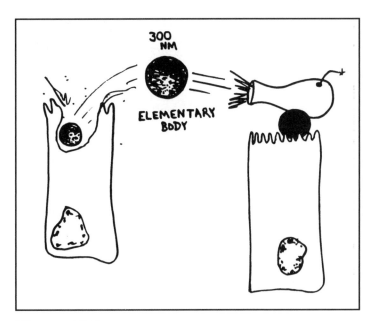

Figure 13-3

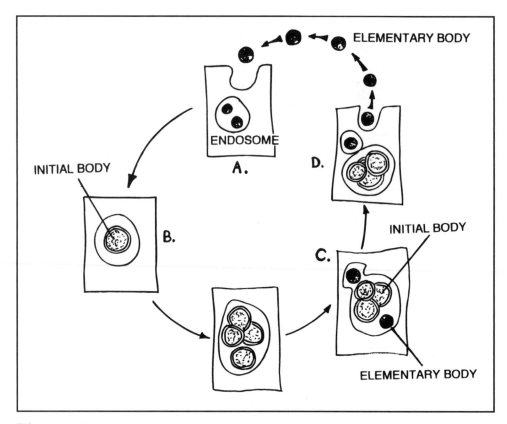

Figure 13-4

Trachoma

Chlamydia trachomatis is responsible for trachoma, a type of chronic conjunctivitis that is currently the leading cause of **preventable blindness** in the world. It is a disease of poverty, prevalent in underdeveloped parts of the world. In the U.S., Native Americans are the group most frequently infected. Children act as the main reservoir, and transmission occurs by hand-to-hand transfer of infected eye secretions and by sharing contaminated clothing or towels. Blindness develops slowly over 10–15 years.

SPECIES	DISEASE
Chlamydia trachomatis serotypes A, B, & C	• Trachoma (a leading cause of blindness in the world)
serotypes D thru K	1. Inclusion conjunctivitis (usually in newborns, contracted in the birth canal) 2. Infant pneumonia 3. Cervicitis 4. Nongonococcal urethritis in men
serotypes L_1, L_2, L_3	• Lymphogranuloma venereum (LGV)
Chlamydophila psittaci	• Atypical pneumonia
Chlamydophila pneumoniae Serogroup TWAR	• Atypical pneumonia

Figure 13-5   CHLAMYDIAL DISEASES

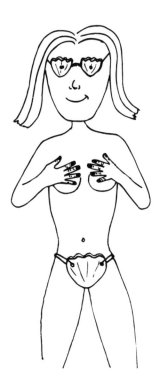

Figure 13-6

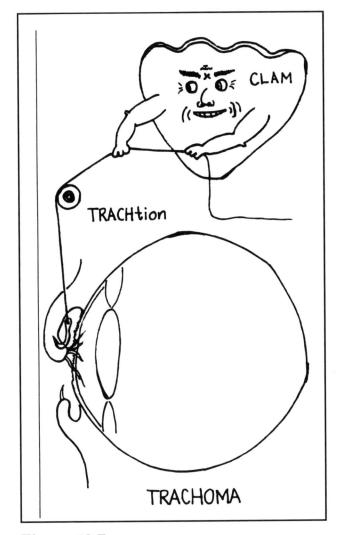

Figure 13-7

Fig. 13-7. The conjunctival infection causes inflammation and scarring. Scar traction (**trach**tion for **trach**-oma) pulls and folds the eyelid inward so that the eyelashes rub against the conjunctiva and cornea, which causes corneal scarring, secondary bacterial infections, and ultimately blindness. Topical treatment is ineffective for this illness. Oral azithromycin is standard first-line therapy.

Inclusion Conjunctivitis

As *Chlamydia trachomatis* is the most common sexually transmitted disease in the U.S., it is not surprising that many babies delivered through birth canals infected with this organism develop **inclusion conjunctivitis**. Conjunctival inflammation with a purulent yellow discharge and swelling of the eyelids usually arises 5–14 days after birth. In the U.S., all newborns are given **erythromycin** eye drops prophylactically.

Diagnosis is made by demonstrating basophilic **intracytoplasmic inclusion bodies** in cells taken from scrapings of the palpebral conjunctival surface. These inclusion bodies are collections of initial bodies in the cytoplasm of the conjunctival cells.

Inclusion conjunctivitis can also occur in adults, usually with an associated genital infection. Inclusion conjunctivitis requires oral therapy. Infants most commonly receive erythromycin solution and adults can take either doxycycline or a macrolide (erythromycin or azithromycin).

Infant Pneumonia

A baby's passage through an infected birth canal may also lead to a chlamydial pneumonia, which usually occurs between 4–11 weeks of life. Initially, the infant develops upper respiratory symptoms followed by rapid breathing, cough, and respiratory distress.

Diagnosis is made clinically, and the diagnosis can be later confirmed by the presence of anti-chlamydial IgM antibodies and/or demonstration of *Chlamydia trachomatis* in clinical specimens. Treat with oral **erythromycin**.

Urethritis

Urethritis, an infection of the urethra, is usually contracted sexually. *Neisseria gonorrhoeae* is the most famous bacterium causing urethritis, but not the most common. Urethritis that is **not** caused by *Neisseria gonorrhoeae* is called nongonococcal urethritis (NGU), and is thought to be the most **common** sexually transmitted disease. NGU is predominantly caused by *Chlamydia trachomatis* and *Ureaplasma urealyticum*.

Many patients with NGU are asymptomatic. Symptomatic patients develop painful urination (dysuria) along with a thin to thick, mucoid discharge from the urethra. It is impossible clinically to differentiate gonococcal urethritis from NGU and they often occur together as a mixed infection. These mixed infections are discovered when patients are treated only with a penicillin family antibiotic and don't get better. Penicillins treat the gonorrhea, but are ineffective against *Chlamydia trachomatis*. Remember that *Chlamydia trachomatis* has no peptidoglycan layer, which is the target for penicillin.

Therefore, all patients diagnosed with urethritis are empirically treated with antibiotics to cover *Neisseria gonorrhoeae*, *Chlamydia trachomatis*, and *Ureaplasma urealyticum*. A commonly used treatment regimen involves a single dose of intramuscular ceftriaxone (a third-generation cephalosporin that is extremely effective against *Neisseria gonorrhoeae*) followed by a 7-day course of oral doxycycline or 1 oral dose of azithromycin (which covers both *Chlamydia trachomatis* and *Ureaplasma urealyticum*). (See Chapter 18, page 181.)

While the patient is on empiric antibiotics, diagnostic tests are performed to determine which organism is responsible. The diagnosis of chlamydial NGU is a bit roundabout because the bacteria are too small to visualize with the Gram stain and cannot be cultured on nonliving media. If the Gram stain reveals polymorphonuclear leukocytes but NO intracellular or extracellular gram-negative diplococci (that is, NO *Neisseria gonorrhoeae*), a diagnosis of NGU is likely. The preferred diagnostic test for gonorrhea and chlamydia in most centers is a nucleic acid amplification test, such as polymerase chain reaction, on an endocervical swab or urine sample.

Cervicitis and Pelvic Inflammatory Disease (PID)

The cervix is a frequent site for *Chlamydia trachomatis* infection. The inflamed cervix appears red, swollen, and has a yellow mucopurulent endocervical discharge. This infection can spread upwards to involve the uterus, fallopian tubes, and ovaries. This infection, which can be caused by both *Chlamydia trachomatis* and *Neisseria gonorrhoeae*, is called **pelvic inflammatory disease (PID)**.

Women with PID often develop abnormal vaginal discharge or uterine bleeding, pain with sexual intercourse (dyspareunia), nausea, vomiting, and fever. The most common symptom is lower abdominal pain. The inflamed cervix, uterus, tubes, and ovaries are very painful. Some medical slang emphasizes this. Women are observed to have the "PID shuffle" (small, widebased steps to minimize shaking of abdomen). With movement of the cervix on bimanual vaginal examination the patient may exhibit the "Chandelier sign" (cervical motion tenderness is so severe that the patient leaps to the chandelier).

PID often results in fallopian tube scarring, which can cause infertility, tubal (ectopic) pregnancy, and chronic pelvic pain. It is estimated that 1 million women suffer from PID every year in the U.S. and 25% of them will become infertile. In one prospective study (Westrom, 1992), tubal occlusion leading to infertility occurred in 8% of women after 1 episode of PID, 19.5% after 2 episodes, and 40% after 3 episodes. Likewise, the risk of ectopic pregnancy and chronic pelvic pain increases with recurrent PID.

Chlamydia trachomatis is particularly dangerous as it often causes asymptomatic or mild PID that goes undiagnosed and untreated, yet can still lead to infertility.

Fig. 13-8. Infected fallopian tubes scar easily, which can result in infertility. The silent sinister **CLAM** (***Chlam***ydia trachomatis) causes asymptomatic PID that can lead to infertility.

A simple shot of **ceftriaxone** and 14 days of oral **doxycycline** will vanquish PID.

(McCormack, 1994)

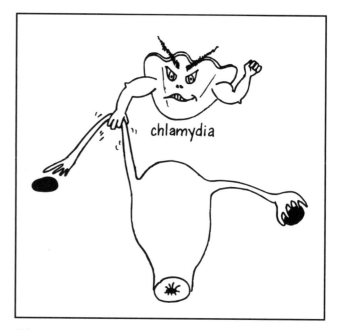

Figure 13-8

Epididymitis

Chlamydial **epididymitis** can develop in men with urethritis and presents clinically as unilateral scrotal swelling, tenderness, and pain, associated with fever.

Other Complications of Chlamydial Infection

Chlamydia trachomatis is also linked to **Reiter's syndrome**, an inflammatory arthritis of large joints, that commonly occurs in young men between the ages of 20 and 40. Inflammation of the eyes (uveitis and conjunctivitis) and urethritis also occur. However, other infectious agents may also precipitate this syndrome.

Fitz-Hugh-Curtis syndrome is an infection of the liver capsule with symptoms of right upper quadrant pain that can occur in men and women. This syndrome is associated with either chlamydial or gonococcal infection.

Lymphogranuloma Venereum

Lymphogranuloma venereum, another sexually transmitted disease caused by *Chlamydia trachomatis*, (serotypes L_1, L_2 and L_3) starts with a painless papule (bump) or ulceration on the genitals that heals spontaneously. The bacteria migrate to regional lymph nodes, which enlarge over the next 2 months. These nodes become increasingly tender and may break open and drain pus (see Chapter 11, page 97).

Chlamydophila psittaci
(Psittacosis)

Chlamydophila psittaci infects more than 130 species of birds, even pet parrots. Humans are infected by inhaling *Chlamydia*-laden dust from feathers or dried-out feces. This infection is an occupational hazard for breeders of carrier pigeons, veterinarians, and workers in pet-shops or poultry slaughterhouses. Infection most commonly results in an atypical pneumonia called **psittacosis**, which occurs 1–3 weeks after exposure.

Atypical Pneumonia

Pneumonia caused by viruses, *Mycoplasma pneumoniae*, and *Chylamydophila* species have frequently been called **atypical pneumonias** because it was felt that they presented clinically and radiographically differently from a typical bacterial pneumonia such as that caused by *Streptococcus pneumoniae*. The classic teaching is that patients infected with atypical organisms present more often with a dry cough, fever, and are less sick appearing than those infected with "typical organisms." Physicians also claim to be able to distinguish

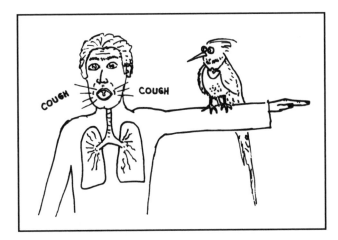

Figure 13-9

between the two classes of pneumonias based on radiographic appearance, with atypical pathogens causing less well-defined infiltrates than typical organisms. For the most part this dogma has been debunked, but you will still hear attendings banter on about atypical versus typical presentations of pneumonia.

Fig. 13-9. A man with atypical pneumonia caused by *Chlamydophila psittaci*. He has a bird with a CLAM necklace, PSITTING on his arm.

Chlamydophila pneumoniae
(strain TWAR)

Chlamydophila pneumonia has a single species, TWAR, which is transmitted from person to person by the respiratory route and causes an atypical pneumonia in young adults worldwide (along with *Mycoplasma pneumoniae*). TWAR is an acronym for its original isolation in **Tai**wan and **A**cute **R**espiratory.

RICKETTSIA

Rickettsia is a small, gram-negative, non-motile, rod-to coccoid-shaped bacterium. It is similar to *Chlamydia* in that they both are the size of large viruses. Both are obligate intracellular energy parasites (they steal ATP). However, *Rickettsia* differs from *Chlamydia* in a number of ways:

1) *Rickettsia* requires an arthropod vector (except for Q fever).

Fig. 13-10. Ricky the riding *Rickettsia* loves to travel. He rides a tick in Rocky Mountain spotted fever, a louse in epidemic typhus, and a flea in endemic typhus.

2) *Rickettsia* replicates freely in the cytoplasm, in contrast to *Chlamydia*, which replicates in endosomes (inclusions).

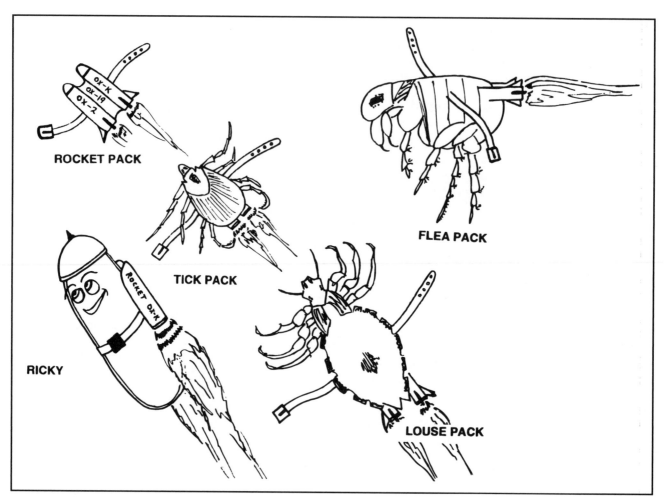

Figure 13-10

3) *Rickettsia* has a tropism for **endo**thelial cells that line blood vessels (*Chlamydia* likes columnar epithelium).

4) They cause different diseases!!! Most *Rickettsia* cause rashes, high fevers, and bad headaches.

Some *Rickettsia* share antigenic characteristics with certain strains of *Proteus vulgaris* bacteria. It is purely coincidental that they have the same antigens. *Proteus* is not involved at all in rickettsial disease. The *Proteus vulgaris* strains that share these common antigens are designated **OX-2, OX-19**, and **OX-K**.

The **Weil-Felix** reaction is a classic test that uses these cross-reacting *Proteus vulgaris* antigens to help confirm a diagnosis of a rickettsial infection. This test is done by mixing the serum of a patient suspected of having a rickettsial disease, with antigens from specific strains of *Proteus vulgaris*. If the serum has antirickettsial antibodies, latex beads coated with *Proteus* antigens will agglutinate, indicating a positive Weil-

Felix test. Comparison of the laboratory results with **Fig. 13-11** can even help distinguish specific rickettsial diseases. For example, when this test is performed on a patient with signs and symptoms of a scrub typhus infection, a negative OX-19 and OX-2 along with a positive OX-K is confirmatory. Unfortunately the Weil-Felix test is neither sensitive nor specific, leading most experts to recommend against its use.

Fig. 13-11. Antigenic differences among the *Rickettsiae.*

Diagnosis of a rickettsial infection can also be made with specific serologic tests documenting a rise in anti-Rickettsial antibody titers over time. These include the indirect immunofluorescence test (IFA), the complement fixation test (CF), and the enzyme-linked immunosorbent assay (ELISA). These tests can specifically identify species and even subspecies.

Therapy for all rickettsial diseases consists primarily of **doxycycline** and **chloramphenicol**.

DISEASE	WEIL-FELIX		
	OX-19	OX-2	OX-K
Rocky Mountain spotted fever	+	+	−
Rickettsial pox	−	−	−
Epidemic typhus	+	−	−
Endemic typhus	+	−	−
Brill-Zinsser disease	+/−	−	−
Scrub typhus	−	−	+
Trench fever	−	−	−
Q fever	−	−	−

Figure 13-11   WEIL-FELIX

Rickettsia rickettsii
(Rocky Mountain Spotted Fever)

Ricky is riding a wood tick. . . .

Fig. 13-12. Rocky Mountain spotted fever presents within a week after a person is bitten by either the wood tick *Dermacentor andersoni* or the dog tick *Dermacentor variabilis*. Both of these ticks transmit the causative organism, *Rickettsia rickettsii*. This disease is characterized by fever, conjunctival redness, severe headache, and a rash that initially appears on the wrists, ankles, soles and palms and later spreads to the trunk. This figure illustrates the spotted Rocky Mountains behind a boy with headache, fever, palmar rash, and tick infestation.

Rocky Mountain spotted fever is more common in the southeastern U.S. tick belt than in the Rocky Mountain region. This disease should be called Appalachian spotted fever, as most cases currently occur in the south Atlantic and south central states such as North Carolina, South Carolina, Tennessee, and Oklahoma. However, cases have been reported in nearly every state.

The organisms proliferate in the endothelial lining of small blood vessels and capillaries, causing small hemorrhages and thrombi. The inflammation and damage to small blood vessels explains the conjunctival redness and skin rash. Although this disease often resolves in about 3 weeks, it can progress to death (especially when antibiotic therapy is delayed).

Since the tick transmits this bacteria during its 6–10 hours of feeding, early discovery and removal of ticks will prevent infection (Spach, 1993).

Rickettsia akari
(Rickettsialpox)

Ricky is riding a mite. . . .

Fig. 13-13. *Rickettsia akari* causes rickettsia**pox** and is transmitted to humans via **mites** that live on house mice. Imagine Ricky, with **pox** marks, playing **Atari** (old type of Nintendo) with his rodent friend **mite**y mouse.

Rickettsialpox is a mild, self-limited, febrile disease that starts with an initial localized red skin bump (papule) at the site of the mite bite. The bump turns into a blister (vesicle) and days later fever and headache develop, and other vesicles appear over the body (similar to chickenpox). Although this disease is self-limiting, there is a dramatic response to **doxycycline**. Elimination of nearby rodents, which can serve as a reservoir for *Rickettsia akari*, is important in preventing this disease.

Rickettsia prowazekii
(Epidemic Typhus)

Ricky is riding a louse. . . .

An **epidemic** is the sudden onset and rapid spread of an infection that affects a large proportion of a population. **Endemic** refers to an infectious disease that exists constantly throughout a population. Two species of *Rickettsia* cause typhus. *Rickettsia prowazekii* causes an epidemic form, while *Rickettsia typhi* is responsible for endemic typhus. Although they have different reservoirs and vectors, these are closely related bacteria that cause a similar disease, and infection with one confers immunity to the other!!

Figure 13-12

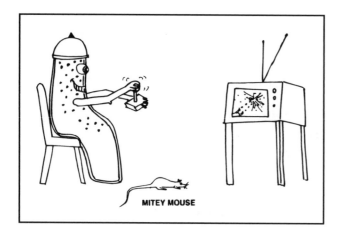

Figure 13-13

Fig. 13-14.   Prowazekii is **Prowar**!!! With war, over-crowding, and poverty, unsanitary conditions prevail

and lice take control, harboring *Rickettsia prowazekii*. The lice transmit the bacteria to humans, causing epidemic typhus.

This disease wiped out a third of Napoleon's army when he advanced on Moscow in 1812, and was responsible for more than 3 million Russian deaths in World War I. The last epidemic in the U.S. occurred more than 70 years ago. Currently, flying squirrels serve as a reservoir in the southern U.S. Sporadic cases occur when lice or fleas from infected squirrels bite humans.

Clinically, **epidemic typhus** is characterized by an abrupt onset of fever and headache following a 2-week incubation period. Small pink macules appear around the fifth day on the upper trunk and quickly cover the entire body. In contrast to Rocky Mountain spotted fever, this rash spares the palms, soles, and face. The patient may become delirious or stuporous. Since *Rickettsia* invade the endothelial cells of blood vessels, there is an increased risk of blood vessel clotting leading

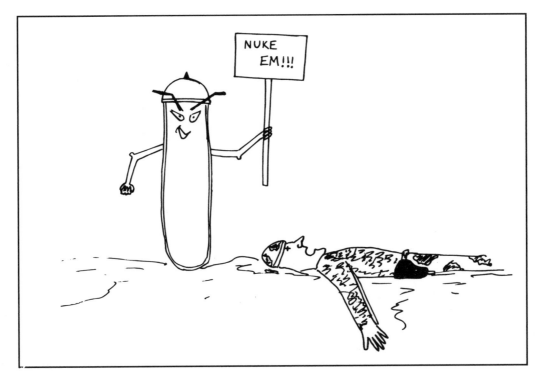

Figure 13-14

to gangrene of the feet or hands. This disease will often resolve by 3 weeks, but occasionally is fatal (especially in older patients).

Diagnosis would be easy during an epidemic. The poor doctor with Napoleon's retreating forces surely became an expert diagnostician of louse-borne typhus! It is the sporadic case in the southern U.S., transmitted from flying squirrels to humans by louse or flea bites, that is unexpected and thus difficult to diagnose. Close contact with the flying squirrel vector should raise suspicion.

Besides tetracycline and chloramphenicol, improved sanitation and eradication of human lice will help control epidemics.

Brill-Zinsser Disease

For those of you who have referred to the Zinsser Microbiology textbook, it is interesting to note that Hans Zinsser is credited with correctly postulating that patients who recovered without antibiotic therapy from epidemic louse-borne typhus could still retain the pathogen *Rickettsia prowazekii* in a latent state. Occasionally, it breaks out of its latent state to produce Brill-Zinsser disease. However, symptoms are usually milder (no skin rash) due to the presence of pre-formed antibodies from the original infection. Diagnosis is made by demonstrating a rapid early rise in Ig**G** titer specific for

Rickettsia prowazekii, rather than a rapid rise in Ig**M**, which occurs in the primary infection.

It is always important to completely eradicate *Rickettsia prowazekii* from your patient with sufficient antibiotic therapy because untreated patients may serve as a reservoir between epidemics.

Rickettsia typhi
(Endemic or Murine Typhus)

Ricky is riding a flea. . . .

Endemic flea-borne typhus is similar to epidemic typhus, yet it is not as severe and does not occur in epidemics. This disease is caused by *Rickettsia typhi*. Rodents serve as the primary reservoir, and the disease is transmitted to humans via the rat flea, *Xenopsylla cheopis*. (This flea was also responsible for transmission of bubonic plague in the past.)

Following a 10-day incubation period, fever, headache, and a flat and sometimes bumpy (maculopapular) rash develop, just as with epidemic typhus. Although this disease is milder than that caused by epidemic typhus, it is still very serious.

Treat with doxycycline or chloramphenicol. Control flea and rat populations. As with the bubonic plague (*Yersinia pestis*), we don't want to just kill the rats because the starving fleas would then all move to bite humans!!!

Figure 13-15

Rickettsia tsutsugamushi
(Scrub Typhus, or Tsutsugamushi Fever)

Rickettsia tsutsugamushi is found in Asia and the southwest Pacific. This disease affected soldiers in the South Pacific during World War II and in Vietnam. *Rickettsia tsutsugamushi* is spread by the bite of larvae (chiggers) of mites. The mites live on rodents, and the larval chiggers live in the soil.

Fig. 13-15. Ricky is now a South Pacific sumo wrestler named Ricky Tsutsugamushi. He is walking in the scrub (scrub typhus) being bitten by chiggers that are on his feet and legs.

After a 2-week incubation period, there is high fever, headache, and a scab at the original bite site. Later a flat and sometimes bumpy (maculopapular) rash develops.

Rickettsia parkeri

Infection with *Rickettsia parkeri* was first established in 2002 in the southeastern coastal United States. The initial patient presented with fever, headaches, eschars, and regional lymphadenopathy.

Rickettsia africae

This organism is responsible for African tick-bite fever (ATBF), an increasingly reported travel-related rickettsiosis. ATBF has been increasingly identified in travelers returning from sub-Sahara Africa as a cause of unexplained fever.

Bartonella quintana
(Trench Fever)

Trench fever is a louse-borne febrile disease that occurred during World War I. The organism responsible for this disease is *Bartonella quintana*. Although it is

Rickettsia-like, it has a different genus name because it is **not an obligate intracellular organism**.

This disease was spread in the trenches by the body louse. Infected soldiers developed high fevers, rash, headache, and severe back and leg pains. After appearing to recover, the soldier would relapse 5 days later. Multiple relapses can occur but fatalities are rare. The organism's species name, quintana, reflects the characteristic 5-day interval between febrile episodes.

Notice the similarities here with epidemic typhus (*Rickettsia prowazekii*-Prowar Ricky). Both achieve epidemic proportions during war, when filth and poor sanitation lead to lice overgrowth.

FILTH = LICE = *Rickettsia prowazekii*
(Epidemic typhus) + *Bartonella quintana*
(trench fever)

Bartonella henselae
(Cat-scratch Disease)

Cat-scratch disease occurs following a cat bite or scratch. A regional lymph node or nodes will enlarge and the patient may develop low-grade fever and malaise. The disease usually resolves within a few months without complications.

A motile, gram-negative rod named *Afipia felis* was originally isolated from affected lymph nodes. However, there is now growing evidence that another bacterium may be the etiologic agent: *Bartonella henselae*. Several studies have now documented high levels of anti-*Bartonella henselae* antibodies in patients with cat-scratch disease.

Both *Bartonella quintana* and *Bartonella henselae* can cause bacteremia, endocarditis, and a syndrome called bacillary angiomatosis, which involves proliferation of small blood vessels in the skin and organs of AIDS patients.

Coxiella burnetii
(Q Fever)

Coxiella burnetii is unique to the *Rickettsia* because, like the gram-positive spore formers (*Clostridium* and *Bacillus*), it has an **endospore** form. This endospore confers properties to the bacteria that differ from other Rickettsiae:

1) **Resistance to heat and drying**: Spores may contaminate milk products so pasteurization temperatures have to be raised to greater than 60°C to kill the endospores.

2) **Extracellular existence**: The spore's resistance allows extended survival outside a host cell. However, like *Chlamydia* and *Rickettsia*, growth and division must occur intracellularly using the host's ATP.

Figure 13-16

3) **Non-arthropod transmission**: *Coxiella burnetii* grows in ticks and cattle. The spores remain viable in dried tick feces deposited on cattle hides, and in dried cow placentas following birthing. These spores are aerosolized and when inhaled cause human disease. Spore inhalation rather than an arthropod bite causes Q fever.

4) **Pneumonia**: Because the spores are inhaled into the lungs, a mild pneumonia similar to that of a *Mycoplasma* pneumonia often develops.

Clinically, abrupt onset of fever and soaking sweats occur 2–3 weeks after infection, along with a pneumonia. This is the only rickettsial disease that causes pneumonia and in which there is NO rash.

Fig. 13-16. Visualize **Carol Burnett** (Coxiella **bur-net**ii) coughing after inhaling the spores from a cowhide and dried placental products in the grass.

Most infections with *Coxiella burnetii* are asymptomatic. *Coxiella burnetii* can also cause granulomatous hepatitis and "culture negative" endocarditis (the diagnosis is most often made by serology).

Human ehrlichiosis

Ehrlichiosis is a tick-borne disease similar to Rocky Mountain Spotted Fever without a rash, caused by *Ehrlichia chaffeensis* (Human Monocytic Ehrlichiosis), *Anaplasma phagocytophilum* (Human Granulocytic Anaplasmosis), or less frequently by a more recently identified bacterium *Ehrlichia ewingii*.

Fig. 13-17. Summary chart of *Chlamydia* and *Rickettsia*.

References

Chapman AS, Bakken JS, Folk SM, et al. Diagnosis and management of tickborne Rickettsial diseases: Rocky Mountain Spotted Fever, Ehrlichiosis, and Anaplasmosis—United States. MMWR 2006;55:1–27.

Freedman DO, Weld LH, Kozarsky PE, et al. Spectrum of disease and relation to place of exposure among ill returned travelers. N Engl J Med 2006;354:119–130.

Parola P, Raoult D. Ticks and tickborne bacterial diseases in humans: An emerging infectious threat. Clin Infect Dis 2001;32:897–928.

Recommended Review Articles:

Dumler JS, Madigan JE, et al. Ehrlichioses in humans: epidemiology, clinical presentation, diagnosis, and treatment. Clin Infect Dis. 2007;45 Suppl 1:S45–51.

Marrie TJ. Q fever pneumonia. Infect Dis Clin North Am. 2010;24(1):27–41.

ORGANISM	RESERVOIR	TRANSMISSION	METABOLISM

Chlamydia trachomatis	• Humans
	• Morphologic note: gram-negative, but lacks peptidoglycan layer and muramic acid
		1. Direct personal contact
		2. Primarily affects the:
		  A. Eyes
		  B. Genitals
		  C. Lungs
		3. Note that trachoma is found in underdeveloped countries, and transmission occurs due to poor hygiene
			LIFE CYCLE
			1. Elementary body (EB): dense spherule that infects cells
			2. Initial (reticulate) body: After EB enters cell, it transforms into an initial body
			  A. Larger & osmotically fragile
			  B. Can reproduce via binary fission
			  C. Requires ATP from the host
			  D. The initial body transforms back into EB, which leaves the cell to infect other cells
			
			  • Note: *Chlamydia* are obligate Intracellular parasites – steal ATP from host with ATP/ADP translocator

Chlamydophila psittaci	• Birds & poultry	1. Bird feces dry out 2. Fecal particles are inhaled, infecting the lungs	Life cycle is similar to *Chlamydia trachomatis*

Figure 13-17   GRAM-NEGATIVE OBLIGATE INTRACELLULAR PARASITES: CHLAMYDIA AND RICKETTSIA

VIRULENCE	CLINICAL	TREATMENT	DIAGNOSTICS
1. Resistant to lysozyme (since their cell wall lacks muramic acid) 2. Prevents phagosome – lysosome fusion 3. Non-motile 4. No pili 5. No exotoxins	*Serotypes A, B, & C:* • Trachoma: causes scarring of the inside of the eyelid, resulting in redirection of the eyelashes onto the corneal surface. This results in corneal scarring and blindness *Serotypes D through K:* 1. Inclusion conjunctivitis (ophthalmia neonatorum) 2. Infant pneumonia 3. Urethritis, cervicitis and pelvic inflammatory disease (PID) in women 4. Nongonococcal urethritis, epididymitis and prostatitis in men *Complications of chlamydial genital tract infections* 1. Sterility, ectopic pregnancy and chronic pain may occur after pelvic inflammatory disease 2. Reiter's syndrome: triad of conjunctivitis, urethritis, and arthritis 3. Fitz-Hugh-Curtis Syndrome: perihepatitis *Serotypes L_1, L_2, & L_3:* • Lymphogranuloma venereum	• Genital and eye infections: 1. Doxycycline (use only for adults) 2. Erythromycin (especially for infants and pregnant woman) 3. Azithromycin • Note: systemic treatment is required for any chlamydial eye infection!! This is especially true for infants, who can develop chlamydial pneumonia following chlamydial conjunctivitis	1. Can *NOT* be grown on artificial media. Can classically be grown in chick yolk sacs. More commonly, *Chlamydia* is cultured in certain cell lines (McCoy cells, for example) 2. For *inclusion conjunctivitis* (ophthalmia neonatorum): Scrapings from the surface of the conjunctiva will show intracytoplasmic inclusion bodies within conjunctival epithelial cells. The inclusion bodies contain glycogen, and thus stain with iodine or giemsa 3. Gram stain of genital secretions will NOT show gram-negative intracellular diplococci 4. Urethritis: most commonly diagnosed by polymerase chain reaction of urethral swab or urine sample. 5. Immunofluorescent slide test: place infected genital or ocular secretions on a slide and stain with fluorescein-conjugated anti-chlamydial antibody 6. Serologic: Examine blood for elevated titers of anti-chlamydial antibodies with complement fixation and immunofluorescence tests 7. *Lymphogranuloma venereum:* A. Serologic tests B. Frei test, which is rarely used, is similar to the PPD skin test for tuberculosis
same	• Psittacosis: a viral-like atypical pneumonia, with fever and dry, non-productive cough (similar to a *Mycoplasma* pneumonia)	1. Doxycycline 2. Erythromycin	1. Serologic: Examine blood for elevated titers of antibodies with complement fixation and immunofluorescence tests 2. Intracytoplasmic inclusion bodies do not stain with iodine

ORGANISM	RESERVOIR	TRANSMISSION	METABOLISM
Chlamydophila pneumoniae (strain TWAR)	• Humans (spread from human to human)	• Respiratory route	Life cycle is similar to *Chlamydia trachomatis*
Rickettsia generalities		• Arthropod vector (except Q fever)	1. Rickettsiae are obligate intracellular parasites: They can not make their own ATP 2. Grow in cytoplasm (in contrast to *Chlamydia*, which replicates in endosomes)
Rickettsia rickettsii	• Dogs, rabbits & wild rodents	1. Wood tick: In Western U.S. *Dermacentor andersoni* 2. Dog tick: In Eastern U.S. *Dermacentor variabilis*	
Rickettsia akari	• House mice	• Mites (which live on the house mice)	
Rickettsia prowazekii	1. Humans 2. Flying squirrels	• Human body louse (*Pediculus corporis*)	
Rickettsia typhi	1. Rats 2. Small rodents	• Rat flea (*Xenopsylla cheopis*)	
Rickettsia tsutsugamushi	1. Rats 2. Shrew 3. Mongooses 4. Birds	• Mite larvae (chiggers)	

Figure 13-17   (continued)

VIRULENCE	CLINICAL	TREATMENT	DIAGNOSTICS
	• Atypical pneumonia: viral-like atypical pneumonia (similar to a *Mycoplasma* pneumonia) in young adults	1. Doxycycline 2. Erythromycin	1. Serologic: Examine blood for elevated titers of antibodies with complement fixation and immunofluorescence tests 2. Intracytoplasmic inclusion bodies do not stain with iodine
1. Non-motile 2. No exotoxins	• Damages *ENDO*thelial cells lining blood vessels	1. Doxycycline 2. Chloramphenicol	1. Culture: chick yolk sac • Can *NOT* be grown on artificial media (except for *Bartonella* species) 2. Serology: identify antibodies against the rickettsial organism 3. Weil-Felix reaction
	• *Rocky Mountain Spotted Fever*: 1. Fever 2. Conjunctival injection (redness) 3. Severe headache 4. Rash on wrists, ankles, soles & palms initially, becomes more generalized later	1. Doxycycline 2. Chloramphenicol	1. Clinical exam 2. Direct immunofluorescent exam of skin biopsy from rash site 3. Serology 4. Weil-Felix reaction: A. Positive OX-19 B. Positive OX-2
	• *Rickettsial Pox*: vesicular rash similar to chicken pox. It resolves over 2 weeks	1. Doxycycline 2. Chloramphenicol	• Weil-Felix reaction negative
	1. *Epidemic Louse-borne Typhus* A. Abrupt onset of fever and headache B. Rash, which spares the palms, soles and face C. Delirium/stupor D. Gangrene of hands or feet 2. *Brill-Zinsser Disease*: A. Reactivation of *Rickettsia prowazekii* B. Mild symptoms C. *NO* rash	1. Doxycycline 2. Chloramphenicol 3. Eradicate human lice	1. Weil-Felix reaction: positive OX-19 2. Serology
	• *Endemic* (or Murine) *Typhus*: fever, headache and rash	1. Doxycycline 2. Chloramphenicol	• Weil-Felix reaction: positive OX-19
	• *Scrub Typhus*: 1. Fever and headache 2. Eschar (scab) at bite site 3. Followed by a rash	1. Doxycycline 2. Chloramphenicol	• Weil-Felix reaction: positive OX-K

ORGANISM	RESERVOIR	TRANSMISSION	METABOLISM
Bartonella quintana	• Humans	• Body louse	• Not an obligate intracellular parasite
Bartonella henselae			• Not an obligate intracellular parasite
Coxiella burnetii	• Cattle, sheep, & goats	*EXCEPTION:* • *No* arthropod vector required. Direct airborne transmission of endospore from cow hide or dried placenta, or via consumption of endospore-contaminated unpasteurized cow milk	*EXCEPTION:* 1. Can grow at a pH of 4.5 within phagolysosomes 2. Has an endospore form
Ehrlichia chaffeensis (HME) *Anaplasma phogocytophilum* (HGA) *Ehrlichia ewingii*	• Deer, dogs, coyotes • Deer, white-footed mouse	• Ticks	

Figure 13-17   (continued)

VIRULENCE	CLINICAL	TREATMENT	DIAGNOSTICS
	1. *Trench Fever*: fever, headache and back pain. It lasts for *5* days and recurs at *5* day intervals 2. Bacteremia, endocarditis, and bacillary angiomatosis	1. Doxycycline 2. Chloramphenicol 3. Azithromycin	1. Serology 2. PCR
	1. Cat-scratch disease 2. Bacillary angiomatosis 3. Bacteremia 4. Endocarditis ("culture negative")	1. Azithromycin 2. Doxycycline	1. Serology 2. PCR
	1. *Q Fever*: fever, headache & viral-like pneumonia. No Rash!!! (This is the only rickettsial disease without a skin rash!!) 2. *Complications*: 1. Hepatitis 2. Endocarditis	1. Doxycycline 2. Erythromycin • Pasteurize milk to 60 °C	1. Complement fixation test demonstrating a rise in antibody 2. PCR
	• Human Ehrlichiosis: similar to Rocky Mountain spotted fever, but rash is rare	1. Doxycycline 2. Rifampin • Resistant to chloramphenicol	1. Rise in acute and convalescent antibody titers 2. Characteristic ehrlichial inclusion bodies are sometimes seen in leukocytes on blood smears 3. PCR

M. Gladwin, W. Trattler, and S. Mahan, *Clinical Microbiology Made Ridiculously Simple* ©MedMaster

CHAPTER 14.   SPIROCHETES

Spirochetes are tiny gram-negative organisms that look like corkscrews. They move in a unique spinning fashion via thin endoflagella called **axial filaments**. Spirochetes are all very slender and tightly coiled. From the inside out, they have a cytoplasm surrounded by an inner cytoplasmic membrane. Like all gram-negative bacteria they then have a thin peptidoglycan layer (cell wall) surrounded by the LPS containing outer lipoprotein membrane. However, 2 things are unique with the spirochetes: 1) Spirochetes are surrounded by an additional phospholipid-rich outer membrane with few exposed proteins; this is thought to protect the spirochetes from immune recognition ("stealth" organisms). 2) Axial flagella come out of the ends of the spirochete cell wall, but rather than protrude out of the outer membrane (like other bacteria shown in **Figure 2-1**), the flagella run sideways along the spirochete under the outer membrane sheath. These specialized flagella are called **periplasmic flagella**. Rotation of these periplasmic flagella spins the spirochete around and generates thrust, propelling them forward. These organisms replicate by transverse fission.

Spirochetes are a diagnostic problem. They cannot be cultured in ordinary media, and although they have gram-negative cell membranes, they are too small to be seen using the light microscope. Special procedures are required to view these organisms, including darkfield microscopy, immunofluorescence, and silver stains. Also, serologic tests help screen for infections with spirochetes.

Spirochetes are divided into 3 genera: 1) *Treponema*, 2) *Borrelia*, and 3) *Leptospira*.

TREPONEMA

Treponemes produce no known toxins or tissue destructive enzymes. Instead, many of the disease manifestations are caused by the host's own immune responses, such as inflammatory cell infiltrates, proliferative vascular changes, and granuloma formation.

Treponema pallidum
(Syphilis)

Treponema pallidum is the infectious agent responsible for the **sexually transmitted** disease syphilis. More than 12 million cases of syphilis occur worldwide. While in the United States the cases per 100,000 persons have decreased dramatically after World War II (450/100,000)

STAGE	CLINICAL
Primary stage	• Painless chancre (ulcer)
Secondary stage	1. Rash on palms and soles 2. Condyloma latum 3. CNS, eyes, bones, kidneys and/or joints can be involved
Latent syphilis	• 25% may relapse and develop secondary stage symptoms again
Tertiary stage	1. Gummas of skin and bone 2. Cardiovascular (aortic aneurysm) 3. Neurosyphilis

Figure 14-1   SYPHILIS: CLINICAL MANIFESTATIONS

to a low today of less than 25/100,000, there was a peak in the late 1980's (crack and HIV epidemics) and a high incidence persists in the southeastern United States. Following a nadir in 2000, syphilis rates again began to rise. This increase in primary and secondary syphilis is due primarily to an increased incidence among men. Today, more than 60% of new cases of syphilis in the U.S. occur in men who have sex with men (MSM), and these cases are often associated with HIV co-infection and high-risk sexual behavior.

Treponema pallidum enters the body by penetrating intact mucous membranes or by invading through epithelial abrasions. Skin contact with an ulcer infected with *Treponema pallidum* (even by the doctor's examining hand) can result in infection. When infection occurs, the spirochetes immediately begin disseminating throughout the body.

If untreated, patients with syphilis will progress through 3 clinical stages, with a latent period between stages 2 and 3.

Fig. 14-1.   Stages of syphilis.

Primary Syphilis

The primary lesion of syphilis is a **painless** chancre that erupts at the site of inoculation 3–**6** weeks after the

Figure 14-2

initial contact. Regional nontender lymph node swelling occurs as well.

Fig. 14-2. The chancre can be described as a firm, ulcerated painless lesion with a punched-out base and rolled edges. It is highly infectious, since *Treponema pallidum* sheds from it continuously. Think of this skin ulcer as a small *Treponema pallidum* resort swimming pool, with thousands of vacationers swimming within. The chancre resolves over 4–**6** weeks without a scar, often fooling the infected individual into thinking that the infection has completely resolved.

Secondary Syphilis

Untreated patients enter the bacteremic stage, or secondary syphilis, often about **6** weeks after the primary chancre has healed (although sometimes the manifestations of secondary syphilis occur while the primary chancre is still healing). In secondary syphilis, the bacteria multiply and spread via the blood throughout the body. Unlike the single lesion of primary syphilis, the second stage is systemic, with widespread rash, generalized lymphadenopathy, and involvement of many organs.

The rash of secondary syphilis consists of small red macular (flat) lesions symmetrically distributed over the body, particularly involving the palms, soles, and mucous membranes of the oral cavity. The skin lesions can become papular (bumpy) and even pustular.

A second characteristic skin finding of the second stage is called **condyloma latum**. This painless, wartlike lesion often occurs in warm, moist sites like the vulva or scrotum. This lesion, which is packed with spirochetes, ulcerates and is therefore extremely contagious.

Skin infection in areas of hair growth results in patchy bald spots and loss of eyebrows.

During the secondary stage, almost any organ can become infected (including the CNS, eyes, kidneys, and bones). Systemic symptoms, such as generalized lymphadenopathy, weight loss, and fever, also occur during this stage.

The rash and condyloma lata resolve over **6** weeks, and this disease enters the latent phase.

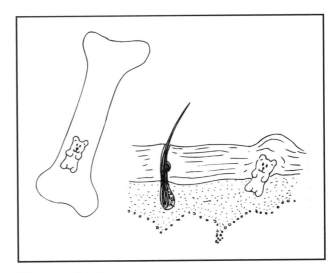

Figure 14-3

Latent Syphilis

In this stage, the features of secondary syphilis have resolved, although serologic tests remain positive. Most patients are asymptomatic during this period, although about 25% will have one or more relapses and develop the infectious skin lesions of secondary syphilis. After 4 years, there are generally no more relapses, and this disease is now considered noninfectious (except in pregnant women, who can still transmit syphilis to their fetus).

About one-third of untreated patients will slowly progress from this stage to tertiary syphilis. The rest will remain asymptomatic.

Tertiary Syphilis

Tertiary syphilis generally develops over **6**–40 years, with slow inflammatory damage to organ tissue, small blood vessels, and nerve cells. It can be grouped into 3 general categories: 1) gummatous syphilis, 2) cardiovascular syphilis, and 3) neurosyphilis.

1) **Gummatous syphilis** occurs 3–10 years after the primary infection in 15% of untreated patients.

Fig. 14-3. **Gummas** (**Gummy bears**) are localized granulomatous lesions which eventually necrose and become fibrotic. These noninfectious lesions are found mainly in the skin and bones. Skin gummas are painless solitary lesions with sharp borders, while bone lesions are associated with a deep gnawing pain. These will resolve with antimicrobial therapy.

2) **Cardiovascular syphilis** occurs at least 10 years after the primary infection in 10% of untreated patients. Characteristically, an aneurysm forms in

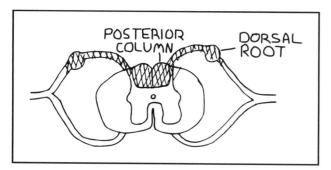

POSTERIOR COLUMN   DORSAL ROOT

Figure 14-4

the ascending aorta or aortic arch. This is caused by chronic inflammatory destruction of the small arterioles (vasa vasorum) supplying the aorta itself, leading to necrosis of the media layer of the aorta. The wall of the aorta splits as blood dissects through the weakened media layer. Aortic valve insufficiency and occlusion of the coronary arteries may also develop as the dissection spreads to involve the coronary arteries. Antimicrobial therapy can NOT reverse these manifestations.

3) **Neurosyphilis** occurs in about 8% of untreated cases. The 5 most common presentations of neurosyphilis are:

a) **Asymptomatic neurosyphilis**: The patient is clinically normal, but cerebrospinal fluid tests positive for syphilis.

b) **Subacute meningitis**: The patient has fever, stiff neck, and headache. Cerebrospinal fluid analysis reveals a high lymphocyte count, high protein, low glucose, and positive syphilis tests.
Note that most bacteria cause an acute meningitis with a high neutrophil count, high protein, and low glucose. *Treponema pallidum* and *Mycobacterium tuberculosis* are two bacteria that cause a subacute meningitis with a predominance of lymphocytes.

c) **Meningovascular syphilis**: The spirochetes attack blood vessels in the brain and meninges (circle of Willis!), resulting in cerebrovascular occlusion and infarction of the nerve tissue in the brain, spinal cord, and meninges, causing a spectrum of neurologic impairments.

d) **Tabes dorsalis**: This condition affects the spinal cord, specifically the posterior column and dorsal roots.

Fig. 14-4. Syphilitic **tabes dorsalis** involves damage to the posterior columns and dorsal roots of the spinal cord. Posterior column damage disrupts vibratory and proprioceptive sensations, resulting in ataxia. Dorsal root and ganglia damage leads to loss of reflexes and loss of pain and temperature sensation.

e) **General paresis** (of the insane): This is a progressive disease of the nerve cells in the brain, leading to mental deterioration and psychiatric symptoms.

The **Argyll-Robertson pupil** may be present in both tabes dorsalis and general paresis. The Argyll-Robertson pupil, caused by a midbrain lesion, constricts during accommodation (near vision) but does not react to light. This is also referred to as the "prostitute's pupil" because the prostitute accommodates but does not react, and is frequently infected with syphilis.

Fig. 14-5. Overview of primary, secondary, latent, and tertiary syphilis. Notice the **rule of sixes**:
Six—Sexual transmission
6 axial filaments
6 week incubation
6 weeks for the ulcer to heal
6 weeks after the ulcer heals, secondary syphilis develops
6 weeks for secondary syphilis to resolve
66% of latent stage patients have resolution (no tertiary syphilis)
6 years to develop tertiary syphilis (at the least)

Congenital Syphilis

Congenital syphilis occurs in the fetus of an infected pregnant woman. *Treponema pallidum* crosses the placental blood barrier, and the treponemes rapidly disseminate throughout the infected fetus. Fetuses that acquire the infection have a high mortality rate (stillbirth, spontaneous abortion, and neonatal death), and almost all of those that survive will develop early or late congenital syphilis.

Early congenital syphilis occurs within 2 years and is like severe adult secondary syphilis with widespread rash and condyloma latum. Involvement of the nasal mucous membranes leads to a runny nose called the "**snuffles**." Lymph node, liver, and spleen enlargement, and bone infection (osteitis—seen on X-ray), are also common afflictions of early congenital syphilis.

Late congenital syphilis is similar to adult tertiary syphilis except that cardiovascular involvement rarely occurs:

1) Neurosyphilis is the same as in adults and eighth nerve deafness is common.

2) Bone and teeth are frequently involved. Periosteal (outer layer of bone) inflammation destroys the cartilage of the palate and nasal septum, giving the nose a sunken appearance called **saddle nose**. A similar inflammation of the tibia leads to bowing called **saber shins**. The upper central incisors are widely spaced with a central notch in each tooth (**Hutchinson's teeth**) and the molars have too many cusps (**mulberry molars**).

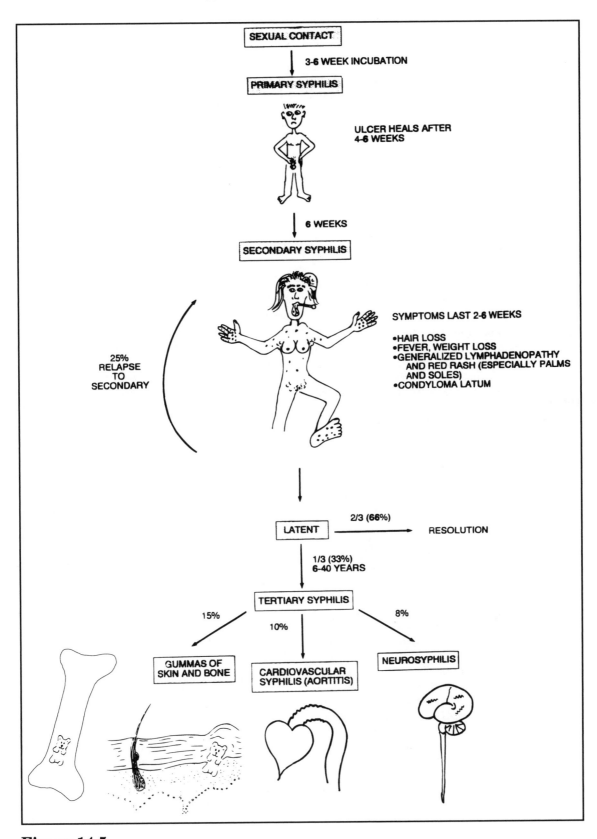

Figure 14-5

3) **Eye disease** such as corneal inflammation can occur.

Interestingly, *Treponema pallidum* infection does not damage the fetus until the fourth month of gestation, so treating the mother with antibiotic therapy prior to this time can prevent congenital syphilis.

Diagnostic Tests for Syphilis

Absolute diagnosis during the first and second stages can be made by direct examination, under darkfield microscopy, of a specimen from the primary chancre, the maculopapular rash, or the condyloma latum. Darkfield microscopy reveals tiny helically-shaped organisms moving in a corkscrew-like fashion.

Since direct visualization of spirochetes is effective only during the **active** stages of primary and secondary syphilis, serologic tests were developed as a screening tool. There are 2 types of serologic screening test: nonspecific and specific.

1) **Nonspecific treponemal tests**: Infection with syphilis results in cellular damage and the release into the serum of a number of lipids, including cardiolipin and lecithin. The body produces antibodies against these antigens. We therefore quantitatively measure the titer of the antibodies that bind to these lipids. If a patient's serum has these antibodies, we suspect that he/she has syphilis. Since invasion of the cerebrospinal fluid (CSF) by syphilis also stimulates an increase of these anti-lipoidal antibodies, we can also perform this test on the CSF to diagnose neurosyphilis. The two most common tests employing this technique are the **Venereal Disease Research Laboratory (VDRL)** and **Rapid Plasma Reagin (RPR)** test.

So why are these tests nonspecific? It is important to realize that 1% of adults without syphilis will also have these antibodies, resulting in a false positive test. For example, false positive tests often occur in patients who are pregnant, have an acute febrile illness such as infectious mononucleosis or viral hepatitis, use intravenous drugs, or following immunization. Therefore, a positive nonspecific test must be confirmed with a specific treponemal antibody test.

2) **Specific treponemal tests**: While the nonspecific tests look for anti-lipoidal antibodies, the specific treponemal tests look for antibodies against the spirochete itself. The **Indirect Immunofluorescent Treponemal Antibody-Absorption (FTA-ABS)** test is the most commonly used specific treponemal test. This test is performed by first mixing the patient's serum with a standardized nonpathogenic strain of *Treponema*, which removes (absorbs) antibodies shared by both *Treponema pallidum* and the nonpathogenic treponemal strains (as nonpathogenic strains of *Treponema* are part of the normal human flora). The remaining serum is then added to a slide covered with killed *Treponema pallidum* (as the antigen). Antibodies that are specific to this organism will subsequently bind, giving a positive result.

Since we all have antibodies to nonpathogenic strains of treponemes, the absorption part of the FTA-ABS test is necessary to cut down on the number of false positives. Only people who have antibodies specific for the pathogenic strain of treponemes will elicit a positive reaction. However, false positives can occur with other spirochetal infections, such as yaws, pinta, leptospirosis, and Lyme disease.

3) Polymerase chain reaction (PCR) detection of bacterial DNA is available.

Treatment

Treponema pallidum is extremely fragile, and can be killed easily with heat, drying, or soap and water. Since syphilis was found to be treatable by raising one's body temperature, patients in the early 1900's were placed in a "fever" box (a closed box in the hot sun, with only the patient's head protruding). Fortunately, the discovery of penicillin provided a less hazardous therapy.

The current drug of choice for syphilis is penicillin (the particular type and dosage of penicillin depends on the stage of the infection). Penicillin can even cross the placenta and cure congenital syphilis. Patients allergic to penicillin can be effectively treated with erythromycin and doxycycline (but doxycycline cannot be used to treat congenital syphilis, as it is toxic to the fetus).

It is important to realize that reinfection can occur. This suggests that antitreponemal antibodies are not protective. Cell-mediated immunity may play a role in the course of syphilis by inducing the regression of the lesions of primary and secondary syphilis.

With adequate treatment, the levels of anticardiolipin antibodies will decrease, while the levels of specific antitreponemal antibodies will remain unchanged. Therefore, a person who is adequately treated will eventually manifest (over months to years) a drop in the VDRL or RPR to nonpositive, while the FTA-ABS will remain positive.

Fig. 14-6. Interpretation of syphilis serology.

Jarisch-Herxheimer Phenomenon

Most patients with syphilis will develop an acute worsening of their symptoms immediately after antibiotics are started. Symptoms include a mild fever, chills, malaise, headache, and muscle aches. The killed organisms release a pyrogen (fever-producing enzyme) that is thought to cause these symptoms. This self-limiting reaction, called the **Jarisch-Herxheimer phenomenon**, may occur with most spirochetes.

VDRL or RPR	FTA-ABS	INTERPRETATION
+	+	• Indicates an active treponemal infection
+	−	• Probably a false positive
−	+	• Successfully treated syphilis
−	−	• Syphilis unlikely, although: 1. Patients with a syphilis infection who also have AIDS may be sero-negative 2. Patient recently infected with syphilis may not have developed an immune response yet

Note: VDRL and RPR are very similar tests.

Figure 14-6   SYPHILIS SEROLOGY INTERPRETATION

Treponema pallidum Subspecies

There are 3 subspecies of *Treponema pallidum* (***endemicum, pertenue***, and ***carateum***) that cause nonvenereal disease (**endemic syphilis, yaws**, and **pinta**, respectively). All 3 subspecies cause skin ulcers and gummas of the skin and bones in children, with the exception of *Treponema carateum*, which only causes skin discoloration (no gummas).

Interestingly, these subspecies are morphologically and genetically identical to *Treponema pallidum*, yet do not cause the sexually transmitted disease syphilis. However, the diseases do share many characteristics with syphilis. Like syphilis, the general pattern of these diseases involves a primary skin papule or ulcer developing at the site of inoculation (usually not the genitals). This is followed by a secondary stage of widespread skin lesions. The tertiary stage is manifested years later by gummas of the skin and bones. Unlike tertiary syphilis, the tertiary stages of the nonvenereal treponemes do not involve the heart or central nervous system.

The antibodies produced by these infections will give a positive VDRL and FTA-ABS. One intramuscular injection of **long-acting penicillin** is curative.

Figure 14-7

Treponema pallidum Subspecies *endemicum*
(Endemic Syphilis: Bejel)

Endemic syphilis occurs in the desert zones of Africa and the Middle East and is spread by sharing drinking and eating utensils. Skin lesions usually occur in the oral mucosa and are similar to condyloma lata of secondary syphilis. Gummas of the skin and bone may develop later.

Treponema pallidum Subspecies *pertenue*
(Yaws)

Yaws, a disease of the moist tropics, spreads from person to person by contact with open ulcers. At the initial site of inoculation a papule appears that grows over months, becoming wartlike and is called the "mother yaw." Secondary lesions appear on exposed parts of the body and years later tertiary gummas develop in the skin and long bones.

Fig. 14-7.   The tertiary lesions in yaws often cause significant disfigurement of the face. Imagine **JAWS** (**Yaws**) taking a bite out of a person's face. These disfiguring lesions on the face, caused by the spirochetes destroying the bone, cartilage and skin, are called **Gangosa**. (A **Gang** of **J(Y)aws**!)

Treponema pallidum Subspecies *carateum*
(Pinta)

Fig. 14-8.   Hispanic person with colored red and blue skin lesions, saying, "Por favor, no **pinta** la cara." **Pinta** is purely a skin disease limited to rural Latin America. After infection by direct contact, a papule develops

Figure 14-8

which slowly expands. This is followed by a secondary eruption of numerous red lesions that turn blue in the sun. Within a year the lesions become depigmented, turning white. These colored lesions look like someone **PAINTED** them on.

BORRELIA

The corkscrew-shaped *Borrelia* are larger than the *Treponema*, and therefore can be viewed under a light microscope with Giemsa or Wright stains.

Borrelia cause Lyme disease (*Borrelia burgdorferi*) and relapsing fever (caused by 18 other species of *Borrelia*). Both of these diseases are transmitted by insect vectors.

Borrelia burgdorferi
(Lyme Disease)

Lyme disease is seen in the Northeast, Midwest and northwestern U.S. This is the most commonly reported tick-borne illness in the U.S.

When walking in the woods during the summer months, you must be careful of the *Ixodes* tick. This tiny creature's bite can transfer the agent for Lyme disease, *Borrelia burgdorferi*. It takes greater than 24 hours of attachment for transfer of the organism, so regular "tick checks" may help prevent infection.

The animal reservoir for *Borrelia burgdorferi* includes the white-footed mouse (as well as other small rodents) and the white-tailed deer. The *Ixodes* ticks pick up the spirochete from these reservoirs and can subsequently transmit them to humans.

Lyme disease has many features that resemble syphilis, although Lyme disease is NOT sexually transmitted. Both of these diseases are caused by spirochetes. The primary stage in both involves a single, painless skin lesion (syphilitic chancre and Lyme's erythema chronicum migrans) that develops at the initial site of inoculation. In both diseases the spirochetes then spread throughout the body, invading many organ systems, especially the skin. Both also cause chronic problems years later (tertiary syphilis and late stage Lyme disease).

Fig. 14-9. Like syphilis, Lyme disease has been divided into three stages: 1) early localized stage, 2) early disseminated stage, and 3) late stage.

Early Localized Stage

The first stage begins about 10 days after the tick bite and lasts about 4 weeks. It consists of just a skin lesion at the site of the tick bite (called **erythema chronicum migrans**) along with a flulike illness, and regional lymphadenopathy.

Fig. 14-10. Erythema chronicum migrans (ECM) starts off as a red (**erythematous**) flat round rash, which spreads out (or **migrates**) over time (**chronicum**). The outer border remains bright red, while the center will clear, turn blue, or even necrose. Visualize a drop of Lyme juice (drawn as drops of spirochetes) landing on the skin and the Lyme acid burning the skin red. With time, the juice spreads out and the erythematous lesion spreads, eventually getting so large that there is not enough juice for the center, so that the center now has normal-looking skin.

Early Disseminated Stage

Fig. 14-11. The early disseminated stage involves the dissemination of *Borrelia burgdorferi* spirochetes to 4 organ systems: the skin, nervous system, heart, and joints. Notice that the Lyme juice (drawn as drops of spirochetes) has begun dripping onto the skin, nervous system, heart, and joints. This stage can occur after or at the same time as the first stage.

The skin lesions in this stage are just ECM again, but this time there are multiple lesions on the body,

STAGE	CLINICAL
Early localized stage (stage 1)	• Erythema chronicum migrans (ECM)
Early disseminated stage (stage 2)	1. Multiple smaller ECM 2. Neurologic: aseptic meningitis, cranial nerve palsies (Bell's palsy), and peripheral neuropathy 3. Cardiac: transient heart block or myocarditis 4. Brief attacks of arthritis of large joints (knee)
Late stage (stage 3)	1. Chronic arthritis 2. Encephalopathy

Figure 14-9   LYME DISEASE: CLINICAL MANIFESTATIONS

and they are smaller (there's just not enough Lyme juice to make them as large as the one in the primary stage).

Borrelia burgdorferi can invade the brain, cranial nerves, and even motor/sensory nerves. Examples include meningitis, cranial nerve palsies (especially of the seventh nerve—a Bell's palsy), and peripheral neuropathies.

Transient cardiac abnormalities occur in about 10% of patients. The most common abnormality is atrioventricular nodal block (heart block), and less commonly myocarditis and left ventricular dysfunction. Since the cardiac lesions usually resolve in a matter of weeks (especially with antibiotic therapy), a permanent pacemaker is often unnecessary.

Migratory joint and muscle pain can also occur. About 6 months after infection attacks of arthritis can occur. Large joints such as the knee become hot, swollen, and painful.

Late Stage

About 10% of untreated patients will develop chronic arthritis that lasts for more than a year. This usually involves 1 or 2 of the large peripheral joints, such as the knee. Interestingly, many of these patients have the B-cell allo antigens HLA-DRB1*0401, HLA-DRB1*0101 and other related alleles.

Like tertiary syphilis, Lyme disease can lead to chronic neurologic damage. An encephalopathy can develop characterized by memory impairment, irritability, and somnolence.

Diagnosis and Treatment

Diagnosis primarily depends on the doctor's recognizing the characteristic clinical findings described above in a person who has been exposed to ticks in an area endemic for Lyme disease.

If the patient presents with ECM, the leading edge of the rash can be biopsied and cultured for *Borrelia burgdorferi*.

As culturing this organism from blood and CSF is very difficult, determination of the levels of anti-*Borrelia burgdorferi* antibodies is often helpful in making a diagnosis. The two most effective techniques are

Figure 14-10

135

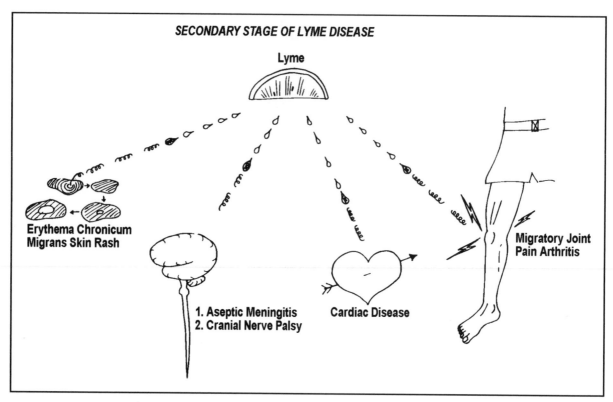

Figure 14-11

enzyme-linked immunosorbent assays (ELISA) and Western immunoblotting.

Doxycycline or **penicillin family** antibiotics are currently the most effective antibiotics for treating this disease.

A vaccine utilizing a recombinant protein of *Borrelia burgdorferi* called OspA was marketed in 1999 but withdrawn in 2002 secondary to poor sales. However, this demonstrated that vaccination was effective for prevention of Lyme disease.

Borrelia recurrentis
(Relapsing Fever)

Of 18 different species of *Borrelia* that can cause relapsing fever, only *Borrelia recurrentis* is transmitted to humans via the body louse (*Pediculus humanus*). The other *Borrelia* species are transmitted by the tick *Ornithodoros*. This tick likes to feed on sleeping campers in the western U.S., especially those who sleep in rodent-infested, rustic mountain cabins.

After the *Borrelia* has been transmitted, via the louse or tick, this bacteria disseminates via the blood. A high fever develops, with chills, headaches and muscle aches. Rash and meningeal involvement may follow. With drenching sweats, the fever and symptoms resolve after 3–6 days. The patient remains afebrile for about 8 days, but then relapses, developing similar features for

another 3–6 days. Relapses will continue to occur, although they will become progressively shorter and milder as the afebrile intervals lengthen.

Antigenic Variation: the Key to Relapsing Fever

Fig. 14-12. "Why the relapses?" you ask. Well, check out our friend, **Boris** the *Borrelia*, who is a master at the art of "antigenic variation." He is initially well camouflaged in blood, but antibodies are soon manufactured by the host's immune system. These antibodies can bind specifically to the *Borrelia* surface proteins and thereby remove the *Borrelia* from the blood. But sneaky Boris rapidly changes his surface proteins, so that the antibodies no longer recognize them. Boris can now safely proliferate without antibody interference, resulting in fever. As soon as the immune system recognizes that there are new foreign proteins in the blood, it churns out a new set of antibodies that are specific for Boris's new surface proteins. But Boris is ready, and quickly changes his surface proteins again. This antigenic variation allows Boris to continue causing relapses for many weeks.

Diagnosis is made by drawing blood cultures (culture on special media) during the febrile periods only (as blood cultures are often negative when the patient is afebrile). A Wright's or Giemsa-stained smear of peripheral blood during febrile periods may reveal

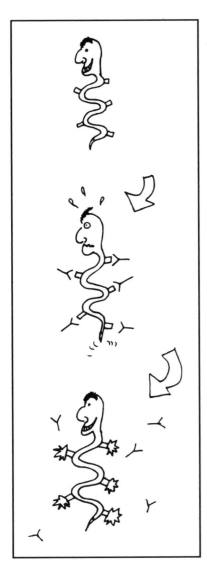

Figure 14-12

the spirochete between red blood cells. Dark-field microscopy is also useful.

Doxycycline or **erythromycin** is the treatment of choice.

LEPTOSPIRA

Leptospira are long, thin aerobic spirochetes that are wound up in a tight coil. They have a hook on one or both ends, giving them an "ice tongs" appearance. Currently *Leptospira* are divided into 2 species. One of them, *Leptospira interrogans*, causes human disease and has been divided by serologic tests into 23 serogroups (sub-groups) and over 250 serovars (sub-subgroups).

Leptospira are found all over the world in the urine of dogs, rats, livestock, and wild animals. These spirochetes can penetrate abraded skin or mucous membranes when humans come in contact with the urine either directly or by swimming in contaminated water (usually swallowed). Outbreaks of leptospirosis have recently been associated with "adventure racing," which often involves swimming through remote rivers and slogging through boggy regions that have been contaminated with leptospira from wild animal urine.

Clinically, there are 2 phases. In the **first** or **leptospiremic phase** the bacteria invade the blood and CSF, causing an abrupt onset of high spiking temperatures, headache, malaise, and severe muscle aches (thighs and lower back). Classically, the conjunctiva are red and the patient experiences photophobia. After about 1 week, there is a short afebrile period and then the fever and earlier symptoms recur. This **second** or **immune phase** correlates with the appearance of IgM antibodies. During the second phase patients may develop meningismus, and the cerebrospinal fluid (CSF) exam reveals an elevated white cell count in most patients.

Leptospira interrogans (classically serogroup *icterohaemorrhagiae*, but can be other serogroups) can cause a more severe illness called **Weil's disease**, or infectious jaundice, which involves renal failure, hepatitis with jaundice, mental status changes, and hemorrhage in many organs.

Diagnosis is made by culturing (on special media) blood and CSF during the first febrile phase. During the second phase and months later the organisms can be cultured from the urine.

The only problem is that treatment should be initiated quickly, before any of the above diagnostic test results are available. To arrive at your diagnosis, you must integrate the clinical history (animal contact or swimming in areas shared by animals), symptoms suggestive of leptospirosis, and lab tests reflecting the affected organs (elevated liver function tests and protein in the urine). Treat patients immediately with either **penicillin** or **doxycycline**.

Newer diagnostic tests include a monoclonal antibody based ELISA to detect *Leptospira* antigens in the urine and Polymerase Chain Reaction (PCR) to detect bacterial DNA in serum, CSF and urine.

Fig. 14-13. Summary of the spirochetes.

Recommended Review Articles:

Dworkin MS, Schwan TG, et al. Tick-borne relapsing fever. Infect Dis Clin North Am. 2008;22(3):449–68.

Lee V, Kinghorn G. Syphilis: an update. Clin Med. 2008 Jun;8 (3):330–3.

Wormser GP, Dattwyler RJ, et al. The clinical assessment, treatment, and prevention of lyme disease, human granulocytic anaplasmosis, and babesiosis: clinical practice guidelines by the Infectious Diseases Society of America. Clin Infect Dis. 2006;43(9):1089–134.

Zetola NM, Klausner JD. Syphilis and HIV infection: an update. Clin Infect Dis. 2007;44(9):1222–8.

SPIROCHETES GRAM-NEGATIVE	RESERVOIR	TRANSMISSION	METABOLISM & MORPHOLOGY	VIRULENCE
Spirochete generalities			1. Multiply by transverse fission 2. Motile: six **axial filaments** wind around the organism between the peptidoglycan layer and the outer cell membrane. Contraction of axial filaments conveys spinning motion	• No exotoxins!!!
Treponema pallidum	• Humans only	• Sexual	1. Microaerophilic 2. Morphology: thick rigid spirals 3. Highly sensitive to elevated temperatures	• Motile
Treponema pallidum subspecies *endemicum*	• Desert zones of Africa and the Middle East	• Sharing of drinking and eating utensils	• Morphologically and serologically indistinguishable from *T. pallidum*	• Motile
Treponema pertenue	• Moist tropical regions	• Person-to-person contact or via flies	• Morphologically, genetically and serologically indistinguishable from *T. Pallidum*	• Motile
Treponema carateum	• Latin America	• Person-to-person contact	• Morphologically and serologically indistinguishable from *T. pallidum*	• Motile

Figure 14-13   SPIROCHETES

CLINICAL	TREATMENT	DIAGNOSTICS
		• Can not culture on artificial media (except for *Leptospira*)

SYPHILIS
A. Primary stage: painless chancre (skin ulcer)
B. Secondary stage:
  1. Rash on palms and soles
  2. Condyloma latum: painless, wartlike lesion which occurs in warm, moist places (vulva or scrotum)
  3. CNS, eyes, bones, kidneys and/or joints can be involved
C. Latent stage: 25% may relapse back to the secondary stage
D. Tertiary stage (33%):
  1. Gummas of skin and bone
  2. Cardiovascular syphilis
  3. Neurosyphilis: may get the Argyll-Robertson pupil
E. Congenital syphilis: contracted in-utero
	1. Penicillin G
	2. Erythromycin
	3. Doxycycline
	• Jarisch-Herxheimer reaction: acute worsening of symptoms after antibiotics are started
		1. Cutaneous lesions examined by dark field microscopy, immunofluorescence, ELISA, or silver stain
		2. Non-specific treponemal test: VDRL; RPR
		3. Specific treponemal test: FTA-ABS, MHA-TP
		• All pregnant women should be screened with VDRL because antibiotic treatment prior to 4 months of gestation prevents congenital syphilis
		4. Polymerase chain reaction (PCR) detection of bacterial DNA is available

BEJEL A. Primary & secondary lesions: occur in oral mucosa B. Tertiary lesions: gummas of skin & bone	• Penicillin	• VDRL and FTA-ABS are positive
YAWS A. Primary and secondary lesions: ulcerative skin lesions near initial site of infection — often looks like condyloma lata B. Tertiary lesions: gummas of skin and bone (resulting in severe facial disfigurement)	1. Azithromycin 2. Penicillin 3. Plastic surgery to correct facial disfigurement	• VDRL and FTA-ABS are positive
PINTA • Flat red or blue lesions which do **NOT** ulcerate	• Penicillin	• VDRL and FTA-ABS are positive

SPIROCHETES GRAM-NEGATIVE	RESERVOIR	TRANSMISSION	METABOLISM & MORPHOLOGY	VIRULENCE
Borrelia burgdorferi	1. White-footed mouse 2. White-tailed deer	• Vector = *Ixodes* ticks 1. *Ixodes scapularis*: East & Midwest 2. *Ixodes pacificus*: West coast	• Microaerophilic	
18 other species of *Borrelia*	• Wild rodents in remote undisturbed areas in the Western U.S.	*Vectors* 1. *Borrelia recurrentis*: louse 2. The other species of *Borrelia*: ticks	• Microaerophilic	1. Antigenic variation: variable expression of outer membrane **Vmp lipoproteins** allows *Borrelia* to escape opsonization and phagocytosis 2. No toxins!!!
Leptospira interrogans 23 serogroups 250 serovars	• Zoonotic (dogs, cats, livestock, and wild animals)	• Direct contact with infected urine or animal tissue. Organisms penetrate broken skin (i.e. on feet) and mucous membranes (swallowing urine-contaminated water)	1. AEROBIC 2. Spiral shaped, with hooks on both ends ("ice tongs") 3. Two axial flagella wrap around and run along the length of the organism under the outer membrane (periplasmic flagella)	

Figure 14-13   (continued)

CLINICAL	TREATMENT	DIAGNOSTICS

LYME DISEASE
A. Early localized stage (stage 1): Erythema chronicum migrans (ECM)
B. Early disseminated stage (stage 2)
  1. Multiple smaller ECM
  2. Neurologic: aseptic meningitis, cranial nerve palsies (Bell's palsy), and peripheral neuropathy
  3. Cardiac: transient heart block or myocarditis
  4. Brief attacks of arthritis of large joints (knee)
C. Late stage (stage 3)
  1. Chronic arthritis
  2. Encephalopathy
	1. Doxycycline
	2. Amoxicillin
	3. Ceftriaxone for neurologic disease
		1. Elevated levels of antibodies against *Borrelia burgdorferi* can be detected by ELISA
		2. Western immunoblotting

RELAPSING FEVER A. Recurring fever about every 8 days B. Fevers break with drenching sweats C. Rash & splenomegaly D. Occasionally meningeal involvement	1. Doxycycline 2. Erythromycin 3. Penicillin G	1. Blood culture during febrile periods 2. Dark field examination of blood drawn during febrile periods 3. Wright's or giemsa - stained peripheral blood smear reveals organism 70% of the time 4. Serologic
1. First phase (leptospiremic): organisms in blood and CSF causes high spiking temperatures, headache and severe muscle aches (thighs and lower back) 2. Second phase (immune): correlates with emergence of IgM and involves recurrence of the above symptoms, often with meningismus (neck pain) 3. **WEIL'S DISEASE**: severe case of leptospirosis with renal failure, hepatitis (and jaundice), mental status changes, and hemorrhage in many organs	1. Penicillin G 2. Doxycycline	1. First week: culture blood or cerebral spinal fluid (on lab media, or by inoculation into animals) 2. Second week to months: culture urine 3. Rarely, dark field microscopy is successful (not recommended) 4. Antibody based ELISA to detect Leptospira antigens in the urine 5. Polymerase Chain Reaction (PCR) to detect bacterial DNA in serum, CSF and urine

M. Gladwin, W. Trattler, and S. Mahan, *Clinical Microbiology Made Ridiculously Simple* ©MedMaster

ACID-FAST BACTERIA

CHAPTER 15. *MYCOBACTERIUM*

The *Mycobacteria* include 2 species that almost everyone has heard of: *Mycobacterium tuberculosis*, which causes tuberculosis, and *Mycobacterium leprae*, which causes leprosy. The *Mycobacteria* also include an expansive group classified as nontuberculous mycobacteria (NTM) which is comprised of over 100 species of organisms. While NTM may not sound familiar, disease due to NTM is more prevalent than tuberculosis and leprosy combined in the United States. *Mycobacteria* are rods with lipid-laden cell walls. This high lipid content makes them **acid-fast** on staining. *Mycobacteria* and *Nocardia* are among a very small group of **acid-fast organisms** and the major two that you need to remember.

In the acid-fast stain, a smear of sputum, for example, is covered with the red stain carbolfuchsin and heated to aid dye penetration. Acid alcohol (95% ethanol and 3% HCl) is poured over the smear, and then a counter-stain of methylene blue is applied. The cell wall lipids of the *Mycobacterium* do not dissolve when the acid alcohol is applied, and thus the red stain does not wash off. So acid-fast organisms resist decolorization with acid alcohol, holding **fast** to their red stain, while bacteria that are not acid-fast lose the red stain and take on the blue.

Fig. 15-1. Visualize a **fast red** sports car to remember that acid-fast organisms stain red.

Figure 15-1

Mycobacterium tuberculosis
(Tuberculosis)

Worldwide, the World Health Organization estimates there are 9 million new cases of tuberculosis and 2 million deaths from tuberculosis annually. In the mid 1980's tuberculosis rates increased in the U.S. for the first time after decades of steady decline. This increase coincided with the emerging HIV/AIDS epidemic. Persons infected with HIV lack the powerful cell-mediated immunity necessary to combat tuberculosis. About $\frac{1}{3}$ of HIV infected persons worldwide also harbor *Mycobacterium tuberculosis*! You will confront this villain again and again in your future career.

This acid-fast bacillus (rod) is an obligate aerobe, which makes sense as it most commonly infects the lungs, where oxygen is abundant. *Mycobacterium tuberculosis* grows very slowly, taking up to 6 weeks for visible growth. The colonies that form lump together due to their hydrophobic lipid nature, resulting in clumped colonies on agar and floating blobs on liquid media.

$$(HO)H_{120}C_{60} - \underset{\underset{OH}{|}}{\overset{\overset{H}{|}}{C}} - \underset{\underset{C_{24}H_{49}}{|}}{\overset{\overset{H}{|}}{C}} - \overset{\overset{O}{||}}{C} - OH$$

Figure 15-2

There is one class of lipid that only acid-fast organisms have and that is involved in mycobacterial virulence—**mycosides**. The terminology is as follows:

1) **Mycolic acid** is a large fatty acid.

Fig. 15-2. The chemical structure of **mycolic acid**, which is a large fatty acid.

2) **Mycoside** is a mycolic acid bound to a carbohydrate, forming a glycolipid.

3) **Cord factor** is a mycoside formed by the union of 2 mycolic acids with a disaccharide (trehalose). This mycoside is only found in virulent strains of *Mycobacterium tuberculosis*. Its presence results in parallel growth of the bacteria, so they appear as cords. Exactly how the virulence occurs is still unknown, but

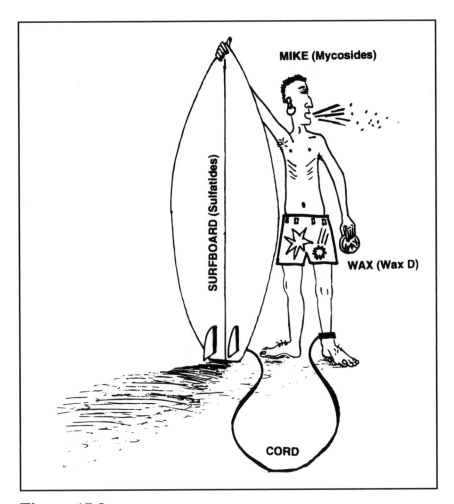

MIKE (Mycosides)

SURFBOARD (Sulfatides)

WAX (Wax D)

CORD

Figure 15-3

experiments show that cord factor inhibits neutrophil migration and damages mitochondria. Its injection into mice results in the release of tumor necrosis factor (TNF or cachectin), resulting in rapid weight loss. Tuberculosis in humans is usually a chronic disease with weight loss that can be mistaken for the cachexia of malignancy. Cord factor might contribute to this weight loss phenomenon.

4) **Sulfatides** are mycosides that resemble cord factor with sulfates attached to the disaccharide. They inhibit the phagosome from fusing with the lysosome that contains bacteriocidal enzymes. The facultative intracellular nature of *Mycobacterium tuberculosis* during early infection may be partly attributable to the sulfatides (see **Fig. 2-7**).

5) **Wax D** is a complicated mycoside that acts as an adjuvant (enhances antibody formation to an antigen) and may be the part of *Mycobacterium tuberculosis* that activates the protective cellular immune system.

Fig. 15-3. To remember the names of the mycosides and their relationship to *Mycobacterium tuberculosis*, picture the surfing dude **Mike** (**myc**osides). He is **WAX-ING** (**wax** D) his **SU**rfboard (**sul**fatides) and has his surfboard **CORD** (**cord** factor) attached to his leg (so as not to lose his stick). Notice Mike has a cough and some weight loss.

Pathogenesis of Tuberculosis

Mycobacterium tuberculosis primarily affects the lung but can also cause disease in almost any other tissue. The way it spreads and damages the body depends on the host's immune response. The organism and the immune system interact as follows:

1) **Facultative intracellular growth**: With the first exposure (usually by inhalation into the lungs), the host has no specific immunity. The inhaled bacteria cause a local infiltration of neutrophils and macrophages. Due to the various virulence factors, the

phagocytosed bacteria are not destroyed. They multiply and survive in the macrophages. The bacteria cruise through the lymphatics and blood to set up camp in distant sites. This period of facultative intracellular existence is usually short-lived because the host rapidly acquires its prime defense against the acid-fast buggers: cell-mediated immunity.

2) **Cell-mediated immunity**: Some of the macrophages succeed in phagocytosing and breaking up the invading bacteria. These macrophages then run toward a local lymph node and present parts of the bacteria to T-helper cells. The sensitized T-cells then multiply and enter the circulation in search of *Mycobacterium tuberculosis*. When the T-cells encounter their antigenic target, they release lymphokines that serve to attract macrophages and activate them when they arrive. These activated macrophages can now destroy the bacteria. It is during this stage that the macrophage attack actually results in local destruction and necrosis of the lung tissue. The necrosed tissue looks like a granular creamy cheese and is called **caseous necrosis**. This soft caseous center is surrounded by macrophages, multinucleated giant cells, fibroblasts, and collagen deposits, and it frequently calcifies. Within this granuloma the bacteria are kept at bay but remain viable. At some point in the future, perhaps due to a depression in the host's resistance, the bacteria may grow again.

PPD Skin Test

Following induction of cell-mediated immunity against *Mycobacterium tuberculosis*, any additional exposure to this organism will result in a localized delayed-type hypersensitivity reaction (type IV hypersensitivity). Intradermal injection of antigenic protein particles from killed *Mycobacterium tuberculosis*, called **PPD** (**P**urified **P**rotein **D**erivative), results in localized skin swelling and redness. Therefore, intradermal injection of PPD will reveal whether or not a person has been **infected** with *Mycobacterium tuberculosis*. This is important because many infected individuals will not manifest a clinical infection for years. When a positive PPD test occurs, you can treat and eradicate the disease before it significantly damages the lungs or other organs.

A person with a positive PPD test is considered to have **latent tuberculosis**. When you have a patient with a low-grade fever and cough, or a patient who has been in contact with people who have tuberculosis (you, for example, after working in the hospital), you will decide to "place a PPD." You inject the PPD intradermally (just barely under the skin so that the skin bubbles up). Macrophages in the skin will take up the antigen and deliver it to the T-cells. The T-cells then move to the skin site, release lymphokines that activate macrophages, and within 1–2 days the skin will become red, raised, and hard. A positive skin test is defined as an area of induration (hardness) that is greater than a pre-defined size after 48 hours (the time it takes for a type IV delayed hypersensitivity reaction to occur). The concept of the test is that the higher risk a person is for having tuberculosis infection, the smaller the size of the induration needs to be! This approach reduces false positives and improves true positives. It is considered positive if ≥ 5mm in persons that are HIV positive or immunosuppressed; ≥10 mm in persons with common risk factors for exposure to TB, such as from being from a high incidence country (any sub-Sahara African country), working in healthcare, prior incarceration, or having a high risk medical condition (diabetes, renal failure); and ≥ 15 mm in all others.

Note that a positive test does not mean that the patient has active tuberculosis; it indicates **exposure** and **infection** to *Mycobacterium tuberculosis* at some time in the past. A positive test is present in persons with active infection, latent infection, and in those who have been cured of their infection.

False positive test: You still must be wary with this test because some people from other countries have had the **BCG** (**b**acillus **C**almette-**G**uerin) **vaccine** for tuberculosis. This vaccine is effective in preventing severe forms of disseminated TB in children, but it provides minimal immunity into later years.

False negative test: Some patients do not react to the PPD even if they have been infected with tuberculosis. These patients are usually **anergic**, which means that they lack a normal immune response due to steroid use, malnutrition, AIDS, etc.

We now have an alternative to the PPD skin tests for screening. A new group of FDA approved blood assays called IGRAS (short for interferon-gamma release assays) measures interferon gamma levels produced in whole blood in response to specific tuberculosis antigens. An advantage of these tests is their relative specificity for *Mycobacterium tuberculosis*; they are not positive in patients with previous BCG vaccination or those with most types of non-tuberculous mycobacteria exposure.

Clinical Manifestations

The first exposure to *Mycobacterium tuberculosis* is called **primary** tuberculosis and usually is a subclinical (asymptomatic) lung infection. Occasionally, an overt symptomatic primary infection occurs.

When an asymptomatic primary infection occurs, the acquired cell-mediated immunity will wall off and suppress the bacteria. These defeated bacteria lie dormant but can later rise up and cause disease. This second infection is called **secondary** or **reactivation** tuberculosis.

For the real number crunchers, here are the statistics: Close contacts, such as household members, of someone with pulmonary tuberculosis have a 30% chance of being infected. Of all the infected persons, about 5% will develop tuberculosis in the next 1 or 2 years and 5% will develop reactivation tuberculosis sometime later in life. So there is a 10% lifetime risk of developing tuberculosis for those infected with *Mycobacterium tuberculosis*.

Primary Tuberculosis

1) *Mycobacterium tuberculosis* is usually transmitted via aerosolized droplet nuclei from the aerosolized respiratory secretions of an adult with pulmonary tuberculosis. This adult will shower the air with these secretions when he coughs, sings, laughs, or talks.

2) The inspired droplets land in the areas of the lung that receive the highest air flow: the middle and lower lung zones. Here there will be a small area of pneumonitis with neutrophils and edema, just like any bacterial pneumonia.

3) Now the bacteria enter macrophages, multiply, and spread via the lymphatics and bloodstream to the regional lymph nodes, other areas of the lungs, and distant organs.

Tuberculosis is a confusing disease because so many different things can happen. As cell-mediated immunity develops, 1) the infection can be contained so that the patient will not even realize he was infected, or 2) it can become a symptomatic disease.

1) **Asymptomatic primary infection**: The cell-mediated defenses kick in, and the foci of bacteria

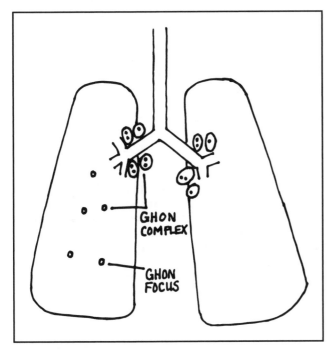

Figure 15-4

become walled off in the caseous granulomas. These granulomas then heal with fibrosis, calcification, and scar formation. The organisms in these lesions are decreased in number but remain viable. Tiny **tubercles** (as the granulomas are called) are often too small to be seen even on chest X-ray. Only a PPD will give the buggers away. Sometimes the chest film will suggest recent infection by showing hilar lymph node enlargement or calcifications.

Fig. 15-4. A calcified tubercle in the middle or lower lung zone is called a **Ghon focus**. A Ghon focus accompanied by perihilar lymph node calcified granulomas is called a **Ghon**, or **Ranke, complex**.

2) **Symptomatic primary tuberculosis** occurs far less frequently, more commonly in children, the elderly, and the immunocompromised (especially HIV infected persons). These groups do not have as powerful a cell-mediated immune system as do healthy adults, so the organisms are not suppressed. When symptoms do occur in primary TB, chest radiographs commonly show enlargement of the mediastinal or hilar lymph nodes, and possibly lower and middle lung infiltrates. Even if untreated, the majority of these persons will control the infection and have healing by encapsulation of the organisms in granulomas, often with calcification. Occasionally, infection will not be contained and the person will become overtly ill with worsening pulmonary and/or disseminated disease. These persons are termed as having primary progressive disease. In severe untreated cases the lung infiltrates will advance to lung necrosis, forming holes in the lungs or cavities. The cavities can become fluid filled and these fluid filled cavities can be visualized on chest radiographs or CT scans. See **Figure 15-5.**

Fig. 15-5. Overt or manifest primary tuberculosis: Large caseous granulomas develop in the lungs or other organs. In the lungs the caseous material eventually liquifies, is extruded out the bronchi, and leaves behind cavitary lesions, shown here with fluid in the cavities (called "cavitary lesions with air-fluid levels" on chest X-ray).

Secondary or Reactivation Tuberculosis

Most adult cases of tuberculosis occur after the bacteria have been dormant for some time. This is called **reactivation** or **secondary tuberculosis**. The infection can occur in any of the organ systems seeded during the primary infection. It is presumed that a temporary weakening of the immune system may precipitate reactivation. Many AIDS patients develop tuberculosis in this manner. HIV infected patients who are infected with *Mycobacterium tuberculosis* have a 10% chance/year of developing reactivation tuberculosis! Based on the estimate that 1/3 of the world's population is latently infected with TB,

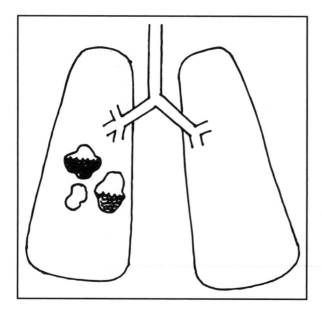

Figure 15-5

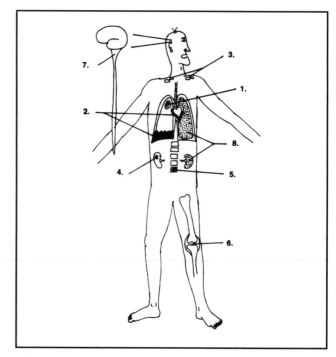

Figure 15-6

one would expect that at least 1/3 of the world's HIV population is also infected.

Risk of reactivation in all persons: 10% for Lifetime!
Risk of reactivation in HIV infected: 10% per year!

Fig. 15-6. The organ systems that can be involved in tuberculosis:

1) **Pulmonary tuberculosis**: This is the most common site of reactivation tuberculosis. The infection usually occurs in the apical areas of the lung around the clavicles. It normally reactivates in the upper lobe because oxygen tension is the highest there, due to decreased pulmonary circulation, and *Mycobacterium tuberculosis* is an aerobic bacterium. Slowly these areas of infection grow, caseate, liquify, and cavitate. Clinically, the patients usually present with a chronic lowgrade fever, night sweats, weight loss, and a productive cough that may have blood in it. This slow erosive infection occurs as the host macrophages and T-cells battle to wall off the bacteria.

2) **Pleural and pericardial infection**: Infection in these spaces results in infected fluid collections around the lung or heart respectively.

3) **Lymph node infection**: Worldwide, this is the most common extrapulmonary manifestation of tuberculosis. The cervical lymph nodes are usually involved. They become swollen, mat together, and drain. Lymph node tuberculosis is called **scrofula**.

4) **Kidney**: Patients will have red and white blood cells in the urine, but no bacteria are seen by Gram stain or grow in culture (remember that *Mycobacterium tuberculosis* takes weeks to grow in culture and are acid-fast). This is referred to as **sterile pyuria**.

5) **Skeletal**: This usually involves the thoracic and lumbar spine, destroying the intervertebral discs and then the adjacent vertebral bodies (**Pott's disease**).

6) **Joints**: There is usually a chronic arthritis of 1 joint.

7) **Central nervous system**: Tuberculosis causes subacute meningitis and forms granulomas in the brain.

8) **Miliary tuberculosis**: Tiny millet-seed-sized tubercles (granulomas) are disseminated all over the body like a shotgun blast. The kidneys, liver, lungs, and other organs are riddled with the tubercles. A chest film will sometimes show a millet-seed pattern throughout the lung. This disease usually occurs in the elderly and in children.

BIG PICTURE: Tuberculosis is usually a chronic disease; it presents slowly with weight loss, low-grade fever, and symptoms related to the organ system infected. Because of its slow course, it may be confused with cancer. Whenever you have an infection of any organ system, tuberculosis will be somewhere on your differential diagnosis list. It is one of the *great imitators*!

Diagnosis

1) **PPD skin test**: This screening test indicates an exposure sometime in the past. Surprisingly, the PPD is negative in up to 1/3 of persons with active pulmonary TB.

2) Chest X-ray: You may pick up an isolated granuloma, Ghon focus, Ghon complex, old scarring in the upper lobes, or active tuberculous pneumonia.

3) Sputum acid-fast stain and culture: When the acid-fast stain or culture are positive, this indicates an active pulmonary infection.

4) Rapid Molecular Detection of MTB in a sputum sample. This is a method of rapid diagnosis which is being embraced in areas with high incidence of MDR tuberculosis. The current most popular platform is the Gene Xpert MTB/RIF which is an automated system that fully integrates processing of the sputum sample with rapid PCR based results, including drug susceptibility for rifampin. It is easy, fast, reliable...but relatively expensive.

The treatment and control of tuberculosis is complicated and will be discussed in the mycobacterial antibiotics chapter (See Chapter 19).

MDR/XDR Tuberculosis

Two events occurred that have raised awareness of the risks posed by multi-drug resistant tuberculosis **(MDR-TB)** and extremely drug resistant tuberculosis **(XDR-TB)**. MDR-TB is defined as resistance to both isoniazid and rifampin. XDR-TB is defined as resistance to isoniazid, rifampin, a flouroquinolone, and an injectable agent (such as an aminoglycoside). In 2005 there was a tuberculosis (TB) outbreak in KwaZula Natal Province, South Africa. There were over 500 culture-confirmed cases of TB, of which 41% were MDR-TB and 10% were XDR-TB. The majority of the persons with XDR-TB had HIV co-infection, and all but one of the 53 confirmed cases of XDR-TB died. Death occurred at a median of 16 days. In July 2007, a young Atlanta attorney was declared to have XDR-TB. Despite this he managed to elude authorities and travel on more than one international flight. Ultimately, he was found to have the less deadly MDR version and he responded to therapy, but the news coverage of this episode raised awareness of this new risk.

Treatment of both MDR- and XDR-TB requires a specialist in TB care and the use of multiple second-line agents for an extended period of time (usually 18–24 months). Just like the old days before effective antibiotics, surgery is often required to remove the focus of infection. As noted above, the failure rate in the treatment of XDR-TB is high! The most common cause of both MDR- and XDR-TB is failure to treat the initial episode of drug-susceptible TB appropriately.

Tuberculosis "Rule of Fives"

• Droplet nuclei are 5 micrometers and contain 5 *Mycobacterium tuberculosis* bacilli.
• Patients infected with *Mycobacterium tuberculosis* have a 5% risk of reactivation in the first 2 years and then a 5% lifetime risk.

• Patients with "high *five*" *HIV* will have a 5 + 5% risk of reactivation per year!

Mycobacterium leprae
(Leprosy, also called Hansen's Disease)

Like *Mycobacterium tuberculosis, Mycobacterium leprae* is an acid-fast rod. It is **impossible** to grow this bacterium on artificial media; it has only been grown in the footpads of mice, in armadillos, and in monkeys. It causes the famous disease leprosy.

The World Health Organizations estimates there are around 2 million persons infected with *Mycobacterium leprae* worldwide, with the majority of cases in 6 countries: India, Brazil, Burma, Indonesia, Madagascar, and Nepal. Every year in the U.S. there are over 100 newly diagnosed cases of leprosy, usually in immigrants. It is unclear why some people are infected and some are not. Many studies have attempted to infect human volunteers, with little success. Infection occurs when a person (who for unknown reasons is susceptible) is exposed to the respiratory secretions or, less likely, skin lesions of an infected individual.

The clinical manifestations of leprosy are dependent on 2 phenomena: 1) The bacteria appear to grow better in cooler body temperatures closer to the skin surface. 2) The severity of the disease is dependent on the host's cell-mediated immune response to the bacilli (which live a facultative intracellular existence, like *Mycobacterium tuberculosis*).

Fig. 15-7. The acid-fast rod *Mycobacterium leprae* is seen here cooling off on an ice cube. Leprosy involves the cooler areas of the body. It damages the **skin** (sparing

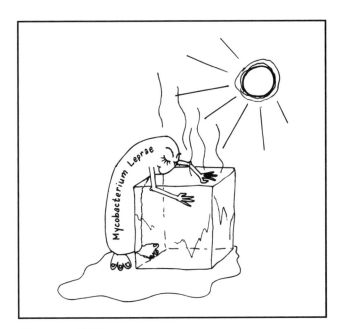

Figure 15-7

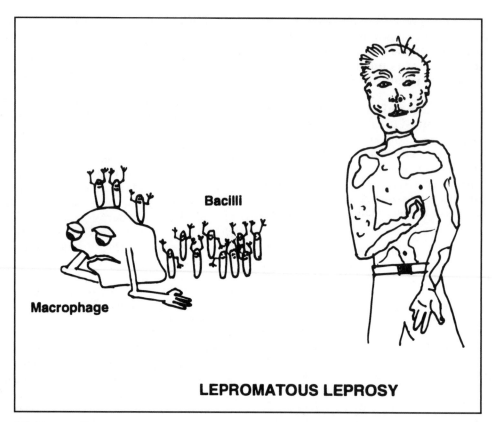

Bacilli

Macrophage

LEPROMATOUS LEPROSY

Figure 15-8

warm areas such as the armpit, groin, and perineum), the superficial **nerves, eyes, nose** and **testes**.

Cell-mediated immunity once again plays an important role in the pathogenesis of this disease. The cellular immunity that limits the spread of the bacteria also causes inflammation and granulomas, particularly in skin and nerves. Clinically, leprosy is broken up into five subdivisions based on the level of cell-mediated immunity, which modulates the severity of the disease:

1) **Lepromatous leprosy** (**LL**): This is the severest form of leprosy because patients canNOT mount a cell-mediated immune response to *Mycobacterium leprae*. It is theorized that defective T-suppressor cells (T-8 cells) block the T-helper cell's response to the *Mycobacterium leprae* antigens.

Fig. 15-8. Lepromatous leprosy (LL): The defeated macrophage is covered with *Mycobacterium leprae* acid-fast rods, demonstrating the very low cellular immunity. The patient with LL cannot mount a delayed hypersensitivity reaction. LL primarily involves the skin, nerves, eyes and testes, but the acid-fast bacilli are found everywhere (respiratory secretions and every body organ). The skin lesions cover the body with all sorts of lumps and thickenings. The facial skin can

become so thickened that the face looks lionlike (hence, **leonine facies**). The nasal cartilage can be destroyed, creating a **saddlenose deformity**, and there is internal testicular damage (leading to infertility). The anterior segment of the eyes can become involved, leading to blindness. Most peripheral nerves are thickened, and there is loss of sensation in the extremities in a glove and stocking distribution. The inability to feel in the fingers and toes leads to repetitive trauma and secondary infections, and ultimately contraction and resorption of the fingers and toes. Lepromatous leprosy will eventually lead to death if untreated.

2) **Tuberculoid leprosy** (**TL**): Patients with TL **can** mount a cell-mediated defense against the bacteria, thus containing the skin damage so that it is not excessive. They will have milder and sometimes self-limiting disease.

Fig. 15-9. Tuberculoid leprosy: The macrophage gobbling up the *Mycobacterium leprae* acid-fast rods demonstrates the high cell-mediated resistance of tuberculoid leprosy. The delayed hypersensitivity reaction is intact, so the lepromin skin test is usually positive. The patient demonstrates **localized superficial**, unilateral **skin** and **nerve** involvement. In this form of leprosy, there are usually only 1 or 2 skin lesions. They are well-defined, hypopigmented, elevated blotches. The area within the

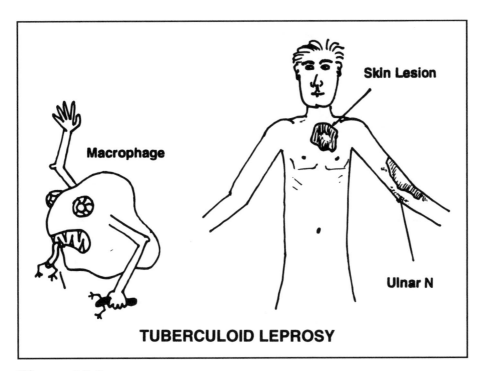

TUBERCULOID LEPROSY

Figure 15-9

	TUBERCULOID	BORDERLINE	LEPROMATOUS
Number of skin lesions	Single	Several	Many
Hair growth on skin lesions	Absent	Slightly decreased	Not affected
Sensation in lesions of the extremities	Completely lost	Moderately lost	Not affected*
Acid fast bacilli in skin scrapings	None	Several	Innumerable
Lepromin skin test	Strongly positive	No reaction	No reaction

* (But a glove and stocking peripheral neuropathy, causing hand and feet numbness, is present!)

Adapted from American Medical Association Drug evaluations, 6th edition, p. 1547.

Figure 15-10   SPECTRUM OF LEPROSY

rash is often hairless with diminished or absent sensation, and enlarged nerves near the skin lesions can be palpated. The most frequently enlarged nerves are those closest to the skin—the greater auricular, the ulnar (above the elbow), the posterior tibial, and the peroneal (over the fibula head). The bacilli are difficult to find in the lesions or blood. Patients are noninfectious and often spontaneously recover.

ACID FAST RODS	MORPHOLOGY	METABOLISM	VIRULENCE	TOXINS
Mycobacterium tuberculosis	1. 40% of total cell dry weight is lipid 2. Composed of mycolic acids 3. Thin rods 4. Non-motile *Remember, mycolic acids are also found in *Nocardia* (which also is acid fast)	1. Aerobic 2. Catalase-positive 3. Slow growth rate	1. Mycosides A. Cord factor: only found in virulent strains (May be responsible for release of tumor necrosis factor (cachectin), causing weight loss) B. Sulfatides: inhibit phagosome-lysosome fusion C. Wax D: acts as an adjuvant 2. Iron siderophore (Mycobactin) 3. Facultative intracellular growth: *M. tuberculosis* can survive and multiply in macrophages *Notice*!!! • Non-motile • No capsule • No attachment pili	*No* exotoxin nor endotoxin. (It has lipopolysac-charide, but *no* Lipid A)

Figure 15-11   ACID FAST BACTERIA

CLINICAL	TREATMENT	DIAGNOSTICS	MISCELLANEOUS

Tuberculosis
A. Primary tuberculosis:
  1. Asymptomatic
  2. Overt disease, involving the lungs or other organs
B. Reactivation or secondary tuberculosis:
  1. Pulmonary
  2. Pleural or pericardial
  3. Lymph node infection
  4. Kidney
  5. Skeletal
  6. Joints
  7. Central nervous system
  8. Miliary tuberculosis
	First line drugs:
	1. Isoniazid (INH)
	2. Rifampin
	3. Pyrazinamide
	4. Ethambutol
	5. Streptomycin
		1. Acid-fast stain of specimen
		2. RAPID CULTURE: Bactec radiometric culture, a liquid broth in a bottle, with radioactive palmitate as a carbon source. Mycobacteria grow and use the carbon, allowing early detection (in 1-2 weeks) even before colonies can be seen.
		3. PPD skin test
		4. IGRA (Interferon gamma release assay)
		5. Chest X-ray
		6. Gene Xpert MTB/Rif (and similar PCR based studies)
			Purified Protein Derivative (PPD) Test
			1. Measure zone of induration:
			  • Positive reaction:
			    1. $\geq$ 5 mm (immuno-compromised host)
			    2. $\geq$ 10 mm (have chronic disease or risk factors for exposure to TB)
			    3. $\geq$ 15 mm (all others)
			2. A positive reaction does not mean active disease.
			3. Can get false negatives in patients with AIDS or malnourished individuals

ACID FAST RODS	MORPHOLOGY	METABOLISM	VIRULENCE	TOXINS
Mycobacterium leprae		1. Catalase-positive 2. Grows best at low temperature 3. Phenolase-positive: converts Dopa into a pigmented product (used for diagnosis)	1. Non-motile 2. Facultative intracellular growth	

NONTUBERCULOUS MYCOBACTERIA	
NAME	**COMMON CLINICAL PRESENTATIONS**

M. avium complex (includes *M. avium* and *M. intracellulare*)	1. In AIDS patients: disseminated infection with fever, weight loss, hepatitis, and diarrhea.
	2. Immunocompetent hosts:
	  a. upper lung cavitary disease in elderly smokers.
	  b. Middle and lower lung nodular and bronchiectatic disease in middle-aged female non-smokers.
	3. Lymphadenitis—most commonly in children.

M. kansasii	1. Pulmonary: upper lung cavitary disease. (Appears similar to tuberculosis.) 2. Disseminated disease (immunocompromised)
M. abscessus	1. Pulmonary disease 2. Skin, soft tissue, and bone disease
M. fortuitum	1. Skin, soft tissue, and bone disease

Figure 15-11   (continued)

CLINICAL	TREATMENT	DIAGNOSTICS	MISCELLANEOUS

Leprosy
A. Lepromatous leprosy (LL):
  1. Low cell-mediated immunity
  2. Organisms found everywhere (organs and blood)
  3. Skin, nerves, eyes and testes involved bilaterally: multiple skin lumps and bumps, leonine facies, saddle nose, peripheral neuropathy, digit absorption, blindness and infertility in men (from testicular damage)
B. Intermediate forms: BL, BB, BT
C. Tuberculoid leprosy (TL):
  1. Intact cell-mediated immunity
  2. Difficult to isolate *M. leprae* from skin or blood
  3. Skin and nerves involved: 1 or 2 superficial unilateral lesions
	1. Rifampin
	2. Dapsone
	3. Clofazimine
	
	• Leprosy reactions (type 1 & type 2) Can occur with treatment (see leprosy drug text for details)
		1. Can *NOT* be grown on artificial lab media; Can only be cultured in certain animals, such as mice foot pads, armadillos or monkeys
		2. Skin or nerve biopsy: will reveal acid-fast bacilli (lepromatous) or granulomas (tuberculoid)
			Lepromin Skin Test
			• Although *not* useful for diagnosis, it allows positioning of patients on the immunologic spectrum

TREATMENT (general overview)	MISCELLANEOUS
1. Disseminated disease in AIDS patients: clarithromycin, rifampin or rifabutin, and ethambutol 2. Pulmonary: clarithromycin, rifampin, ethambutol 3. Lymphadenitis: excisional surgery	1. Common cause of Fever of Unknown Origin (FUO) in AIDS patients 2. Most common cause of NTM lung disease.
Isoniazid, Rifampin, Ethambutol	Second most common cause of NTM pulmonary disease in the U.S.
1. Pulmonary disease: usually requires surgery combined with antibiotics for cure (need susceptibilities to guide therapy). 2. Macrolides (clarithromycin, azithromycin), combined with intravenous agents (amikacin, cefoxitin, or imipenem).	- Rapid grower: usually grows in culture in <7 days.
Two agents with *in vitro* activity: amikacin, ciprofloxacin, sulfonamides, clarithromycin, etc.	1. Rapid grower 2. Common laboratory contaminant 3. Associated with contaminated foot baths.

ATYPICAL MYCOBACTERIA	
NAME	**CLINICAL**
M. chelonae	1. Skin, soft tissue, and bone disease 2. Disseminated disease (immunocompromised) 3. Keratitis – associated with contact use
M. marinum	Skin, soft tissue, and bone disease ("Fish Tank Granuloma")
M. ulcerans	"Buruli ulcers": progressive necrotic skin ulcerations

Figure 15-11   (continued)

The 3 remaining categories represent a continuum **between** LL and TL. They are called **borderline lepromatous (BL), borderline (BB)**, and **borderline tuberculoid (BT)**. The skin lesions of BL will be more numerous and have a greater diversity of shape than those of BT.

The **lepromin skin test** is similar to the PPD used in tuberculosis. It measures the ability of the host to mount a delayed hypersensitivity reaction against antigens of *Mycobacterium leprae*. This test is more prognostic than diagnostic and is used to place patients on the immunologic spectrum. It makes sense that TL patients would have a positive cell-mediated immune response and thus a positive lepromin skin test, while LL patients, who cannot mount a cell-mediated immune response, have a negative response to lepromin.

See Chapter 19 for information about the treatment of leprosy.

Fig. 15-10. The spectrum of leprosy.

NONTUBERCULOUS MYCOBACTERIA

Nontuberculous mycobacteria (NTM) are an expansive group of organisms that are ubiquitous in the soil and water. Healthy immunocompetent persons rarely develop disease despite continued, likely daily, exposure. The incidence of disease due to these organisms has been increasing, likely due to increased awareness and improved laboratory diagnosis. NTM can cause a broad range of disease from asymptomatic colonization to a chronic disabling pneumonia.

A clinician's first exposure to NTM is likely to be caring for AIDS patients with disseminated **Mycobacterium avium-complex (MAC)** disease. This is a very common opportunistic infection in persons with AIDS and CD4 T cell counts <50 cells/mm^3. These patients often present with unexplained fevers, weight loss, diarrhea, and general malaise, with an elevation of alkaline phosphatase on their routine labs. Diagnosis is confirmed by showing growth in mycobacterial blood cultures. These patients generally respond well to appropriate antibiotic therapy and by starting antiretroviral therapy (ART) for HIV.

MAC is also the most common cause of NTM lung disease. It usually presents in one of two ways: 1) as upper lung cavitary disease, predominantly in male smokers, or 2) as lower and middle lung involvement with bronchiectasis and nodular infiltrates in middle aged non-smoking women. In this group it is felt that these women have some as yet undefined underlying predisposition. Treatment of pulmonary MAC disease is long and arduous, requiring an average of 18 months of therapy with a macrolide (clarithromycin, azithromycin) based regimen.

NTM can cause pulmonary disease, lymphadenitis, skin lesions, bone and joint infections, and more. See **Figure 15-11** for a broad overview of the most commonly observed NTM organisms.

Fig. 15-11. Summary of *Mycobacteria*.

References

Britton WJ, Lockwood DNJ. Leprosy. The Lancet. 2004; 363: 1209–1219.

Boehme CC, Nabeta P, Hillemann D, et al. Rapid molecular detection of tuberculosis and rifampin resistance. NEJM 2010;363:1005–15.

Recommended Review Articles:

Diagnosis and Treatment of Disease Caused by Nontuberculous Mycobacteria: The Official Statement of the American Thoracic Society. American Journal of Critical Care Medicine 2007;175:1–50.

Ma Z, Lienhardt C, et al. Global tuberculosis drug development pipeline: the need and the reality. Lancet. 2010;375(9731):2100–9.

Maartens G, Wilkinson RJ. Tuberculosis. Lancet. 2007;370(9604):2030–43.

Schluger NW, Burzynski J. Recent advances in testing for latent TB. Chest. 2010 Dec;138(6):1456–63.

TREATMENT	MISCELLANEOUS
Two agents with *in vitro* activity: tobramycin, clarithromycin, linezolid, imipenem, amikacin.	Usually responds well to treatment.
Usually with two agents: clarithromycin, ethambutol, rifampin.	Common in fresh and salt water.
Surgical debridement often combined with clarithromycin and rifampin.	Found in tropical rain forests

M. Gladwin, W. Trattler, and S. Mahan, *Clinical Microbiology Made Ridiculously Simple* ©MedMaster

BACTERIA WITHOUT CELL WALLS

CHAPTER 16. *MYCOPLASMA*

The Mycoplasmataceae are the tiniest free-living organisms capable of self-replication. They are smaller than some of the larger viruses. Mycoplasmataceae are unique bacteria because they lack a peptidoglycan cell wall. Their only protective layer is a cell membrane, which is packed with sterols (like cholesterol) to help shield their cell organelles from the exterior environment. Due to the lack of a rigid cell wall, Mycoplasmataceae can contort into a broad range of shapes, from round to oblong. They therefore cannot be classified as rods or cocci.

The lack of a cell wall explains the ineffectiveness of antibiotics that attack the cell wall (penicillin, cephalosporin), as well as the effectiveness of the antiribosomal antibiotics erythromycin and tetracycline.

There are 2 pathogenic species of Mycoplasmataceae, ***Mycoplasma pneumoniae*** and ***Ureaplasma urealyticum***.

Fig. 16-1. Mycoplasmataceae surrounded only by a cell membrane, padded with sterols. Penicillin and cephalosporin fail to tear down the cell membrane, while they successfully destroy the cell wall of a nearby gram-positive *Streptococcus*.

Mycoplasma pneumoniae

Mycoplasma pneumoniae causes a mild, self-limited bronchitis and pneumonia. It is the number one cause of bacterial bronchitis and pneumonia in teenagers and young adults. Following transmission via the respiratory route, this organism attaches to respiratory epithelial cells with the help of protein P1 (an adhesin virulence factor). After a 2–3 week incubation period, infected patients will have a gradual onset of fever, sore throat, malaise, and a persistent dry hacking cough. This is referred to as **walking pneumonia**, because clinically these patients do not feel very sick.

Chest X-ray reveals a streaky infiltrate, which usually looks worse than the clinical symptoms and physical exam suggest. Most symptoms resolve in a week, although the cough and infiltration (as seen on X-ray) may last up to 2 months. Although *Mycoplasma* is a bacterium, the nonproductive cough and the streaky infiltrate on the chest X-ray are more consistent with a viral (atypical) pneumonia (see Chapter 13, page 115).

For unclear reasons, up to 7% of patients infected with *Mycoplasma pneumonia* can develop **erythema multiforme** or **Stevens-Johnson syndrome**, a severe skin reaction characterized by erythematous vesicles and bullae over the mucocutaneous junctions of mouth, eyes and skin.

Diagnostic tests include:

1) **Cold agglutinins**: Patients infected with *Mycoplasma pneumonia* can develop monoclonal IgM antibodies directed at a common red blood cell antigen called the "I" antigen, which appears to be modified (making it antigenic) with infection. These antibodies bind to the red cells and cause them to agglutinate at 4 °C. These antibodies are thus called cold agglutinins. They develop by the first or second week of the *Mycoplasma pneumoniae* infection, peak 3 weeks after the onset of the illness, and slowly decline over a few months.

You can perform this simple test at the bedside. Put the patient's blood in a nonclotting tube. After placing this tube on ice, the blood will clump together if the patient has developed the cold agglutinin antibodies. Amazingly, when you lift the tube out of the ice, the clumped blood will unclump as it warms in the palm of your hand.

2) **Complement fixation test**: The patient's serum is mixed with glycolipid antigens prepared from *Mycoplasma*. A fourfold rise in antibody titer between acute and convalescent samples is diagnostic of a recent infection.

3) **Sputum culture**: Mycoplasmataceae (both *M. pneumoniae* and *U. urealyticum*) can be grown on artificial media. These media must be rich in cholesterol and contain nucleic acids (purines and pyrimidines). After 2–3 weeks, a tiny dome-shaped colony of *Mycoplasma* will assume a "fried-egg" appearance. Cultured colonies of *Mycoplasma pneumoniae*, the most significant human pathogen in this genus, do not form a halo. Its colonies have a round bumpy appearance likened to a **mulberry**.

4) **Mycoplasma DNA probe**: Sputum samples are mixed with a labeled recombinant DNA sequence homologous to that of the mycoplasma. The recombinant probe will label mycoplasma DNA if present.

5) Mycoplasma DNA can be detected in sputum samples by **polymerase chain reaction (PCR)**

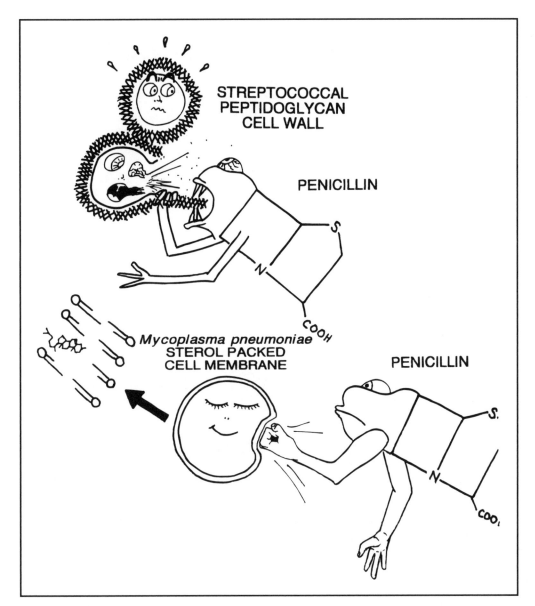

Figure 16-1

Upper respiratory tract infection need not be treated but if walking pneumonia develops, treatment will shorten the course and severity of the illness. Again, because the *Mycoplasma* have no cell wall, the β-lactam antibiotics do not work. The mainstays of treatment for *Mycoplasma pneumoniae* are the **macrolides** (azithromycin, clarithromycin), **tetracyclines** (doxycycline) and **quinolones** (ciprofloxacin, levofloxacin). We call these drugs **"atypical coverage"** since they cover the atypical bacteria *Mycoplasma, Legionella*, and *Chlamydia*, which in addition to viral pneumonia all cause atypical pneumonia (atypical pneumonia was so named because the penicillins did not work for these pneumonias).

Ureaplasma urealyticum
(T-strain *Mycoplasma*)

Hold on!!! Why isn't this second species of Mycoplasmataceae called *"Mycoplasma"*? The man who named this tiny organism didn't want you to ever forget that *Ureaplasma* loves swimming in urine and produces urease to break down urea (so it is "urea-lytic"!). It is sometimes referred to as a **T**-strain

CELL WALL-LESS BACTERIA	MORPHOLOGY	METABOLISM	VIRULENCE	TOXINS
Mycoplasma pneumoniae (Eaton's Agent)	1. **NO Cell Wall** 2. Pleomorphic: can appear round to oblong shaped. 3. Smallest bacteria capable of growth & reproduction outside a living cell (smaller than some viruses: .1–.2 microns) 4. Motile (glides)	1. Requires **CHOLESTEROL** for membrane formation 2. Facultative anaerobe	• Protein P1: adheres to epithelial cells of the respiratory tract	NONE
Ureaplasma urealyticum	1. **NO Cell Wall** 2. Pleomorphic	1. Requires cholesterol 2. Urease: metabolizes urea into ammonia and CO_2)		NONE

Figure 16-2 MYCOPLASMA

Mycoplasma, as it produces **T**iny colonies when cultured.

Ureaplasma urealyticum is part of the normal flora in 60% of healthy sexually active women and commonly infects the lower urinary tract, causing urethritis. Urethritis is characterized by burning on urination (dysuria) and sometimes a yellow mucoid discharge from the urethra. *Neisseria gonorrhoeae* and *Chlamydia trachomatis* are the other 2 bacteria that cause urethritis (see Chapter 13, page 114).

Ureaplasma urealyticum can be identified by its ability to metabolize urea into ammonia and carbon dioxide.

Fig. 16-2. Summary of the Mycoplasmataceae.

Recommended Review Articles:

Burstein GR, Zenilman JM. Nongonococcal urethritis—a new paradigm. Clin Infect Dis. 1999;28 Suppl 1:S66–73.

Loens K, Goossens H, Ieven M. Acute respiratory infection due to Mycoplasma pneumoniae: current status of diagnostic methods. Eur J Clin Microbiol Infect Dis. 2010;29(9):1055–69.

CLINICAL	TREATMENT	DIAGNOSTICS	MISCELLANEOUS

1. Tracheobronchitis
2. Walking pneumonia (also called atypical pneumonia): fever with a dry, non-productive hacking cough
	1. **Macrolides** (azithromycin, clarithromycin)
	2. **Tetracyclines** (doxycycline)
	3. **Quinolones** (ciprofloxacin, levofloxacin)
	
	• Penicillin and cephalosporins do **NOT** work, as *Mycoplasma* does not have a cell wall
		1. Cold agglutinins
		2. Complement fixation test
		3. Culture: Takes 2-3 weeks
		   A. Requires cholesterol and nucleic acids.
		   B. Add penicillin to inhibit growth of contaminating bacteria
		   C. Dome-shaped colonies with "fried egg" appearance or "mulberry" appearance (in the case of *Mycoplasma pneumoniae*)
		4. Rapid identification tests: sputum can be tested with DNA probes (nucleic acid hybridization). PCR of sputum samples.
			1. Chest X-ray will show patchy infiltrates that look worse than physical exam and clinical symptoms suggest
			2. Disease usually occurs in children, adolescents, and young adults

• Non-gonococcal urethritis: burning on urination, with a yellow mucoid discharge from the urethra	1. Erythromycin 2. Tetracycline	1. Requires cholesterol and urea for growth 2. Colonies are extremely tiny (thus called **T-strain**)	• T-Form *Mycoplasma* (T = Tiny)

ANTI-BACTERIAL MEDICATIONS

CHAPTER 17.   PENICILLIN-FAMILY ANTIBIOTICS

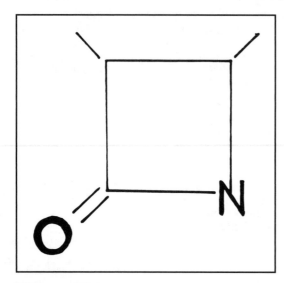

Figure 17-1

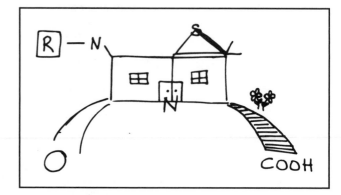

Figure 17-3

Fig. 17-3. Penicillin home. This looks like a house with a new room built on the side. Notice the funky antennae that run the groovy sound system. You will see later that changing the antennae, adding another antenna, or building a basement will create new types of penicillin with differing spectrums of activity and potencies.

Mechanism of Action

The penicillins don't just slow the growth of bacteria, they kill bacteria. They are therefore bacteri**cidal**.

You will recall (see Chapter 1) that both gram-positive and gram-negative bacteria possess peptidoglycans in their cell walls. These are composed of repeating disaccharide units cross-linked with amino-acids (peptides). The enzyme that catalyzes this linkage is called a **transpeptidase**.

The penicillin must evade the bacterial defenses and penetrate the outer cell-wall layers to the inner cytoplasmic membrane, where the transpeptidase enzymes are located. In gram-negative bugs, the penicillin must pass through channels known as **porins**. Then the **penicillin beta-lactam ring binds to and competitively inhibits the transpeptidase enzyme**. Cell wall synthesis is arrested, and the bacteria die. Because penicillin binds to transpeptidase, this enzyme is also called the **penicillin-binding protein**.

To be effective the beta-lactam penicillin must:

1) Penetrate the cell layers.
2) Keep its beta-lactam ring intact.
3) Bind to the transpeptidase (penicillin-binding protein).

Since its introduction during World War II, **penicillin** has provided a safe and effective treatment for a multitude of infections. Over time, many bacteria have designed ways to defeat penicillin. Fortunately, scientists have continued to develop new types of penicillins, as well as other antibiotics that are able to overcome most of the bacterial defenses.

Fig. 17-1.   This simple-looking box is a beta-lactam ring. All penicillin-family antibiotics have a beta-lactam ring. For this reason they are also called the **beta-lactam antibiotics**.

Fig. 17-2.   Penicillin has another ring fused to the beta-lactam ring.

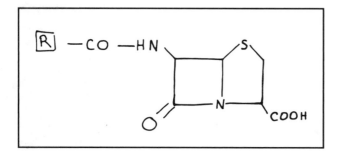

Figure 17-2

Resistance to Beta-Lactam Antibiotics

Bacteria defend themselves from the penicillin family in 4 ways. Gram-positive bacteria and gram-negative bacteria use different mechanisms:

1) One way that gram-negative bacteria defend themselves is by preventing the penicillin from penetrating the cell layers by altering the porins. Remember that gram-negative bacteria have an outer lipid bilayer around their peptidoglycan layer (see Chapter 1). The antibiotic must be the right size and charge to be able to sneak through the porin channels, and some penicillins cannot pass through this layer. Because gram-positive bacteria do not have this perimeter defense, this is not a defense that gram-positives use.

2) Both gram-positive and gram-negative bacteria can have beta-lactamase enzymes that cleave the C-N bond in the beta-lactam ring.

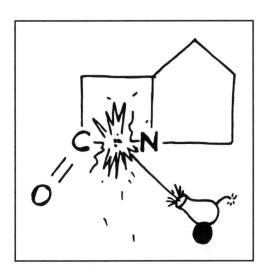

Figure 17-4

Fig. 17-4. Beta-lactamase enzyme (depicted here as a cannon) cleaves the C-N bond.

Gram-positive bacteria (like *Staphylococcus aureus*) **secrete** the beta-lactamase (called **penicillinase** in the secreted form) and thus try to intercept the antibiotic outside the peptidoglycan wall.

Gram-negative bacteria, which have beta-lactamase enzymes bound to their cytoplasmic membranes, destroy the beta-lactam penicillins locally in the periplasmic space.

3) Bacteria can alter the molecular structure of the transpeptidase so that the beta-lactam antibiotic will not be able to bind. Methicillin-resistant *Staphylococcus aureus* (MRSA) defends itself in this way, making it resistant to **ALL** of the penicillin family drugs.

4) Both gram positive and gram negative bacteria may also develop the ability to actively pump out the beta-lactam before it can bind to the transpeptidase enzyme. This is called an "efflux" pump.

Adverse Effects

All penicillins can cause anaphylactic (allergic) reactions. An acute allergic reaction may occur from minutes to hours and is IgE-mediated. Bronchospasm, urticaria (hives), and anaphylactic shock (loss of ability to maintain blood pressure) can occur. More commonly, a delayed rash appears several days to weeks later.

All of the penicillin family antibiotics can cause diarrhea by destroying the natural GI flora and allowing resistant pathogenic bacteria (such as *Clostridium difficile*) to grow in their place.

Types of Penicillin

There are 5 types:

1) **Penicillin G**: This is the original penicillin discovered by Fleming, who noted that the mold *Penicillium notatum* produced a chemical that inhibited *Staphylococcus aureus* Penicillin was first used in humans in 1941.

2) **Aminopenicillins**: These penicillins offer better coverage of gram-negative bacteria.

3) **Penicillinase-resistant penicillins**: This group is useful against beta-lactamase (an enzyme that destroys beta-lactam rings) producing *Staphylococcus aureus*.

4) **Anti-Pseudomonal penicillins** (including the carboxypenicillins, ureidopenicillins, and monobactams): This group offers even wider coverage against gram-negative bacteria (including *Pseudomonas aeruginosa*).

5) **Cephalosporins**: This is a widely used group of antibiotics that have a beta-lactam ring, are resistant to beta-lactamase, and cover a broad spectrum of gram-positive and gram-negative bacteria.

Many bacteria produce cephalosporinases, making them resistant to many of these drugs.

Penicillin G

Fig. 17-5. Penicillin G is the original G-man of the penicillins. There are oral dosage formulations of Penicillin G, but it is usually given intramuscularly (IM) or intravenously (IV). It is usually given in a crystalline form to increase its half-life.

Many organisms have now developed resistance to the old G-man because he is sensitive to beta-lactamase enzymes. But there are a few notable times when the G-man is still used:

1) Pneumonia caused by *Streptococcus pneumoniae*. (However, resistant strains are common.)

Penicillin V is an **oral form of penicillin**. It is acid *stable* in the stomach. It is commonly given for streptococcus pharyngitis caused by group A beta-hemolytic streptococcus since it can be taken orally.

Figure 17-5

Aminopenicillins
(Ampicillin and Amoxicillin)

These drugs have a **broader spectrum** than Penicillin G, hitting more gram-negative organisms. This enhanced gram-negative killing is attributable to better penetration through the outer membranes of gram-negative bacteria and better binding to the transpeptidase. However, like penicillin G, the aminopenicillins are still inhibited by penicillinase.

The gram-negative bacteria killed by these drugs include *Escherichia coli* and the other enterics (*Proteus, Salmonella, Shigella*, etc.). However, resistance has developed: 30% of *Haemophilus influenzae* and many of the enteric gram-negative bacteria have acquired penicillinase and are resistant.

Note that the aminopenicillins are one of the few drugs effective against the gram-positive enterococcus (see **Fig. 17-17**).

Both ampicillin and amoxicillin can be taken **orally**, but amoxicillin is more effectively absorbed orally so you will frequently use it for outpatient treatment of bronchitis, otitis media, and sinusitis. It is the drug of choice for infections caused by *Listeria*.

IV ampicillin is commonly used with other antibiotics such as the aminoglycosides (gentamicin) for broad gram-negative coverage. In the hospital you will become familiar with the "**Amp-gent**" combo! Patients with serious urinary tract infections are often infected with a gram-negative enteric or enterococcus. Amp-gent offers broad empiric coverage until cultures reveal the exact organism responsible.

Penicillinase-Resistant Penicillins

Methicillin, nafcillin, and oxacillin are penicillinase-resistant drugs that can kill *Staphylococcus aureus*. These are usually given IV.

Methicillin was highly efficacious against staphylococcal infections, but because of the occurrence of interstitial nephritis, its use has been discontinued in the United States. You will still hear its name used frequently in reference to sensitivity testing (e.g. Methicillin Resistant *Staphylococcus aureus*).

Figure 17-6

Fig. 17-6. This picture will help you remember the names of the IV beta-lactamase resistant penicillins: I **met** a **na**sty **ox** with a beta-lactam ring around its neck.

Nafcillin is the drug of choice for serious *Staphylococcus aureus* infections, such as cellulitis, endocarditis, and sepsis.

Fig. 17-7. The clocks (**clox**) were ticking. It was only a matter of time before the **oral beta-lactamase resistant** penicillins were discovered: **Clox**acillin and di**clox**acillin.

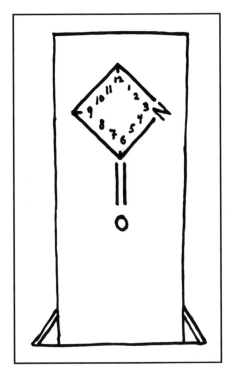

Figure 17-7 THE *CLOX* WERE TICKING

These drugs are not good against gram-negative organisms. They are used for gram-positive bacteria, especially those that produce penicillinase (*Staphylococcus aureus*).

When a patient has an infected skin wound (cellulitis, impetigo, etc.), you know he most likely has *Staphylococcus aureus* or group A beta-hemolytic streptococcus. Treating with Penicillin G, V, or ampicillin would not cover penicillinase-producing *Staphylococcus aureus*. Treating with one of these penicillinase-resistant agents will, and if you give him one of the oral agents he can go home on oral antibiotics. You won't have to take care of him around the **clock**!!!

Beware!!!! Methicillin resistant *Staphylococcus aureus* (**MRSA**) is now prevalent both in hospital acquired infections and infections acquired in the community (community acquired MRSA). The Penicillinase-resistant penicillins are still the drugs of choice for *Staphylococcus aureus* infections once MRSA has been excluded. (see Chapter 5)

Anti-Pseudomonal Penicillins
(Carboxypenicillins and Ureidopenicillins)

This group of penicillins has expanded gram-negative rod coverage, especially against the difficult-to-destroy *Pseudomonas aeruginosa*. They are also active against anaerobes (*Bacteroides fragilis*) and many gram positives.

Fig. 17-8. *Pseudomonas*, which can cause a devastating pneumonia and sepsis, is resistant to many antibiotics. It is so crafty and sneaky that we need James Bond to help with its elimination. Bond is fortunate to have three excellent weapons for his task. He has his pick of a **car** (with special weapons and gadgets), a specially trained **tick** that can home in on its target and suck out the life of the target, or a megaton **pipe** bomb:

Carboxypenicillins: **Tic**arcillin and **Car**benicillin
Ureidopenicillins: **Pip**eracillin and mezlocillin.

These drugs are sensitive to penicillinases, and thus most *Staphylococcus aureus* are resistant. Carbenicillin has certain disadvantages such as lower activity and thus the need for high dosages; high sodium load; platelet dysfunction; and hypokalemia. The parenteral form is currently not available for use in the United States. Replacement with ticarcillin or a ureidopenicillin has reduced these problems and provided antibacterial activity.

Beta-Lactamase Inhibitors
(Clavulanic Acid, Sulbactam, and Tazobactam)

These penicillanic acid derivatives are inhibitors of beta-lactamase. They can be given in combination with penicillins to create a beta-lactamase resistant combination:

Amoxicillin and clavulanic acid = Augmentin (trade name)
Ticarcillin and clavulanic acid = Timentin (trade name)
Ampicillin and sulbactam = Unasyn (trade name)
Piperacillin and tazobactam = Zosyn (trade name)

These drugs provide broad coverage against the beta-lactamase producing gram-positives (*Staphylococcus aureus*), gram-negatives (*Haemophilus influenza*), and anaerobes (*Bacteroides fragilis*).

The Cephalosporins

There are now more than 20 different kinds of cephalosporins. How do you become familiar with so many antibiotics? Do not fear. This chapter will teach you how to master these drugs!

Fig. 17-9. The cephalosporins have 2 advantages over the penicillins:

1) The addition of a new basement makes the beta-lactam ring much more resistant to beta-lactamases (but now susceptible to cephalosporinases!).

2) A new R-group side chain (another antenna-cable TV if you will) allows for double the manipulations in the lab. This leads to all kinds of drugs with different spectrums of activity.

There are 3 major generations of cephalosporins: first, second, and third. These divisions are based on their activity against gram-negative and gram-positive organisms.

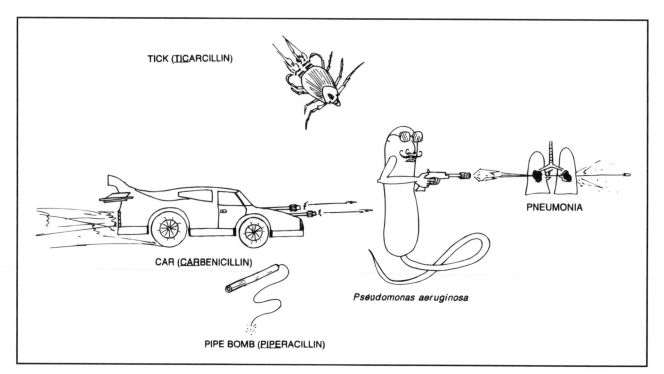

Figure 17-8

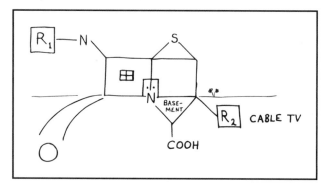

Figure 17-9

Fig. 17-10. With each new generation of cephalosporins, the drugs are able to kill an increasing spectrum of gram negative bacteria.

At the same time, the newer cephalosporins are less effective against the gram-positive organisms. The *Streptococci* and *Staphylococci* are most susceptible to first-generation cephalosporins.

Note that **MRSA** (**M**ethicillin **R**esistant **S**taphylococ-cus **aureus**) is resistant to all cephalosporins because it has changed the structure of its penicillin binding protein (transpeptidase). The Enterococci (including *Streptococcus faecalis*) are also resistant to cephalosporins.

Fig. 17-11. MRSA and the Enterococci are resistant to the cephalosporins.

A new cephalosporin has been classified as a *fourth-generation* antibiotic because it has great gram-negative coverage like the third generation but also has very good gram-positive coverage.

The Names! How do you remember which cephalosporin is in which group!!??!! The most important thing is to remember the trends and then you can look in any pocket reference book for the specific drugs in each group. However, it is nice to be familiar with the names of individual drugs, and exams often expect you to be able to recognize them. Here is an easy, although imperfect, way to learn many of them.

First-Generation

Almost all cephalosporins have the sound **cef** in their names, but the first generation cephalosporins are the only ones with a **PH**. To know the first-generation cephalosporins, you first must get a **PH.D.** in **PH**armacology.

ce**ph**alothin
ce**ph**apirin   (Exceptions: ce**faz**olin, cefadroxil)
ce**ph**radine
ce**ph**alexin

Ce**faz**olin is an important first-generation drug that doesn't have a PH. Don't let this **faz**e you! Cefazolin is also the only IV first generation cephalosporin available in the United States.

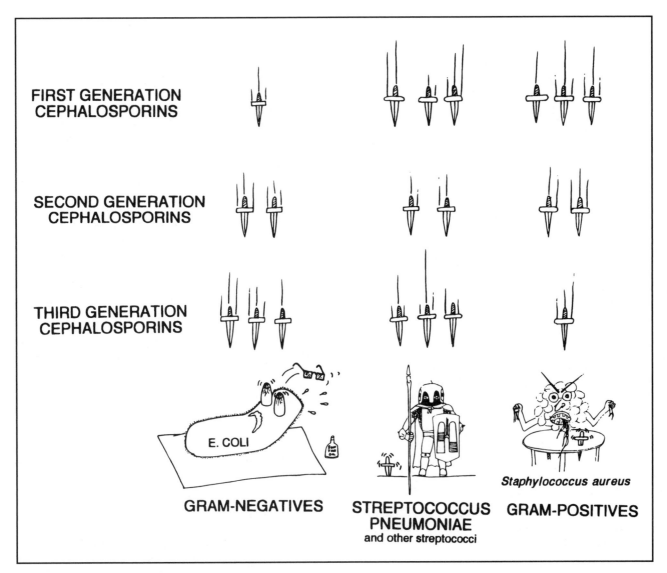

FIRST GENERATION CEPHALOSPORINS

SECOND GENERATION CEPHALOSPORINS

THIRD GENERATION CEPHALOSPORINS

E. COLI

GRAM-NEGATIVES

STREPTOCOCCUS PNEUMONIAE
and other streptococci

Staphylococcus aureus

GRAM-POSITIVES

Figure 17-10

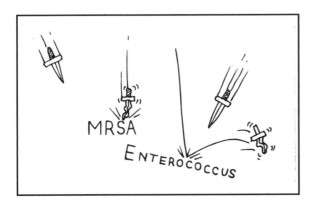

MRSA

ENTEROCOCCUS

Figure 17-11

Second-Generation

Fig. 17-12. Second-generation cephalosporins have **fam, fa, fur, fox,** or **tea,** in their names. After you get your **PH**.D., you would want to gather your **fam**ily to celebrate! The **FAM**ily is gathered, some wearing **FUR** coats, and your **FOX**y cousin is drinking **TEA** in a toast to your achievement.

cefamandole
cefaclor
cefuroxime   (Exceptions: cefmetazole, cefonicid, cefprozil)
cefoxitin
cefotetan (pronounced ce-fo-tea-tan)

165

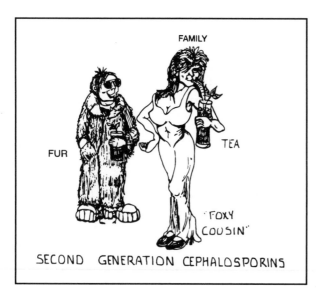

FAMILY

FUR

TEA

"FOXY COUSIN"

SECOND GENERATION CEPHALOSPORINS

Figure 17-12

Third-Generation

TRI for third (you know, triglycerides, etc.). Most of the third-generation cephalosporins have a **T** (for tri) in their names.

ceftriaxone
ceftazidime*   (Exceptions: cefixime, cefdinir,
                 cefpodoxime, cefditoren)
cefotaxime
ceftizoxime
ceftibuten

Note that cefotetan (**tea**) is a second generation drug.
*ceftazidime also comes co-formulated with a beta-lactamase inhibitor (avibactam)

Fourth-Generation

There is only one:
cefepime
and it is the only cephalosporin with a **fep** in its name. (**Cefpirome** is also within this class, but it is not available in the U.S.)

Fifth-Generation

Ceftaroline is the only cephalosporin with activity against MRSA (methicillin-resistant *staphylococcus aureus*). In late 2014, a newer, fifth-generation cephalosporin combined with a beta-lactamase inhibitor was introduced: ceftolozane-tazobactam.

Adverse Effects

Ten percent of patients who have allergic reactions to penicillin will also have a reaction to cephalosporins.

Such allergic reactions are the same as with penicillin: an acute IgE-mediated reaction or the more common rash, which usually appears weeks later.

"When do we use these antibiotics?"

1) **First-generation cephalosporins**: Recall **Fig. 17-10** showing the excellent gram-positive coverage. First-generation cephalosporins are used as alternatives to penicillin for staphylococcal and streptococcal infections when penicillin cannot be tolerated (allergy). Surgeons love to give these drugs **before surgery** to prevent infection from the skin.

2) **Second-generation cephalosporins**: This group covers more of the gram-negative rods than the first-generation cephalosporins. Cefuroxime has good coverage against both *Streptococcus pneumoniae* and *Haemophilus influenzae*. This makes it an ideal agent for community-acquired bacterial pneumonia when the sputum is negative and you don't know what the organism is. (*Streptococcus pneumoniae* and *Haemophilus influenzae* are common causes of community-acquired pneumonia.) Cefuroxime is also good for sinusitis and otitis media, which are often caused by *Haemophilus influenzae* or *Moraxella catarrhalis*.

Anaerobic coverage: Three second-generation cephalosporins cover anaerobic bacteria, such as *Bacteroides fragilis*. These can be used for intra-abdominal infections, aspiration pneumonias, and colorectal surgery prophylaxis, all of which involve anaerobic contamination from the GI tract. These 3 drugs are cefotetan, cefoxitin, and cefmetazole.

Fig. 17-13. Some of the second-generation cephalosporins kill anaerobic bacteria, such as *Bacteroides fragilis*. Study the picture of a **fox** (cefoxitin) who **met** (cefmetazole) an **anaerobic bug** for **tea** (cefotetan).

3) **Third-generation cephalosporins**: These are used primarily for inpatient treatment of community acquired pneumonia, meningitis, and pyelonephritis (urinary tract infection that has gone on to involve the kidneys).

The fourth generation cefepime is sometimes called an extended spectrum 3rd-generation cephalosporin. Think of him as the same but with a little added muscle against gram-positives and the terrible *Pseudomonas aeruginosa*.

Ceftazidime and cefepime are the only cephalosporins that are effective against *Pseudomonas aeruginosa*. So when you encounter the "impossible-to-kill" *Pseudomonas*: Give it the **Taz**, the **Fop**, and the **Fep**!

Ceftriaxone and cefotaxime both have excellent CSF penetration and cover the common bacteria that frequently cause meningitis. Ceftriaxone is the drug of choice in adults with meningitis, while cefotaxime is the first-line drug in neonates and children. (Ceftriaxone may interfere with bilirubin metabolism in neonates,

SECOND GENERATION ANAEROBIC-COVERING
CEPHALOSPORINS

Figure 17.-13

hence the choice of cefotaxime). Ceftriaxone is also given IM for gonorrhea, as more *Neisseria gonorrhoea* have become resistant to penicillin and tetracycline.

Ceftaroline, a 5th generation player, is indicated for community acquired pneumonia and skin infections, but its real beauty is its activity against MRSA. Ceftolozane-tazobactam has improved gram negative and anaerobic coverage due to the addition of the beta-lactamase inhibitor making it ideal for complicated abdominal infections.

Carbapenems

The carbapenems (**imipenem, meropenem, doripenem, and ertapenem**) constitute one of the newer classes of antibiotics and they have some of the broadest coverage! You will see the members of this class used with increasing frequency as bacterial resistance to the earlier generation antibiotics increases. Carbapenems are resistant to beta-lactamases, including the newer Extended Spectrum Beta-Lactamases (referred to in the clinic as the **ESBL's**). Begin by learning about **imipenem** . . .

Tell yourself that you are a pen. Read: "**I'm a pen**."

Now picture the pen crossing out all the bacteria that are difficult to treat. The pen (imipenem) can terminate almost all of them.

Imipenem has the broadest antibacterial activity of any antibiotic known to man!!! It kills gram-negatives, gram-positives, and anaerobes (even tough guys like *Pseudomonas aeruginosa*). Some bacteria that

are still resistant to this drug include our enemy MRSA, some *Pseudomonas* species, and bacteria without peptidoglycan cell walls (*Mycoplasma*).

Imipenem is stable to beta-lactamases. Because it is very small, it can pass through porin channels to the periplasmic space. There it can interact with transpeptidase in a similar fashion as the penicillins and cephalosporins. Unfortunately, with heavy use of this antibiotic some bacterial strains have developed new enzymes that can hydrolyze imipenem, and some gram-negative bacteria have squeezed down their porin channels to prevent its penetration.

The normal kidney has a dihydropeptidase that breaks imipenem down, so a selective enzyme inhibitor of this dihydro peptidase is given with imipenem. The inhibitor is **cilastatin**.

Carbapenems can cause allergic reactions similar to those of penicillin, with about about a 10% cross reactivity. Imipenem is notable for making seizures more likely by lowering one's seizure threshold. (Avoid this drug in patients with a history of seizures, meningitis, prior strokes, or evidence of brain masses).

Clinical note: On the wards imipenem is called a "decerebrate antibiotic" because you don't have to think about what bacteria it covers. It covers almost everything!!!

Meropenem and the lastest newcomer, **doripenem**, are carbapenems that are as powerful as **imipenem**. In general, they can be used interchangeably. Meropenem and doripenem are stable against dihydropeptidase, so cilastatin is not needed. Meropenem and doripenem have less potential for causing seizures relative to imipenem.

Ertapenem, a newer carbanem, has the advantage of requiring only once daily I.V. administration. It has become the drug of choice for the empiric coverage of severe diabetic foot infections (usually polymicrobic). Ertapenem differs from the other carbapenems in that it does not cover *Pseudomonas aeruginosa*.

Warning!!! Even these super antibiotics can be overcome. Newly identified carbapenemase producing bacteria are being found. Many of these very resistant organisms cannot be cured! More on this later...

Aztreonam

Aztreonam is a magic bullet for gram-negative aerobic bacteria!!! It is a beta-lactam antibiotic, but it is different in that it is a **mono**bactam. It only has the beta-lactam ring, with side groups attached to the ring. It does not bind to the transpeptidases of gram-positive or anaerobic bacteria, **only** to the transpeptidase of **gram-negative bacteria**.

Fig. 17-14. A TREE (AzTREonam) has fallen through the center of our house, leaving only the square portion (beta-lactam ring) standing, and letting all the air in (aerobic). You can imagine if this happened to

Figure 17-14

1. Antipseudomonal penicillins
   A. Ticarcillin
   B. Timentin (ticarcillin & clavulanate)
   C. Piperacillin
   D. Zosyn (piperacillin & tazobactam)
   E. Carbenicillin (no longer produced in U.S.)

2. Third generation cephalosporins
   A. Ceftazidime
   B. Cefoperazone (no longer produced in U.S.)

3. Fourth generation cephalosporins
   A. Cefepime

4. Carbapenems
   A. Imipenem
   B. Meropenem
   C. Doripenem

5. Aztreonam

6. Ciprofloxacin

7. Aminoglycosides
   A. Amikacin
   B. Gentamicin
   C. Tobramycin

8. Polymixins

Figure 17-15   ANTIBIOTICS THAT COVER *PSEUDOMONAS AERUGINOSA*

1. Penicillins with beta-lactamase inhibitor
   A. Augmentin (Amoxicillin & clavulanate)
   B. Timentin (Ticarcillin & clavulanate)
   C. Unasyn (Ampicillin & sulbactam)
   D. Zosyn (Pipericillin & tazobactam)

2. Second generation cephalosporins
   A. Cefoxitin
   B. Cefotetan
   C. Cefmetazole

3. Imipenem, Meropenem, Doripenem, and Ertapenem

4. Chloramphenicol

5. Clindamycin

6. Metronidazole

7. Moxifloxacin

8. Tigecycline

Figure 17-16   ANTIBIOTICS THAT COVER THE ANAEROBES (INCLUDING *BACTEROIDES FRAGILIS*)

Methicillin-resistant *Staphylococcus aureus* (MRSA)	1. Vancomycin 2. Linezolid 3. Daptomycin 4. Quinupristin/dalfopristin 5. Tigecycline 6. Ceftaroline
Methicillin-resistant *Staphylococcus epidermidis*	
Vancomycin-resistant *Enterococci* (VRE)	1. Linezolid 2. Daptomycin 3. Tigecycline

Figure 17-17   ANTIBIOTICS THAT COVER THE DIFFICULT-TO-KILL GRAM-POSITIVE BACTERIA

your house it would be a **negative** (gram) experience. Aztreonam kills gram-negative aerobic bacteria.

Aztreonam kills the tough hospital-acquired, multidrug resistant, gram-negative bacteria, including *Pseudomonas aeruginosa.*

Data suggest there is little cross-reactivity with the bicyclic beta-lactams, so we can use this in penicillin-allergic patients!

Clinical notes: Because this antibiotic only kills gram-negative bugs, it is used (much like the aminoglycosides) along with an antibiotic that covers gram-positives. The resulting combinations give powerful broad-spectrum coverage:

vancomycin + aztreonam
clindamycin + aztreonam

Fig. 17-15. Antibiotics that cover *Pseudomonas aeruginosa.*

Fig. 17-16. Antibiotics that cover anaerobic bacteria, including *Bacteroides fragilis.*

Fig. 17-17. Antibiotics that cover the difficult-to-kill gram-positive bacteria: *methicillin*-resistant *Staphylococcus aureus* (MRSA), methicillin-resistant *Staphylococcus epidermidis* and Vancomycin-Resistant *Enterococci* (VRE).

Fig. 17-18. Summary of the penicillin (beta-lactam) family antibiotics.

References

Fish DN, Singletary TJ. Meropenem, a new carbapenem antibiotic. Pharmacotherapy 1997; 17:644–669.

Fraser KL, Grossman RF. What new antibiotics to offer in the outpatient setting. Sem Resp Infect 1998; 13:24–35.

Gilbert DN, Moellering RC, Eliopoulos GM, Sande MA. The Sanford Guide to Antimicrobial Therapy 2008. 38th edition. Antimicrobial Therapy Inc. Sperryville VA, 2008.

Mandell GL, Bennett JE, Dollin R, eds. Principles and Practice of Infectious Diseases; 4th edition. New York: Livingstone 1995.

Owens RC, Nightingale CH, et al. Ceftibuten: An overview. Pharmacotherapy 1997; 17:707–720.

Rockefeller University Workshop. Special report: multiple-antibiotic-resistant pathogenic bacteria. N Engl J Med 1994; 330:1247–1251.

Recommended Review Articles:

Bush K, Macielag MJ. New ß-lactam antibiotics and ß-lactamase inhibitors. Expert Opin Ther Pat. 2010;20(10):1277–93.

NAME	MECHANISM OF ACTION	PHARMOKINETICS
		PENICILLINS
Penicillin G Aqueous (crystalline) penicillin G Procaine penicillin G Benzathine penicillin G	• Competitive inhibitor of the transpeptidase enzyme; Inhibits bacterial cell wall synthesis	• Can not survive passage through the stomach • Aqueous penicillin G: intravenous (IV) • Procaine and benzathine penicillin G: intramuscular (IM) • PO penicillin G.
Penicillin V	• Same	• Oral
Amino penicillins Ampicillin Amoxicillin	• Same	1. Ampicillin: IV or oral 2. Amoxicillin: Oral (better oral absorption than ampicillin)
Penicillinase-resistant penicillins (IV) Methicillin Nafcillin Oxacillin	• Same	IV
Penicillinase-resistant penicillins (Oral) Cloxacillin Dicloxacillin	• Same	• Oral
Antipseudomonal penicillins Carbenicillin Ticarcillin Piperacillin	• Same	1. IV-Ticarcillin and Piperacillin 2. PO-Carbenicillin (rarely used)
Combination of penicillin with beta-lactamase inhibitors: Amoxicillin + clavulanate (Augmentin) Ticarcillin + clavulanate (Timentin) Ampicillin + sulfbactam (Unasyn) Piperacillin + tazobactam (Zosyn)	• Same	1. Augmentin: oral 2. Unasyn, Timentin, and Zosyn: IV.
	CEPHALOSPORINS	

First Generation
1. Cephalothin
2. Cephapirin
3. Cephradine
4. Cephalexin
5. Cefazolin
6. Cefadroxil
	• Competitive inhibitor of the transpeptidase enzyme; Inhibits bacterial cell wall synthesis	1. Oral
		   a. Cephalexin
		   b. Cefadroxil
		2. IV
		   a. Cephalothin
		   b. Cephapirin
		   c. Cefazolin
		3. Oral or IV: Cephradine

Figure 17-18   PENICILLIN FAMILY ANTIBIOTICS (chart continued on page 172)

ADVERSE EFFECTS	THERAPEUTIC USES	MISCELLANEOUS

1. Allergy (due to presence of preformed IgE)
  A. Anaphylactic shock
  B. Urticaria (Hives)
  C. Rash
2. Delayed rash 1–2 weeks later
3. Superinfections: *Clostridium difficile* can overrun the colon, causing pseudomembranous enterocolitis
	1. *Streptococci pneumoniae*
	2. Group A beta-hemolytic streptococci (*Streptococcus pyogenes*)
	3. *Neisseria meningitidis*
	4. *Treponema pallidum* (syphilis)
	5. *Pasteurella multocida*
	6. *Listeria monocytogenes*
	7. *Actinomyces israelii*
		• Bacteria "cidal"
		• *Strategies of bacteria resistance*:
		  1. Prevent entrance of penicillin
		  2. Enzymatically cleave the beta lactam ring (with a beta lactamase enzyme)
		  3. Alter the structure of the transpeptidase enzyme

	1. Strep throat caused by group A beta-hemolytic streptococci (*Streptococcus pyogenes*) 2. Covers all organisms that penicillin G does	
	1. Broader gram-negative coverage than the above penicillins 2. Covers the enterococci (group D streptococci)	
	• Used for skin infections when penicillinase-producing *Staphylococcus aureus* is a possible pathogen	• **MET** a **NA**sty **OX**
	• Used for skin infections when penicillinase-producing *Staphylococcus aureus* is a possible pathogen	• The clocks (**clox**) were ticking

	1. Use when *Pseudomonas aeruginosa* is a possible pathogen
	2. Anaerobic coverage
		• James Bond's weapons:
		  1. **Car**
		  2. **Tick**
		  3. **Pipe** bomb

	1. Very broad coverage; can be used with hospital-acquired pneumonias 2. Anaerobic coverage 3. Timentin and Zosyn cover *Pseudomonas*	

1. Allergy (due to presence of preformed IgE)
  A. Anaphylactic shock
  B. Urticaria (Hives)
  C. Rash
2. Delayed rash 1–2 weeks later
  • Note: 5–10% of patients with allergy to penicillin will also
	1. Excellent gram-positive bacteria coverage
	2. Excellent for skin infections
		1. *Strategies of bacteria resistance*: cleaving the beta lactam ring (with a beta lactamase enzyme)
		2. The first generation cephalosporins are the only ones with a **PH** in their name

NAME	MECHANISM OF ACTION	PHARMOKINETICS
First Generation (continued)		4. Renal excretion
Second Generation 1. Cefamandole 2. Cefaclor 3. Cefuroxime 4. Cefoxitin 5. Cefotetan 6. Cefmetazole 7. Cefonicid 8. Cefprozil	• Same	1. Oral: Cefaclor, Cefprozil 2. Oral or IV: Cefuroxime 3. IV: the rest 4. Renal excretion

Third Generation
1. Ceftriaxone
2. Ceftazidime (+/− avibactam)
3. Cefotaxime
4. Ceftizoxime
5. Cefixime
6. Cefoperazone
7. Cefpodoxime
8. Ceftibuten
9. Cefepime
   (a fourth-generation)
	• Same	1. Oral:
		   a. Cefixime
		   b. Cefpodoxime
		   c. Ceftibuten
		   d. Cefdinir
		   e. Cefditoren
		2. IV: the rest
		3. Renal excretion

Fourth Generation Cefepime	• Same	1. Intravenous 2. Renal excretion
Fifth Generation 1. Ceftaroline 2. Ceftolozane-tazobactam	• Same	1. Intravenous 2. Renal excretion

Carbapenems
Imipenem (co-formulated with cilastatin)
Meropenem
Doripenem
Ertapenem
	1. Imipenem: inhibits bacterial cell wall synthesis
	2. Cilastatin:
	   A. Inhibits an enzyme in the kidneys that metabolizes imipenem (thus increasing its half life)
	   B. Protects the kidney from toxicity caused by imipenem
		1. IV or IM
		2. Renal excretion

Aztreonam (a monobactam)	• Inhibits bacterial cell wall synthesis	1. IV or IM 2. Renal excretion

Figure 17-18   (continued)

ADVERSE EFFECTS	THERAPEUTIC USES	MISCELLANEOUS
have a reaction to cephalosporins 3. Superinfections: *Clostridium difficile* can overrun the colon, causing pseudomembranous enterocolitis		

1. Allergy
2. Superinfection
3. Cephalosporins with the methyl-thio-tetrazole (MTT) side chain (includes **cefamandole, cefmetazole, & cefotetan**):
  A. Interferes with the synthesis of vitamin K dependent clotting factors, resulting in poor coagulation
  B. May interfere with the metabolism of alcohol, resulting in the accumulation of acetaldehyde, which causes nausea and vomiting
	1. Covers more gram-negatives than the first generation
	2. Cefotetan, cefoxitin and cefmetazole: anaerobic coverage
		• The **FAM**ily is gathered, some wearing **FUR** coats, and your **FOX**y cousin is drinking **TEA** in a toast to your achievement

1. Allergy 2. Superinfection 3. Cephalosporins with the methyl-thio-tetrazole (MTT) side chain (as mentioned above)	1. Ceftazidime, cefoperazone, and cefepime have antipseudomonal activity 2. Ceftriaxone has excellent penetration into the cerebrospinal fluid. So excellent choice for meningitis	• Most of the third generation cephalosporins have a 'T' (for tri) in their names
1. Allergy 2. Superinfection	1. Health care associated pneumonia 2. Neutropenic fever 3. Pseudomonal infections	• Poor anaerobic coverage
1. Allergy 2. Superinfection	Ceftaroline- MRSA, skin and soft tissue infections, pneumonia Ceftolozane-tazobactram- Abdominal infections* and complicated urinary tract infections	• *often combined with metronidazole
1. Nausea/vomiting (when infused rapidly) 2. Individuals allergic to penicillin are at high risk to be allergic to imipenem 3. Seizures	*Broad spectrum* 1. Gram-positives 2. Gram-negatives (Ertapenem does not cover *Pseudomonas*) 3. Anaerobes 4. Does not cover methicillin-resistant *Staphylococcus aureus* (MRSA)	• I'm a pen crossing out all bacteria "decerebrate antibiotic"
• Minimal cross-reactivity with penicillins	• Gram-negative organisms only	• Magic bullet for gram-negatives

M. Gladwin, W. Trattler, and S. Mahan, *Clinical Microbiology Made Ridiculously Simple* ©MedMaster

CHAPTER 18. ANTI-RIBOSOMAL ANTIBIOTICS

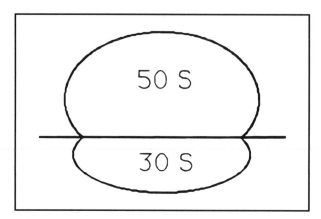

Figure 18-1

Figure 18-2

All cells depend on the continued production of proteins for growth and survival. Translation of mRNA into the polypeptides that make up these proteins requires the use of ribosomes. Antibiotics that inhibit ribosomal action would thus inhibit cellular growth and survival. Since we only want to inhibit the growth of pathogenic bacterial cells during an infection and not our own cells, we are fortunate that bacteria actually have a different type of ribosome than we do. We can exploit this difference by specifically inhibiting the ribosomes of bacteria, while sparing the function of our own ribosomes. Bacterial ribosomes are smaller than ours. While we have an 80S particle, the bacterial ribosome consists of a 70S particle that has 2 subunits: the 50S (large) and the 30S (small). (Surprisingly, 50S + 30S = 70S)

Fig. 18-1. The bacterial ribosome.

There are 5 important types of antibiotics that inhibit the function of the bacterial ribosome. Three of them inhibit the large 50S subunit, and the other two inhibit the small 30S subunit.

Here's how you can remember these 5 drugs:

Fig. 18-2. Convert the ribosome to home plate and picture a baseball player sliding into home. The ball is fielded by the catcher, who makes a **CLEan TAG** and the player is out!!! Here is what **CLEan TAG** helps you remember:

C for **Ch**loramphenicol and **C**lindamycin
L for **L**inezolid

E for **E**rythromycin
T for **T**etracycline and **T**igecycline
AG for **A**mino**g**lycosides

Note that the word **CLEan** lies over the base and the word **TAG** beneath the base. This corresponds to the ribosomal subunit that these drugs inhibit: CLEan inhibits the 50S; TAG inhibits the 30S.

Fig. 18-3. To remember which of these are orally absorbed, we have drawn boxes around the CLEan TAG on the ribosome. Notice that the boxes do not extend around the aminoglycosides (AG). We now draw a cake with one-fourth missing—the same quadrant that is missing above. You can eat three quarters of the cake. The fourth piece (representing the aminoglycoside quadrant) is missing, as this is the one anti-ribosomal antibiotic that cannot be absorbed orally. The aminoglycoside must be given IM or IV for systemic treatment of infections. (Tigecycline, the newest tetracycline derivative, is also only available in IV formulation).

Chloramphenicol
(The "Chlorine")

This drug has an amazing spectrum of activity. It is one of the few drugs (like Imipenem) that **kills most clinically important bacteria**. It is like pouring "chlorine" on the organisms. Gram-positive, gram-negative, and **even anaerobic bacteria** are susceptible. It is one of the handful of drugs that can kill the anaerobic *Bacteroides fragilis*.

Clinical Uses

Because of its rare but severe side effects, this otherwise excellent drug is used only when there is no alternate antibiotic, and thus the benefits far outweigh the risks:

1) It is used to treat bacterial meningitis, when the organism is not yet known and the patient has severe allergies to the penicillins, including the cephalosporins. The wide spectrum of activity of chloramphenicol and excellent penetration into the CSF will protect this patient from the devastating consequences of meningitis.

2) Young children and pregnant women who have Rocky Mountain spotted fever cannot be treated with tetracycline due to the side effects of tetracycline discussed on page 178. Chloramphenicol then becomes the drug of choice.

Note: In under-developed countries this drug is **widely used**. It only costs pennies and covers everything. Third world nations do not have the luxury of expensive alternative drugs available in the U.S.

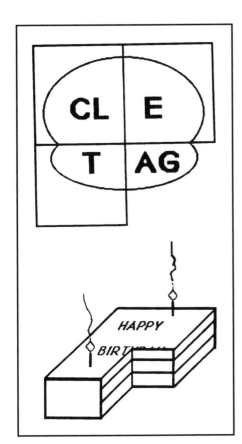

Figure 18-3

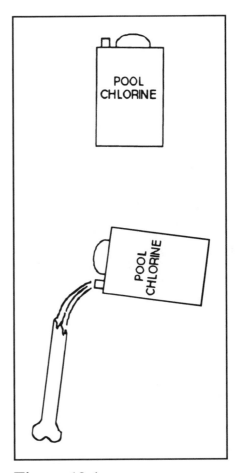

Figure 18-4

Adverse Effects

Fig. 18-4.   Picture a can of chloramphenicol chlorine. Now picture the chlorine being poured down the shaft of a long bone. You can well imagine that the bone marrow would dissolve. This drug is famous for 2 types of bone marrow depression. The first is dose-related and reversible, and often only causes an anemia. The second type wipes out the bone marrow irreversibly and is usually fatal. This is called **aplastic anemia**. Aplastic anemia caused by chloramphenicol is extremely rare, occurring in only 1:24,000 to 1:40,000 recipients of the drug.

Fig. 18-5.   Now picture a baby who leaps into a freshly chlorinated pool. The baby crawls out of the pool, and the chlorine has turned the baby's skin gray (**Gray Baby Syndrome**). Neonates, especially preemies, are unable to fully conjugate chloramphenicol in the liver or excrete it through the kidney, resulting in very high blood levels. Toxicity occurs with vasomotor collapse (shock), abdominal distention, and cyanosis, which appears as an ashen gray color.

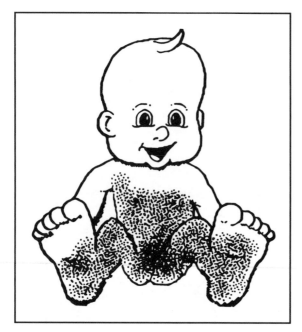

Figure 18-5

Clindamycin

Clinical Uses

This drug is NOT useful against gram-negative bugs. So what is it good for? Many gram-positive bugs are inhibited. So what? What else?!!!?

Anaerobic infections! This is another of the rare handful of antibiotics that cover anaerobes (including *Bacteroides fragilis*). Surgeons use clindamycin along with an aminoglycoside for penetrating wound infections of the abdomen, which may occur with bullet and knife trauma. When the GI tract is perforated, it releases its contents of gram-negative and anaerobic bugs into the sterile peritoneal cavity. The aminoglycosides cover the aerobic gram-negative organisms, and clindamycin covers the anaerobes.

Clindamycin is also used for infections of the female genital tract, such as septic abortions, as there are a lot of anaerobes there. Oral preparations of clindamycin and vaginal cream are alternatives to metronidazole for the treatment of bacterial vaginosis. Topical clindamycin solution is also useful in the treatment of acne vulgaris and rosacea (adult acne).

Clindamycin is often combined with a beta-lactam (such as penicillin) or vancomycin for the treatment of toxic shock syndrome associated with Group A streptococcus or *Staphylococcus aureus*. By inhibiting the ribosome and protein translation, clindamycin is able to turn off toxin production and thereby lessen the clinical severity of this life threatening condition.

Clindamycin is also becoming the go-to drug in uncomplicated skin and soft tissue infections where there is a high suspicion for *Staphylococcus aureus* or group A streptococcal infection. Most of the community-acquired skin and soft tissue infections we are seeing currently are caused by CA-MRSA and are usually susceptible to clindamycin.

Clindamycin is often given orally for months to outpatients who develop an anaerobic aspiration pneumonia. Alcoholics who have a seizure and aspirate or individuals with very poor dentition who aspirate can develop a polymicrobial lung abscess. This abscess slowly eats away at the lung to produce a lung cavity. These patients can present with weight loss, chronic low-grade fevers, night sweats and expectoration of foul smelling sputum. **Clinical pearl:** The lung is the only organ that can contain an abscess that does not need surgical drainage. Chronic treatment with Clindamycin or another agent that covers anaerobic bacteria (see **Fig. 17-16**) will do the trick.

Adverse Effects

You **must** know this: Clindamycin can cause **Pseudomembranous Colitis!!!!!**

When you give a patient clindamycin, or another potent antibiotic for that matter, it will destroy the natural flora of the GI tract. *Clostridium difficile*, if resistant to clindamycin, will grow like crazy and secrete its exotoxin in the colon. This exotoxin causes epithelial cell death and colonic ulcerations that are covered with an exudative membrane; thus the name pseudomembranous colitis. These patients often present with a severe diarrhea. Stool cultures yielding *Clostridium difficile* or titers of toxin found in the stool can help establish a diagnosis.

Note: While clindamycin was first identified as the cause of pseudomembranous colitis, it is noteworthy that other antibiotics also cause this condition. In fact **most cases are now caused by the penicillin family drugs** because they are prescribed more frequently.

To treat pseudomembranous colitis, you must give oral vancomycin or metronidazole. Vancomycin passes through the GI tract without being absorbed and is therefore highly concentrated upon reaching the colon. The high concentration can overwhelm and kill *Clostridium difficile*. Metronidazole is less expensive and is now the preferred agent, because use of oral vancomycin may contribute to vancomycin resistant enterococcus!

A highly toxigenic strain of *Clostridium difficile* was identified in 2005 with confirmed cases in Canada and several US states. This strain produces excessive amounts of toxin and has led to higher rates of morbidity and mortality than in the past. Several patients

Figure 18-6

developed disease without any identifiable risk factors (recent hospitalization, recent antibiotics, etc.).

Fig. 18-6. Visualize a VAN (vancomycin) and a METRO (metronidazole) cruising down the GI tract. They run over the ulcerative potholes of pseudomembranous colitis and kill the offending *Clostridium difficile.*

Linezolid and Tedizolid
(Oxazolidinones)

Clinical Uses

Linezolid and Tedizolid are used to stamp out resistant gram positive bugs. Imagine Linezolid, the Godzilla Lizard, wreaking havoc on the unsuspecting bugs. Both block the 50S ribosomal subunit and thus have activity against gram positive organisms, including those resistant to other antimicrobials. They are often used for infections due to *Staphylococcus aureus* (including MRSA) and against Vancomycin resistant *enterococcus* (VRE). Linezolid has a broad range of uses including healthcare associated pneumonia and skin and soft tissue infections. Linezolid and Tedizolid are available in IV and oral formulations, making them an attractive option for home therapy.

Adverse Effects

Oxazolidinones as a class can cause bone marrow suppression, including thrombocytopenia, anemia and neutropenia. They can also precipitate symptoms of serotonin syndrome (a potentially life-threatening drug reaction) if used with SSRI antidepressants or MAO inhibitors. **Avoid use in patients on antidepressants**.

Macrolides and Ketolide

This class of antibiotics which includes **erythromycin** ("A Wreath"), **azithromycin, clarithromycin**, and a newer agent, **telithromycin** (actually a ketolide), all inhibit bacterial ribosomal function at the 50S subunit. In general this class of medications is well tolerated and has a broad spectrum of activity against gram positive, some gram negatives, and "atypical" bacterial pathogens such as *Legionella, Chlamydia pneumoniae*, and *Mycoplasma*.

Erythromycin, previously the drug of choice for outpatient treatment of community acquired pneumonia, has largely been replaced in this role by azithromycin, clarithromycin, and now telithromycin, all of which have excellent oral absorption with fewer side effects.

Erythromycin, azithromycin, clarithromycin are all commonly used as second line agents for skin and soft tissue infections not due to Methicillin Resistant *Staphylococcus aureus* (MRSA). They are also frequently used for upper respiratory tract infections (sinusitis, otitis media, and bronchitis) and for coverage of "atypical" bacterial pathogens such as *Legionella, Chlamydia pneumoniae*, and *Mycoplasma*, that are associated with community acquired pneumonia. Azithromycin is marketed in a "Z-pack" which you will undoubtably see if you have any patient contact! *Streptococcus pneumoniae* is becoming more resistant to the first and second generation macrolides (erythromycin, azithromycin, and clarithromycin). Azithromycin has also been used as an alternative for the treatment of syphilis, but resistance is quickly emerging. Azithromycin and clarithromycin are both used in the treatment and prevention of atypical mycobacterial infections.

Telithromycin, a newer macrolide, has a similar spectrum of action as the older macrolides, but it has efficacy against most macrolide resistant *Streptococcus pneumoniae* species. Avoid Telithromycin in persons with **myasthenia gravis**, as they can develop acute respiratory failure. It recently received a "black box warning" due to this complication. Imagine a man with droopy eyelids due to **myasthenia gravis**. His eyes are so droopy he is unable to "**TELL IF** you **THRO**w a ball at him". The fear of you hurting him "**takes his breath away**". (mnemonic courtesy of Dr. Michael Waring)

Adverse Effects

Macrolides are among the safest antibiotics; think of a pretty wreath (erythromycin) compared to the nasty chlorine (chloramphenicol). The few side effects include:

1) Common and dose-dependent abdominal pain (GI irritation) resulting from intestinal peristalsis. Erythromycin is the worst culprit. In fact, erythromycin is intentionally prescribed for patients with gastric dysmotility (such as diabetic gastroparesis) due to its ability to induce peristalsis.

2) Rare cholestatic hepatitis. Imagine a wreath slipping into the bile duct and blocking flow.

Figure 18-7

3) Prolonged QT syndrome (primarily with erythromycin). You must be careful prescribing this medication to patients with a history of arrhythmias.

Fig. 18-7. Erythromycin was previously the drug of choice for Legionnaires' disease. (It has since been replaced by azithromycin or a new generation fluoroquinolone). The heroic French foreign legionnaire has died in a desert battle. In his honor, **a wreath** is laid by his grave. Notice the tomb stone is in the shape of a cross to help you remember that erythromycin covers gram-positive organisms (and don't forget atypicals! Tombstone courtesy of Dr. Cornejo, U. of Colorado).

Tetracycline/Doxycycline
("The Tet Offensive")

Tetracycline chelates with cations in milk and milk products, aluminum hydroxide, Ca++, and Mg++. When it is chelated, it will pass through the intestine without being absorbed. **Doxycycline** is a tetracycline that chelates cations poorly and is thus better absorbed with food. IV **tetracycline** is no longer available.

Clinical Uses of Doxycycline

This drug is used for all the diseases you would expect a young soldier in the **Tet offensive** to get by crawling around in the jungle and mingling with prostitutes on leave:

1) Venereal diseases caused by *Chlamydia trachomatis*.

2) Walking pneumonia caused by *Mycoplasma pneumoniae* (used as an alternative to erythromycin).

3) Animal and tick-borne diseases caused by *Brucella* and *Rickettsia* (see the ticks on the soldier's pants in **Fig. 18-8**).

4) Doxycycline also works wonders for acne.

Adverse Effects

Fig. 18-8. Picture a Vietcong soldier involved in the **Tet offensive** to help remember these important side effects:

1) This soldier is naturally very nervous as the Tet offensive involved waves of soldiers running into 20th century American fire power. So he has **GI irritation** with nausea, vomiting, and diarrhea. This is a common side effect.

2) A grenade has blown up near him, burning his skin like a sunburn. Notice the rays of light going from the explosion to his face. **Phototoxic dermatitis** is a skin inflammation on exposure to sunlight.

3) Shrapnel has struck his kidney and liver: **renal and hepatic toxicity**. These adverse effects are rare and usually occur in pregnant women receiving high doses by the intravenous route.

4) Note the dark **discolored teeth** of the soldier. This drug will chelate to the calcium in the teeth and bones of babies and children under age 7, resulting in brown teeth and **depressed bone growth**. Don't give the drug to pregnant women or their baby's teeth will look like those of the soldier.

TIGECYCLINE

This tetracycline derivative, is the first of a new class of antibiotics, the glycyclines, with a similar broad spectrum of activity as the tetracyclines, but a reduced propensity to induce resistance. Tigecycline is only available in I.V. formulation and is currently marketed for use in complicated skin and soft tissue infections and for the empiric treatment of intra-abdominal infections. It has activity against MRSA and VRE. Its main side-effect is gastrointestinal upset (very common)! In late 2010 the FDA issued a drug safety communication regarding increased mortality risk associated with the use of tigecycline compared to that of other drugs used to treat a variety of serious infections. What this nebulous warning means I am not sure, but it leads me (C.S.M.) to choose alternative antibiotics when possible.

Aminoglycosides
(A Mean Guy)

Aminoglycosides must diffuse across the cell wall to enter the bacterial cell, so they are often used with penicillin, which breaks down this wall to facilitate diffusion.

Figure 18-8

Clinical Uses

In general, aminoglycosides kill aerobic gram-negative enteric organisms (the enterics are the bugs that call the GI tract home, such as *E. coli* and company). The aminoglycosides are among the handful of drugs that kill the terrible *Pseudomonas aeruginosa*!!!

Most aminoglycosides end with -**mycin**:

1) **Streptomycin** is the oldest one in the family. Many bugs are resistant to it.

2) **Gentamicin** is the most commonly used of all the aminoglycosides. It is combined with penicillins to treat in-hospital infections. There are also many bacterial strains resistant to this drug.

3) **Tobramycin** is good against the terrible *Pseudomonas aeruginosa*.

4) **Amikacin** does not end with **mycin** (sorry). Maybe that is to set it apart. It has the broadest spectrum and is good for hospital-acquired (nosocomial) infections that have developed resistance to other drugs while doing time in the hospital.

5) **Neomycin** has very broad coverage but is toxic, so it can only be used topically for skin infections.

6) **Netilmicin** is used for preoperative coverage for GI surgery. This drug is given orally before GI surgery. It cruises down the GI tract, without being absorbed, killing the local inhabitants. This prevents spilling of organisms during surgery into the sterile peritoneal cavity.

Figure 18-9

Adverse Effects

Here's how we will remember the side effects: Picture this huge boxer, **a mean guy** (Aminoglycoside), and now check out these pictures:

Fig. 18-9. In the **eighth** round **A MEAN GUY** delivers a crushing right hook to his opponent's **ear**, hurling him off balance, ears ringing and head spinning (eighth cranial nerve toxicity: vertigo, hearing loss). The hearing loss is usually irreversible.

Fig. 18-10. With his opponent off balance, **A mean guy** surges upward with a savage left hook into his right side, pulverizing his **kidney** (renal toxicity). Aminoglycosides are renally cleared and can damage the kidney. This can be reversible, so always follow a patient's BUN and creatinine levels, which increase with kidney damage.

Fig. 18-11. The opponent drops to the floor, out cold in a complete **neuromuscular blockade**, unable to move a muscle, or even breathe. This curare-like effect is rare.

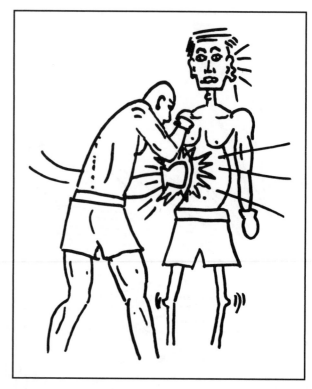

Figure 18-10

Note: These side effects occur if the dose is very high, so when using these in the hospital, the drug level in the blood is checked after steady state levels have been achieved (usually after the third dose). With appropriate blood levels, these agents are generally safe.

Quinupristin/dalfopristin

Quinupristin/dalfopristin is an old antibiotic in the streptogramin class that has found new life with the recent emergence of drug resistant organisms. It inhibits bacterial protein synthesis by binding the 50S ribosomal subunit. This is a nasty medication and should only be used when absolutely necessary.

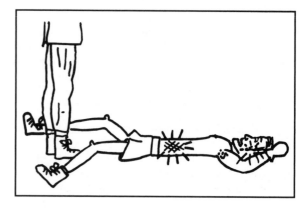

Figure 18-11

Quinupristin/dalfopristin is primarily active against gram positive organisms. Its main use is in life threatening infections with Vancomycin Resistant *Enterococcus faecium* infections (VRE), but not Vancomycin Resistant *Enterococcus faecalis*. (UM, for faeci *UM*, yes, IS, for faecal *IS*, no!!). It can also be used in complicated skin infections with *Staphylococcus aureus* (including MRSA) and Group A *Streptococcus*.

Remember that we said it was a nasty medication . . . . . side effects are common and include hyperbilirubinemia (3–35%), pain at the infusion site (~40%) and arthralgias/myalgias (~40%). Basically, save this medication for when you have no other choice.

Spectinomycin
(Spectacular Spectinomycin)

This drug has a name that sounds like an aminoglycoside, but it is different structurally and biologically. Its mechanism is similar in that it acts on the 30S ribosome to inhibit protein synthesis, but exactly how is not known. Group this with the aminoglycosides in the CLEan TAG mnemonic (see **Fig. 18-2**) to remember its action, but note that it is NOT an aminoglycoside. It is given as an IM injection.

Clinical Uses

Spectinomycin is used to treat gonorrhea, caused by *Neisseria gonorrhoeae*, as an alternative to penicillin and tetracycline (doxycycline), since many strains are resistant to these drugs.

Fig. 18-12.   Mr. Gonorrhoeae, resistant to tetracycline and penicillin.

Fig. 18-13.   Spectacular spectinomycin treats resistant *Neisseria gonorrhoeae*.

Let's briefly review the treatment of gonococcal urethritis (gonorrhea) since this will incorporate a lot of the drugs we have studied.

"Mr. Gonorrhoeae"

Figure 18-12

A patient presents with burning on urination and a purulent penile discharge. When you Gram stain the discharge, you see tiny red (gram-negative) kidney-shaped diplococci inside the white blood cells. Now what? There are many penicillinase-producing and tetracycline-resistant *Neisseria gonorrhoeae*, but you still have a few antibiotics to chose from:

1) **Ceftriaxone** (a third generation cephalosporin): Give one shot IM in the butt! Also give azithromycin 1 gm orally to combat against growing cephalosporin resistance in *gonorrhoeae* and to get the *Chlamydia trachomatis* that is hiding in the background in 50% of cases of urethritis. Seven days of doxycycline is an alternative to azithromycin. Or:
2) **Spectinomycin**: Give one shot in the butt! (along with doxycycline for the *Chlamydia*).

Adverse Effects

Infrequent and minor. Spectinomycin does NOT cause the vestibular, cochlear, and renal toxicity that the aminoglycosides do.

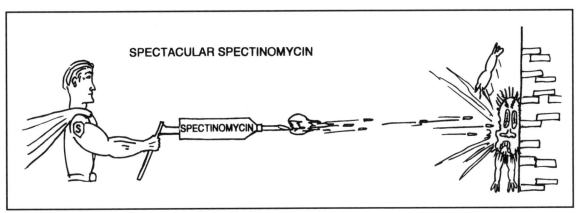

SPECTACULAR SPECTINOMYCIN

SPECTINOMYCIN

Figure 18-13

NAME	MECHANISM OF ACTION	PHARMOKINETICS	ADVERSE EFFECTS

Chloramphenicol	• Binds to 50S ribosomal subunit, & inhibits protein synthesis	1. Oral or IV
		2. Metabolized & completely inactivated in the liver
		3. Metabolites excreted in urine
			1. **Bone marrow depression**:
			  A. Dose related anemia
			  B. Aplastic anemia (rare, but often fatal)
			2. **Gray Baby Syndrome** (40% fatal): cyanosis, vomiting, green stools & vasomotor collapse (This is caused by the accumulation of unmetabolized chloramphenicol, since the neonatal liver has yet to synthesize sufficient metabolic enzymes)

Clindamycin (derivative of lincomycin)	• Binds to 50S ribosomal subunit, & inhibits protein synthesis	1. Oral or IV 2. Excreted from bile and urine	• **Pseudomembranous colitis**: destroys the normal intestinal flora, which allows *Clostridium difficile* to grow and secrete its toxin, causing a bloody diarrhea. Treat with oral vancomycin or metronidazole • Pseudomembranous colitis can also be caused by other antibiotics, such as the penicillins (ampicillin)
Oxazolidinones Linezolid Tedizolid	• Binds to 50S ribosomal subunit, & inhibits protein synthesis	1. Oral or IV 2. Metabolized partially in the liver	1. Bone marrow suppression (thrombocytopenia, anemia, and neutropenia) 2. Serotonin syndrome
Macrolides Erythromycin, Azithromycin, Clarithromycin	• Binds to 50S ribosomal subunit & inhibits protein synthesis	Well absorbed orally. (Erythro and Azithro also in IV)	1. GI upset due to stimulation of gastric motility 2. Rare cholestatic jaundice 3. Prolonged QT syndrome
Telithromycin (a ketolide)	• Binds to 50S ribosomal subunit & inhibits protein synthesis	1. Oral formulation	1. GI upset. 2. QT prolongation—don't give in patients with arrythmias or on anti-arrhythmics
• Tetracycline • Doxycycline • Minocycline • Demeclocycline	• Binds to 30S ribosomal subunit, & inhibits protein synthesis	1. Oral absorption from the stomach and small intestine (however, absorption is severely impaired by food, milk, Ca^{++} & Mg^{++} salts)	1. GI irritation: nausea, vomiting and diarrhea 2. **Phototoxic Dermatitis** (often get a skin rash) 3. Renal & hepatic toxicity (with high doses)

Figure 18-14   ANTI-RIBOSOMAL DRUGS (chart continued on page 184)

THERAPEUTIC USES	MISCELLANEOUS

• Wide spectrum of activity: kills gram-positives, gram-negatives and anaerobes (but its toxicity can be lethal)
• Generally, it is only used for:
  1. Bacterial meningitis in infants who are known to have severe allergies to penicillin and cephalosporin
  2. Rickettsial infections in children and pregnant women (since tetracycline should be avoided in children)
	• Think "chlorine": wide spectrum, but toxic

1. Anaerobes:
  A. For wounds which penetrate the abdomen
  B. For anaerobic infections of the female genital tract
2. Gram-positive organisms, if the patient has severe allergies to penicillin and cephalosporin
3. *Toxoplasma gondii*: use clindamycin in combination with pyrimethamine
4. Toxic shock syndrome: due to Group A *Streptococcus* and *Staphylococcus aureus*	

1. Healthcare associated pneumonia
  2. Complicated skin and soft tissue infections
  3. *Staph aureus* pneumonia
  4. Infections due to: Methicillin Resistant *Staphylococcus Aureus* (MRSA) and Vancomycin Resistant *Enterococcus* (VRE)
	• **Avoid with antidepressants**

1. Outpatient treatment of upper and lower tract respiratory infections 2. Atypical organisms—*Legionella, Mycoplasma, Chlamydia*	• Metabolized by Cytochrome P450
• Community acquired pneumonia	1. Primarily metabolized by CYP450 2. Do not use in patients with **myasthenia gravis**!
1. *Rickettsia* 2. *Chlamydia* (erythromycin is equally effective) 3. *Mycoplasma pneumoniae* 4. *Entamoeba histolytica* 5. Spirochetes:	• Democlocycline: used primarily in the US for the treatment of the Syndrome of Inappropriate ADH secretion (SIADH) rather than for its antibacterial effect.

NAME	MECHANISM OF ACTION	PHARMOKINETICS	ADVERSE EFFECTS
• Tetracycline • Doxycycline • Minocycline • Demeclocycline (Continued)		2. IV formulations available 3. Concentrates in liver and undergoes extrahepatic circulation 4. Excretion: A. Urine: tetracycline B. Stool: doxycycline	4. **Fanconi Syndrome**: occurs with ingestion of outdated drug; Results in renal tubular dysfunction, which can lead to renal failure 5. Superinfections (like *Clostridium difficile* induced pseudomembranous colitis) 6. **Teratogenic**: depresses bone growth in fetus, by chelating Ca^{++} and therefore decreasing Ca^{++} serum levels 7. Discolors teeth and stains bone at site of bone calcification!
Tigecycline (a glycycline)	• Binds *30s* ribosomal subunit	1. IV only	1. GI irritation-nausea, vomiting, diarrhea- (very common!) 2. Other side effects similar to tetracyclines
Aminoglycosides	• Binds to *30S* ribosomal subunit, & inhibits protein synthesis	1. IV or IM (Not oral) 2. Diffuses across cell wall of microbes, so synergistic with penicillin (since penicillin breaks down cell walls, so that the aminoglycoside works better) 3. **Crosses CNS only if meninges are inflamed**! 4. Not metabolized 5. Excreted renally	1. Vestibular and auditory **ototoxicity** (due to cranial nerve 8 damage) 2. Nephrotoxicity 3. Neuromuscular blockade: muscle paralysis and apnea
Quinupristin/ dalfopristin (streptogramin)	• Inhibits *50S* ribosomal subunit	1. IV only	Common: 1. Hyperbilirubinemia 2. Infusion site pain and inflammation 3. Myalgia/arthralgia
Spectinomycin	• Binds to *30S* ribosomal subunit, & inhibits protein synthesis	1. IM 2. Excreted in urine unmetabolized	• *NO* serious toxicity

Figure 18-14 (continued)

Fig. 18-14. Summary of anti-ribosomal antibiotics.

References and Recommended Reading

Eckmann C, Dryden M. Treatment of complicated skin and soft-tissue infections caused by resistant bacteria: value of linezolid, tigecycline, daptomycin and vancomycin. Eur J Med Res. 2010; 15(12):554–63.

Gilbert, DN. Aminoglycosides. In: Principles and Practice of Infectious Diseases, 6th ed, Mandell, GL, Bennett, JE, Dolin, R (Eds), Churchill Livingstone, New York 2005. p. 328.

McDonald, LC, Killgore, GE, Thompson, A, et al. An epidemic, toxin gene-variant strain of *Clostridium difficile*. N Engl J Med 2005; 353:2433.

THERAPEUTIC USES	MISCELLANEOUS
A. *Borrelia* and *Leptospira* B. *Treponema pallidum* (second choice behind penicillin) 6. Brucella (second choice behind Bactrim) 7. Nocardia (second or third choice) 8. Facial acne	
1. Complicated skin and soft tissue infections 2. Intra-abdominal infections 3. Covers Methicillin Resistant *Staphylococcus Aureus* (MRSA) and Vancomycin Resistant *Enterococcus* (VRE)	1. Use associated with increased mortality 2. Similar in structure to tetracyclines
1. Aminoglycosides are effective against gram-negative enteric organisms 2. Also effective against: A. Tularemia B. Yersinia pestis C. Brucellosis D. *Mycobacterium tuberculosis*	1. *Streptomycin*: oldest member of family; many bugs are resistant! 2. *Gentamicin*: most commonly used of all aminoglycosides 3. *Tobramycin*: good against *Pseudomonas aeruginosa* 4. *Amikacin*: has the broadest spectrum 5. *Neomycin*: used topically, as it is very toxic. Also very broad spectrum 6. Netilmicin
1. Complicated skin infections with Group A *strep* and S. *aureus* 2. Life-threatening bacteremia with Vancomycin Resistant *Enterococcus* (VRE) (*Enterococcus faecium* only)	• A "nasty" medication
• Gonorrhea (as an alternative to penicillin) • Not effective against *Treponema pallidum* (syphilis) or *Chlamydia*	

M. Gladwin, W. Trattler, and S. Mahan, *Clinical Microbiology Made Ridiculously Simple* ©MedMaster

Severe *Clostridium difficile*-associated disease in populations previously at low risk—four states, 2005. MMWR Morb Mortal Wkly Rep 2005; 54:1201.

Tigecycline (tygacil). Med Lett Drugs Ther 2005; 47:73

Warny, M, Pepin, J, Fang, A, et al. Toxin production by an emerging strain of *Clostridium difficile* associated with outbreaks of severe disease in North America and Europe. Lancet 2005; 366:1079.

Zuckerman, JM. Macrolides and ketolides: azithromycin, clarithromycin, telithromycin. Infect Dis Clin North Am 2004; 18:621

CHAPTER 19. ANTI-TB AND ANTI-LEPROSY ANTIBIOTICS

TREATMENT OF TUBERCULOSIS

This chapter will cover the first-line anti-tuberculosis antibiotics and the logical approach to their use.

The first-line drugs, in order of their frequency of use, are:

Isoniazid (INH)	"**I saw** a
Rifampin	**R**ed
Pyrazinamide	**Pyr**e—BURNING THE LIVER"
Ethambutol	
Streptomycin	

Fig. 19-1. Isoniazid ("I saw"), *R*ifampin ("red"), and *Pyr*azinamide ("pyre"), are first-line antituberculosis antibiotics that can cause liver damage ("burning the liver").

When it comes to tuberculosis, you will encounter 2 populations of patients: 1) those with active tuberculosis and 2) those with a reactive PPD skin test, representing a latent infection. These 2 populations are treated very differently.

Treatment of Active Tuberculosis

A patient presents with dyspnea, fever, productive cough, and night sweats that have lasted 2 months, along with upper lobe consolidation on chest X-ray. Acid-fast bacilli are identified from a sputum sample.

A patient with active pulmonary or extra-pulmonary disease should be treated for 6 months with a rifampin based regimen. Usually this consists of a 2 month intensive phase with isoniazid, rifampin, ethambutol, and pyrazinamide, followed by 4 months of isoniazid and rifampin (**remember 4 for 2, then 2 for 4**). Patients with cavitary pulmonary tuberculosis and sputum cultures that are still positive after 2 months of treatment should have their second phase of treatment with isoniazid and rifampin extended from 4 to 7 months. Many authorities treat patients with tuberculous meningitis, tuberculous spondylitis (bone infection or Pott's disease), or tuberculous arthritis for 12 months. **This makes sense ... sequestered sites require longer therapy.**

Treatment of PPD Reactors

These persons may have latent *Mycobacterium tuberculosis* in their bodies and might develop a reactivation tuberculosis. Treatment of PPD reactors is thus preventive.

There are three important steps in the diagnosis of latent tuberculosis:

1) **Is there sufficient risk of latent tuberculosis infection to warrant screening with a PPD skin test?** The following groups have the highest risk of latent tuberculosis and subsequent reactivation:

- Persons at increased risk of exposure to infectious cases (recent close contact, health care workers)
- Persons at increased risk of tuberculosis infection (foreign-born persons from countries with a high prevalence of tuberculosis, homeless persons, persons in long-term care facilities or prisons)

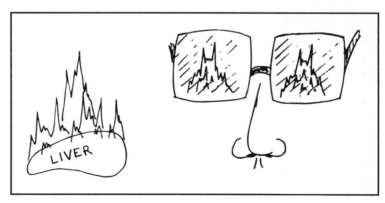

Figure 19-1

• Persons at increased risk of reactivation and development of active disease (HIV infection, injection drug-users, kidney disease, diabetes, malnourishment, on immunosuppressive therapy or with cancers)

2) **Is the PPD test positive?** The goal of testing is to identify persons at greatest risk of reactivation so that the beneficial effects of treatment outweigh the risks of side effects. The precise details of what constitutes a positive PPD test and who to treat is discussed below under **Risk of Reactivation Tuberculosis.**

3) **If the PPD is positive, have we ruled out active disease?**

• It is imperative to exclude active tuberculosis with a chest radiograph and if abnormal to perform a sputum AFB stain and culture. This is a vital step because if we treated active tuberculosis with only one agent such as INH, resistance would rapidly develop.

If the answer to all these three questions is **YES, YES, YES**, then treat the latent tuberculosis: Isoniazid is usually given for 9 months for treatment of latent tuberculosis. Taking rifampin for 4 months or a combination of once weekly isoniazid and rifapentine for 3 months are approved alternatives. **(9 months for latent TB is the key)**

The main serious complication of these regimens is **hepatotoxicity!** While as many as 10–20% of patients treated with INH develop liver enzyme abnormalities (an increase in ALT or AST liver enzymes), the development of symptomatic hepatitis is very uncommon. While a classic US Public Health Service study (Kopanoff et al. Am Rev Respir Dis 1978) suggested that as many as 2.3% of patients older than 50 years of age developed hepatitis while taking isoniazid, more recent studies suggest that this risk is only 1 out of 1000 persons (Nolan et al. JAMA 1999). For this reason we now treat with INH if treatment is indicated, regardless of age. One still must be cautious in treating PPD reactors with underlying liver disease or who are heavy users of alcohol. Liver function tests should be checked at baseline and periodically during therapy in patients with chronic liver disease, heavy use of alcohol, and in pregnancy. While receiving treatment for latent TB infection, patients should be advised to immediately report symptoms of nausea, jaundice, diarrhea, or abdominal pain. **(Don't be mellow: Call the doctor if you turn yellow!)**

Risk of Reactivation Tuberculosis

Factors that increase the risk of reactivation include recent PPD conversion, having fibrotic scars on chest X-ray, exposure to household members with active tuberculosis, and being immunosuppressed. The larger the skin reaction to PPD the more likely the test is a true positive, representing *M. tuberculosis* infection. Because these factors increase the risk of reactivation of tuberculosis, they will tip the scales towards **treatment**.

The Center for Disease Control and American Thoracic Society have used these concepts to formulate the following recommendations for treating patients with a 9 months course of prophylactic isoniazid:

GREATEST RISK OF REACTIVATION: Treat the following patients if **PPD ≥ 5 mm**:

1) Persons with HIV infection.
2) Persons with fibrotic changes on chest X-ray compatible with old healed tuberculosis.
3) Close contacts of persons with newly diagnosed active tuberculosis. Note: In this case (especially with children), even if the PPD is negative, treat for 3 months, then repeat the PPD. If at that time it is < 5 mm, the isoniazid may be discontinued.
4) patients with organ transplants, and other immunosuppressed patients (receiving the equivalent of ≥ 15 mg of prednisone per day for greater than one month).

MODERATE RISK: Treat these patients if **PPD ≥ 10 mm**:

1) Persons with medical conditions that lower the immune system, like diabetes, prolonged steroid or immunosuppressive treatment, renal failure, and others.
2) Recent arrivals (<5 years) from countries with a high prevalence of tuberculosis.
3) Persons who inject drugs.
4) Residents and employees of high-risk settings, such as homeless shelters, long-term care facilities, prisons, and other health care facilities.
5) Recent PPD conversion within a 2-year period.
6) Health care workers.

LOW RISK: Treat these patients if **PPD ≥ 15 mm**:

Persons with no known risk factors (although these patients should not have been screened with a PPD in the first place, as only patients with a risk of tuberculosis infection should be screened).

Isoniazid-Resistant Organisms

Resistance to INH and the other antibiotics is developing. Resistant organisms are most common in persons previously treated for tuberculosis and should also be suspected in persons from Africa, Asia, or South America; homeless persons and others exposed to resistant organisms; and those whose sputum cultures remain positive after 2 months of treatment.

1) Culture susceptibility testing should follow the initiation of treatment.

2) If resistance is suspected, 4 or more first-line drugs should be used (INH, rifampin, pyrazinamide, ethambutol, or streptomycin).

3) If resistance develops, never add a single new antibiotic; always add **two**. This will insure that the resistant *M. tuberculosis* will be unable to develop further resistance.

Note that there are now **multiple** resistant *M. tuberculosis* organisms that may require 4–5 different antibiotics.

DOT the I's and Cross the T's to Prevent Resistance!!!

DOT or **D**irectly **O**bserved **T**herapy: Health care providers in outpatient settings have their patients come into the clinic to receive their medications under **direct observation to ensure adherence**. Numerous studies have now documented that this strategy can decrease resistance and increase treatment efficacy.

ANTI-TB ANTIBIOTICS

Isoniazid, Rifampin, and Pyrazinamide:

1) **All cause hepatotoxicity**: Patients on these medications must understand the symptoms of hepatitis so they can report to a doctor immediately should they develop. Mild elevations of liver enzymes can be expected to occur in 15–20% of patients on isoniazid, but should these levels exceed 3–5× the upper limit of normal, the drugs should be discontinued.

2) **All are absorbed orally**: This is very important since these must be administered for 6–9 months. They must be orally absorbed!!!

3) **All penetrate into most tissues**: They must reach the center of caseous granulomas.

Isoniazid (INH)
("I Saw")

This is a great antibiotic because it is inexpensive, absorbed orally, and bacteriocidal. Were it not for the toxicity, it would be perfect!!!

INH interferes with the biosynthesis of the mycolic acid component of the cell wall of the Mycobacteria.

Adverse Effects

1) What do you think? **Hepatotoxicity**!!! Alcoholics beware! Alcohol increases the metabolism of INH by the liver, which increases the risk of developing hepatitis and decreases the therapeutic effect.

2) INH increases the urinary excretion and depletion of pyridoxine (vitamin B_6), which is needed for proper nerve function. INH will thus lead to decreased B_6 levels,

characterized by **peripheral neuropathy**, rash, and anemia. Many Docs routinely give B_6 vitamins with INH.

Rifampin
("Red")

Think **R**ifampin:

1) **R**ed: Body fluids such as **urine**, feces, saliva, sweat, and tears are colored a bright red-orange color by rifampin. This is not harmful to the patient, but patients must be made aware of this or they will discontinue the medicine in a panic.

2) **R**NA: Rifampin inhibits the DNA-dependent RNA polymerase of the *Mycobacterium tuberculosis* bugs.

Adverse Effects

1) Hepatitis (much less than INH).

2) Rifampin induces the cytochrome P450 enzyme system (also called the microsomal oxidase system, or MOS), so many other drugs are gobbled up by the spruced-up MOS. This results in decreased half-lives of certain drugs in patients taking rifampin. Some examples:

   a) Coumadin (an anticoagulant): Blood-thinning effect will be reduced.
   b) Oral contraceptives: Women can get pregnant and get breakthrough bleeding!
   c) Oral hypoglycemics and corticosteroids are less effective.
   d) Anticonvulsants such as phenytoin (seizures!).

Rifabutin

Rifabutin is very similar to rifampin in structure, antibacterial activity, metabolism, and adverse reactions. It is commonly used in the treatment of *Mycobacterium avium-intracellulare* (MAI).

The same drug-drug interactions of rifampin should be considered for rifabutin however, rifabutin induces cytochrome P450 less than rifampin. Thus, rifabutin is often used in HIV patients with tuberculosis who are on a protease inhibitor as part of their HIV antiviral regimen.

Rifapentine

Rifapentine is a long acting anti-tubercular medication similar to rifampin in its antibacterial activity. It too is an inducer of cytochrome P450. It has similar side effects as the other rifamycins in its class. It recently gained approval to be used in combination with isoniazid as part of a once weekly regimen for 12 weeks for the treatment of latent TB. Additionally, it is an option as part of a multi-drug regimen that can be given twice weekly in the treatment of fully drug susceptible active tuberculosis in non-HIV infected persons.

Pyrazinamide
("Pyre")

The mechanism of action of pyrazinamide is not known. Pyrazinamide is very important for its ability to kill slowly replicating tubercle bacilli. It is a key component of short course therapy. Without pyrazinamide in the treatment regimen, therapy must be extended to at least 9 months.

Adverse Effects

Pyrazinamide is hepatotoxic (no kidding?!) This medicine is usually given for no more than 2 months to avoid liver toxicity. Avoid it in pregnancy (unknown effect on fetus). Pyrazinamide commonly causes **joint pain** and may precipitate flares of gout.

Ethambutol
("Ethane-Butane Torch")

Adverse Effects

Fig. 19-2. The main side effect of ethambutol is a dose-dependent, reversible, ocular toxicity. Think of an **ethane-butane** flame torch, torching an eye. The ocular toxicity is manifested by:

1) Decreased visual acuity with loss of central vision (central scotomata).
2) Color vision loss.

Ethambutol is not used in young children because they are not able to report vision deterioration. Adults are tested for visual acuity and color perception at regular intervals. Many doctors instruct their patients to read the fine newspaper print everyday as a self exam.

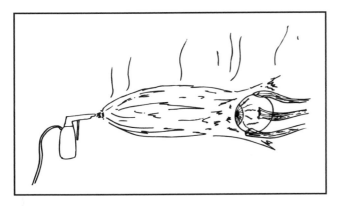

Figure 19-2

Streptomycin

Streptomycin is in the aminoglycoside family, which inhibits protein synthesis at the **30S** ribosomal subunit, and is given IM or IV. It is ototoxic and nephrotoxic (see Chapter 18). Avoid it in pregnant women (can cause congenital deafness).

Fixed-Dose Combinations

Fixed-dose combinations are available as **Rifamate** (isoniazid and rifampin) and **Rifater** (isoniazid, rifampin, and pyrazinamide). Such combinations are strongly encouraged for adults who are self-administering their medications because they may enhance adherence, reduce the risk of inappropriate monotherapy, and prevent drug resistance.

Second-line Drugs

These can be used when multiple antibiotics are needed for the treatment of multi-drug resistant *Mycobacterium tuberculosis*.
Para-aminosalicyclic acid
Capreomycin sulfate
Cycloserine
Ethionamide
Kanamycin
Amikacin (aminoglycoside)
Quinolones such as levofloxacin
Linezolid
Bedaquiline*
*Approved in December 2012 for MDR TB

Treatment of LEPROSY

Three drugs are used in the treatment of leprosy: dapsone, rifampin, and clofazimine.

The **Rap Zone** of **Dapsone**

Where's your ears?
    There on the floor.
Where's your hand?
    I left it on the door.
Come on Doc, look and see,
    these rappin' clowns got leprosy.

Hey, you clown!
    You left your nose in the car,
Hey, you clown!
    You left your toes in a bar,
Call the doc, to the **Rap Zone**,
    Your first-line drug remains **dapsone**.

And if you dance to this stupid rappin',
watch your feet as they start a stampin',

There's only one thing to help your **dancin**,
Time to reach for the drug, **rifampin**.
Their peelin' **clown faces** look really lean.
They're healing faster than can be seen,
As long as they stay close to **clofazimine**.

Severe cases of leprosy should be treated with rifampin, dapsone, and clofazimine for a minimum of 2 years and until patients are acid-fast bacilli negative.

Less severe cases are treated with rifampin and dapsone for 6 months.

See anti-tuberculosis medications (page 188), for more on rifampin. See the sulfa drugs (Chapter 20) for more on dapsone.

Clofazimine

Fig. 19-3.   A **clown-faced** clown climbs a DNA double helix stairway. His outfit is colored red and black:

1) Clofazimine works by binding to the DNA of *Mycobacterium leprae*. It also has anti-inflammatory actions that are helpful in treating the **leprosy reactions**.

2) Clofazimine is a red-colored compound, and when it deposits in the skin and conjunctiva, it colors these tissues red. Any place on the body where there is a leprosy lesion, the skin will appear tan to black. Note the clown's red and black outfit.

Leprosy Reactions

Fifty percent of patients treated for leprosy develop a leprosy reaction. There are 2 types (1 and 2)

Figure 19-3

NAME	MECHANISM OF ACTION	PHARMOKINETICS
ANTI-TUBERCULOSIS DRUGS		
Isoniazid (INH)	• Interferes with the biosynthesis of the mycolic acid component of the cell wall of Mycobacterium	1. Oral, IM, or IV 2. Penetrates into all fluids & tissues 3. Metabolized in liver by acetylation (a small population in U.S. are "slow acetylators") 4. Excreted via urine 5. Increases urinary excretion of pyridoxine (vitamin B_6)

Figure 19-4   ANTIBIOTICS FOR MYCOBACTERIA

and both are immune-mediated, possibly in response to the increase in dead organisms with treatment. The reactions involve inflammation of the nerves, testicles, eyes, joints, and skin (erythematous nodules).

Type 1 reactions occur only in borderline patients (BT, BB, BL), and almost always occur during the first year of treatment. The skin lesions of leprosy typically swell, becoming more edematous, and occasionally ulcerate. Neuritis can also occur, leading to sensory or motor nerve loss. The type 1 reaction is thought to be a delayed hypersensitivity reaction to the dead bacilli. When this reaction occurs, patients can be treated with prednisone. It is important that you do NOT withdraw the anti-leprosy drugs if a leprosy reaction occurs.

Type 2 reaction (called **Erythema Nodosum Leprosum**) is associated with borderline lepromatous (BL) and lepromatous leprosy (LL). Commonly, a painful nodular rash erupts in a previously normal-appearing area of skin, along with a high fever. Neuritis, orchitis, arthritis, iritis, and lymphadenopathy can occur as well. The type 2 reaction is thought to be an immune complex-mediated reaction involving the deposition of the immune complexes in tissues followed by complement activation. These patients can also be treated with prednisone or clofazimine. However, the treatment of choice is **thalidomide**. This is one of the few uses of thalidomide that is condoned in the U.S. because it is a potent teratogen (also used in the treatment of multiple myeloma). Again, the anti-leprosy antibiotics are NOT to be withdrawn!

Fig. 19-4. Summary of antibiotics for *Mycobacteria*.

References

American Thoracic Society, CDC, and IDSA. Treatment of Tuberculosis. MMWR. 2003; 52(RR11): 1–77.

Hopewell PC, Bloom BR. Tuberculosis and other Mycobacterial Diseases. In: Murray JF, Nadel JA, eds. Textbook of Respiratory Medicine. 2nd ed. Philadelphia: W.B. Saunders Co. 1994: 1094–1160.

U.S. Department of Health and Human Services, Division of Tuberculosis Elimination, Centers for Disease Control and Prevention, American Thoracic Society. Core Curriculum on Tuberculosis: What the Clinician Should Know. 4th Edition. Atlanta, Georgia, 2000.

Recommended Reading:

Daley CL. Update in tuberculosis 2009. Am J Respir Crit Care Med. 2010; 181(6):550–5.

Forno C, Häusermann P, et al. The difficulty in diagnosis and treatment of leprosy. J Travel Med. 2010; 17(4):281–3.

LoBue PA, Enarson DA, Thoen TC. Tuberculosis in humans and its epidemiology, diagnosis and treatment in the United State. Int J Tuberc Lung Dis. 2010; 14(10):1226–32.

Ma Z, Lienhardt C, et al. Global tuberculosis drug development pipeline: the need and the reality. Lancet. 2010; 375 (9731):2100–9.

Sterling T, Villarino M, et al. Three months of rifapentine and isoniazid for latent tuberculosis infection. NEJM. 2011; 365(23):2155–66.

ADVERSE EFFECTS	THERAPEUTIC USES	MISCELLANEOUS
1. Hepatotoxicity: A. Risk of hepatitis increases with age B. Increase risk of hepatitis when alcohol is consumed 2. Can induce pyridoxine (vitamin B_6) deficiency, resulting in pellagra (which is manifested as peripheral neuritis, rash and anemia)	1. Prophylaxis for tuberculosis (used alone) 2. For active tuberculosis (use in combo with other drugs)	1. No beer, wine or liquor, as alcohol increases the risk of developing hepatitis 2. Vitamin B_6 (pyridoxine) supplements are often given to avoid deficiency of pyridoxine 3. Monitor hepatic enzyme level 4. Combo drugs: A. Rifamate: Isoniazid & rifampin B. Rifater: Isoniazid, rifampin, & pyrazinamide

NAME	MECHANISM OF ACTION	PHARMOKINETICS
Rifampin	• Inhibits DNA dependent RNA polymerase	1. Oral 2. Penetrates into all fluids & tissues 3. Metabolized via microsomal oxidase system (MOS): induces MOS, thereby increasing its own metabolism as well as the metabolism of other drugs 4. Excreted via liver
Pyrazinamide	1. Unknown mechanism 2. This drug is an analog of nicotinamide	1. Oral 2. Renal excretion
Ethambutol	1. Unknown mechanism 2. Metal chelator	1. Oral 2. Can cross blood brain barrier 3. Excreted unchanged in urine & feces
Rifabutin	• Rifabutin inhibits DNA-dependent RNA polymerase in susceptible strains of *Escherichia coli* and *Bacillus subtilis* but not in mammalian cells. It is not known whether rifabutin inhibits DNA-dependent RNA polymerase in MAC	1. Oral 2. Induces Cytochrome P450 and may alter blood levels of other medications (such as antiretroviral drugs)
Rifapentine	• Inhibits DNA dependent RNA polymerase	1. Oral 2. Hepatic metabolism 3. Long half life
Streptomycin (an aminoglycoside)	• Binds to 30S ribosomal subunit and inhibits protein synthesis	• Streptomycin can be administered IM or IV
ANTI-LEPROSY DRUGS		
Sulfones 1. Dapsone 2. Sulfoxone	• PABA antagonist (similar mechanism as sulfonamides). Results in blockage of dihydrofolic acid (DHF) synthesis, a precursor to tetrahydrofolic acid (TH4), which is crucial to the synthesis of purines. This results in inhibition of bacterial DNA synthesis	1. Oral 2. Absorbed from GI tract via enterohepatic circulation. 3. Metabolized in liver - via acetylation 4. Excreted in urine
Clofazimine	1. Binds to DNA 2. Anti-inflammatory actions are helpful for treating the leprosy reactions.	• Oral
Rifampin: see above		

Figure 19-4   (continued)

ADVERSE EFFECTS	THERAPEUTIC USES	MISCELLANEOUS

1. Asymptomatic jaundice, elevated liver enzymes
2. Urine, sweat & tears become RED-ORANGE color
	1. This drug is used for both tuberculosis and leprosy
	2. Also used prophylactically for persons exposed to patients ill with *N. meningitidis*
	3. Sometimes used for:
	  A. Legionella pneumophila
	  B. *Staph. aureus* endocarditis
		• Increases metabolism of (and thus decreases half-life) of:
		1. Coumadin
		2. Corticosteroids
		3. Oral contraceptives – careful!!!
		4. Oral hypoglycemics
		5. Digoxin
		6. Methadone

1. Hepatotoxic!! 2. Gout (inhibits uric acid secretion, thus increasing uric acid levels) (Go put out the pyre)	• Mycobacterium tuberculosis	Do not use in pregnancy

1. Dose related, bilateral, ocular toxicity that is usually reversible
  A. Decreased visual acuity
  B. Color vision loss
  C. Loss of central vision
    (central scotoma)
	• Mycobacterium tuberculosis	• Only first line drug that is bacterio*static*

1. Possible kidney and liver effects 2. Bone marrow suppression 3. Rash, fever 4. Uveitis (inflammation of the eye). 5. Orange discoloration of urine, sweat, tears and even soft contact lenses	• Rifabutin is used in combination with other drugs for prevention and treatment of Mycobacterium avium or in M. intracellulare which comprise M. avium complex (MAC)	1. MAC is related to tuberculosis (TB), but no one anti-TB drug works against MAC 2. Care must be taken when rifabutin is used with other medications that are metabolized by the cytochrome P450 system
1. Hepatitis 2. Hypersensitivity reaction	• Mycobacterium tuberculosis	• Induces P450
• Vestibular & ototoxic	• Mycobacterium tuberculosis	Do not use in pregnancy
1. Skin rash, drug fever 2. Bone marrow suppression causing agranulocytosis (low neutrophils) 3. Leprosy reactions may occur with treatment (see text)	• Mycobacterium leprae	• Mycobacterium leprae develops resistance rapidly
• Red and black skin discolorations	1. Mycobacterium leprae 2. Leprosy reaction	• Resistance develops slowly!

M. Gladwin, W. Trattler, and S. Mahan, *Clinical Microbiology Made Ridiculously Simple* ©MedMaster

CHAPTER 20.   MISCELLANEOUS ANTIBIOTICS

THE FLUOROQUINOLONE ANTIBIOTICS

Ciprofloxacin and Family

This group of antibiotics is expanding. The fluoroquinolones have become as large and important a group as the penicillins and cephalosporins. The reason for this is that they are **safe**, achieve high blood levels with **oral** absorption, and **penetrate** extremely well into tissues.

Fig. 20-1. All the antibiotics in the fluoroquinolone family have the common ending -**FLOXACIN**. To help you remember some facts about this drug family, think of a crazy naked group of partyers, a **FLOCK OF SINNERS**. They are all **gyrating** their hips as they party

and dance. The fluoroquinolones act by inhibiting **DNA gyrase**, resulting in the breakage of the bacterial DNA structure.

In the basic quinolone structure, the main feature that distinguishes the fluoroquinolones from their predecessor, nalidixic acid, is the addition of a fluorine group, hence the name fluoroquinolone. The evolution of quinolone structures now allow their classification into first, second, third and fourth generations, with nalidixic acid solely in the first generation. Classification by generation is available on page 198.

Resistance to the Fluoroquinolones

Like all antibiotics, with this excessive use, resistant organisms are rapidly spreading. What makes this particularly disconcerting is that the resistance is against

Figure 20-1

all the fluoroquinolones. Resistance is caused by a point mutation in the bacterial DNA gyrase subunits. This powerful new family of antibiotics should be used carefully to reduce the spread of resistance.

Adverse Effects

There are very few:

1) Some patients experience GI irritability (nausea, vomiting, belly pain, diarrhea), as occurs with erythromycin and doxycycline. The **flock of sinners** often vomit after their excessive drinking.

2) The drugs damage cartilage in animals. Avoid them in children, as children have more cartilage. More recent data suggests that fluoroquinolones appear safe in pregnancy. (Antimicrob Agents Chemother 1998;42:1336–9)

3) The fluoroquinoles, ciprofloxacin in particular, have been associated with tendonitis and tendon rupture (most commonly involving the Achilles tendon).

4) CNS side effects are rare: headache, restlessness and insomnia.

5) Through disruption of the normal bowel flora, fluoroquinolone use increases one's risk of developing *Clostridium difficile* colitis and diarrhea. A retrospective evaluation of a recent outbreak of hypervirulent *Clostridium difficile*-associated diarrhea in Quebec found fluoroquinolone use to be the strongest associated risk factor.

6) Gatifloxacin in its oral form was associated with hyper- and hypoglycemia. The pill form has been removed from the market and, currently, only gatifloxacin ophthalmic solution is available for the treatment of bacterial conjunctivitis.

Ciprofloxacin, unlike third and fourth generation quinolones seems to inhibit gamma-aminobutyric acid (GABA) and therefore can cause seizures in patients with renal insufficiency or when coadministered with drugs that decrease renal blood flow (e.g. nonsteroidal anti-inflammatory drugs).

Pharmacokinetics

These drugs undergo enterohepatic circulation (excreted in the bile and reabsorbed in the intestine) and then are excreted via the kidney. So drug levels are high in the stool and kidneys. They also penetrate well into bone and prostate. Finally, they achieve high intracellular levels.

High drug level in target tissue
+
ability to kill organism
=
clinical utility.

Clinical Uses

1) Ciprofloxacin has **poor gram-positive coverage**: It does poorly against *Streptococcus pneumoniae* but has activity against *Staphylococcus aureus* (also against anthrax).

2) Most fluoroquinolones **do not cover anaerobes**. NOTE: Some of the new fluoroquinolones have increasing *Staphylococcus* and *Streptococcus* coverage, and even show promise for anaerobic coverage.

So what are these drugs good for??

3) **Gram-negatives**!!! including:

a) The multi-resistant *Pseudomonas aeruginosa*. They are used to treat patients with cystic fibrosis, who are all colonized with this evil bug.

b) The Enterobacteriaceae (Enterics: bugs in the GI tract) except anaerobes. They are used for diarrhea caused by enterotoxigenic *Escherichia coli*, *Salmonella, Shigella*, and *Campylobacter*. One can take these pills to treat and prevent traveler's diarrhea. High intestinal drug level + Enteric coverage = Treatment of diarrhea.

c) Complicated urinary tract infections (UTIs) caused by resistant strains of Enterobacteriaceae (*Pseudomonas*, etc.); prostatitis and epididymitis. High renal, prostatic drug levels + coverage of gram-negatives = Treatment of UTIs.

Fig. 20-2. The quinolones (floxacin) achieve high renal levels and can be used for UTIs.

Figure 20-2

d) Gram-negative facultative intracellular organisms, such as *Legionella, Brucella, Salmonella,* and *Mycobacterium.* High intracellular concentration + gram-negative coverage = Treatment of *Legionella, Brucella, Salmonella,* and *Mycobacterium.*

The newer generation quinolones have been engineered to fill in the "holes" in ciprofloxacin's coverage. Levofloxacin has expanded gram positive coverage (*Streptococcus pneumoniae, Enterococcus faecalis,* group A,B,C,G Streptococci). It is indicated primarily for community acquired pneumonia and skin infections.

Moxifloxacin has been dubbed a "respiratory quinolone" due to its improved activity against *Streptococcus pneumoniae,* as well as improved anaerobic coverage. Moxifloxacin is so good against anaerobes that it has recently been approved for the empiric treatment of intra-abdominal infections.

Despite the many "improvements" in the newer quinolones **ciprofloxacin** retains the best activity against *Pseudomonas aeruginosa* and is probably the only one that should be used for this indication (with the possible exception of levofloxacin).

Because they also cover the atypical bacteria (*Legionella, Mycoplasma,* and *Chlamydia*) they are good choices for community acquired pneumonia.

Several other fluoroquinolones have come and gone due to adverse side-effects. In addition to gatifloxacin (removed due to hyper- and hypoglycemia), trovafloxacin was withdrawn due to hepatic toxicity, grepafloxacin due to increased cardiac events, and sparfloxacin due to photosensitivity and prolonged Q-T.

VANCOMYCIN

This IV antibiotic, a glycopeptide, has a critical role in the treatment of infectious diseases. It is the opposite of aztreonam, which covers **all** gram-negative bugs. **Vancomycin covers ALL GRAM-POSITIVE bugs!!!**

Even MRSA (methicillin resistant *Staphylococcus aureus*)!!

Even *Enterococcus* (both *faecalis* and *faecium* species).

Even multi-resistant *Staphylococcus epidermidis* (in infections of indwelling intravenous catheters).

It is also used to treat endocarditis caused by *Streptococcus* and *Staphylococcus* in penicillin-allergic patients.

Fig. 20-3. A **VAN** with a + on its side (an ambulance van) is driving out of some **IV tubing**. It is about to run over an **ear** ("D" of D-alanine) and hit a **peptidoglycan cell wall**. The VAN is being driven by an Indian, the **red man**. This picture helps us remember that **VANCOMYCIN** is given IV, kills gram-positive bugs by inhibiting peptidoglycan production, and causes the **red man syndrome**. In the latter, which follows rapid infusion of vancomycin, there is often a nonimmunologic

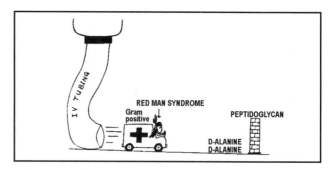

Figure 20-3

release of histamine, resulting in a red rash of the torso and itching skin. Slow infusion over an hour or antihistamine premedication can prevent this problem.

Vancomycin inhibits the biosynthesis of the gram-positive peptidoglycan at a step earlier than penicillin. It complexes with D-alanine D-alanine to inhibit transpeptidation. Like penicillin, it acts **synergistically** with the aminoglycosides.

Vancomycin is **not absorbed orally**. We take advantage of this in the treatment of *Clostridium difficile* pseudomembranous colitis. Vancomycin is taken orally, cruises down the GI tract unabsorbed, and kills the *Clostridium difficile*!!

With the extensive use of vancomycin, new strains of multiple drug-resistant gram-positive organisms have emerged (see Chapter 4). With the emergence of vancomycin-resistant gram-positive organisms, in particular *Enterococcus* and *Staphylococcus aureus*, alternative agents are being developed and frequently used. **Quinopristin/dalfopristin**, **linezolid**, and **tigecycline** are examples of newer agents used in the battle against resistant gram-positive bugs.

MORE GLYCOPEPTIDES

Telavancin, Dalbavancin, Oritavancin

The more recently developed glycopeptides- Telavancin, Dalbavancin, and Oritavancin- have a similar mechanism of action to vancomycin, inhibiting cell wall synthesis by binding to **D-alanine D-alanine** to block transpeptidation. Telavancin has an additional lipophilic side chain, which contributes to bacterial killing via disruption of cell permeability. It has the advantage of still being effective against *Staphylococcus aureus* with intermediate resistance to vancomycin. Unfortunately, telavancin has been associated with poor outcomes when used in persons with poor renal function. Dalbavancin and oritavancin are both used for skin and soft tissue infections and have the advantage of once weekly dosing due to their long half-lives.

DAPTOMYCIN

This is the sole agent in a newer class of antibiotics, a cyclic lipopetide. It kills gram positive bacteria by altering the microbe cell-membrane electrical charge and transport. Similar to vancomycin, it has no activity against gram negative bacteria. It is active against Methicillin Resistant *Staphylococcus aureus* (MRSA) and Vancomycin Resistant *Enterococcus* (VRE)!!!! Be sure to monitor your patients CPK levels while on daptomycin as one of its main side effects is myopathy. It has also been associated with eosinophilic pneumonia. Eosinophilic pneumonia presents as an acute febrile illness with pulmonary infiltrates and an increase in eosinophils in broncheoalveolar fluid. It resolves by stopping daptomycin and giving steroids.

ANTIMETABOLITES

Trimethoprim and Sulfamethoxazole

Nucleotide and DNA formation require tetrahydrofolate (TH4). Bacteria make their own TH4 and use Para Amino Benzoic Acid (PABA—you know, the stuff in sunscreen) to make part of the TH4. People don't make TH4; we get it as a vitamin in our diet.

Fig. 20-4. The Sulfa drugs look like PABA.

When you give a person one of the sulfa drugs (e.g., sulfamethoxazole), the bacteria use it, thinking it's PABA. (There isn't much room for higher cortical neurons in a single-celled creature.) The sulfa drug competitively inhibits production of TH4. Since our cells don't make TH4, it doesn't affect us, but it affects the bacteria.

TH4 gives up carbons to form purines and other metabolic building blocks. After giving up a carbon, it becomes dihydrofolate (TH2) and must be reduced back to TH4 by the enzyme **dihydrofolate reductase**. Trimethoprim looks like the dihydrofolate reductase of bacteria and competitively inhibits this reduction. This inhibits bacterial DNA formation.

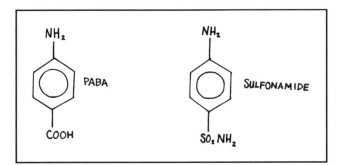

Figure 20-4

The big picture here is that trimethoprim (TMP) and sulfamethoxazole (SMX) act synergistically to kill many gram-positive and gram-negative bacteria. They both inhibit TH4 production but at different steps.

Pharmacokinetics

Oral absorption: Just imagine eating a big chunk of rotten egg which smells like sulfur.

Excretion: Because it is excreted in the urine, this is a good drug for urinary tract infections. To remember this, think of the pungent smell of sulfur, that rotten egg smell, and then think of the pungent smell of urine when you walk by a Porta-Potti at a construction site. Make this smell association, and you won't forget.

Adverse Effects

Adverse effects are rare in persons without AIDS. They include nausea, vomiting, diarrhea, and skin rashes (drug eruptions). Approximately half of persons with AIDS develop adverse effects on TMP/SMX, including skin rashes and bone marrow suppression.

Giving TMP/SULFA to a patient on warfarin blood thinner is very dangerous! This drug interaction increases warfarin levels, resulting in a high risk of bleeding.

Clinical Uses

TMP/SMX has no anaerobic coverage, but does have a wide gram-negative and gram-positive coverage (and even covers some Protozoans). Study the following mnemonic **TMP S**MX:

T (Tree): Respiratory tree. TMP/SMX covers *Streptococcus pneumoniae* and *Haemophilus influenzae*. It is good for otitis media, sinusitis, bronchitis, and pneumonia, which are frequently caused by these bugs.

M (Mouth): Gastrointestinal tract. TMP/SMX covers gram-negatives that cause diarrhea such as *Shigella*, *Salmonella*, and *Escherichia coli*.

P (PEE): Genitourinary tract. TMP/SMX covers urinary tract infections, prostatitis, and urethritis caused by the Enterics (*Escherichia coli* and clan).

SMX (**S**yndrome): AIDS. TMP/SMX covers *Pneumocystis carinii* pneumonia (PCP). It is given to **prevent PCP** when CD4+ T-cell counts drop below 200–250. More than 60% of PCP infections are being prevented with this prophylactic intervention! It is also given intravenously in high doses for active pneumonia.

In addition to *Pneumocystis carinii*, other protozoans covered by TMP/SMX are *Toxoplasma gondii* and *Isospora belli*.

Fig. 20-5. Summary of the miscellaneous antibiotics.

NAME	MECHANISM OF ACTION	PHARMOKINETICS
Fluoroquinolones **First generation** • Nalidixic acid **Second generation** • Norfloxacin • Ciprofloxacin • Enoxacin • Lomefloxacin • Ofloxacin • Levofloxacin **Third generation** • Gatifloxacin **Fourth generation** • Moxifloxacin • Gemifloxacin	• Inhibits the enzyme **DNA Gyrase**. This results in the breakage of the bacterial DNA structure, and inhibition of DNA synthesis	1. Oral or IV 2. Enterohepatic circulation results in high concentration within stool) 3. Excellent tissue penetration 4. Renal excretion (results in high urinary levels)
Glycopeptides Vancomycin Telavancin Dalbavancin Oritavancin	• Inhibits biosynthesis of the gram-positive peptidoglycan a step earlier than penicillin. Specifically: inhibits transpeptidation of D-alanine. • Telavancin also disrupts cell membrane potential and changes cell permeability due to its lipophilic side chain.	1. If administered IV: excreted renally 2. If administered orally: A. Not absorbed from GI tract B. Attains high concentration in stool
Daptomycin (CUBICIN)	1. Binds to the gram-positive bacterial cell membrane in a calcium (Ca^{2+})-dependent manner without penetrating the cytoplasm, leading to rapid depolarization of membrane potential. 2. Depolarization causes rapid inhibition of bacterial DNA, RNA & protein synthesis, which results in bacterial cell death.	1. IV administration 2. 92% protein bound 3. Not extensively metabolized 4. Renal excretion: • Adjust dose in renal disease

Figure 20-5   MISCELLANEOUS ANTIBIOTICS (chart continued on page 200)

ADVERSE EFFECTS	THERAPEUTIC USES	MISCELLANEOUS

1. Gastrointestinal symptoms
2. Damage to cartilage in animals. So it is not used in children or pregnant women
3. Achilles Tendonitis
4. CNS: headache, insomnia, restlessness
5. Disruption of bowel flora and increased risk of *Clostridium difficile* diarrhea
	• Covers the gram-negative bacteria exceptionally well
	  1. Ciprofloxacin is indicated for the coverage of *Pseudomonas aeruginosa* (although resistance is increasing!)
	  2. Diarrhea caused by enteric organisms (*Salmonella, Shigella, Campylobacter* or *E. coli*) as high levels are attained in stool
	  3. Urinary tract infections: High renal and prostate concentrations
	  4. Chronic bone infections (osteomyelitis): covers *Pseudomonas, Staphylococcus aureus* or *Enterobacteriaceae*
	  5. Covers gram-negative facultative intracellular organisms, including *Legionella, Brucella, Salmonella,* and atypical *Mycobacteria*
	  6. Newer generation fluoroquinolones: Expanded gram-positive coverage (*Streptococcus pneumoniae, Staphylococcus aureus,* and *Enterococcus faecalis*) and atypical bacteria coverage (*Legionella, Mycoplasma,* and *Chlamydia*) make them good choices for community acquired pneumonia
	  7. Moxifloxacin is indicated for the empiric coverage of intra-abdominal infections due to its broad spectrum of activity, including anaerobes.
		• Resistance develops by point mutations of the DNA gyrase enzyme.

• When administered IV:
  1. Hearing loss is rare
  2. "Red man syndrome": Get red, pruritic rash on torso. This occurs with rapid IV infusion of vancomycin, which stimulates histamine release; simply slow down infusion to prevent
	1. Covers all gram-positive organisms, including exceptionally resistant organisms such as:
	  A. MRSA (methicillin-resistant *Staphylococcus aureus*)
	  B. *Enterococcus*
	  C. Multidrug resistant *Staphylococcus epidermidis.*
	2. Pseudomembranous colitis caused by *Clostridium difficile* (administer orally)
	3. Useful for the treatment of gram-positive organisms in patients who are allergic to penicillin and cephalosporin
		1. There is increasing resistance to vancomycin.
		2. Telavancin retains activity against *S. aureus* with intermediate susceptibility to vancomycin.
		3. Dalbavancin and Oritavancin given IV once weekly

1. Potential for myopathy: monitor baseline CPK enzyme levels and weekly thereafter 2. Eosinophilic pneumonia	• Broad gram-positive coverage, including organisms resistant to methicillin & vancomycin	1. New antibiotic class: cyclic lipopeptide 2. Not indicated for the treatment of pneumonia due to low lung penetration and decreased activity in the presence of pulmonary surfactant.

NAME	MECHANISM OF ACTION	PHARMACOKINETICS

Trimethoprim/ sulfamethoxazole (TMP/SMX): called **Bactrim**	• Together, these two drugs inhibit the synthesis of tetrahydrofolate (TH4), which is a crucial cofactor for the synthesis of purines (nucleic acids). Inhibition of TH4 production will therefore block DNA synthesis
	  1. Sulfamethoxazole looks like PABA. It competitively inhibits conversion of PABA to dihydrofolate (DHF)
	  2. Trimethoprim inhibits the enzyme DHF reductase, blocking conversion of DHF to TH4
	• Animal cells do not synthesize TH4. They require folate in their diet since they can not synthesize TH4. Therefore, TMP/SMX does not block mammalian DNA synthesis
		1. Good oral absorption
		2. Can also be given intravenously
		3. Metabolized in the liver
		4. Renal excretion

Figure 20-5   (continued)

References

Khaliq Y, Zhanel GG. Fluoroquinolone-associated tendionopathy: A critical review of the literature. Clin Infect Dis. 2003;36:1404–1410.

Gilbert DN, Moellering RC, Eliopoulos GM, Sande MA. The Sanford Guide to Antimicrobial Therapy 2005. 35th Edition. Antimicrobial Therapy Inc., Hyde Park VT 2005.

Pepin J, Saheb N, Coulombe M, et al. Emergence of fluoroquinolones as the predominant risk factor for *Clostridium difficile*-associated diarrhea: A cohort study during an epidemic in Quebec. Clin Infect Dis. 2005;41:1254–1260.

Tedesco KL, Rybak MJ. Daptomycin. Pharmacotherapy 2004; 24:41.

Recommended Review Articles:

Nailor MD, Sobel JD. Antibiotics for gram-positive bacterial infections: vancomycin, teicoplanin, quinupristin/dalfopristin, oxazolidinones, daptomycin, dalbavancin, and telavancin. Infect Dis Clin North Am. 2009; 23(4):965–82.

Pappas G, Athanasoulia AP, Matthaiou DK, Falagas ME. Trimethoprim-sulfamethoxazole for methicillin-resistant Staphylococcus aureus: a forgotten alternative? J Chemother. 2009; 21(2):115–26.

ADVERSE EFFECTS	THERAPEUTIC USES	MISCELLANEOUS
1. GI: nausea, vomiting and diarrhea 2. Skin rashes 3. Bone marrow suppression; primarily in patients infected with AIDS 4. Do not use in pregnancy, as it causes increased bilirubin levels in the fetus and reduces TH4 • Folate deficiency: can increase neural tube defects in first trimester 5. Patients with low folate levels can get macrocytic anemia. Coadministering folinic acid will prevent the anemia without affecting its antibacterial effect	• Wide gram-positive and gram-negative coverage (but no anaerobic coverage) • Uses: **TMP S**MX 1. **T** (**T**ree): Respiratory tree. TMP/SMX covers *Streptococcus pneumoniae* and *Haemophilus influenzae*. It is good for otitis media, sinusitis, bronchitis, and pneumonia, which are frequently caused by these bugs 2. **M** (**M**outh): Gastrointestinal tract TMP/SMX covers gram-negatives that cause diarrhea such as *Shigella, Salmonella,* and *E. coli* 3. **P** (**P**EE): Genitourinary tract. TMP/SMX covers urinary tract infections, prostatitis and urethritis caused by the Enterics, *N. gonorrhoeae* and *Chlamydia* 4. **S**MX (**S**yndrome): AIDS. TMP/SMX covers *Pneumocystis carinii* pneumonia (PCP). It is given to **prevent PCP** when CD_4^+ T-cell counts drop below 200–250. More than 60% of PCP infections are being prevented with this prophylactic intervention! It is also given intravenously in high doses for active pneumonia 5. In addition to *Pneumocystis carinii*, other protozoans covered by TMP/SMX are *Toxoplasma gondii* and *Isospora belli* 6. Nocardia	• Other anti-folate drugs: 1. Dapsone 2. Sulfadiazine

M. Gladwin, W. Trattler, and S. Mahan, *Clinical Microbiology Made Ridiculously Simple* ©MedMaster

PART 2. FUNGI

CHAPTER 21. THE FUNGI

As "budding" doctors in the modern world of AIDS, organ transplantation, and modern chemotherapy, you will treat an unprecedented number of immunocompromised patients. With their lowered cell-mediated immunity, there is a dramatic increase in the incidence of virtually every fungal infection! You will commonly see fungi that used to be exceedingly rare.

Fungi are eucaryotic cells, which lack chlorophyll, so they cannot generate energy through photosynthesis. They do require an aerobic environment. After discussing the following crucial terms, we will discuss the categories of fungi pathogenic to humans.

Yeast: Unicellular growth form of fungi. These cells can appear spherical to ellipsoidal. Yeast reproduce by budding. When buds do not separate, they can form long chains of yeast cells, which are called **pseudohyphae**. Yeast reproduce at a slower rate than bacteria.

Hyphae: Threadlike, branching, cylindrical, tubules composed of fungal cells attached end to end. These grow by extending in length from the tips of the tubules.

Molds (also called **Mycelia**): Multicellular colonies composed of clumps of intertwined branching hyphae. Molds grow by longitudinal extension and produce spores.

Spores: The reproducing bodies of molds. Spores are rarely seen in skin scrapings.

Dimorphic fungi: Fungi that can grow as either a yeast or mold, depending on environmental conditions and temperature (usually growing as a yeast at body temperatures).

Saprophytes: Fungi that live in and utilize organic matter (soil, rotten vegetation) as an energy source.

FUNGAL MORPHOLOGY

Certain morphologic characteristics serve as virulence factors as well as targets for antifungal antibiotics.

Cell membrane: The bilayered cell membrane is the innermost layer around the fungal cytoplasm. It contains sterols (sterols are also found in the cell membranes of humans as well as the bacteria *Mycoplasma*). **Ergosterol** is the essential sterol in fungi, while cholesterol is the essential sterol in humans. The majority of antifungal agents work by disrupting ergosterol, either by binding to it and thereby punching holes in the fungal cell wall (**amphotericin B** and **nystatin**), or by interfering with ergosterol synthesis (the **azoles** and **echinocandins**).

Cell wall: Surrounding the cell membrane is the cell wall, composed mostly of carbohydrate with some protein. Fungal cell walls are potent antigens to the human immune system.

Capsule: This is a polysaccharide coating that surrounds the cell wall. This antiphagocytic virulence factor is employed by *Cryptococcus neoformans*. The capsule can be visualized with the **India ink stain**.

Fig. 21-1. It is helpful to organize the human fungal diseases by the depth of the skin that they infect.

SUPERFICIAL FUNGAL INFECTIONS

Pityriasis versicolor and **tinea nigra** are extremely superficial fungus infections, whose primary manifestation is pigment change of the skin. Neither of these cause symptoms and will only come to your attention because skin pigment change is noted!!! Both are named for their respective skin manifestations:

Pityriasis **versicolor** (multicolored)
Tinea **nigra** (black colored)

1) **Pityriasis versicolor** (also called **tinea versicolor**) is a chronic superficial fungal infection which leads to **hypo**pigmented or **hyper**pigmented patches on the skin. With sunlight exposure the skin around the patches will tan, but the patches will remain white. This infection is caused by *Malassezia furfur*.

2) **Tinea nigra** is a superficial fungal infection that causes dark brown to black painless patches on the soles of the hands and feet. This infection is caused by *Exophiala werneckii*.

Diagnosis of both infections is based on microscopic examination of skin scrapings, mixed on a slide with potassium hydroxide (KOH). This will reveal hyphae and spherical yeast, as the KOH digests nonfungal debris. *Malassezia* can look like spaghetti (hyphae) with meatballs (spherical yeast).

Treatment of both consists of spreading dandruff shampoo containing selenium sulfide over the skin. This is an inexpensive and effective treatment. The topical antifungal imidazoles can also be used.

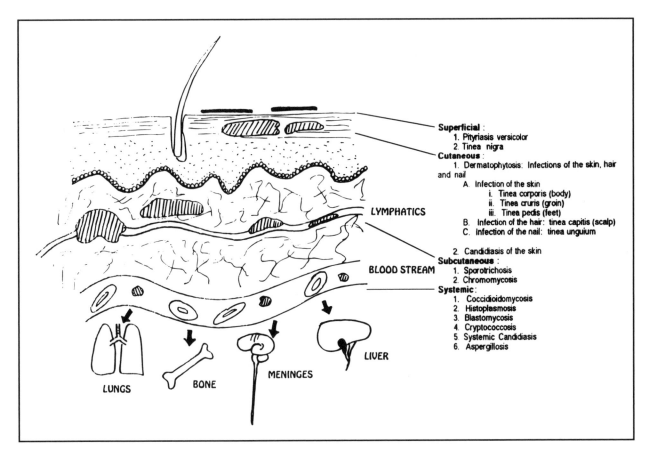

LYMPHATICS

BLOOD STREAM

LUNGS   BONE   MENINGES   LIVER

Superficial :
1. Pityriasis versicolor
2. Tinea nigra

Cutaneous :
1. Dermatophytosis: Infections of the skin, hair and nail
   A. Infection of the skin
      i. Tinea corporis (body)
      ii. Tinea cruris (groin)
      iii. Tinea pedis (feet)
   B. Infection of the hair: tinea capitis (scalp)
   C. Infection of the nail: tinea unguium
2. Candidiasis of the skin

Subcutaneous :
1. Sporotrichosis
2. Chromomycosis

Systemic :
1. Coccidioidomycosis
2. Histoplasmosis
3. Blastomycosis
4. Cryptococcosis
5. Systemic Candidiasis
6. Aspergillosis

Figure 21-1

CUTANEOUS FUNGAL INFECTIONS OF THE SKIN, HAIR, AND NAILS

The Dermatophytoses

Dermatophytoses are a category of cutaneous fungal infections caused by more than 30 species of fungi. The dermatophytic fungi live in the dead, horny layer of the skin, hair, and nails (cutaneous layer seen in **Fig. 21-1**). These fungi secrete an enzyme called keratinase, which digests keratin. Since keratin is the primary structural protein of skin, nails, and hair, the digestion of keratin manifests as scaling of the skin, loss of hair, and crumbling of the nails.

The common dermatophytes include *Microsporum*, *Trichophyton*, and *Epidermophyton*.

1) **Tinea corporis** (body): Following invasion of the horny layer of the skin, the fungi spread, forming a ring shape with a red, raised border. This expanding raised red border represents areas of active inflammation with a healing center. This is appropriately called **ringworm**, since it looks like a ring-shaped worm under the skin.

2) **Tinea cruris (jock itch)**: Patients develop itchy red patches on the groin and scrotum.

3) **Tinea pedis (athlete's foot)**: This infection commonly begins between the toes, and causes cracking and peeling of the skin. Infection requires warmth and moisture, so it only occurs in those wearing shoes.

4) **Tinea capitis** (scalp): This condition primarily occurs in children. The infecting organisms grow in the hair and scalp, resulting in scaly red lesions with loss of hair. The infection appears as an expanding ring.

5) **Tinea unguium (onychomycosis)** (nails): The nails are thickened, discolored, and brittle.

To diagnose a dermatophyte infection:

1) Dissolve skin scrapings in potassium hydroxide (KOH). The KOH digests the keratin. Microscopic examination will reveal branched hyphae.

2) Direct examination of hair and skin with **Wood's light** (ultraviolet light at a wavelength of 365 nm). Certain species of *Microsporum* will fluoresce a brilliant green.

The first-line drugs for treatment of dermatophytoses are the topical imidazoles. The skin should be kept dry and exposed to the drying effects of the air (nudity has its advantages!!). More involved dermatophyte infections of the skin and nails can be treated with oral agents such as terbinafine, and the azoles (primarily fluconazole and itraconazole).

Candida albicans

The last type of cutaneous fungal infection is caused by **Candida albicans**. *Candida* can infect the mouth (oral thrush), groin (diaper rash), and the vagina (*Candida* vaginitis). It can also cause opportunistic systemic infections. All these infections will be discussed in more detail later in this chapter (page 208).

SUBCUTANEOUS FUNGAL INFECTIONS

Subcutaneous fungal infections gain entrance to the body following trauma to the skin. They usually remain localized to the subcutaneous tissue or spread along lymphatics to local nodes. These fungi are normal soil inhabitants and are of low virulence.

Sporothrix schenckii
(Sporotrichosis)

Fig. 21-2. Spore tricks. *Sporothrix schenckii* is a dimorphic fungi commonly found in soil and on plants (rose

thorns and splinters). **Sporotrichosis**, the disease, is an occupational hazard for gardeners. Following a prick by a thorn contaminated with *Sporothrix schenckii*, a subcutaneous nodule gradually appears. This nodule becomes necrotic and ulcerates. The ulcer heals, but new nodules pop up nearby and along the lymphatic tracts up the arm.

Microscopic examination of this fungus reveals yeast cells that reproduce by budding. Culture at 37°C reveals yeast, while culture at 25 °C reveals branching hyphae (dimorphism). Treat with itraconazole, fluconazole, or oral potassium iodide. So if you are going to **POT** roses you might buy some **potassium** iodide!

Phialophora and Cladosporium
(Chromoblastomycosis)

Fig. 21-3. Visualize a **chrome-plated (chromo)** fungus blasting **cauliflower warts** on the skin to help you remember the disease **Chromoblastomycosis**. It is a subcutaneous infection caused by a variety of copper-colored soil saprophytes (*Phialophora* and *Cladosporium*) found on rotting wood. Infection occurs following a puncture wound. Initially, a small, violet wartlike lesion develops. Over months to years, additional violet-colored wartlike lesions arise nearby. Clusters of these lesions resemble cauliflower. Skin scrapings with KOH reveal **copper-colored sclerotic bodies**. Treat with itraconazole and local excision.

Figure 21-2

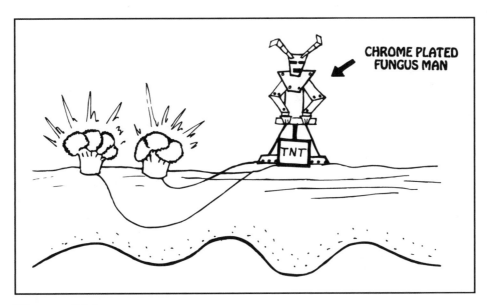

Figure 21-3

SYSTEMIC FUNGAL INFECTIONS

Three fungi that cause systemic disease in humans are *Histoplasma capsulatum*, *Blastomyces dermatitidis*, and *Coccidioides immitis*. All 3 are **dimorphic fungi**. They grow as mycelial forms, with spores, at 25 °C on Sabouraud's agar. At 37 °C on blood agar, they grow in a yeast form. This **dimorphism** plays a part in human infection. In their natural habitat (the soil) they grow as mycelia and release spores into the air. These spores are inhaled by humans and at the "human temperature" of 37 °C they grow as yeast cells.

Geography

Fig. 21-4. *Histoplasma* and *Blastomyces* are endemic to the vast areas that drain into the Mississippi River. Visualize a fungi pilot firing a rocket that **HITS** and **BLASTS** a hole in the Mississippi River.

Fig. 21-5. *Coccidioides* is endemic to the southwestern U.S. (Arizona, New Mexico, southern California) and northern Mexico. Visualize Mr. Fungus as he **COCKS** his pistol in the old **SOUTHWEST**.

Knowledge of these geographic areas is important clinically. For example, *Coccidioides* has become the

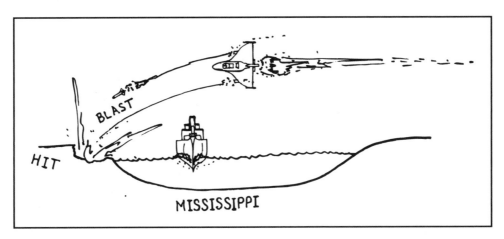

Figure 21-4

Figure 21-5

second most common opportunistic infection in AIDS patients who have resided in Arizona. So a sick AIDS patient with a history of previous residence in the Southwest would raise suspicions.

Mechanism of Disease

All 3 fungi have a similar disease mechanism. Notice the parallels to tuberculosis.

Like *Mycobacterium tuberculosis* the 3 fungi are acquired by inhalation. However, unlike *Mycobacterium tuberculosis*, the fungal infections are inhaled as a spore form and are never transmitted from person to person. Rather, the spores are aerosolized from soil, bird droppings, or vegetation. Like *Mycobacterium tuberculosis*, once inhaled, local infection in the lung is followed by bloodstream dissemination. In most infected persons the fungi are destroyed at this point by the cell-mediated immune system. Antigenic preparations called **coccidioidin** and **histoplasmin** are like the PPD of *Mycobacterium tuberculosis*: when injected intradermally in a previously exposed person they yield a delayed type hypersensitivity reaction which results in localized swelling within 24–48 hours.

The 3 fungi have 3 clinical presentations:

1) **Asymptomatic**: The majority of cases are asymptomatic or mild respiratory illnesses that go unreported.
2) **Pneumonia**: A mild pneumonia can develop with fever, cough, and chest X-ray infiltrates. Like

tuberculosis, granulomas with calcifications can follow resolution of the pneumonia.

A small percentage of persons will develop a severe pneumonia, and an even smaller group will progress to a chronic cavitary pneumonia, marked by weight loss, night sweats, and low-grade fevers, much like a chronic tuberculosis pneumonia.

3) **Disseminated**: Rarely, the hematogenously spread fungi can actually cause disseminated disease, such as meningitis, bone lytic granulomas, skin granulomas that break down into ulcers, and other organ lesions. This disseminated form commonly occurs in the immunocompromised host.

All 3 are best diagnosed by obtaining a biopsy of the affected tissue: bronchoscopic biopsy of lung lesions, skin biopsy, etc. The tissue can be examined with silver stain for yeast or can be grown on Sabouraud's agar or blood agar. **The tissue is the issue**!!!! Skin tests are not very helpful for diagnosis, as many people have been previously exposed asymptomatically and will have a positive test anyway. Serologic tests can be helpful (complement fixation, latex agglutination). A urine histoplasma antigen test is also used to assist with diagnosis.

Acute pulmonary histoplasmosis and coccidioidomycosis usually require no treatment, as the infection is mild. For chronic or disseminated disease, itraconazole or amphotericin B is often required for months! All *Blastomyces* infections require aggressive amphotericin B or itraconazole treatment.

Summary of *Histoplasma, Blastomyces,* and *Coccidioides*

LIKE TUBERCULOSIS
Inhaled, primary infection in the lung.
Asymptomatic, mild, severe,
  or chronic lung infections.
Lung granulomas, calcifications,
  and/or cavitations.
Can disseminate hematogenously
  to distant sites.
Skin test like PPD.

UNLIKE TUBERCULOSIS
No person-to-person transmission.
Fungi with spores, NOT acid-fast
bacteria.

Fig. 21-6. To remember that *Coccidioides, Blastomyces,* and *Histoplasma* all are inhaled as spores and cause disease in the lungs, skin, bones, and meninges, study **Cowboy Fungus**. He has **spore** bullets, **cocks** his gun, then **blasts** and **hits** the lung, skin, bone, and meninges.
Histoplasma capsulatum:
  Nonencapsulated despite its name.

Figure 21-6

Present in bird and bat droppings, so outbreaks of pneumonia occur when cleaning chicken coops or spelunking (cave exploring).

Blastomyces dermatitidis:
Fungi are isolated from soil and rotten wood.
The rarest systemic fungal infection.
When it does cause infection, it is rarely asymptomatic or mild. Most cases present as chronic disseminated disease with weight loss, night sweats, lung involvement, and skin ulcers. Blastomyces is the hardest to get and the hardest to have!

The b**LAST** to get,
No **BLAST** to have!!!

Coccidioides immitis:
Commonly causes a mild pneumonia in normal persons in the southwestern U.S.
Common opportunistic infection in AIDS patients from that area.

Cryptococcus neoformans and *Cryptococcus gattii*
(Cryptococcosis)

There are over 30 species of *Cryptococcus* but only two are known to cause disease in humans, *Cryptococcus neoformans* and *Cryptococcus gattii*. Although endemic to Papua New Guinea and Northern Australia *Cryptococcus gattii* was first reported in the United States in 1999 in the Pacific Northwest and is considered an emerging infectious disease. There were over 100 *C. gattii* confirmed infections in the United States between 2004 and 2011.

Cryptococcus neoformans is widespread and the more common species to cause infection in humans. *Cryptococcus* is a polysaccharide encapsulated yeast (not dimorphic) similar to the previous 3 fungi in that it is inhaled into the lungs and the infection is usually asymptomatic. It differs in that the major manifestation is not pneumonia but rather meningoencephalitis.

This fungus is found in nature, especially in pigeon droppings. Following inhalation and local lung infection, often asymptomatic, the yeast spreads via the blood to the brain. Most cases (3/4) occur in immunocompromised persons. In fact, almost 10% of AIDS patients develop cryptococcosis.

A subacute to chronic meningitis develops in cryptococcosis with headache, nausea, confusion, staggering gait, and/or cranial nerve deficits. Fever and meningismus can be mild. Cryptococcal meningitis is fatal without treatment, because cerebral edema progresses to eventual brainstem compression.

Cryptococcus can also cause pneumonia, skin ulcers, and bone lesions like the other systemic fungi.

The key to diagnosis is doing a lumbar puncture and analyzing the cerebrospinal fluid. An India ink stain shows yeast cells with a surrounding halo, the polysaccharide capsule. This test is positive half of the time. A more sensitive test is the cryptococcal antigen test, which detects cryptococcal polysaccharide antigens. Culture will confirm the diagnosis.

Fig. 21-7. AIDS and cryptococcosis.
The usual treatment of crytococcal meningitis in persons with AIDS is amphotericin B and flucytosine initially for 2 weeks followed by fluconazole for at least

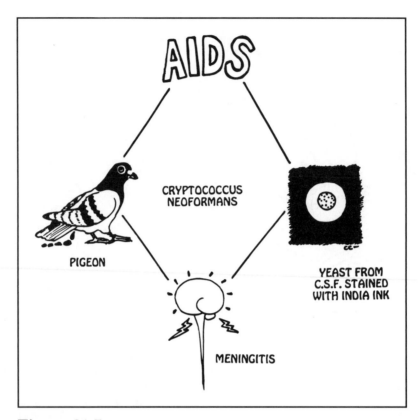

Figure 21-7

8 weeks followed by chronic suppressive therapy until CD4 counts have risen above 100. Non-meningeal disease in immunocompetent patients can often be treated with long courses of oral fluconazole alone.

Candida albicans
(Candidiasis)

As a physician you will see this yeast everywhere. It is given out like **CANDY** to humanity: women with vaginitis, babies with diaper rash, AIDS patients, and the list goes on. In the **normal host**, *Candida albicans* causes 3 infections that are cutaneous. In the **immunocompromised patient**, it can cause any of the 3 cutaneous infections, as well as cause invasive systemic disease.

In Normal Hosts

1) **Oral thrush**: Patches of creamy white exudate with a reddish base cover the mucous membranes of the mouth. These are difficult to scrape off with a tongue blade. Swish and spit preparations of nystatin or amphotericin B, or merely sucking on imidazole candies will resolve this infection.

2) **Vaginitis**: There are 20 million cases of "yeast infection" every year in the U.S. alone! Women develop

Candida vaginitis more frequently when taking antibiotics, oral contraceptives, or during menses and pregnancy. The symptoms are vaginal itching and discharge (thick copious secretions wetting the underwear). Speculum examination reveals inflamed vaginal mucosa and patches of cottage cheese-appearing white clumps affixed to the vaginal wall. Imidazole vaginal suppositories are helpful. Alternatively, a single dose of oral fluconazole is very effective.

3) **Diaper rash**: Warm moist areas under diapers and in adults between skin folds (under breasts for example) can become red and macerated secondary to *Candida* invasion.

In Immunocompromised Patients

4) **Esophagitis**: Extension of thrush into the esophagus causes burning substernal pain worse with swallowing. *Candida* does not infect the esophagus in immune-competent persons!

5) **Disseminated**: *Candida* can invade the bloodstream and virtually every organ. When systemic candidiasis is suspected, the retina must be examined with the ophthalmoscope. Multiple white fluffy candidal patches occasionally may be visualized. **Clinical Pearl**: Since *Candida* is normal flora, it is often cultured from the urine, sputum, and stool. These can represent

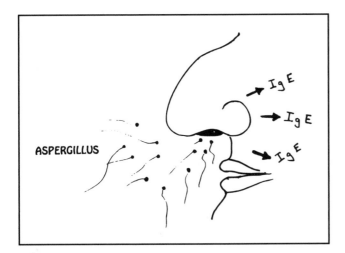

Figure 21-8

contaminants. However, isolation from the blood is never normal and must be respected!!!

Diagnosis is made with KOH preparation of skin scrapings, or with stains and cultures of biopsied tissue or blood. New non-culture methods to diagnose systemic candidiasis include: a blood test for beta-D-glucan, a component of fungal cell walls; the T2Candida assay which detects amplified DNA using magnetic resonance technology, and a PCR assay. Each of these new tests has pros and cons and has not replaced cultures as the gold standard.

Systemic infection requires intravenous therapy with **amphotericin B, fluconazole**, or an echinocandin (**caspofungin or micafungin**).

Aspergillus
(Aspergillosis)

Aspergillus species, of which *Aspergillus fumigatus* is the most common human pathogen, cause three major types of diseases in humans: an allergic reaction in the airways called **Allergic Bronchopulmonary Aspergillosis (ABPA)**, an infection in preformed lung cavities, called an **Aspergilloma**, and invasive infection of lung, called **invasive aspergillosis**.

ASPIRATION of *ASPERGILLUS* = ASTHMA

Fig. 21-8. Inhalation of aspergillus spores can cause a disease called **Allergic Bronchopulmonary Aspergillosis (ABPA)**. The spores of *Aspergillus* mold are floating in the air everywhere. Some persons develop an asthma-type reaction to these spores. They have a type 1 hypersensitivity reaction (IgE-mediated immediate allergic reaction) with bronchospasm, increase in IgE antibodies, and blood eosinophilia. They also manifest a type 4 reaction (delayed type cell-mediated allergic reaction) with cell-mediated inflammation and lung infiltrates. Systemic corticosteroids and oral itraconazole are an effective treatment.

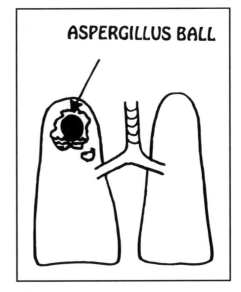

Figure 21-9

Fig. 21-9. Persons with lung cavitations from tuberculosis or malignancies can grow an aspergillus fungal ball in the cavity, called an **aspergilloma**. This ball can be large (as big as a golf ball) and requires surgical removal.

Immunocompromised hosts can develop invasive pneumonias and disseminated disease. Bloody sputum may occur, due to blood vessel wall invasion by *Aspergillus* hyphae.

Invasive aspergillosis usually occurs in immunocompromised hosts such as patients who are neutropenic after chemotherapy or patients on high dose steroids for the treatment of graft versus host disease. Patients with end-stage AIDS and CD4 counts less than 50 cells/microL can sometimes develop this as well. The disease presents as a slowly progressive, often asymptomatic (or with low grade fevers) pneumonia, characterized by multiple nodular infiltrates on the chest CT scan. This disease has a very high mortality rate, and treatment with the most powerful antifungal agents is required to save your patient's life. Treatment options include voriconazole, amphotericin preparations, and caspofungin.

Aspergillus and other fungi produce toxins that cause liver damage and liver cancer. These toxins are called **mycotoxins**. The toxin produced by *Aspergillus* is called the **aflatoxin**. This has worldwide significance since *Aspergillus* grows ubiquitously, contaminating peanuts, grains, and rice. The fact that half of the cancers south of the Sahara desert in Africa are liver cancers and 40% of screened foods contain aflatoxins suggests that this is a real threat.

Mucormycosis

Mucormycosis (zygomycosis) is an opportunistic disease caused by several molds in the order Mucorales of the phylum Zygomycota. The primary pathogens are included in the generas of *Rhizopus, Rhizomucor,* and *Mucor*. These

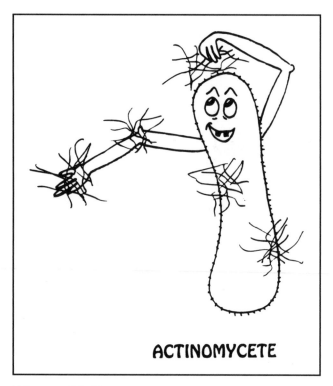

ACTINOMYCETE

Figure 21-10

molds are found everywhere in the environment. Persons at risk for infection are those who develop profound **acidosis**, such as **diabetics**, the immunocompromised, burn patients, and persons taking the iron chelator **deferoxamine**. Rhinocerebral and pulmonary involvement is most common. The molds aggressively invade the sinuses, cranial bones, and blood vessels. Treatment requires aggressive surgical debridement and anti-fungal therapy with amphotericin or posaconazole.

THE FUNGI-LIKE BACTERIA: ACTINOMYCETES AND NOCARDIA

Fig. 21-10. Actinomycetes are bacteria **acting like fungi.** These are procaryotic organisms and are truly bacteria. They are discussed here because they frequently grow in the form of mycelia and are water and soil saprophytes. The 2 that cause human disease are *Actinomyces* and *Nocardia*, both of which are gram-positive rods.

Actinomyces Israelii

There are **4 concepts** you should know about this organism:

1) It is a gram-positive, beaded, filamentous anaerobic organism that grows as normal flora in the mouth and GI tract.

NAME	RESERVOIR	MORPHOLOGY	CLINICAL
Malassezia furfur		• "Spaghetti and meatballs"	***Pityriasis versicolor*** • Hypo or hyperpigmented patches on the skin; surrounding skin darkens with sunlight while the patches remain white
Exophiala werneckii		• Brown-pigmented, branched, septate hyphae and budding yeast cells	***Tinea nigra*** • Dark brown to black patches on the soles of the hands or feet
1. *Microsporum* species 2. *Trichophyton* species 3. *Epidermophyton floccosum*	• Depending on the particular species: 1. Soil 2. Animals 3. Humans		***Dermatophytosis*** 1. Tinea corporis (body): "ringworm" 2. Tinea cruris (groin): "jock itch" 3. Tinea pedis (feet): "athlete's foot" 4. Tinea capitis (scalp) 5. Tinea unguium (nail): "Onychomycosis"
Sporothrix schenckii	• Found on rose thorns		***Sporotrichosis*** 1. Subcutaneous nodule gradually appears at site of thorn prick 2. This nodule becomes necrotic and ulcerates 3. This ulcer heals, but new nodules pop up nearby along the lymphatic tracts

Figure 21-11 FUNGI

2) It causes eroding abscesses following trauma to the mucous membranes of the mouth or GI tract. The infection is named according to the area of the body through which the abscess erodes: **cervicofacial actinomycosis, abdominal actinomycosis**, and **thoracic actinomycosis**.

3) When examined under the microscope, the pus draining from the abscess reveals yellow granules, called **sulfur granules**. These are not composed of sulfur but of microcolonies of *Actinomyces* and cellular debris.

4) Treatment of this gram-positive bacterium is with **penicillin G** and surgical drainage.

Note: *Actinomyces* and *Nocardia* are **both** filamentors, beaded, branching gram-positive organisms, but **only** *Actinomyces* forms sulfur granules and **only** *Nocardia* is acid-fast.

Nocardia asteroides

Nocardia forms weakly gram-positive, **partially acid-fast beaded branching thin filaments**! It is not considered normal flora. Infections with *Nocardia* are frequently misdiagnosed as tuberculosis because it is acid-fast and it causes the same disease process. Like *Mycobacterium tuberculosis, Nocardia* is inhaled and grows in the lung to produce lung abscesses and cavitations. Erosion into the pleural space can occur, as well as blood-born dissemination, resulting in abscesses in the brain and other organs. Immunocompromised patients, especially those taking steroids, are particularly at risk for *Nocardia* infection. Treatment is with **trimethoprim** and **sulfamethoxazole**.

Treatment of *Actinomyces* and *Nocardia* is a **SNAP!**
Sulfa for
Nocardia
Actinomyces give
Penicillin

Fig. 21-11. Summary of the fungi.

References:

Harris JR, et al. Cryptococcus gattii in the United States: Clinical Aspects of Infection with an Emerging Pathogen. Clin Infect Dis 2011;53:1188–95.

Ostrosky-Zeichner L, Alexander BD, Kett DH, et al. Multicenter clinical evaluation of the (1 → 3) beta-D-glucan assay as an aid to diagnosis of fungal infections in humans. Clin Infect Dis 2005;41:654–9.

www.doctorfungus.org

Recommended Review Articles:

Pfaller MA, Diekema DJ. Epidemiology of invasive candidiasis: a persistent public health problem. Clin Microbiol Rev. 2007 Jan; 20(1):133–63.

TREATMENT	DIAGNOSIS	MISCELLANEOUS	LOCATION
1. Dandruff shampoo (containing selenium sulfide) 2. Topical imidazole	**Potassium hydroxide** (KOH) prep: reveals short, curved, unbranched hyphae with spherical yeast cells (looks like "spaghetti and meatballs")		SUPERFICIAL
1. Dandruff shampoo (containing selenium sulfide) 2. Topical imidazole	• KOH prep: brown-pigmented, branched, septate hyphae and budding yeast cells		SUPERFICIAL
1. Topical imidazole 2. Oral griseofulvin is used for tinea unguium and tinea capitis 3. Oral terbinafine	1. KOH: branched hyphae 2. Wood's light: certain species of *Microsporum* will fluoresce under ultraviolet light	• Secretes the enzyme keratinase, which digests keratin	CUTANEOUS
1. Itraconazole 2. Fluconazole 3. Oral potassium iodide	*Dimorphic* 1. Culture at 25 °C: will grow branching hyphae 2. Culture at 37 °C: will grow yeast cells		SUBCUTANEOUS

NAME	RESERVOIR	MORPHOLOGY	CLINICAL
1. *Phialophora verrucosa* 2. *Cladosporium carrionii* 3. *Fonsecaea* species	• These copper-colored soil saprophytes can be found on rotting wood	• Sclerotic bodies: copper colored cells	**Chromoblastomycosis** Following a puncture wound, a small, violet wart-like lesion develops. With time, clusters of these skin lesions can develop (resembling cauliflower)
Coccidioides immitis	1. Desert areas of the southwestern United States and northern Mexico 2. Respiratory transmission	*Dimorphic*: 1. Mycelial forms with spores at 25 °C 2. Yeast forms at 37 °C	**Coccidioidomycosis** 1. Asymptomatic (in most persons) 2. Pneumonia 3. Disseminated: can affect the lungs, skin, bones, and meninges • Note: A small percentage of individuals with this infection will develop painful erythematous nodular lesions called **erythema nodosum**
Histoplasma capsulatum	1. Mississippi valley 2. Present in bird and bat droppings 3. Respiratory transmission	*Dimorphic*: 1. Mycelial forms with spores at 25 °C 2. Yeast forms at 37 °C 3. No capsule (despite its name)	**Histoplasmosis** 1. Asymptomatic (in most persons) 2. Pneumonia: lesions calcify, which can be seen on chest X-ray (may look similar to tuberculosis) 3. Disseminated: can occur in almost any organ, especially in the lung, spleen, or liver
Blastomyces dermatitidis	1. Mississippi River valley extending North to the Great Lakes 2. Resides in soil or rotten wood 3. Respiratory transmission	*Dimorphic* 1. Mycelial forms with spores at 25°C 2. Yeast forms at 37°C	**Blastomycosis** 1. Asymptomatic (uncommon) 2. Pneumonia: lesions rarely calcify 3. Disseminated (most common): present with weight loss, night sweats, lung involvement, and skin ulcers 4. Cutaneous: skin ulcers
Cryptococcus neoformans *Cryptococcus gattii*	1. Found in pigeon droppings 2. Respiratory transmission 1. Found in soil	1. Polysaccharide capsule 2. Yeast form only (Not dimorphic!)	**Cryptococcosis** 1. Subacute or chronic meningitis A. Headache B. Fever C. Vomiting D. Neurologic or mental status changes 2. Pneumonia: usually self-limited 3. Skin lesions: look like acne
Candida albicans	1. Normal inhabitant of the skin, mouth and gastrointestinal tract 2. Not found in blood!	• Pseudohyphae and yeast	**Candidiasis in a normal host** 1. Oral thrush 2. Vulvovaginal candidiasis 3. Cutaneous A. Diaper rash B. Rash in the skin folds of obese individuals

Figure 21-11   (continued) (chart continued on page 214)

TREATMENT	DIAGNOSIS	MISCELLANEOUS	LOCATION
1. Itraconazole 2. Local excision	• Skin scrapings with KOH prep reveal copper-colored cells, called sclerotic bodies		SUBCUTANEOUS

1. Amphotericin B
2. Itraconazole
3. Fluconazole
	1. Biopsy of affected tissue: lung biopsy, skin biopsy, etc.
	  A. Silver stain or KOH prep
	  B. Culture on Sabouraud's agar
	2. Serology
	3. Skin test (tests for exposure only)
		Common opportunistic infection in AIDS patients from the southwest United States	SYSTEMIC

1. Itraconazole
2. Amphotericin B (in immunocompromised patients)
	1. Lung biopsy
	  A. Silver stain specimen
	  B. Culture on Sabouraud's agar will reveal hyphae at 25 °C, and yeast at 37 °C.
	2. Serology
	3. Skin test (tests for exposure only)
	4. Urine antigen test.
		• Can survive intracellularly within macrophages	SYSTEMIC

1. Itraconazole
2. Ketoconazole
3. Amphotericin B
	1. Biopsy of affected tissue: lung biopsy, skin biopsy, etc.
	  A. Silver stain specimen
	  B. Culture on Sabouraud's agar
	2. Serology
	3. Skin test (tests for exposure only)
		1. b**LAST** to get
		2. **No BLAST** to have
			SYSTEMIC

1. Amphotericin B and flucytosine (is superior to amphotericin B alone) 2. Fluconazole	1. **India-ink** stain of cerebrospinal fluid (CSF): observe encapsulated yeast 2. Cryptococcal antigen test of CSF: detects polysaccharide antigens 3. Fungal culture	• Most cases occur in immunocompromised persons	SYSTEMIC
• The choice of antifungal agent depends on on the area involved and its severity. Possible choices include: 1. Thrush- oral fluconazole, nystatin swish and spit, and clotrimazole candies 2. Cutaneous infection: topical imidazole or oral fluconazole.	1. KOH stain of specimen 2. Silver stain of specimen 3. Blood culture: growth must be respected 4. Blood assay for beta-D-glucan. 5. T2Candida- blood assay for amplified candida DNA 6. PCR		CUTANEOUS or SYSTEMIC (Normal host, or opportunistic)

NAME	RESERVOIR	MORPHOLOGY	CLINICAL

Candida albicans (continued)			***Candidiasis in an immunocompromised host***
			• Thrush, vaginitis and/or cutaneous, plus:
			  1. Esophageal:
			    A. Retrosternal chest pain
			    B. Dysphagia
			    C. Fever
			  2. Disseminated candidiasis: acquired by very sick hospitalized patients, resulting in multi-organ system failure
			  3. Chronic mucocutaneous candidiasis

1. *Aspergillus fumigatus* 2. *Aspergillus flavus* 3. *Aspergillus niger*	1. Everywhere (frequent lab contaminant) 2. Aspiration of *Aspergillus* = asthma	• Branching septated hyphae	***Aspergillosis*** 1. **Allergic bronchopulmonary aspergillosis** (IgE mediated): asthma type reaction with shortness of breath and high fever 2. **Aspergilloma** (fungus ball): associated with hemoptysis (bloody cough) 3. **Invasive aspergillosis**: necrotizing pneumonia. May disseminate to other organs in immunocompromised patients 4. **Aflatoxin** consumption (produced by *Aspergillus flavus*) can cause liver damage and liver cancer
1. *Rhizopus* 2. *Rhizomucor* 3. *Mucor*	• Saprophytic molds	• Broad, non-septated, branching hyphae	***Mucormycosis*** 1. Rhinocerebral (associated with diabetes): starts on nasal mucosa and invades the sinus and orbit 2. Pulmonary mucormycosis
THE FUNGI-LIKE BACTERIA			

Actinomyces israelii	• Part of the normal flora of the mouth and gastrointestinal tract	1. Gram-positive rods
		2. **Anaerobic** bacteria
		3. Grow as branching chains or beaded filaments
			• Eroding abscesses of the mouth, lung or gastrointestinal tract, classified as:
			  1. Cervicofacial actinomycosis
			  2. Thoracic actinomycosis
			  3. Abdominal actinomycosis

Nocardia asteroides	1. Never part of the normal flora 2. Respiratory transmission	1. Gram-positive rods 2. Partially **acid-fast:** due to mycolic acids in the cell wall 3. Aerobic 4. Grow as branching chains or beaded filaments	1. Pneumonia 2. Formation of abscesses in the lung, kidney, and central nervous system

Figure 21-11 (continued)

TREATMENT	DIAGNOSIS	MISCELLANEOUS	LOCATION
3. Esophageal candidiasis (most common in HIV): fluconazole or caspofungin. 4. Systemic candidiasis: intravenous amphotericin B, fluconazole, or caspofungin. 5. Chronic mucocutaneous candidiasis: ketoconazole or fluconazole			

1. Allergic bronchopul-monary aspergillosis: treat with cortico-steroids
2. Aspergilloma: removal via surgery
3. invasive aspergillosis: treat with voriconazole, amphotericin B, or possibly caspofungin. (very high mortality)
	A. Allergic bronchopulmonary aspergillosis:
	  1. High level of Ige and IgG against aspergillis
	  2. Sputum culture
	  3. Wheezing patient and chest X-ray with fleeting infiltrates
	  4. Increased level of eosinophils
	  5. Skin test: immediate hypersen-sitivity reaction
	B. Aspergilloma: diagnose with chest X-ray or CT scan
	C. Invasive aspergillosis: sputum examination and culture
		• Aflatoxins contaminate peanuts, grains and rice	

1. Aggressive surgical debridement 2. IV Amphotericin or posaconazole as an oral alternative	1. Biopsy 2. Black nasal discharge	• This disease is rapidly fatal	
1. Penicillin G 2. Surgery	1. Examine tissue or pus from infection site, and look for "sulfur granules" 2. Anaerobic culture	• Yellow "sulfur granules": microcolonies of *Actino-myces* and cellular debris	
• Trimethoprim/ sulfamethoxazole	1. Gram stain 2. Modified acid fast stain: Decolorize with 1% sulfuric acid instead of acid alcohol. 3. Aerobic culture	• *Nocardia* infections usually occur in immunocompro-mised patients	

CHAPTER 22. ANTIFUNGAL ANTIBIOTICS

The number of fungal infections has risen dramatically with the increase in patients who are immunocompromised from AIDS, chemotherapeutic drugs, and organ transplant immunosuppressive drugs. For this reason you will frequently use the handful of antifungal antibiotics available.

Ergosterol is a vital part of the cell membranes of fungi but is not found in the cell membranes of humans, which contain cholesterol. Most antifungal agents bind more avidly to ergosterol than to cholesterol, thus more selectively damaging fungal cells than human cells. By binding to or inhibiting ergosterol synthesis, they increase the permeability of the cell membranes, causing cell lysis.

We can divide antifungal agents into 6 major groups based on their mechanism of action:

1) **Polyenes**: This group includes **amphotericin B** and its newer formulations. Amphotericin is the granddaddy of all antifungal agents. It forms a complex with **ergosterol** and disrupts the fungal plasma membrane, leading to leakage of the cytoplasmic contents and fungal cell death. Amphotericin covers the majority of medically important fungi. It must be given intravenously due to poor oral absorption. It is also used intrathecally (directly into the cerebrospinal fluid) and may be used as a bladder wash for fungal cystitis (bladder infection).

2) **Anti-metabolites**: There is only one agent in this class, flucytosine.

3) **Azoles**: These agents inhibit the fungal cytochrome P450 3A-dependent C 14-demethylase which converts lanosterol to ergosterol, leading to the depletion of ergosterol in the fungal cell membrane. This class contains multiple agents such as **clotrimazole, fluconazole, itraconazole, voriconazole, posaconazole,** and **isavuconazole**. This class of anti-fungals is used extensively due its favorable side-effect profile.

4) **Glucan synthesis inhibitors**: Otherwise known as **echinocandins**, these newer agents, which include **caspofungin, micafungin,** and **anidulafungin**, are given intravenously with few side effects. They inhibit fungal cell wall synthesis by inhibiting the enzyme 1,3 D-glucan synthase.

5) **Allylamines**: This class consists primarily of topical antifungal agents, with the exception of **terbinafine**, which is topical and oral. These agents inhibit ergosterol biosynthesis via inhibition of squalene oxidase.

6) **Others**: This catch-all group includes **griseofulvin**, a novel antifungal which prevents fungal cell division by disruption of the fungal cell mitotic spindles.

Amphotericin B

The classic antifungal antibiotic is amphotericin B. Most species of fungi are susceptible to it, and although it has many side effects, it remains a first line option for many serious systemic fungal infections:

Systemic *Candida* infections.

Cryptococcal meningitis: used in combination with flucytosine.

Severe pneumonia and extrapulmonary Blastomycosis, Histoplasmosis, and Coccidioidomycosis.

Invasive Aspergillosis.

Invasive Sporotrichosis.

Mucormycosis

Adverse Effects

On the wards this drug is referred to as ampho*terrible* and *Awful*tericin because of its numerous adverse effects:

1) **Renal toxicity**: (Poor Mr. Kidney!) There is a dose-dependent azotemia (increase in BUN and creatinine reflecting kidney damage) in most patients taking this drug. This is **reversible** if the drug is stopped. The creatinine level must be followed closely, and if it becomes too high (creatinine > 3), the dosage may have to be lowered, terminated, or switched to alternate day regimens.

2) **Acute febrile reaction**: A shaking chill (rigors) with fever occurs in some people after IV infusion. This is a **common** side effect.

3) **Anemia**.

4) **Inflammation of the vein** (phlebitis) at the IV site.

These side effects are important because they are very common. In fact, when amphotericin B is given in the hospital, it is usually given with aspirin or acetaminophen to prevent the febrile reaction. Daily BUN and creatinine levels are drawn to monitor kidney function. You can see that these side effects are important in day-to-day clinical management! Newer lipid and liposomal preparations of Amphotericin B cause less nephrotoxicity.

Fig. 22-1. Properties of **amphotericin B**, the "**amphibian terrorist**": intravenous drug delivery; fungicidal by binding to ergosterol in the fungal cell membrane, causing membrane disruption and osmotic lysis of the cell; nephrotoxicity.

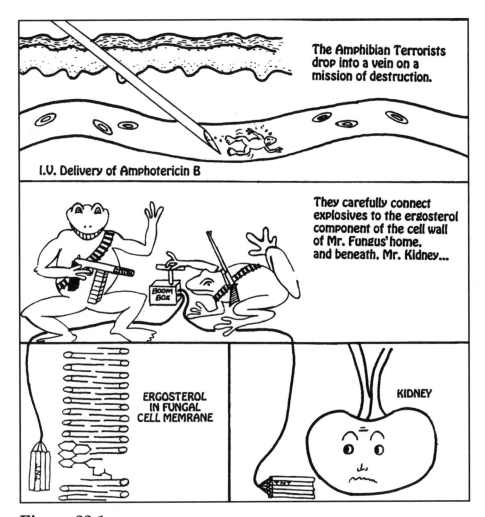

The Amphibian Terrorists drop into a vein on a mission of destruction.

I.V. Delivery of Amphotericin B

They carefully connect explosives to the ergosterol component of the cell wall of Mr. Fungus' home. and beneath. Mr. Kidney...

ERGOSTEROL IN FUNGAL CELL MEMRANE

KIDNEY

Figure 22-1

To speed amphotericin's travel through the kidneys and decrease renal toxicity, hydration with normal saline is used commonly with traditional amphotericin B. This hydration is generally not required with the newer preparations of amphotericin B. Electrolyte replacement is another important adjunct of amphotericin therapy because amphotericin causes increases in urinary excretion of potassium, magnesium and bicarbonate.

Newer preparations of amphotericin B are available that add different lipids (fats!) to the traditional (old fashioned) amphotericin B deoxycholate. The addition of the lipid decreases the nephrotoxicity of the drug, making it less *Amphoterrible*.

Amphotericin B colloidal dispersion (ABCD: Amphocil): Ampho B + cholesterol sulfate. Rigors still occur but nephrotoxicity is reduced.

Amphotericin B lipid complex (ABLC): Ampho B + dimyristoylphosphatidylglycerols and dimyristoylphosphatidylcholines. Rigors still occur but there is less nephrotoxicity.

Liposomal amphotericin B (Ambisome): A unilamellar liposome containing a mixture of 1 molecule of amphotericin B surrounded by a coating of nine molecules of lipid (soy lecithin, cholesterol, and distearoylphosphatidylglycerol), like a coated jaw breaker. There is little nephrotoxicity or rigors.

Some hospitals make their own concoction by adding amphotericin B deoxycholate to Intralipid (parenteral fat for intravenous feeding) in a mixture of 1–2 mg amphotericin B per ml lipid. Less nephrotoxicity is seen, but once again we do not yet know enough about antifungal efficacy.

Flucytosine

Flucytosine is rarely used alone because of rapid development of resistance. Think of it as the tag team wrestling partner of amphotericin B. Amphotericin B busts holes in the cell membranes and flucytosine enters and inhibits DNA/RNA synthesis.

Most fungi are resistant to flucytosine, but *Cryptococcus* and *Candida* are the exceptions. Flucytosine use is mostly limited to the treatment of cryptococcal meningitis, in conjunction with Amphotericin B.

Adverse Effects

1) **Bone marrow depression**, resulting in leukopenia and thrombocytopenia. Remember that most antimetabolite type drugs will do this (methotrexate, sulfa drugs, 5-fluorouracil, etc.).
2) **Nausea, vomiting, diarrhea**. This again is common with the antimetabolites, such as the chemotherapeutic drugs.

The reason for these adverse effects is that the drugs damage DNA during its formation in rapidly dividing cells such as bone marrow and GI epithelial cells.

The Azole Family

The azole family may be classified into 2 groups of drugs: the imidazoles and the triazoles.

IMIDAZOLES	TRIAZOLES
Ketoconazole	Fluconazole
Miconazole	Itraconazole
Clotrimazole	Voriconazole
	Posaconazole
	Isavuconazole

The azoles inhibit the cytochrome P-450 enzyme system, which is involved in ergosterol synthesis. The depletion of ergosterol disrupts the permeability of the fungal cell membrane.

These drugs are active against a **broad** spectrum of fungi.

Clotrimazole and **miconazole** are too toxic for systemic use and for this reason, are primarily used for **topical fungal infections**, including pityriasis versicolor, cutaneous candidiasis, and the dermatophytosis (tinea pedis, corporis, etc.). Clotrimazole troches (like candies) are sucked to treat oral *Candida* (thrush), and clotrimazole vaginal suppositories treat *Candida* vaginitis.

Ketoconazole, fluconazole, itraconazole, voriconazole, posaconazole, and **Isavuconazole** are tolerated orally and have many important uses for **systemic fungal infections**.

Ketoconazole

Ketoconazole, one of the imidazoles, has a fairly broad spectrum of activity against many fungi, but it has been largely replaced by the newer, more effective, less toxic triazoles.

Adverse Effects

1) **GI**: Nausea, vomiting, and anorexia, all common.

2) **Hepatotoxicity**: This is usually seen as a temporary rise of hepatic enzymes but on rare occasions can lead to hepatic necrosis. Follow enzymes when on this drug.
3) **Inhibition of testosterone synthesis**: Ketoconazole inhibits the cytochrome P-450 system, which is important in testosterone synthesis. The result is gynecomastia, impotence, decreased sex drive (libido), and decreased sperm production.
4) Adrenal suppression.

Fluconazole

Fluconazole is one of the triazoles; it is less toxic and has broader antifungal coverage than ketoconazole. Fluconazole is one of the most heavily used antifungals. On your medicine rotations you will see it prescribed often!! It is used primarily for susceptible candida infections (primarily *Candida albicans*), both for superficial and disseminated infections. In HIV/AIDS patients with low CD4 counts fluconazole is used to treat oral thrush (a candida infection) and it is also used in the treatment of **cryptococcal** meningitis (primarily after an initial course of IV amphotericin).

The big picture with fluconazole is that it kills *Candida albicans* very well:

1) Studies comparing it to amphotericin B in the treatment of systemic *Candida albicans* infection (in non-neutropenic patients) demonstrated equivalent efficacy.
2) A single dose of fluconazole very effectively clears candida vaginitis.

Itraconazole

Itraconazole is useful in the treatment of chromoblastomycosis, histoplasmosis, coccidioidomycosis, blastomycosis, sporotrichosis, and indolent cases of aspergillosis. The main problem with this drug is poor oral absorption. Taking it with acid drinks such as orange juice or colas enhances absorption (need low pH). A new IV formulation has also been developed to avoid poor absorption.

Voriconazole

Voriconazole has a **Voracious** appetite for fungi!!! Voriconazole is an exciting newer arrival to our antifungal armamentarium. It has broad activity against multiple fungi, similar to fluconazole and itraconazole. Its niche at this point is its superior activity against invasive aspergillus and against fluconazole resistant candida species (non-albicans). Voriconazole was found to be at least equivalent to Amphotericin versus invasive aspergillus and not nearly as toxic!!! Voriconazole has both I.V. and oral formulations. Toxic effects

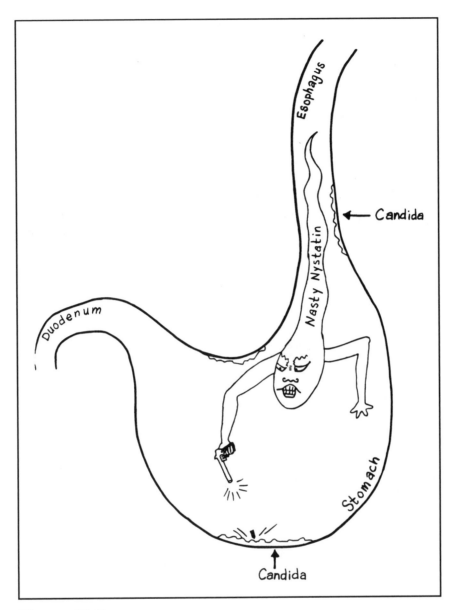

Figure 22-2

of voriconazole include transient visual changes in 30% of patients, and occasional hepatotoxicity and rash. Unique to its IV formulation is a component called cyclodextrin which is renally excreted; therefore the IV formulation of voriconazole should not be used in patients with creatinine clearance rates of <50.

Posaconazole and Isavuconazole

These are two of the new triazoles. They have a similar spectrum of activity as voriconazole so they can be used against the terrible aspergillus, but with a few potential twists:

Posaconazole and isavuconazole have activity against Mucormycosis, one of the saprophytic molds that can cause devastating disease and is very difficult to treat. Similar to voriconazole, they are available in both IV and oral forms.

Glucan Synthesis Inhibitors (Echinocandins)

Caspofungin, micafungin, and anidulafungin are the three FDA approved agents in this class. They are both given by the intravenous route and have a similar range of activity.

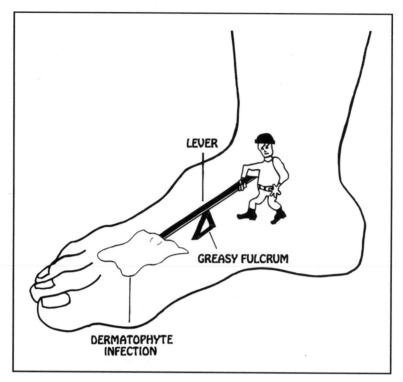

Figure 22-3

NAME	MECHANISM OF ACTION	PHARMOKINETICS
Amphotericin-B (a polyene antibiotic)	• Punches holes in ergosterol: This increases membrane permeability, resulting in cell death	1. **No** oral absorption: so give it IV 2. Does **NOT** cross blood brain barrier: so must give intrathecally to reach the cerebrospinal fluid (CSF) 3. Excreted via biliary tract and kidney (clearance is not affected by kidney dysfunction)
Flucytosine	• Converted to 5-fluorouracil, which inhibits fungal DNA & RNA synthesis	1. Oral absorption 2. Excreted in urine 3. Penetrates CSF well!!!
Ketoconazole (an Imidazole)	• Blocks ergosterol synthesis by inhibiting the cytochrome P_{450} enzymes. This causes depletion of ergosterol, resulting in disruption of the permeability of the cell membrane	1. Oral absorption 2. Absorbed better at low pH - so worse absorption when taken with antacids or H_2 blockers 3. Extensive hepatic metabolism

Figure 22-4   ANTI-FUNGAL DRUGS

These antifungals are great against all species of candida and also as an option for salvage of patients with invasive aspergillosis failing conventional therapy.

Other Antifungal Drugs

Nystatin

Nystatin, like amphotericin B, binds to ergosterol, increasing the permeability of the cell membrane and causing cell lysis.

Think "Nasty Nystatin" because this drug is too toxic to take parenterally (intravenously). It is only used topically on the skin and mucous membranes. Also, since it is not absorbed from the gastrointestinal tract, oral nystatin can be used to treat oral and esophageal infections with yeast or fungi. You will order nystatin on the wards as **Nystatin, Swish and Swallow** for treatment of oral, esophageal, and gastric candidiasis. It is also given topically for vaginal candidiasis.

Fig. 22-2. Nasty Nystatin cruises down the esophagus killing fungi on the wall of the esophagus. In one end and out the other!

Griseofulvin

Fig. 22-3. Visualize **griseofulvin** as a **greasy fulcrum** used to lever the dermatophyte plaques off the skin. It inhibits fungal growth by disrupting spindle formation, thus preventing **mitosis**. Note the worker peeling fungus off your **"toe,"sis**.

Griseofulvin deposits in keratin precursor cells in the skin, hair, and nails, where it inhibits the growth of fungi in those cells. Note that it does not kill the fungi; it just inhibits their growth (static rather than cidal). The uninfected drug-infiltrated keratin precursor cells mature and move outward toward the keratinized layer. As the older, infected cells fall off with normal cell turnover, this translates into a slow cure of skin fungus.

Adverse effects of griseofulvin are uncommon. They include headache, nausea, vomiting, photosensitivity, and mental confusion, in addition to bone marrow suppression (leukopenia and neutropenia).

Potassium Iodide

Potassium iodide is used to treat sporotrichosis. Remember that you get sporotrichosis from pricking

ADVERSE EFFECTS	THERAPEUTIC USES	MISCELLANEOUS
1. Nephrotoxic (reversible) 2. Acute febrile reaction 3. Anemia 4. Phlebitis at IV site	• Severe systemic fungal infections: 1. Systemic Candida infections 2. Cryptococcal meningitis 3. Excellent for blastomycosis, histoplasmosis and coccidioides 4. Invasive aspergillosis 5. Invasive sporotrichosis 6. Mucormycosis	1. Monitor BUN and creatinine levels daily to follow kidney dysfunction 2. Newer lipid and liposomal preparations are less nephrotoxic

1. Bone marrow suppression
  A. leukopenia
  B. thrombocytopenia
2. Nausea, vomiting and diarrhea
	1. Cryptococcal meningitis (in combination with amphotericin B)
	2. Candidal endocarditis (in combo with amphotericin B)
		• The reason for these adverse effects is that flucytosine inhibits DNA synthesis, which occurs in rapidly dividing cells such as bone marrow cells & GI epithelial cells

1. Nausea, vomiting and anorexia
2. Hepatotoxic
3. Inhibits CYP$_{450}$ system, resulting in decreased androgen & testosterone synthesis:
  A. Gynecomastia
  B. Impotence
  C. Decreased sex drive
  D. Decreased sperm production
4. Rash/pruritus
	• Chronic mucocutaneous candidiasis	

NAME	MECHANISM OF ACTION	PHARMOKINETICS
Miconazole and clotrimazole (these are imidazoles)	• Blocks ergosterol synthesis by inhibiting the cytochrome P_{450} enzymes	1. Topical usage 2. Oral (not absorbed systemically)
Fluconazole (a triazole)	• Blocks ergosterol synthesis by inhibiting the cytochrome P_{450} enzymes	1. Oral absorption 2. Can also be administered IV
Itraconazole (a triazole)	• Blocks ergosterol synthesis by inhibiting the cytochrome P_{450} enzymes	1. Oral absorption 2. Metabolized and excreted via liver 3. Available IV
Voriconazole	• Blocks ergosterol synthesis by inhibiting the cytochrome P_{450} enzymes	1. Oral absorption 2. Can also be administered IV
Posaconazole Isavuconazole	1. Blocks ergosterol synthesis by inhibiting the cytochrome P450 enzymes	1. Oral and IV
Echinocandins (caspofungin, micafungin, and anidulafungin)	1. Inhibit fungal cell wall synthesis by inhibiting 1,3 D-glucan synthase	1. Intravenous
Terbinafine	Blocks ergosterol synthesis by inhibiting squalene epoxide	1. Oral 2. Topical

Figure 22-4   (continued)

ADVERSE EFFECTS	THERAPEUTIC USES	MISCELLANEOUS
• Low toxicity when used topically	A. *Topical fungal infections* 1. Tinea versicolor 2. Cutaneous candidiasis 3. Dermatophytosis B. Oral troches for thrush (oral candidiasis) C. Vaginal suppositories for candida vaginitis	
• Less toxic than ketoconazole No interference with testosterone synthesis 1. Nausea 2. Skin rash 3. Headache	1. Oral, vaginal and esophageal *Candida* 2. Alternative to amphotericin B for treatment of: A. Systemic candidiasis B. Cryptococcal meningitis C. Pulmonary & extrapulmonary coccidioidomycosis (but fluconazole is not used to treat coccidioides meningitis)	
• Less toxic than ketoconazole No interference with testosterone synthesis 1. Nausea 2. Skin rash 3. Headache	A therapeutic option for the following fungal infections: 1. Blastomycosis 2. Histoplasmosis 3. Coccidioidomycosis 4. Sporotrichosis 5. Chromomycosis 6. Invasive aspergillosis	Absorption enhanced by taking with acid drinks (such as O.J. or colas)
1. Photophobia 2. Rash 3. Liver enzyme increases	Broad activity against multiple fungi. Its primary use at this time is for infections with: 1. Aspergillus 2. Fluconazole resistant candida 3. Antifungal prophylaxis in bone marrow transplant recipients	Do not use IV formulation in patients with decreased renal function (creatinine clearance <50)
Few side-effects in initial trials	Broad antifungal activity including Mucormycosis	
1. Well tolerated 2. Infusion may cause flushing due to histamine release	1. A good candida drug (including fluconazole resistant species) 2. Indicated for salvage of aspergillus infections	
1. Rare gastrointestinal side effects 2. Rare rash 3. Rare reversible agranulocytosis	1. Primarily used for dermatophyte infections – has largely replaced griseofulvin in treatment of onychomycosis (fungal nail infections)	

NAME	MECHANISM OF ACTION	PHARMOKINETICS
Nystatin	• Punches holes in ergosterol: This increases membrane permeability, resulting in cell death	1. Not absorbed from GI tract. Oral administration results in "topical" treatment along the GI tract 2. Apply topically to skin and vaginal infections 3. Too toxic to give IV
Griseofulvin	• Inhibits mitosis of cells, by disrupting spindle formation	1. Oral absorption 2. Absorbed better with fatty foods. 3. Deposits in keratin 4. Excreted in feces unchanged
Potassium Iodide		

Figure 22-4   (continued)

your finger in the garden. "You get Sporotrichosis while **Pott**ing plants." If the infection becomes systemic, amphotericin B or itraconazole is better.

Terbinafine

Terbinafine is a newer oral fungicidal agent that blocks fungal cell wall synthesis. It blocks ergosterol synthesis by inhibiting the formation of squalene epoxide from squalene. Terbinafine tends to accumulate in nails, and is therefore useful for tinea unguium (onychomycosis). It also appears useful in the treatment of tinea pedis, tinea capitis, and tinea corporis. Since it is not metabolized by the cytochrome p450 system (as are the azole antifungals), there is little potential for drug-drug interactions.

Tavaborole and Efinaconazole

These are newer topical antifungal agents that are FDA approved for use against toenail fungus (onychomycosis) due to *Trichophyton*. Their efficacy is still less than desired, with cure rates of <20%.

Fig. 22-4.   Summary of the anti-fungal drugs.

Reference

Andriole VT. Current and future antifungal therapy: new targets for antifungal therapy. International Journal of Antimicrobial Agents 2000;16:317–21.

Bennett JE. Antifungal Agents. In: Mandell GL. Bennett JE, Dolin R, eds. Principles and Practice of Infectious Diseases. 4th edition. New York: Churchill Livingstone 1995; 401–410.

Gupta AK, Tomas E. New antifungal agents. Dermatologic Clinics 2003; 21(3).

Jackson CA, et al. Drug therapy: Oral azole drugs as systemic antifungal therapy. N Engl J Med 1994;330:263–272.

Sanford JP, Gilbert DN, et al. Guide to antimicrobial therapy 1994. Antimicrobial Therapy, Inc, Dallas Texas; 1994.

www.doctorfungus.org

Recommended Review Articles:

Chen SC, Playford EG, Sorrell TC. Antifungal therapy in invasive fungal infections. Curr Opin Pharmacol. 2010;10(5): 522–30.

Bennett JE. Echinocandins for candidemia in adults without neutropenia. N Engl J Med. 2006;355(11):1154–9.

ADVERSE EFFECTS	THERAPEUTIC USES	MISCELLANEOUS
• Highly toxic if given IV	1. Oral, esophageal or gastric candidiasis (oral administration) 2. Vaginal candidiasis (apply topically)	• Order as "nystatin, swish and swallow"
1. Headache, nausea, vomiting, photosensitivity and mental confusion 2. Bone marrow suppression	• Dermatophytosis of the skin, hair and nails	• Works very slowly!!!
• Skin rash	• Cutaneous sporotrichosis	"You get Sporotrichosis while *Potting* plants"

M. Gladwin, W. Trattler, and S. Mahan, *Clinical Microbiology Made Ridiculously Simple* ©MedMaster

PART 3. VIRUSES

CHAPTER 23. VIRAL REPLICATION AND TAXONOMY

Fig. 23-1. Imagine that a virus is a specially designed military spaceship with a super-resistant outer shell. However, the engineers forgot to build a fuel tank for their spaceship. It must therefore drift around until it encounters a satisfactory planet. When it lands, it transfers the military cargo off the ship. This cargo is designed to take control of all factories and then instruct the factories to begin building copies of the original spacecraft (without fuel tanks). When the replica ships have been constructed, they will depart in mass, leaving the planet in complete ruins.

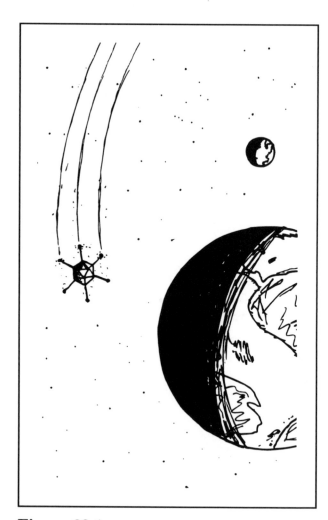

Figure 23-1

Viruses have these unique characteristics:

1) They are energy-less. They float around until they come in contact with an appropriate cell.

2) They are basic life forms composed of a protein coat, called a **capsid**, that surrounds genetic material. Viruses do not have organelles or ribosomes. Certain viruses are further enclosed by an external lipid bilayer membrane that surrounds the capsid and may contain glycoproteins. Some viruses also carry some structural proteins and enzymes inside their capsid.

3) The genetic material is **either** DNA **or** RNA. **Never both**!! The genetic material contains instructions to make millions of clones of the original virus.

4) Replication of the genetic material occurs when the virus takes control of the host cell's synthetic machinery. Viruses contain all of the genetic information, but not the enzymes, needed to build millions of replicas of the original virus.

VIRAL MORPHOLOGY

They say we learn best by doing, so let's study viral structure by making a virus, starting from the **nucleic acid** inside and proceeding to the **capsid** and **envelope**.

Nucleic Acid

Fig. 23-2. Viruses are classified as either DNA or RNA viruses. So we have two choices for our virus: DNA or RNA. Of course nothing is quite that simple. The nucleic acid strands can be single-stranded, double-stranded, linear, or looped, in separate segments or one continuous strand. The nucleic acid sequences can encode a simple message or encode hundreds of enzymes and structural proteins.

RNA Viruses

Let us choose RNA for the core of our virus. There are 2 types of RNA viruses: positive (+) stranded and negative (−) stranded.

The **POSITIVE** (+) means that the RNA is JUST LIKE a messenger RNA (mRNA). When a positive (+) stranded RNA virus enters a host cell, its RNA can immediately be translated by the host's ribosomes into protein.

When negative (−) stranded RNA viruses enter a cell, they are not able to begin translation immediately.

Figure 23-2

They must first be transcribed into a positive (+) strand of RNA (like mRNA). To do this, negative (−) stranded RNA viruses must carry, in their capsid, an enzyme called **RNA-dependent RNA polymerase**, which will carry out the transcription of the negative (−) strand into positive (+). Human cells do not have an RNA-dependent RNA polymerase, so negative (−) standed viruses must carry their own.

Fig. 23-3.   Translation of positive (+) and negative (−) RNA viruses.

Two special RNA viruses deserve mention here:

1) Retroviruses, of which HIV is a member, are unique because of their ability to incorporate into the host genome.

2) Reoviridae, including rotavirus, is unique as they are the only viruses with a double stranded RNA genome.

Fig. 23-4.   The RNA of the retroviruses is transcribed in a reverse fashion ("retrograde") into DNA! To do this, these viruses carry a unique enzyme called **reverse transcriptase**.

DNA Viruses

Fig. 23-5.   Unlike RNA, DNA cannot be translated directly into proteins. It must first be transcribed into mRNA, with subsequent translation of the mRNA into structural proteins and enzymes.

Most DNA viruses have **both** a negative (−) strand and a positive (+) strand. Here is the confusing part: The negative (−) strand refers to the strand of DNA that is read, while the positive (+) strand is ignored. Parvoviruses are the exception, and have a single stranded DNA genome.

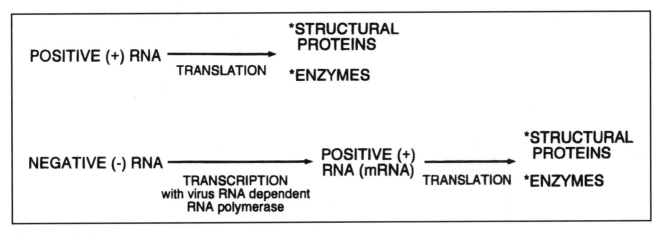

Figure 23-3

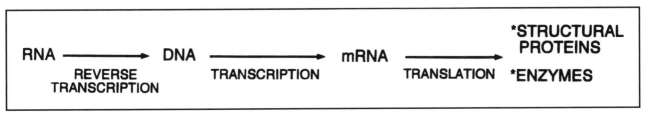

Figure 23-4

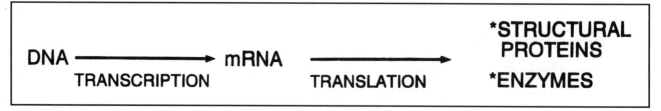

Figure 23-5

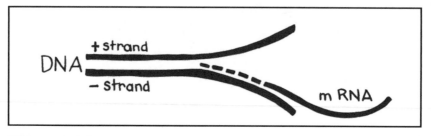

Figure 23-6

Fig. 23-6. Unlike positive (+) stranded RNA, which is translated directly into proteins, the negative (-) strand of DNA is used as the template for transcription into mRNA.

Capsids

Now that we know about the different types of nucleic acids, let's build a structure to house the genome. First we will build a capsid. There are two types of capsids: icosahedral and helical.

Icosahedral Symmetry Capsids

Fig. 23-7. Take 1 or more polypeptide chains and organize them into a globular protein subunit. This will be the building block of our structure and is called a **capsomer**.

Fig. 23-8. Arrange the capsomers into an equilateral triangle.

Figure 23-7

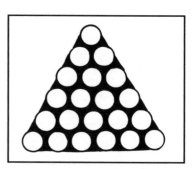

Figure 23-8

Fig. 23-9. Place 20 triangles together to form an icosahedron.

Package the DNA or RNA inside the icosahedral capsid!

Helical Symmetry Capsids

Fig. 23-10. In helical symmetry the protein capsomers are bound to RNA (always RNA because only RNA viruses have helical symmetry) and coiled into a helical nucleoprotein capsid. Most of these assume a spherical shape except for the rhabdoviruses (rabies virus), which have a bullet-shaped capsid.

Envelope

Fig. 23-11. Now that we have made an icosahedral capsid with a nucleic acid (RNA or DNA) inside and a coiled helical nucleocapsid (RNA), let us cover the

Figure 23-9

structure with a lipid bilayer membrane. Viruses acquire this membrane by budding through the host cell nuclear or cytoplasmic membrane and tearing off a piece of the membrane as they leave. There may be various glycoproteins embedded in their cell membranes.

Viruses that do not have membranes are referred to as **naked** or nonenveloped. Those with membranes are referred to as **enveloped**.

Finished Product

Fig. 23-12. The appearance of the complete viruses and their approximate sizes compared to the bacterium *E. coli*, and the very small bacteria, *Chlamydia, Rickettsia*, and *Mycoplasma pneumoniae*.

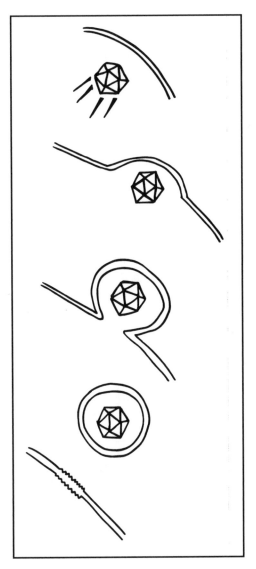

Figure 23-11

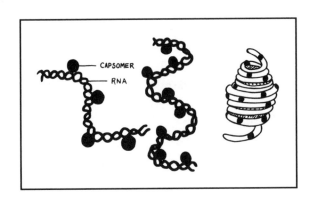

Figure 23-10

CLASSIFICATION

Viruses are classified according to their:

1) Nucleic acid:
   Type of nucleic acid: DNA, RNA
   Double- vs. single-stranded
   Single or segmented pieces of nucleic acid
   Positive (+) or negative (−) stranded RNA
   Complexity of genome
2) Capsid:
   Icosahedral
   Helical

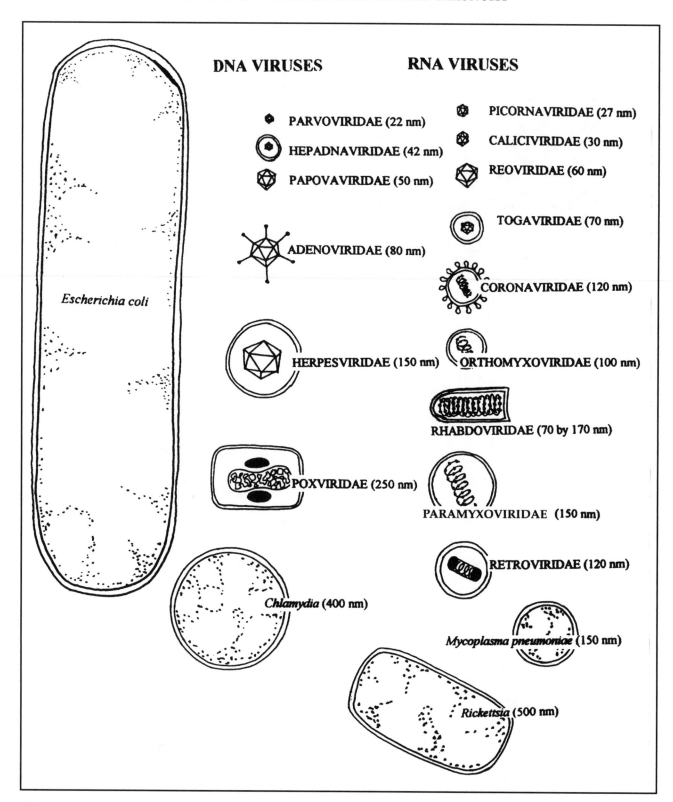

Figure 23-12

3) Envelope:
   Naked
   Enveloped
4) Size:
   The diameter of the helical capsid viruses
   The number of capsomers in icosahedral capsids

We will now go over the virus families and the characteristics that separate them.

DNA Viruses

These are sometimes referred to as the **HHAPPP**y viruses:

Herpes
Hepadna
Adeno
Papova
Parvo
Pox

Most DNA viruses are double-stranded, show icosahedral symmetry, and replicate in the nucleus (where DNA customarily replicates).
Two DNA viruses break these rules:

1) **Parvoviridae**: This virus is so simple that it only has a single strand of DNA. It is as simple as playing a **ONE PAR** hole in golf.
2) **Poxviridae**: This virus is at the opposite end of the spectrum and is extremely complex. Although it does have double-stranded DNA, the DNA is complex in nature, coding for hundreds of proteins. This virus does not have icosahedral symmetry. The DNA is surrounded by complex structural proteins looking much

like a box (**POX IN A BOX**). This virus replicates in the **cytoplasm**.

Three of the DNA viruses have envelopes:
Herpes Hepadna Pox
Three are naked: A woman must be naked for the **PAP** smear exam.
PApova Adeno **PA**rvo

Fig. 23-13.   The DNA viruses.

RNA Viruses

There are certain generalities about RNA viruses, most of which are the opposite of DNA viruses.
Most RNA viruses are single-stranded (half are positive [+] stranded, half negative [−]), enveloped, show helical capsid symmetry, and replicate in the cytoplasm:

Toga	Orthomyxo
Corona	Paramyxo
Retro	Rhabdo
Picorna	Bunya
Calici	Arena
Reo	Filo
Flavi	

Exceptions:

1) Reoviridae are double-stranded.
2) Three are nonenveloped: Picorna, Calici, and Reoviridae.
3) Five have icosahedral symmetry: Reo, Picorna, Toga, Flavi, Calici (Rhabdo has helical symmetry but is shaped like a bullet).
4) Two undergo replication in the nucleus: Retro and Orthomyxo.

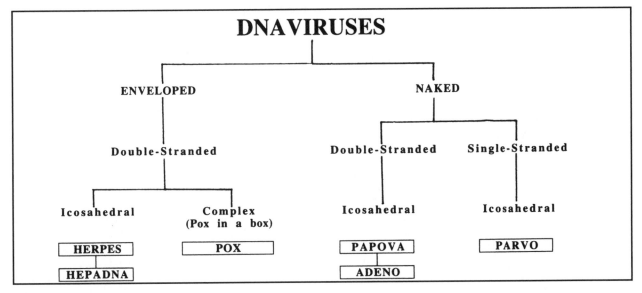

Figure 23-13   DNA VIRUSES

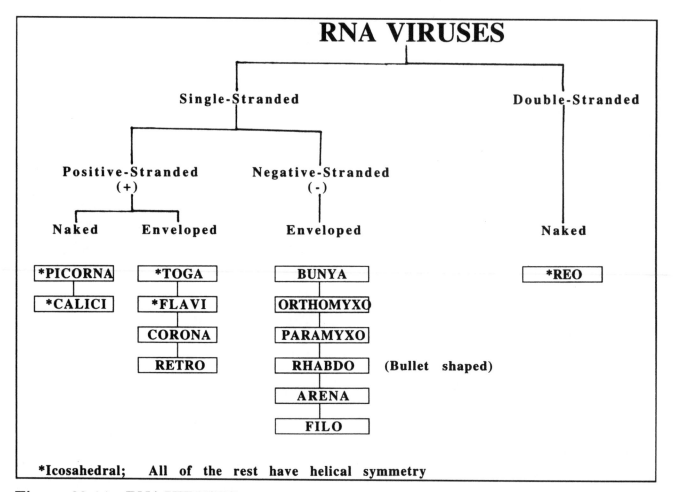

RNA VIRUSES

Figure 23-14 RNA VIRUSES

Fig. 23-14. The RNA viruses.

Here are two catchy mnemonics to remember the (+) and (−) stranded RNA viruses. (Courtesy of John Marcel, Georgetown U. School of Medicine)

(+) stranded: "The **Calci**fied old Emperor **Pico** is wearing his Crown and **Toga** and is eating **Flav**orful grapes from a **Retro** bowl" (Calici, Pico, Corona, Toga, Flavi, and Retroviridae).

(−) stranded: "**Old Pete's Rab**id dog **Filo** fights Paul **Buny**on in the **Arena**" (Orthomyxo, Paramyxo, Rhabdo, Filo, Bunya, and Arenaviridae)

VIRAL REPLICATION

Viruses cannot reproduce on their own. They must invade a cell, take over the cell's internal machinery and instruct the machinery to build enzymes and new viral structural proteins. Then they copy the viral genetic material enough times so that a copy can be placed in each newly constructed virus. Finally, they leave the host cell.

In order for viruses to reproduce, they must complete these 4 steps:

1) Adsorption and penetration.
2) Uncoating of the virus.
3) Synthesis and assembly of viral products (as well as inhibition of the host cell's own DNA, RNA and protein synthesis).
4) Release of virions from the host cell (either by lysis or budding).

Adsorption and Penetration

Fig. 23-15. The viral particle binds to the host cell membrane. This is usually a specific interaction in which a viral encoded protein on the capsid or a glycoprotein embedded in the virion envelope binds to a host cell membrane receptor. Unlike the bacteriophage virion (see Chapter 3 on Bacterial Genetics), which injects its DNA, these viruses are completely internalized, capsid and nucleic acid. This internalization occurs by endocytosis or by fusion of the virion envelope with the host cell membrane.

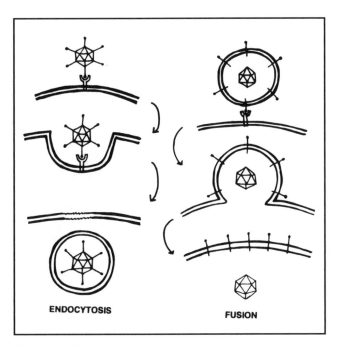

Figure 23-15

Uncoating

The nucleic acid is released from the capsid into the nucleus or cytoplasm.

Transcription, Translation, Replication

RNA Viruses

These viruses usually undergo transcription, translation, and replication in the cytoplasm.

Positive stranded RNA viruses are the equivalent of preformed messenger RNA (mRNA). As soon as they invade the cell they are ready for translation. These viruses immediately use the host cell's ribosomal proteins and enzymes to translate their positive RNA into an **RNA dependent RNA polymerase** to make negative stranded copies of their RNA for replication.

Negative stranded RNA viruses have a bit of a problem. They cannot translate into protein because they are a negative strand (copy of mRNA) so they need to carry with them in the virion a **viral RNA dependent RNA polymerase** to first make a positive strand copy which can then be translated into viral proteins.

Fig. 23-16. Positive (+) stranded RNA virus replication. Positive strand RNA viruses first have to make the RNA dependent RNA polymerase by protein translation of their positive strand of RNA (which is like mRNA).

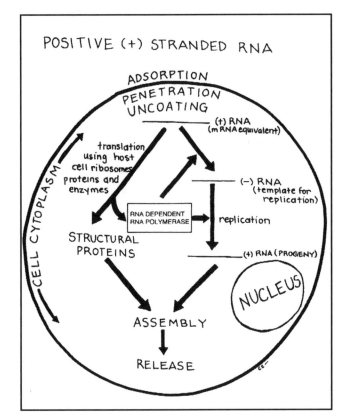

Figure 23-16   POSITIVE (+) STRANDED RNA VIRUS REPLICATION

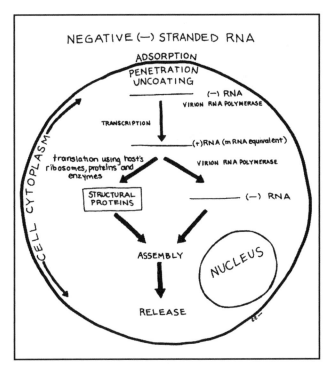

Figure 23-17   NEGATIVE (−) STRANDED RNA VIRUS REPLICATION

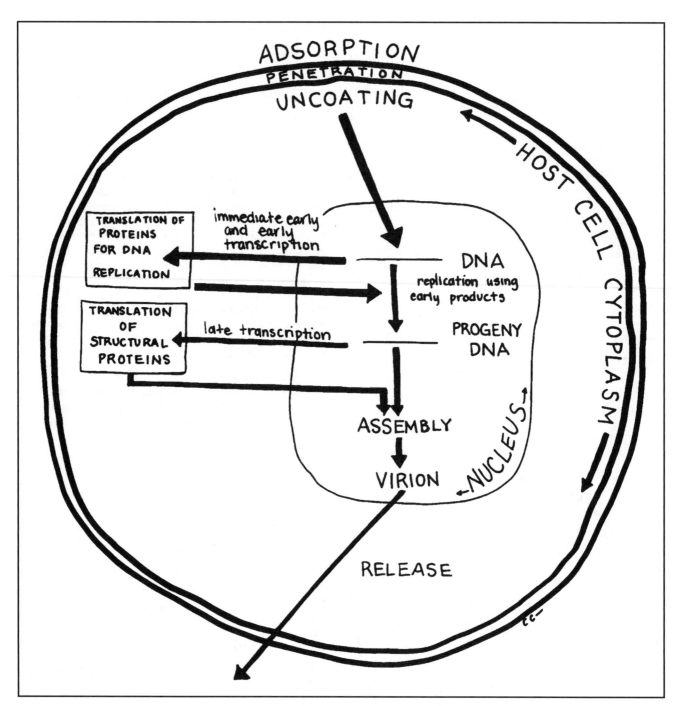

Figure 23-18   DNA VIRUS REPLICATION

The newly synthesized polymerase can then make the negative strand and positive strand RNA for replication of RNA progeny.

Fig. 23-17.   Negative (−) stranded RNA virus replication. The virus uncoats, releases a virion associated

RNA polymerase, and must first transcribe the negative (−) strand to a positive (+) strand (using the RNA polymerase). The positive (+) strand then acts like mRNA and undergoes both transcription and translation.

Retroviruses deserve special mention here. Unlike other single stranded positive RNA, these viruses do not undergo immediate translation. Instead, they carry a preformed RNA dependent DNA polymerase (reverse transcriptase), and are transcribed into DNA which can then be incorporated directly into the host genomic DNA. The DNA can then be transcribed into mRNA to make viral proteins, or RNA that will form the genome of the budding viruses.

DNA Viruses

Transcription and replication usually occur in the nucleus.

Fig. 23-18. DNA virus replication. DNA viruses tend to be more genetically complex than RNA viruses. Thus, viral transcription is divided into immediate early, early, and late transcription. Another important concept is that DNA viruses act in a similar fashion as our own genome. Segments of DNA are transcribed in the nucleus into mRNA, are spliced and processed, and the mRNA then moves to the cytoplasm (endoplasmic reticulum) where translation occurs.

Immediate early and Early: The initially transcribed mRNA here encodes enzymes and proteins needed for DNA replication and for further transcription of late mRNA.

Late: The mRNA is usually transcribed after viral DNA replication has begun and is transcribed from progeny DNA. The capsid structural proteins are synthesized from the late mRNA genome.

Assembly and Release

The structural proteins and genome (RNA or DNA) assemble into the intact helical or icosahedral virion. The virion is then released.

Naked virions: The cell may lyse and release the virions, or the virions may be released by reverse phagocytosis (exocytosis).

Enveloped virions: The newly formed naked virion acquires its new "clothing" by budding through the Golgi apparatus, nuclear membrane, or cytoplasmic membrane, tearing off a piece of host cell lipid bilayer as it exits.

HOST CELL OUTCOME

Death: With the viral infection, the host cell's own function shuts down as the cell is commandeered for virion replication. This can result in cell death.

Transformation: Infection can activate or introduce oncogenes. This results in uncontrolled and uninhibited cell growth.

Latent infection: The virus can survive in a sleeping state, surviving but not producing clinically overt infection. Various factors can result in viral reactivation.

Chronic slow infection: Some viruses will cause disease only after many years, often decades, of indolent infection.

Fig. 23-19. Summary of virus morphology.

NUCLEIC ACID	SYMMETRY	PRESENCE OR ABSENCE (NAKED) OF ENVELOPE	PHYSICAL STATE OF NUCLEIC ACID
RNA	ICOSAHEDRAL	NAKED	SS NONsegmented
			SS NONsegmented
			DS SEGMENTED (11)
		ENVELOPED	SS NONsegmented
			SS NONsegmented
	HELICAL	ENVELOPED	SS NONsegmented
			SS SEGMENTED (3)
			SS SEGMENTED (8)
			SS NONsegmented
			SS NONsegmented
			SS NONsegmented
			SS SEGMENTED (2)
	COMPLEX	COMPLEX COAT	SS DIPLOID (2 identical copies of + stranded RNA)

Figure 23-19   VIRAL MORPHOLOGY

POSITIVE (+) OR NEGATIVE (−) STRANDED	FAMILY	SPECIFIC PATHOGENIC VIRUSES (OR DISEASES CAUSED)
+	*PICORNA* viridae	Polio virus Coxsackie A & B virus ECHO virus Hepatitis A virus Rhino virus New enteroviruses
+	*CALCI* viridae *Hepe*viridae	Norwalk virus Hepatitis E virus
Double stranded	*REO* viridae	Rota virus
+	*TOGA* viridae	Mosquito borne encephalitis (WEE, EEE, VEE) Rubivirus (rubella)
+	*FLAVI* viridae	Yellow fever virus Dengue virus St. Louis encephalitis Japanese encephalitis Hepatitis C virus
+	*CORONA* viridae	Respiratory illness (cold)
−	*BUNYA* viridae	California encephalitis virus Rift Valley fever virus Sandfly fever virus Hantavirus
−	*ORTHOMYXO* viridae	Influenza virus (types A, B & C)
−	*PARAMYXO* viridae	Para-influenza virus Respiratory syncytial virus Mumps Measles Metapneumovirus
−	*RHABDO* viridae	Rabies virus
−	*FILO* viridae	Marburg virus (acute hemorrhagic fever) Ebola virus (acute hemorrhagic fever)
−	*ARENA* viridae	Lymphocytic choriomeningitis virus Lassa virus
+ • Note: RNA reverse transcribed to DNA using reverse transcriptase enzyme	*RETRO* viridae	Human immunodeficiency virus (HIV) types I and II HTLV types I and II

NUCLEIC ACID	SYMMETRY	PRESENCE OR ABSENCE (NAKED) OF ENVELOPE	PHYSICAL STATE OF NUCLEIC ACID
DNA	ICOSAHEDRAL	NAKED	*SS* LINEAR
			DS CIRCULAR
			DS LINEAR
		ENVELOPED	DS LINEAR
			DS CIRCULAR
	COMPLEX	COMPLEX ENVELOPE	DS LINEAR

* Note: Delta virus (causes hepatitis) is an incomplete RNA virus. It needs the coinfection with **hepatitis B virus** to cause disease

Figure 23-19   (continued)

POSITIVE (+) OR NEGATIVE (−) STRANDED		FAMILY	SPECIFIC PATHOGENIC VIRUSES (OR DISEASES CAUSED)
		PARVO viridae	Erythema infectiosum Transient aplastic anemia crisis
		PAPOVA viridae	Human papilloma virus BK polyomavirus JC polyomavirus
		ADENO viridae	Childhood respiratory illness ("cold") Epidemic keratoconjunctivitis
		HERPES viridae	Herpes simplex virus types 1 & 2 Varicella-zoster virus Cytomegalovirus Epstein-Barr virus Human Herpesvirus 6 (roseola)
		HEPADNA viridae	*Hepatitis B virus (see note)
		POX viridae	Smallpox Vaccinia Molluscum contagiosum

M. Gladwin, W. Trattler, and S. Mahan, *Clinical Microbiology Made Ridiculously Simple* ©MedMaster

CHAPTER 24. ORTHOMYXOVIRIDAE AND PARAMYXOVIRIDAE

ORTHOMYXO virus	ORDINARY
Influenza virus	• The Flu; Pneumonia in at risk groups

PARAMYXO VIRUS	PARADE OF DISEASES
Parainfluenza virus & Respiratory syncytial virus	1. Bronchiolitis, viral pneumonia, croup, in children 2. Cold/Flu in adults
Metapneumovirus	• Upper and lower respiratory tract infections. Primarily in young children and older adults.
Mumps virus	• Mumps: parotitis, testicular inflammation
Measles (rubeola) virus	• Measles: prodrome, Koplik's spots, rash, encephalitis

Figure 24-1

These 2 viral families have similar structures and the ability to adsorb to glycoprotein receptors, particularly in the upper respiratory tract. The **or**thomyxoviridae are all influenza viruses, which cause the "**OR**dinary flu." **Para**myxoviridae also replicate in the upper respiratory tract and can produce influenza-like illness, but they also produce a **PARA**de of distinctly different diseases. The paramyxoviridae include parainfluenza virus, mumps, measles, metapneumovirus and respiratory syncytial virus.

Fig. 24-1. The orthomyxoviridae and paramyxoviridae.

ORTHOMYXOVIRIDAE

Approximately 20% of the entire world population gets infected with the influenza virus each year! In the United States the infection occurs as **epidemic influenza** (an outbreak in a city, states, or entire country) each winter, typically between late December and early March, and up to 10–40% of people can be infected in certain communities during these outbreaks. Almost everyone has experienced the flu: high fever, chills which can become frank shaking chills, headaches, malaise (feeling really bad!) and myalgias (muscle aches). The common cold-like viral upper respiratory symptoms develop as well, including dry cough, sore throat, and rhinorrhea (runny clogged nose), but the painful muscle aches, high fevers and headaches really set this apart from the run of the mill cold. **Pandemic influenza** are much more severe outbreaks that involve much of the world and are caused by the emergence of a new influenza A virus to which we have no current immunity.

Influenza can cause a primary pneumonia, particularly with the severe infections seen in the influenza pandemics, or weaken the immunity to promote a delayed secondary bacterial pneumonia or otitis media, typically with *Staphylococcus aureus*, *Haemophilus influenzae* or *Streptococcus pneumoniae*.

Owing to its high infectivity and spread via small-particle respiratory aerosols, influenza epidemics hit a community like a tsunami: Within two weeks of infecting a community (abrupt illness after a 1–2 day incubation) the first sign of an epidemic is that kids start to miss school or are diagnosed with pneumonia and otitis media. Then the infection hits the adults, causing missed work followed by later hospital admissions for secondary pneumonia. This is followed by death, which usually occurs in the elderly, the immunocompromised and in persons with chronic lung disease (emphysema or lung fibrosis). While death is rare in previously healthy children and adults during **epidemic influenza**, as many as 20–40,000 people still die each year in the United States from complications (namely pulmonary) of influenza infection. The story is much worse during **pandemic influenza** which will be discussed in more detail at the end of this section.

Virion Structure and Pathogenesis

Understanding the structure of this virus will be important to you as a physician.

The ability to produce epidemics and susceptibility to antibody immunity, vaccination, and the antiviral drugs called **neuraminidase inhibitors**, all depend on the viral ultrastructure. You will see at the end of this section that the paramyxoviridae have a similar structure with a few small changes (making it oh so easy to learn!).

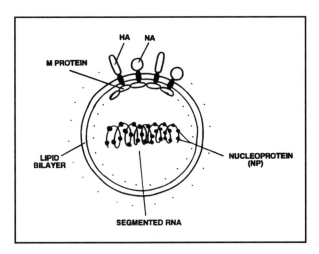

Figure 24-2

Fig. 24-2. The orthomyxoviridae are spherical virions. At the virion center lie 8 segments of negative (−) stranded RNA put together with a protein (**nucleocapsid protein – NP**) into a helical symmetry capsid. Surrounding the nucleocapsid lies an outer membrane studded with long glycoprotein spikes. There are 2 distinct types of glycoprotein: one with **Hemagglutinin Activity** (**HA**) and one with **Neuraminidase Activity** (**NA**). Anchoring the bases of each of these spikes on the inside of the viral lipid bilayer are membrane proteins (**M-proteins**).

Hemagglutinin (HA)

Fig. 24-3. Hemagglutinin (HA) can attach to host sialic acid receptors. Sialic acid receptors are present on

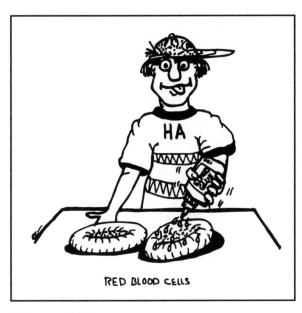

Figure 24-3

the surface of erythrocytes, so viruses with HA glycoproteins cause heme-agglutination when mixed with red blood cells.

Host cell sialic acid receptors also exist on upper respiratory tract cell membranes, and HA binding to these receptors activates fusion of the host cell membrane with the virion membrane, resulting in dumping of the viral genome into the host cell. So HA is needed for **adsorption**! Antibodies against HA will block this binding and prevent infection.

Neuraminidase (NA)

Neuraminic acid is an important component of mucin, the substance covering mucosal epithelial cells and forming an integral part of the host's upper respiratory defense barrier. As the name implies, neuraminidase (NA) cleaves neuraminic acid and disrupts the mucin barrier, exposing the sialic acid binding sites beneath.

Neuraminidase is also critical for the release of the newly formed virion from the infected host cell. As new viruses are assembled, they bud out of the host membrane, and after the budding is complete, the virion's hemagglutinin is bound to a host cell sialic acid-containing receptor. The neuraminidase then cleaves this sialic acid receptor to allow the formed virus to escape and infect a new cell.

Fig. 24-4. Viral NA and HA act as tag team wrestling partners, wrestling down the host's defenses. NA cleaves the cell mucin barrier, while HA fuses to the cell's sialic acid residues, enabling viral adsorption and penetration.

Again, antibodies and drugs, called the neuraminidase inhibitors, directed against the neuraminidase are protective.

Influenza Serology and Epidemiology

There are 3 types of influenza virus: A, B, and C. These types have many strains separated by antigenic differences in HA and NA. Type A infects humans, other mammals (swine, etc.), and birds. Type B and C have only been isolated from humans.

When looking at the disease influenza, 2 questions about epidemiology arise:

Q: If antibody to the NA and HA are protective, why do we continually get **epidemics** of the bothersome flu, with fever, chills, myalgias, arthralgias, headache, and other miseries?

A: Antigenic Drift: During viral replication mutations can occur in the HA or NA, leading to changes in the antigenic nature of these glycoproteins. This is termed **antigenic drift** because the changes are small, just a little drift of the sailboat in the water. The resulting new strains are only partially attacked by our immune system, resulting in milder disease in adults who have previously acquired

Figure 24-4

antibodies. Major mutational changes usually result in altered codon reading frames and a nonviable virus.

Q: We all think that this is a pesky but mild self-limiting disease. It can cause pneumonia and more serious disease in the elderly, but usually it resolves without complications in 3 to 7 days. So why have there been devastating **pandemics** of influenza throughout history, as in 1918? From 1918 to 1919 a pandemic of influenza swept across the world, killing 549,000 people in the United States, 12.5 million people in India alone, and more than 20 million people (21 million reported) worldwide.

A: Antigenic Shift: Now we are really **shift**ing gears. We are taking the boat mentioned above and airlifting it to a mountain in the Himalayas. With antigenic shift there is a complete change of the HA, NA, or both. This can only occur with influenza type A because the mechanism involves the trading of RNA segments between animal and human strains. When 2 influenza types co-infect the same cell, undergo replication and capsid packaging, RNA segments can be mispackaged into another virus. This virus now wields a new HA or NA glycoprotein that has never been exposed to a human immune system

anywhere on the planet. So the entire human population would be susceptible, leading to devastating pandemics.

The new HA and NA antigens are given number subscripts to differentiate them. The pandemic of 1889 was caused by a virus with an H2 hemagglutinin, the pandemic of 1900 was caused by a new virus with H3 hemagglutinin; in 1918 a swine flu virus transferred its HA to a human virus and so was called Hswine hemagglutinin (HSW). This pandemic was also called the Spanish Flu and resulted in 21 million deaths recorded world-wide, with 549,000 deaths in the United States. All in all there have been an estimated 31 pandemics described since 1580; the pandemics recorded since 1989 and their HA and NA composition are shown on page 243. It is the emergence of a new virus to which the majority of the population has little or no immunity that leads to a pandemic. Often these novel viruses emerge via the combination of avian and human viruses. Examples of avian influenza that have jumped the species barrier from birds to humans are the H5N1 virus which emerged in 1997 and the H7N9 virus in 2013. More on "Bird Flu" later...

Figure 24-5

More recently, in 2009, we had a new pandemic caused by **H1N1**, again of swine flu origin. Discovery of this most recent antigenic shift led to great concern and mass vaccination campaigns, but in the end we did not see higher rates of mortality than in traditional flu seasons. H1N1 virus is now a human seasonal flu virus that also circulates in pigs. It still causes human infection, but now much of the world's population has some degree of immunity. At least for the current generation further infections with H1N1 will be largely due to "**antigenic drift**" rather than due to the original "**antigenic shift**" that led to the pandemic of 2009. H1N1 has been incorporated into the most recent seasonal flu vaccines (2011–2012).

Global pandemics:

1889:	H2N2
1901:	H3N8
1918:	H1N1: "Spanish flu"—highly pathogenic strain—with high mortality (estimated 21 million deaths worldwide)
1947:	H1N1*
1957:	H2N2* "Asian flu"—illness but low mortality
1968:	H3N2* "Hong Kong flu"—illness but low mortality
1977:	H1N1*
1997:	H5N1 (avian influenza pandemic; great risk for human pandemic)
2009:	H1N1*

*Notice also that some strains caused a second pandemic as a new unexposed population grows to adulthood.

Fig. 24-5. Antigenic drift and shift: For influenza epidemics and pandemics to occur in people other than children (who do not yet have antibodies), the antigens in the NA and HA must somehow **change**, through antigenic drift and antigenic shift.

Complications of Influenza

Even the normal yearly flu can cause complications. The elderly and immunocompromised suffer more serious illness as the virus spreads to the lower respiratory tract, resulting in pneumonia. The viral infection also lowers the host defenses against many **bacteria**. Secondary bacterial pneumonias by *Staphylococcus aureus, Streptococcus pneumoniae*, and others are common and the physician must follow patients (especially the elderly) closely until complete resolution of their illness. New fevers or failure to improve means danger!!

Fig. 24-6. Study the figure of the child with the **crown** (the **Rey**- Spanish for king) and the lightning bolts around his **head** and **liver**. Children given aspirin when they have influenza or varicella (chicken pox) can develop a severe liver and brain disease called **Reye's Syndrome**. It is not yet known why this occurs. **Give acetaminophen for fever in children, no aspirin!!!!**

Diagnostic tests for influenza fall into 4 broad categories:

1. Virus isolation: Culture of the virus allows for genetic and antigenic analysis
2. Detection of viral proteins: New one hour tests help guide the choice of antiviral agents

Figure 24-6

3. Detection of viral nucleic acid (RNA) in clinical material is available by reverse transcription followed by PCR (very sensitive method)

4. Serological diagnosis: 4-fold increase in specific antibody levels over 2 weeks

Treatment and Control

Influenza viruses are grown in mass quantities in chick embryos, which are then inactivated, purified, and used as vaccines. Scientists carefully choose 3 strains that are circulating in the population or expected to cause an outbreak in the next season. These vaccines have variable success depending on the accuracy of the "guesswork." The vaccines should be given to the elderly, immunocompromised patients, and health-care workers.

A live-attenuated influenza vaccine that can be given as a nasal spray has been approved for use in previously healthy people from ages 5–49 years. This is called the cold-adapted influenza vaccine-trivalent (CAIV-T).

There are **four drugs** available for the treatment and prophylaxis of influenza virus infections:

The **adamantanes** (**amantadine** and **rimantadine**) are **M2 ion channel** inhibitors; inhibition of the influenza A M2 protein blocks acidification of the interior of the virion required for normal viral uncoating inside the cell. These agents are only effective against influenza A (see Chapter 30, **Fig. 30–1** and **Fig 30–6**). While these drugs can be given prophylactically to prevent the flu and also shorten the duration of active infection, they do cause some CNS side effects like anxiety or confusion. Most problematic, these drugs are associated with the rapid emergence of drug-resistant isolates of influenza A that are genetically stable and infectious. In fact, *the majority of the recent circulating strains of influenza A virus, including H5N1 and the more recent H1N1 strain are resistant to the adamantanes.*

The **neuraminidase inhibitors** (**zanamivir** and **oseltamivir**) interfere with the release of the progeny virus from the infected host cell and unlike the adamantanes, which only works on influenza A, are effective against all strains of influenza. Because the release of new progeny viruses requires neuraminidase to cleave the host cell sialic acid receptor, the inhibition of this enzyme prevents release and only one round of infection is possible, limiting the severity of the infection. These drugs are actually mimics of the sialic acid receptor substrate for the active catalytic site of the neuraminidase. Zanamivir is given by dry powder inhalation and oseltamivir is used orally. If these medications are given to adults or children with influenza infection within 36–48 hours of the onset of symptoms, they decrease the illness by 1–2 days and reduce the severity of the infection (and also appear to prevent secondary bacterial pneumonia and otitis media). The earlier in the course of illness they are given, the more effective they are. These drugs are also 70–90% effective in preventing infection when given prophylactically after exposure to an infected close contact or for seasonal prophylaxis. Side effects to these medicines are rare and so is the development of drug resistance, although resistance to oseltamivir has developed in viral isolates from children. A newer neuroaminidase inhibitor, **peramivir**, is available in intravenous form only and received emergency use authorization by the FDA in 2009 for use in persons with **H1N1** infection who were unable to tolerate, or who were not responding to oseltamivir or zanamivir.

The next pandemic? Bird Flu- H5N1, H7N9, or something else?

In 1997 an avian influenza A virus referred to as H5N1 crossed the bird-human species barrier in Asia and ultimately caused hundreds of human fatalities and was felt to pose a major pandemic threat. Cases increased dramatically in 2003, raising fears of a new pandemic. The pandemic has yet to occur. Instead there has been sporadic transmission since 2003 leading to more than 600 confirmed human cases in 15 countries (primarily in Asia and Africa) with a greater than 50% mortality rate. Human-to-human transmission has remained limited. To date, most of the human cases have occurred in either poultry workers, close house-hold contacts, and in health care workers in close contact with an infected person, suggesting that human-to-human transmission is still limited and requires heavy exposure. A typical exposure history includes plucking and preparing diseased chickens or ducks, handling fighting cocks, or playing with asymptomatic ducks. While there has been no evidence to date of human-to-human transmission via small particle respiratory aerosols, with every case there is an opportunity for a mutation that makes the virus more virulent and capable of aerosolized human-to-human transmission, an event necessary for a true influenza pandemic similar to the infamous 1918 pandemic.

Clinical Manifestations of H5N1

Following an incubation period of 2–4 days after exposure (up to 8-days), most patients develop a high fever and a typical flu-like illess (headache, myalgias – muscle aches, diarrhea, abdominal pain, vomiting, sore throat, rhinorrhea) with lower respiratory symptoms such as cough, shortness of breath, and sputum production. Unlike the run of the mill epidemic influenza, almost all patients that are infected with influenza H5N1 then develop a clinical pneumonia with diffuse patchy infiltrates on chest radiogram which progress to consolidation with air-bronchograms in more than one lung zone. At the time of hospitalization, this is typically a primary viral pneumonitis without secondary bacterial infection.

This rapidly progresses to the Acute Respiratory Distress Syndrome (ARDS or non-cardiogenic pulmonary edema, which is defined by low blood oxygen levels, infiltrates in multiple parts of the lung, and no evidence of heart failure). 50% of infected people die and almost 90% of those younger than 15 years of age die. While most infections tend to be milder in children and infants, this is sadly not the case for influenza H5N1.

Diagnosis of H5N1

Viral culture or Reverse Transcriptase Polymerase Chain Reaction (RT-PCR) of viral RNA from pharyngeal or nasal washings is used to identify the virus. Virus can also be detected in feces. Immunofluorescence test for antigen with antibody directed against H5 or a four-fold rise in H5-specific antibody titers can also be used to document infection.

Treatment of H5N1

The mainstay of therapy is mechanical ventilation in an effort to keep the patient alive until the immune system and neuraminidase inhibitors, like oseltamivir, can work. Typically, broad spectrum standard antibiotics are given to prevent secondary bacterial pneumonia. While it is not clear that oseltamivir is effective after the viral pneumonia develops, most patients have been treated with this drug. There is some evidence that treatment early in the course of the illness is effective. A detailed list of current drugs and resistance:

1) **Oseltamivir**- This drug is used for primary prevention and treatment and is given orally at 75 mg twice daily for five days (once a day for 7–10 days for prophylaxis) in adults and weight adjusted dosing for children. In severe cases the dose is typically doubled and given for up to 10 days. Unfortunately, high level resistance has developed secondary to a viral mutation with substitution of a single amino acid in the N1 neuraminidase (Histidine 274 to Tyrosine). Up to 16% of children with regular influenza A develop this variant and this mutation has now been identified in infected children with H5N1.

2) **Zanamivir** is an inhaled neuraminidase inhibitor that still works in the oseltamivir-resistant strains of H5N1. Further clinical investigation of this agent and of a new neuraminidase inhibitor, **peramivir**, are on-going.

3) **Amantadine** and **Rimantadine** – Unfortunately, the influenza A (H5N1) is highly resistant to these drugs so they have no clinical utility against the avian flu.

H7N9

On April 1, 2013 the World Health Organization reported 3 human infections with a novel influenza A (H7N9) virus in China. This new virus, similar to H5N1, is of avian origin. The number of new cases of H7N9 peaked in April 2013 and then declined, likely related to closure of live bird markets and increased public awareness. There have been spikes in H7N9 cases annually since 2013 coinciding with influenza season. There have been over 500 laboratory confirmed cases. There are no commercially available vaccines for H5N1 or H7N9, although a H5N1 vaccine was FDA approved in 2007 and is stockpiled by the US federal government for distribution if needed.

PARAMYXOVIRIDAE

The structure of paramyxoviridae is very similar to that of the orthomyxoviridae. The differences are that:

1) The negative (−) stranded RNA is in a single strand, not segmented.
2) HA and NA are a part of the same glycoprotein spike, not 2 different spikes.
3) They possess a **fusion (F) protein** (not present in the orthomyxoviridae) that causes the infected host cells to fuse together into **multinucleated giant cells** (syncytial cells similar to those caused by herpesviridae and retroviridae infection).

There are 5 paramyxoviridae that cause human disease: parainfluenza virus, respiratory syncytial virus, metapneumovirus, mumps virus, and measles virus. Before reading about each, let's examine the **big picture**:

1) **Think lungs**: All adsorb to and replicate in the upper respiratory tract. Respiratory syncytial virus, metapneumovirus, and parainfluenza virus all cause lower respiratory infections (pneumonia) in children and upper respiratory tract infections (bad colds) in adults.
2) **Think kids**: Most infections occur in children.
3) **Think viremia**: The viral infection results in dissemination of virions in the blood to distant sites. Mumps and measles reproduce in the upper respiratory tract and spread hematogenously to distant organs. Mumps can produce local parotid and testes infection (parotitis and orchitis), and measles can produce a severe systemic febrile illness. Brain infection (encephalitis) can occur with both mumps and measles.

Parainfluenza Virus

The parainfluenza virus causes upper respiratory infection in adults ranging from cold symptoms such as rhinitis, pharyngitis, and sinus congestion, to bronchitis and flu-like illness. Children, elderly, and the immunocompromised also suffer from lower respiratory tract infections (pneumonia).

Croup is a parainfluenza infection of the larynx and other upper respiratory structures (laryngotracheobronchitis) that occurs in children. Swelling of these structures produces airway narrowing. Stridor (a wheezing sound) and a barking cough (like a seal) occur as air moves through the narrowed upper airways.

Respiratory Syncytial Virus (RSV)

RSV is so-named because it causes respiratory infections and contains an F-protein that causes formation of multinucleated giant cells (syncytial cells). This virus differs from the rest of its kin by lacking both the HA and NA glycoproteins.

RSV is the number one cause of pneumonia in young children, especially in infants less than 6 months of age. The virus is highly contagious with outbreaks occurring in winter and spring. The treatment of RSV infection is less than ideal, with ribavirin studies showing conflicting results. Efforts have therefore focused on prevention. RSV infection can be prevented in a high percentage of cases with **palivizumab**, which is a monoclonal antibody against RSV that is produced by a recombinant DNA method. A blood-derived product, **serum RSV immune globulin (RSVIG)**, previously was used alone or in combination with ribavirin in seriously ill patients but is no longer available.

Previously infected persons are not entirely immune, but the subsequent infections are usually limited to the upper respiratory tract.

Metapneumovirus

Metapneumovirus was first isolated in 2001 and subsequently has been determined to be the second most common etiology of lower respiratory infection in young children. Children infected with this virus tend to be slightly older than those infected with RSV, 1 year old versus <6 months. Most illness occurs in the winter and early spring with illness ranging from bronchiolitis (~50%), croup (~20%), to pneumonia (<10%) in children. Disease has also been increasingly found in older adults who may merely develop cold symptoms or alternatively may develop more severe lower respiratory tract disease. Diagnosis is most commonly made from reverse transcriptase PCR studies performed on respiratory secretions and nasopharyngeal swabs. Treatment is supportive.

Mumps Virus

Mumps virus replicates in the upper respiratory tract and in regional lymph nodes and spreads via the blood to distant organs. Infection can occur in many organs, but the most frequently involved is the parotid gland.

Fig. 24-7. About 3 weeks after initial exposure to mumps virus the parotid gland swells and becomes painful. The testes are also frequently infected. About 25% of infected males who have reached puberty can

develop orchitis. The testes enlarge and stretch the capsule, resulting in intense pain. Infertility is a rare complication. Meningitis and encephalitis can also occur, the former being more common and less severe.

There is only one antigenic type, and a live attenuated viral vaccine is a part of the trivalent measles-mumps-rubella (MMR) vaccine.

Measles Virus

Due to the effectiveness of the MMR vaccine, there were only 216 cases reported in the U.S. between 2001 and 2003. However, the disease continues to have worldwide impact with about 1 million deaths annually.

Fig. 24-8. The clinical manifestations of measles (also called **rubeola**).

Exposure

Measles virus is highly contagious and spreads through nasopharyngeal secretions by air or by direct contact. The virus multiplies in the respiratory mucous membranes and in the conjunctival membranes. Incubation lasts for 2 weeks prior to the development of rash.

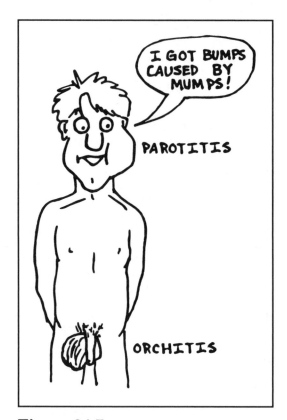

Figure 24-7

Prodrome

Fig. 24-9.   Measles prodrome. Prior to the appearance of the rash, the patient suffers from prodromal illness with conjunctivitis, swelling of the eyelids, photophobia, high fevers to 105° F, hacking cough, rhinitis, and malaise (feels cruddy).

Koplik's Spots

Fig. 24-10.   Koplik's spots. A day or 2 before the rash, the patient develops small red-based lesions with blue-white centers in the mouth. Think of a cop licking a red-white-blue lollipop.

Rash

The measles rash is red, flat to slightly bumpy (maculopapular). It spreads out from the forehead to the face, neck, and torso, and hits the feet by the third day.

Fig. 24-11.   As the measles rash spreads downward, the initial rash on the head and shoulders coalesces. The rash disappears in the same sequence as it developed. Visualize a can of measles-brand red paint being poured over a patient's head. The paint is thicker over the head and shoulders and drips completely off in 6 days.

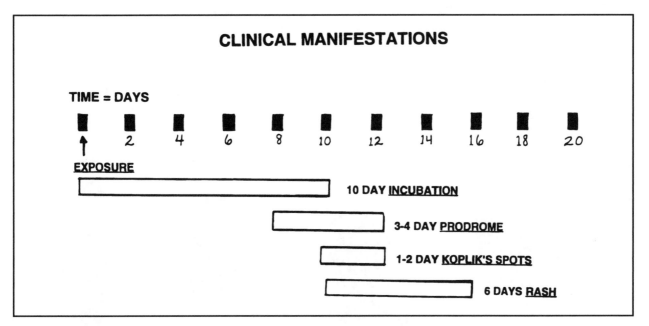

Figure 24-8

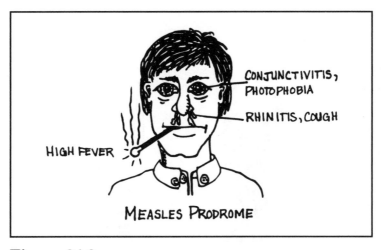

Figure 24-9

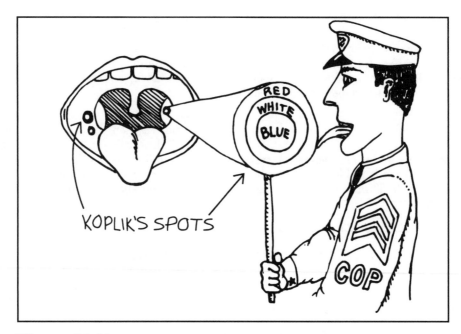

Figure 24-10

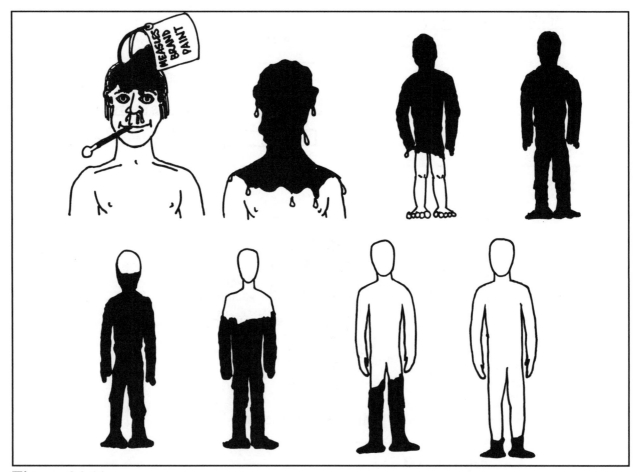

Figure 24-11

Complications

Like mumps, the measles virus disseminates to many organ systems and can damage those sites, causing pneumonia, eye damage, heart involvement (myocarditis), and the most feared complication, encephalitis. Encephalitis is rare, but 10% of patients who develop this will die.

Infection with measles during pregnancy does not cause birth defects but has been associated with spontaneous abortion and premature delivery. In fact, measles in pregnant women results in fetal death in 20% of cases.

Subacute sclerosing panencephalitis (SSPE) is a slow form of encephalitis caused by measles virus. Many years after a measles infection the child or adolescent may have slowly progressing central nervous system disease, with mental deterioration and incoordination.

The MMR vaccine, which contains live attenuated measles virus, is preventative.

Fig. 24-12. Summary of the orthomyxoviridae and paramyxoviridae.

Recommended Review Articles:

Jefferson T, Demicheli V, et al. Antivirals for influenza in healthy adults: systematic review. Lancet. 2006;367(9507): 303–13.

Li Q, et al. Preliminary Report: Epidemiology of the Avian Influenza A (H7N9) Outbreak in China. N Eng J Med 2013;1–11.

Moscona A. Drug therapy: Neuraminidase inhibitors for influenza. N Eng J Med 2005;353:1363–73.

The Writing Committee of the WHO Consultation on Clinical Aspects of Pandemic (H1N1) 2009 Influenza, Bautista E, Chotpitayasunondh T, et al. Clinical aspects of pandemic 2009 influenza A (H1N1) virus infection. N Engl J Med. 2010; 362(18):1708–19.

The Writing Committee of the World Health Organization Consultation on Human Influenza A/H5. Current concepts: Avian Influenza (H5N1) infection in humans. N Eng J Med 2005;353:1374–85

Williams JV, et al. Human Metapneumovirus and Lower Respiratory Tract Disease in Otherwise Health Infants and Children. N Eng J Med 2004;350:443–50.

The Writing Committee of the World Health Organization Consultation on the Clinical Aspects of Pandemic (H1N1) 2009 Influenza. N Eng J Med 2010;362:1708–19.

NAME	MORPHOLOGY	VIRULENCE FACTORS
ORTHOMYXO VIRUS		
Influenza type A: human and animal strain type B: human only strain type C: human only strain	1. Negative (−) single-stranded RNA 2. Segmented (7–8) 3. Lipid containing envelope 4. Helical symmetry 5. Replicates in the nucleus! (Retroviruses are the only other type of RNA viruses that replicate in the nucleus)	1. **Hemagglutinin (HA) glycoprotein**: binds to red blood cells. Also binds to cells of the upper respiratory tract. The HA is then cleaved into two pieces (HA1 & HA2) by host cell proteases, which allows HA to activate fusion. The viral RNA is then dumped into these cells. 2. **Neuraminidase (NA) glycoprotein**: breaks down neuraminic acid, an important component of mucin
PARAMYXO VIRUS		
Parainfluenza	1. Negative (−) single-stranded RNA 2. Unsegmented 3. Lipid containing envelope 4. Helical symmetry 5. Replicates in the cytoplasm	1. Glycoproteins with combined HA and NA activity 2. **F-protein** (Fusion protein): results in multinucleated giant cells (called syncytial cells)
Respiratory syncytial virus	• Same as above	1. **F-protein** 2. **NO** HA nor NA glycoproteins
Metapneumovirus	• Same as above	• Integrin alpha-V-beta receptor allows infection of respiratory tract epithelial cells.
Mumps	• Same as above	1. Glycoproteins with combined HA and NA activity 2. **F-protein**

Figure 24-12   ORTHOMYXOVIRIDAE AND PARAMYXOVIRIDAE

CLINICAL	TREATMENT & PREVENTION	MISCELLANEOUS
• The **Flu**: Fever, runny nose, cough, myalgias arthralgias, etc. • *Complications* 1. Secondary bacterial pneumonias in the elderly 2. **Reyes Syndrome** in children who use aspirin; get liver and brain disease 3. Increased mortality in the elderly and in those with underlying pulmonary and cardiac disease.	1. Vaccine: contraindicated in egg allergies. (vaccine grown in eggs) 2. Amantadine & Rimantidine: prevent viral uncoating of influenza A 3. Zanamivir (inhaled) & Oseltamivir (oral) are neuraminidase inhibitors. Can shorten course of influenza A and B.	1. *Antigenic drift*: small mutations, resulting in minor changes in the antigenicity of HA or NA. This results in **epidemics** of the common flu 2. *Antigenic shift* (only occurs with influenza type A): reassortment. Major changes of the HA or NA (including acquisition of animal HA or NA). This results in devastating influenza **pandemics** 3. Avian influenza viruses such as H5N1 and H7N9 pose great risk for human pandemics.
1. Upper respiratory tract infection in adults: bronchitis, pharyngitis, rhinitis 2. Viral pneumonia in children, elderly and immuno-compromised 3. **Croup**: Children develop a barking cough due to infection and swelling (narrowing) of the larynx 4. Bronchiolitis in children	• Supportive	
1. Most common cause of pneumonia in infants less than 6 months of age 2. Acute otitis media occurs in up to 33% of children with RSV illness	1. Palivizumab: a monoclonal antibody against RSV that is produced by a recombinant DNA. It is given intramuscularly. Indicated for prophylaxis in premature infants (less than 32 weeks) or infants younger than 2 years with severe chronic lung disease 2. Ribavirin	
• Upper and lower respiratory tract infections in young children and older adults.	• Supportive	• Diagnose with RT-PCR of respiratory samples.
Mumps 1. Parotid gland swelling (painful) 2. Testicular inflammation (very painful) 3. Meningitis 4. Encephalitis	• Prevention: **MMR** vaccine: 1. **M**easles 2. **M**umps (live attenuated) 3. **R**ubella	• Only one antigenic type. Therefore, the vaccine is protective

NAME	MORPHOLOGY	VIRULENCE FACTORS
PARAMYXO VIRUS		
Measles (rubeola)	• Same as above	1. HA, but no NA 2. **F-protein**

Figure 24-12   (continued)

CLINICAL	TREATMENT & PREVENTION	MISCELLANEOUS

Measles
1. **Prodrome**: high fever, hacking cough and conjunctivitis
2. **Koplik's** spots: small red based blue-white centered lesions in the mouth
3. **Rash**: from head, then to neck & torso, then to feet. As the rash spreads, it coalesces
4. Complications:
  A. Pneumonia, eye damage, myocarditis and encephalitis
  B. 20% risk of fetal death if acquired by a pregnant woman early in her pregnancy
  C. **Subacute Sclerosing Panencephalitis**: slow form of encephalitis that occurs many years after a measles infection.
	Prevention: **MMR** vaccine:
	1. **Measles** (live attenuated)
	2. **Mumps**
	3. **Rubella**
		• Biopsy of rash or Koplik's spots reveals multinucleated giant cells

M. Gladwin, W. Trattler, and S. Mahan, *Clinical Microbiology Made Ridiculously Simple* ©MedMaster

CHAPTER 25. HEPATITIS VIRIDAE

Viral hepatitis is an infection of the liver hepatocytes by viruses. There are 5 known viruses that primarily infect the liver.

The 5 RNA viruses are:

1) Hepatitis A virus (HAV).
2) Hepatitis C virus (HCV), which was previously called NON-A NON-B until it was isolated.
3) Hepatitis D virus (HDV).
4) Hepatitis E virus (HEV).
5) Hepatitis G virus.

There is 1 DNA virus called:

1) Hepatitis B virus (HBV)

Hepatitis A and E are both transmitted via the fecal-oral route, while the rest are transmitted via blood-to-blood (parenteral) contact. Just as A and E are at both ends of ABCDE, so they are transmitted by elements of both ends of the GI tract. **A** = Anal, **E** = Enteric, **BCD** = **BlooD**.

We will now discuss the clinical disease hepatitis and then cover each virus in more detail.

VIRAL HEPATITIS

Viral hepatitis can be a sudden illness with a mild to severe course followed by complete resolution. This is called **acute viral hepatitis** and can be caused by all of these viruses. Hepatitis can also have a prolonged course of active disease or silent asymptomatic infection termed **chronic viral hepatitis**. The parenterally (blood-to-blood) transmitted HBV, HCV, and HDV can cause chronic hepatitis.

1) **Acute viral hepatitis** has a variable incubation period, depending on the virus type. The growth of the virus first results in systemic symptoms much like the flu, with fatigue, low-grade fever, muscle/joint aches, cough, runny nose, and pharyngitis. One to two weeks later the patient may develop jaundice as the level of bilirubin, which is normally cleared by the liver, rises. As the virus grows in the hepatocytes, these liver cells necrose (die). The hepatocytes produce enzymes that are released during cell death. These are the liver-function enzymes aspartate aminotransferase (AST), alanine aminotransferase (ALT), gamma-glutamyl transpepti-

dase (GGT), and alkaline phosphatase. Elevated blood levels of these liver enzymes help establish the diagnosis of hepatitis.

So about 2 weeks into the illness, the patient is often jaundiced, has a painful enlarged liver and high blood levels of liver-function enzymes.

The details of how to determine which virus is causing the hepatitis will be discussed with each virus.

2) **Chronic viral hepatitis** is more difficult to diagnose because the patient is often asymptomatic with only an enlarged tender liver and mildly elevated liver-function enzyme levels.

The Pattern of Liver Enzyme Elevation

Different diseases result in different patterns of liver-function enzyme elevation. For example, viral hepatitis usually causes the transaminases ALT and AST to elevate to very high levels, while GGT, alkaline phosphatase, and bilirubin are only mildly elevated. As the disease progresses, bilirubin levels rise higher. A gallstone in the bile duct causes the opposite to occur: bilirubin, alkaline phosphatase, and GGT rise higher than ALT and AST. The reason for this is as follows.

Fig. 25-1. The hepatocytes produce AST and ALT. The cells that line the bile canaliculi produce alkaline phosphatase and GGT. The bile canaliculi carry bilirubin.

Fig. 25-2. With viral hepatitis there is cell necrosis as the virus takes over the cell machinery. The hepatocytes rupture, resulting in the release of AST and ALT. Some pericanalicular cells will also be destroyed. So we will see very high AST and ALT with a little elevation of alkaline phosphatase and GGT. As the infection worsens, the liver swells and the canaliculi narrow, resulting in a backup of bilirubin into the blood. So with viral hepatitis the ALT and AST are very high, and the bilirubin, alkaline phosphatase, and GGT rise later in the course.

Fig. 25-3. A stone blocking the bile duct would result in:

1) Inability to excrete the bilirubin, resulting in high blood levels of bilirubin.
2) The backup results in increased alkaline phosphatase and GGT synthesis by the canalicular cells. Imagine the backup of pressure popping the cells, resulting in the release of alkaline phosphatase and GGT (not the true mechanism but a good memory tool).

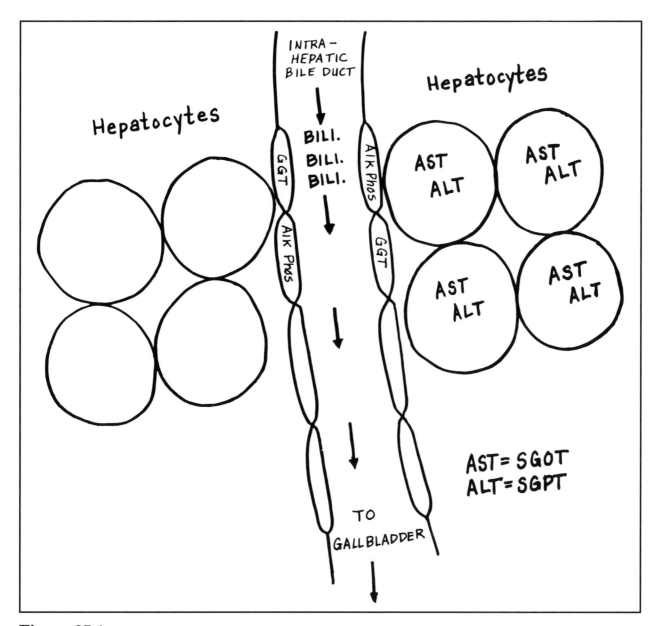

Figure 25-1

So with an obstructive process (stone in bile duct), AST and ALT would only be minimally elevated; alkaline phosphatase, GGT, and bilirubin would be very elevated.

Hepatitis A Virus (HAV)

HAV has a naked icosahedral capsid with a positive (+) single-stranded RNA nucleic acid. It is in the family Picornaviridae, and as is the case with most of this family it is transmitted by the fecal-to-oral route (**HAV** = **A**nus).

Epidemiology

HAVe you washed your hands??? About 2000 cases of hepatitis A infection are reported each year in the U.S., and there are many more infections that are asymptomatic or unreported. In fact, 40% of Americans living in urban centers have serologic evidence of prior infection, but only 5% remember the infection. Overall, the incidence of new infections has been declining in the U.S. with the advent of widespread vaccination. Outbreaks often occur secondary to fecal-to-hand-to-mouth contact. Examples of this include an infected food handler

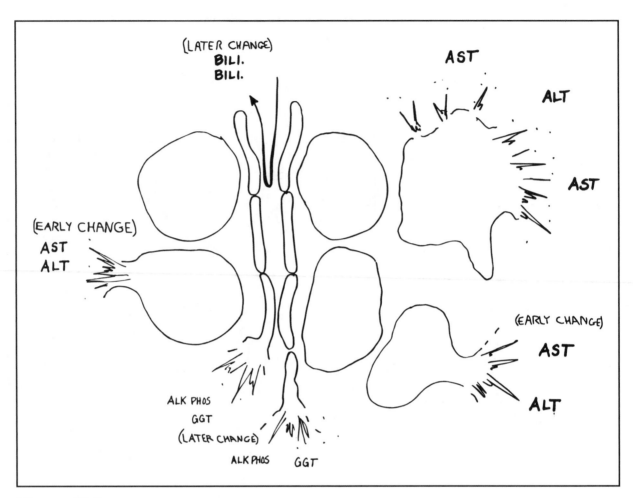

Figure 25-2

contaminating food after poor hand washing, persons ingesting fecally contaminated drinking water, or close person-to-person contact in institutions such as day care centers. There is a 15–40 day incubation period (about 1 month) before the patient develops acute hepatitis as described earlier.

Young children are the most frequently infected, and they have a milder course than do adults, often without developing jaundice or even symptoms. At the other end of the spectrum, a small percentage (1–4%), usually adults, will develop fulminant (severe) hepatitis. However, death from HAV is very rare (1%).

Serology

Serologic tests can help establish the diagnosis. The HAV capsid is antigenic, resulting in the host production of anti-HAV IgM and later, the anti-HAV IgG. A patient with active infection will have anti-HAV IgM detectable in the serum. Anti-HAV IgG

indicates old infection and no active disease. This antibody lasts indefinitely and is protective, which means that it will protect against future infection with HAV.

Fig. 25-4. Hepatitis A infection: A time line of clinical symptoms and antibody development.

Fig. 25-5. Hepatitis A serology.

Treatment

Inactivated Hepatitis A vaccine is recommended for adults at high risk of HAV infection, such as travelers. Beginning in 2005, Hepatitis A vaccine was incorporated into the recommended routine pediatric vaccination schedule. If a person has been exposed, pooled immune serum globulin will prevent or decrease the severity of infection, if given early during incubation. Pooled immune serum globulin is obtained by ethanol fractionation from the plasma of hundreds of donors.

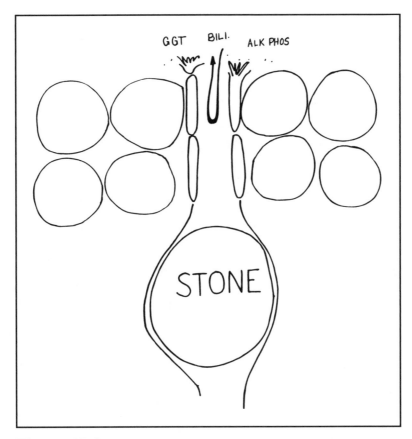

Figure 25-3

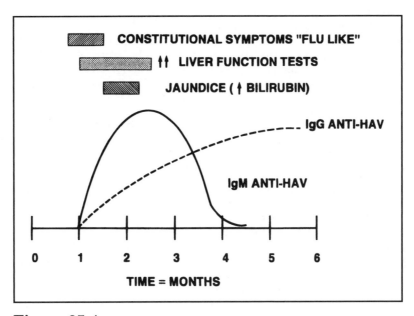

Figure 25-4

	IgM ANTI-HAV	IgG ANTI-HAV
Acute infection	+	−
Old infection ("immune to HAV")	−	+
Incubation *or* no infection	−	−

Figure 25-5 SEROLOGY OF HEPATITIS A

Since anti-HAV IgG is present in about 40% of the population, there will be antibody in the pooled immune serum globulin that will inactivate the virus. Once infection is established, treatment is only supportive.

Hepatitis B Virus (HBV)

You will have an intimate relationship with HBV throughout your career. Why is this? In an infected patient, this virus lives in all human body fluids (semen, urine, saliva, blood, breast milk . . .). As a physician, you will come into contact with patients who harbor this virus. Since hospital workers are considered to be an at-risk group, you will receive immunization against HBV. So you will touch this virus frequently and will actually harbor antibodies against it.

HBV = Big and Bad

HBV is very different from HAV. It is a **Big** (42 NM) virus with an enveloped icosahedral capsid and double-stranded circular DNA.

Fig. 25-6. The intact virus is called the **Dane particle** (**B**ig like a Great Dane) and looks like a sphere under electron microscopy. Notice that the Dane particle has an envelope and an icosahedral capsid studded with protein spikes. In its core is a double-stranded DNA with associated DNA polymerase enzyme.

When looking at infected blood with electron microscopy, you will notice the Dane particle spheres, described above, as well as longer filamentous structures. These filamentous structures (as seen under electron microscopy) are composed of the envelope and some capsid proteins that have disassociated from the intact virion. This part of the virus is called the **hepatitis B surface antigen** (**HBsAg**) and is of critical importance because antibodies against this component (**anti-HBsAg**) are protective. Having anti-HBsAg means the patient is **immune** against HBV.

Removing HBsAg leaves the viral core, which is called **hepatitis B core antigen** (**HBcAg**) and is also antigenic. However, antibodies against the core (**anti-HBcAg**) are not protective (do not result in immunity).

During active infection and viral growth, a soluble component of the core is released. This is called **HBeAg**. This antigen is a cleavage product of the viral core structural polypeptide. HBeAg is found dissolved in the serum, and is a **marker for active disease and a highly infectious state**. Pregnant mothers with HBeAg in their blood will almost always transmit HBV to their offspring (90% transmission rate), whereas mothers who have no HBeAg will rarely infect the neonate (10% transmission rate).

Epidemiology

HBV is present in human body fluids and is transmitted from blood-to-blood contact. This non-oral transmission is called **parenteral** transmission. Transmission from an infected patient can occur by needle sharing, accidental medical exposures (needle sticks, blood spray, touching blood with unprotected hands), sexual contact, blood transfusions, perinatal transmission, etc. This virus is extremely contagious.

Pathogenesis

Another reason that HBV is a **B**ad dude is that unlike hepatitis A, which can only cause an acute hepatitis, HBV can cause acute and chronic hepatitis. The following are disease states caused by *BIG BAD HBV*:

1) **Acute hepatitis**.
2) **Fulminant hepatitis**: Severe acute hepatitis with rapid destruction of the liver.
3) **Chronic hepatitis**:

   a) **Asymptomatic carrier**: The carrier patient never develops antibodies against HBsAg (anti-HBsAg) and harbors the virus without liver injury. There are an estimated 200 million carriers of HBV in the world.
   b) **Chronic-persistent hepatitis**: The patient has a low-grade "smoldering" hepatitis.
   c) **Chronic active hepatitis**: The patient has an acute hepatitis state that continues without the normal recovery (lasts longer than 6–12 months).

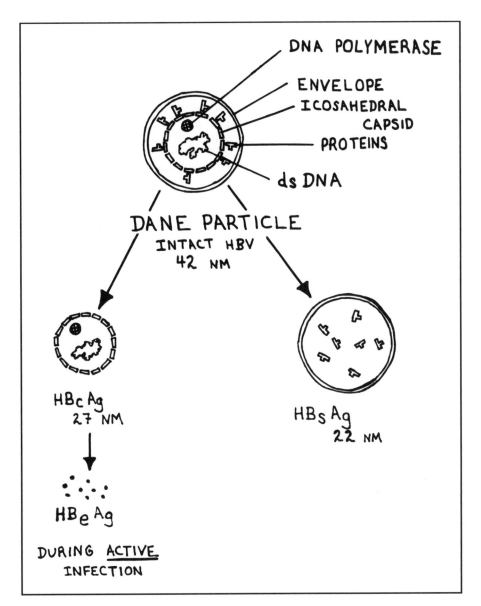

DNA POLYMERASE

ENVELOPE

ICOSAHEDRAL
CAPSID
PROTEINS

ds DNA

DANE PARTICLE
INTACT HBV
42 NM

HBc Ag
27 NM

HBe Ag

DURING ACTIVE
INFECTION

HBs Ag
22 NM

Figure 25-6

4) **Co-infection with hepatitis delta virus (HDV):** See HDV section.

Liver injury appears to occur from a cell-mediated immune system attack on HBV. Viral antigens on the surface of infected hepatocytes are targets for cytotoxic T-cells. Immune complexes of antibody and HBsAg can deposit in tissues and activate the immune system, resulting in arthritis, as well as skin and kidney damage. Patients who have immunosuppressed states, such as malnutrition, AIDS, and chronic illness, are more likely to be asymptomatic carriers because their immune system does not attack.

Complications

Primary hepatocellular carcinoma is a complication of HBV. With chronic infection the HBV DNA becomes incorporated into the hepatocyte DNA and triggers malignant growth. There is a 200X increase in the risk of developing primary hepatocellular carcinoma in HBV carriers as compared to noncarriers.

Infection with HBV can result in permanent liver scarring and loss of hepatocytes. This is called **cirrhosis**.

Serology

Serologic tests help establish HBV infection. The many antigens and antibodies are simpler than they seem, as follows:

1) **HBsAg**: The presence of HBsAg always means there is LIVE virus and infection, either acute, chronic, or carrier. When anti-HBsAg develops, HBsAg disappears and the patient is protected and immune.

    a) **HBsAg = DISEASE (chronic or acute)**
    b) **Anti-HBsAg = IMMUNE, CURE, NO ACTIVE DISEASE!!!**

2) **HBcAg**: Antibodies to HBcAg are not protective but we can use them to understand how long the infection has been ongoing. With acute illness we will see IgM anti-HBcAg. With chronic or resolving infection IgG anti-HBcAg will develop.

    a) **IgM anti-HBcAg = NEW INFECTION**
    b) **IgG anti-HBcAg = OLD INFECTION**

3) **HBeAg**: The presence of HBeAg connotes a high infectivity and active disease. Presence of anti-HBeAg suggests lower infectivity.

    a) **HBeAg = HIGH INFECTIVITY, virus going wild!**
    b) **anti-HBeAg = LOW INFECTIVITY**

Fig. 25-7. Time course of acute HBV infection, with complete resolution and immunity.

Fig. 25-8. Time course of chronic HBV infection, with failure to develop the protective anti-HBsAg antibodies.

Treatment

Prevention and control of hepatitis B involves:

1) Serologic tests on donor blood to remove HBV-contaminated blood from the donor pool.

2) Active immunization: The vaccine is a recombinant vaccine. The gene coding for **HBsAg** is cloned in yeast and used to produce mass quantities of HBsAg, used as the vaccine. There is no risk of developing disease from the vaccine because it contains only the surface envelope and proteins (HBsAg = no DNA or capsid). The HBV vaccine is now given to all infants at birth, 2, 4, and 15 months; it is also given as 3 injections to adolescents and high-risk adults (health care workers, IV drug users, etc.).

3) Anti-viral agents for treatment of chronic active or persistent HBV infection:

    a) Interferons (either interferon alfa or pegylated-interferon alfa) suppress HBV DNA levels and lead to seroconversion of HBeAg in around 35% of patients with chronic Hepatitis B infection. The advantages of interferons are that they are given for a finite period of time (as short as 4 months) and there is little evidence of drug resistance. The drawbacks include cost, less than ideal efficacy, and frequent side effects.

    b) Nucleoside analogs (lamivudine, adefovir, entecavir, and telbivudine) or a nucleotide analog (tenofovir) are a frequently used alternative in the United States. They have the advantages of oral delivery, few side effects, and potent inhibition of viral replication. Seroconversion of HBeAg is similar to

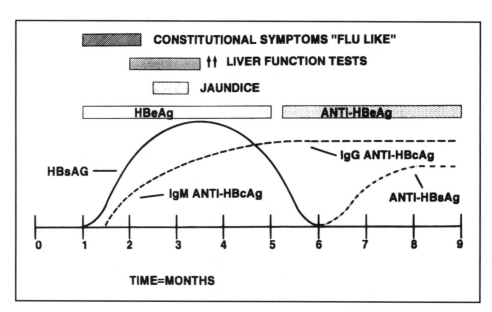

Figure 25-7

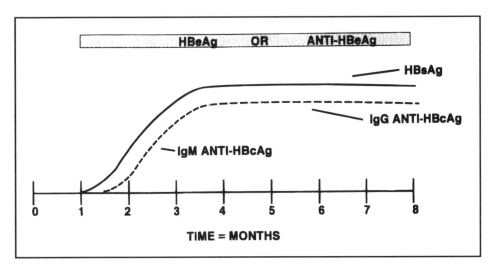

Figure 25-8

the interferons (~30–35%). Drawbacks include the development of drug resistance (particularly with lamivudine), the need for long term or indefinite treatment, and expense.

c) Current therapy of hepatitis B usually entails a combination of the above listed agents (a nucleoside/nucleotide analog combined with pegylated-interferon or a combination of nucleoside analogs).

The main problem with all available therapies for Hepatitis B is that they fail to fully eradicate HBV. As a result, relapse of hepatitis is always a possibility even in the face of apparently successful therapy.

Fig. 25-9. Serology of the medical student vaccinated with HBV recombinant vaccine.

Fig. 25-10. Hepatitis B serology.

Hepatitis Delta Virus (HDV)

This RNA virus is transmitted parenterally and **can only replicate with the help of HBV**. The delta virus helical nucleocapsid actually uses HBV's envelope, HBsAg. HDV steals the clothes from HBV and can only cause infection with the HBsAg coat.

Fig. 25-11. Notice the HBV envelope and proteins surround the HDV helical nucleocapsid. Next to it is a conceptual figure of the letter **D** in a big **B**.

HBV + HDV = Big Bad Dude

Without Big Bad HBV, HDV is just a **dud** and is not infectious. Hepatitis **D** is **D**efective and requires Hep **B** as a **B**uddy (mnemonic thanks to Katherine Putz). Infection occurs in 2 ways:

1) **Co-infection**: HBV and HDV both are transmitted together parenterally (IV drug use, blood transfusions, sexual contact, etc.) and cause an acute hepatitis similar to that caused by HBV. Antibodies to HBsAg will be protective against both, ending the infection.

2) **Superinfection**: HDV infects a person who has chronic HBV infection (like the 200 million worldwide HBV carriers). This results in acute hepatitis in a patient already chronically infected with HBV. This HDV infection is often severe, with a higher incidence of fulminant hepatitis, cirrhosis, and a greater mortality (5–15%). The patient with chronic HBV cannot make Anti-HBsAg and so remains chronically infected with both HBV and HDV.

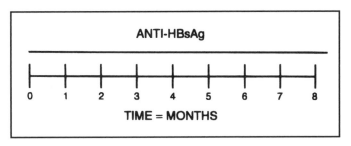

Figure 25-9

	HBsAg	Anti-HBsAg	HBeAG	Anti-HBeAg	Anti-HBcAg
Acute HBV	+	–	+	–	IgM
Chronic HBV High infectivity	+	–	+	–	IgG
Chronic HBV Low infectivity	+	–	–	+	IgG
Recovery	–	+	–	+	IgG
Immunized	–	+	–	–	–

Figure 25-10   SEROLOGY OF HEPATITIS B

Serology is currently not very helpful for diagnostic purposes because IgM and IgG anti-HDV are in the serum for only a short period. As there is no treatment, control of HBV infection is currently the only way to protect against HDV.

Hepatitis C Virus (HCV)

Hepatitis C is an emerging disease that became publicized in the mid-1990's. Originally termed "non-A, non-B hepatitis", hepatitis C is the leading cause of chronic hepatitis in the United States. Sero-surveys show 1.5% of Americans are seropositive for HCV. Unlike hepatitis B where only about 10% of persons exposed to the virus go on to develop chronic infection, with HCV up to 85% of those with exposure and acute infection go on to develop chronic hepatitis. Chronic hepatitis C is the leading indication for liver transplantation in the United States today.

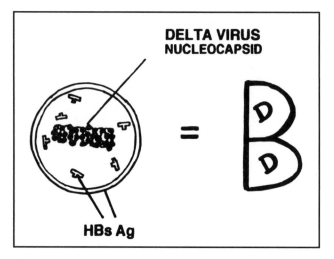

DELTA VIRUS NUCLEOCAPSID

HBs Ag

Figure 25-11

Transmission

HCV is an enveloped icosahedral RNA virus in the flavivirus family. It has a simple genomic structure with just 3 structural and 5 nonstructural genes. There are several genotypes (1, 2, and 3, but genotype 1 is the most common in the United States). These genotypes vary by geographic isolation and they respond variably to antiviral therapy. HCV is transmitted parenterally, with the primary means of infection being via injection drug use.

Manifestations

The incubation period for hepatitis C is 6 to 12 weeks. Acute infection is usually asymptomatic, although a small percentage of patients will display classic symptoms of acute hepatitis (fever, nausea, muscle aches, fatigue, right upper quadrant pain, and jaundice). Up to 85% of patients will go on to develop chronic hepatitis. Of these around 20% will develop cirrhosis.

Diagnosis

Infection with HCV is diagnosed by testing for anti-HCV antibodies that are detectable within 6 to 8 weeks after exposure and remain positive thereafter. A positive test is usually confirmed by recombinant immunoblot assay (RIBA) or by measuring the HCV viral RNA.

Treatment

This is one of the most exciting areas in infectious diseases! We finally have effective treatments for chronic hepatitis C infection. HCV can be cured. Unlike hepatitis B which incorporates in the host DNA, making definitive cure elusive, hepatitis C is an RNA virus which hangs out in the cytoplasm of infected liver cells. It hijacks the cell's machinery to replicate itself, but since it is not incorporated into the host DNA, it can be fully eradicated. Until recently, our armamentarium to fight hepatitis C was pitiful. We tortured patients with 48 weeks of **interferon** and **ribavirin** and had less than 50% cure rates overall. Things got a little better in 2011 when **telaprevir** and **boceprevir** (NS3 protease inhibitors) were approved, but interferon and ribavirin were still required and response rates still only were 60–80% successful.

Fast forwarding to 2015, there has been an explosion of new treatments with many more to come. Several of the new "interferon-free regimens" require only 12 weeks of therapy, have minimal side–effects, and have **cure rates of greater than 90%**! First line regimens for genotype 1 disease include: **ledipasvir/sofosbuvir; paritaprevir/ritonavir/ombitasvir/dasabuvir/riba virin** ; and **sofosbuvir/simeprivir**. The biggest drawback is cost. These meds are very pricey. We are talking insurance companies refusing to pay expenses ($50,000–100,000 for a treatment course). Despite the short term expense, the long term gains – decrease in liver failure and liver cancer – should be great.

Hepatitis E Virus (HEV)

HEV hepatitis is often referred to as **non-A hepatitis** because it shares similarities with HAV. HEV is also transferred by the fecal-oral route, frequently with the consumption of fecally contaminated water during monsoon flooding. The **E** stands for **Enteric** (fecal-oral). It is endemic to Asia, India, Africa, and Central America.

Hepatitis G Virus

Hepatitis G is an RNA virus in the *Flavivirus* family. It is transmissible by transfusion & parenteral routes. It has not conclusively shown to cause liver disease.

Fig. 25-12. Summary of hepatitis viruses.

References

Bhattacharya D, Thio CL. Review of Hepatitis B therapeutics. Clin Infect Dis 2010;51:1201–7.

Butt AB, Kanwal F. Boceprevir and Telaprevir in the management of Hepatitis C Virus-infected patients. Clin Infect Dis 2012;54:96–104.

Doo EC, Ghany MG. Hepatitis B virology for clinicians. Clin Liver Dis. 2010;14(3):397–408.

Hoofnagle JH. Hepatitis B-preventable and now treatable. New Engl J Med 2006;354:1074–76.

Hoofnagle JH, Seeff LB. Peginterferon and ribavirin for chronic hepatitis C. N Engl J Med. 2006;355(23):2444–51.

Lau GKK, Piratvisuth T, Luo KX, Marcellin P, Thongsawat S, Cooksley G, et al. Peginterferon alfa-2a, lamivudine, and the combination for HBeAg-positive chronic hepatitis B. New Engl J Med 2005;352:2682–95.

Lok AS. The maze of treatments for hepatitis B. New Engl J Med 2005;2743–46.

Strader DB, Wright T, Thomas DL, Seef LB. Diagnosis, management, and treatment of hepatitis C. Hepatology 2004; 39:1147.

Zhao S, Tang L, et al. Comparison of the efficacy of lamivudine and telbivudine in the treatment of chronic hepatitis B: a systematic review. Virol J. 2010;7:211.

NAME	MORPHOLOGY	TRANSMISSION	CLINICAL
Hepatitis A	• Picorna viridae 1. Positive (+) single-stranded **RNA** 2. No envelope (naked) 3. Icosahedral capsid	Fecal-oral	• **Acute viral hepatitis**: fever, jaundice, and a painful enlarged liver A. 1% develop fulminant hepatitis B. Never becomes chronic
Hepatitis B	• Hepadna viridae 1. Double-stranded circular **DNA** 2. Envelope 3. Icosahedral capsid 4. **Dane particle** (intact virus) includes: A. Envelope B. Capsid associated proteins C. Capsid D. Core (DNA + protein enzymes) 5. Hepatitis B **surface** antigen (HBsAg) A. Envelope B. Capsid associated proteins 6. Hepatitis B core antigen (HBcAg) A. Double stranded DNA B. DNA polymerase enzyme C. Capsid • Disassociation of the Dane particles leaves HBsAg and HBcAg 7. Hepatitis B antigen (HBeAg): soluble component of the core, which is a marker for active disease.	1. Blood transfusion 2. Needle sticks 3. Sexual 4. Across the placenta	1. **Acute viral hepatitis** 2. **Fulminant hepatitis**: severe acute hepatitis with rapid destruction of the liver 3. **Chronic hepatitis** (10%) A. Asymptomatic carrier B. Chronic persistent hepatitis C. Chronic active hepatitis 4. Coinfection or superinfection with Hepatitis Delta virus (HDV) *Complications*: 1. Primary hepatocellular carcinoma 2. Cirrhosis
Hepatitis C	1. Probably a Flavivirus 2. Single-stranded **RNA** 3. Enveloped icosahedral capsid	1. Blood transfusion 2. Needle sticks 3. Sexual 4. Across the placenta	**Acute viral hepatitis** • Up to 85% develop chronic hepatitis • 20% will develop cirrhosis • Increased risk of developing primary hepatocellular carcinoma.

Hepatitis D	1. Incomplete RNA virus-only infective with the help of hepatitis B virus
	2. Helical nucleocapsid that requires the hepatitis B envelope (HBsAg) to be infectious
		1. Blood transfusion
		2. Needle sticks
		3. Sexual
		4. Across the placenta
			1. **Coinfection**: HBV and HDV are acquired at the same time, and cause an acute hepatitis. Anti-HBV antibodies help cure infection
			2. **Superinfection**: HDV infects a patient with chronic hepatitis B who can not manufacture Anti-HBsAg antibodies.
			    Complications:
			    A. Fulminant hepatitis
			    B. Cirrhosis

Hepatitis E	1. In family of Hepeviridae 2. Single-stranded **RNA** 3. No envelope (naked)	• Fecal-oral	• **Hepatitis** (like hepatitis A)
Hepatitis G	• Flavivirus	1. Transfusion 2. Needle sticks	It has not been conclusively shown to cause liver disease.

Figure 25-12   HEPATITIS VIRIDAE

TREATMENT	SEROLOGY	MISCELLANEOUS
1. Pooled immune serum globulin 2. Supportive care 3. New HAV vaccine	1. **Anti-HAV IgM** = Active disease 2. **Anti-HAV IgG** = Old; No active disease. Protective against repeated infection	
1. Prevention: Hepatitis B recombinant vaccine is now given to all infants and adolescents in the U.S. at birth, 1–2 months and 6–18 months. It is also given to adolescents and high-risk individuals 2. Screen blood to remove HBV-contaminated blood from the donor pool 3. Treatment options: interferons (alfa and pegylated-interferon alfa); nucleoside analogs (lamivudine, adefovir, entecavir, and telbivudine); nucleotide analog (tenofovir); or combinations of the above	• **HBsAg** = Disease (Acute or chronic) • **Anti-HBsAg** = Immunity: provides protection against repeat infection • **IgM anti-HBcAg** = New infection • **IgG anti-HBcAg** = Old infection • **HBeAg** = High infectivity • **Anti-HBeAg** = Low infectivity	1. Only hepatitis B carries a DNA polymerase enzyme within the virion 2. Liver injury occurs from a cell-mediated immune system attack on HBV
• Combination therapy with HCV direct acting antivirals (refer to http://www.hcvguidelines.org for most recent guidelines)	• Screening: anti-HCV antibodies	1. HCV is leading cause for liver transplantation 2. Genotype 1 most common in United States
• Control of HBV infection is currently the only way to protect against HDV	• Serology is not very helpful, since detectable titers of IgM and IgG anti-HDV are present only fleetingly	
		• Responsible for epidemics of hepatitis in Asia. Very rare in the United States
		• Some studies have shown that co-infection with HGV might actually slow the progression of HIV disease!

M. Gladwin, W. Trattler, and S. Mahan, *Clinical Microbiology Made Ridiculously Simple* ©MedMaster

CHAPTER 26. RETROVIRIDAE, HIV, AND AIDS

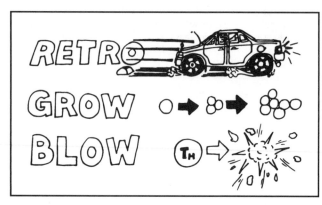

Figure 26-1

The retroviridae are a large group of RNA viruses that infect animals and humans.

Fig. 26-1. There are 3 **big concepts** unique to the retroviridae: **RETRO, GROW,** and **BLOW.**

1) **Retro**: Most RNA viruses (negative- or positive-stranded), enter the host cell and act as mRNA or are transcribed into mRNA. The retroviridae are different. They carry with them a unique enzyme called **reverse transcriptase**. This enzyme is an RNA-dependent DNA polymerase that converts the viral RNA into DNA. This viral DNA has unique "sticky" ends that allow it to integrate into the host's own DNA, as do transposons.

2) **Grow**: Retroviruses can cause cancer in the cells they infect. Genes called **oncogenes** in humans and animals can cause the malignant transformation of normal cells into cancer cells. Some retroviruses carry oncogenes in their genome. Inactive oncogenes in animals and humans are called **proto-oncogenes**. These are genetic time bombs waiting for activation, which can occur by carcinogen-induced DNA mutation or by retrovirus infection.

3) **Blow**: Some retroviridae are cytotoxic to certain cells, blowing them up. The most notable is the human immunodeficiency virus (HIV) which destroys the T-helper lymphocytes it infects. This ultimately results in devastating immunodeficiency.

In this chapter we will discuss: 1) oncogenes; 2) the human retroviridae, using HIV as a model for retroviridae structure and genetics; and 3) HIV infection and 4) the Acquired Immunodeficiency Syndrome (AIDS) caused by HIV.

ONCOGENES

What Is Cancer?

Normal cells have strict growth control. They divide an exact number of times with exact timing, depending where in the body they are located. When normal cells are grown on a plate of nutrient media, they form a single layer and stop dividing when they touch each other. This is called **contact inhibition**. Malignant cells lose contact inhibition and pile up in vitro. They become immortal, dividing continuously as long as there is nutrient supply. In the body this uncontrolled growth causes physical damage by local expansion, steals nutrients from other cells, produces abnormally high levels of hormones or other cell products, and can infiltrate areas where other cells grow (such as the bone marrow).

How Do Oncogenes Cause Cancer?

The normal cell has receptor proteins in the cell membrane that regulate cell growth. Growth factors (**mitogens**) bind to these receptors to regulate growth. Examples of these receptors are the insulin receptor, the Epidermal Growth Factor (EGF) receptor, and the Platelet Derived Growth Factor (PDGF) receptor. Mitogen stimulation of these receptors causes intracellular phosphorylation of tyrosine. Phosphotyrosine then acts as an intracellular growth messenger.

Fig. 26-2. The Rous sarcoma virus oncogene (*src*), encodes a transmembrane protein that also phosphorylates tyrosine, **but at ten times the normal rate!!!** The *erb*-B oncogene that causes cancer in chickens encodes a protein similar to the EGF receptor. Other oncogenes encode proteins that are similar to the PDGF and insulin receptors!

Still other oncogenes encode proteins that act at the nuclear level to promote uncontrolled expression of growth genes.

Retroviridae and Oncogenes

Because most of the retroviridae isolated to date cause either leukemia or sarcoma in their vertebrate

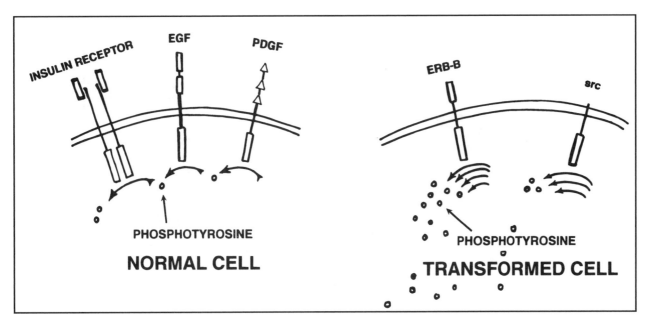

Figure 26-2

hosts, they are sometimes called **leukemia sarcoma viruses**. Some retroviruses cause cancer directly (acute) by integrating an intact oncogene into the host DNA. Others cause cancer indirectly (nonacute) by activating a host proto-oncogene.

Acute Transforming Viruses

Fig. 26-3. Some retroviridae, the acute transforming viruses, carry intact oncogenes within their viral genome, which when integrated into the host DNA cause malignant transformation. This integration is facilitated by the "sticky" ends and an enzyme called **integrase**.

The acute transforming viruses were discovered in 1911 when Peyton Rous injected cell-free filtered material from a chicken tumor into another chicken. The chicken subsequently developed tumor. The causative agent is now known to be a retrovirus called the **Rous sarcoma virus** which possesses within its DNA an intact oncogene called *src*. When the Rous sarcoma virus infects a cell, it reverse transcribes its RNA into DNA. The DNA is integrated into the host's genome by integrase. Once integrated the *src* gene is expressed, causing malignant transformation.

Where Do Viral Oncogenes Come From?

Studies have shown that normal host DNA has sequences that are homologous to viral oncogenes but are still inactive (proto-oncogenes). These genes most likely are involved in cell growth regulation. Somehow, during normal viral infection and integration, a mistake in splicing occurs and a virus "captures" a proto-oncogene. This

proto-oncogene ultimately becomes activated in the virus so the virus now carries an oncogene.

The oncogene gene sequence is so long that most acute transforming viruses have lost their own RNA critical for viral replication. These viruses are called **defective acute transforming viruses** and require a coinfecting virus to cause cancer.

The Rous sarcoma virus is the only known acute transforming virus that is non-defective. It has the full RNA genome needed for replication and also carries an accessory *src* oncogene.

Non-acute Transforming Viruses

Fig. 26-4. Other retroviridae, the non-acute transforming viruses, activate host cell proto-oncogenes by integrating viral DNA into a key regulatory area. These viruses do not carry oncogenes and thus have room for the full genome necessary for viral replication.

HUMAN RETROVIRUSES

By the mid-1970's, retroviruses had been discovered in many vertebrate species, including apes. The hypothesis that humans may also be infected with retroviruses led to a search that ultimately resulted in the isolation of a retrovirus from the cell lines and blood of patients with adult T-cell leukemia. This virus is called human T-cell leukemia virus (HTLV-I).

HTLV-I has now been linked to a paralytic disease that occurs in the tropics (Caribbean islands) called **tropical spastic paraparesis**. HTLV-I induced

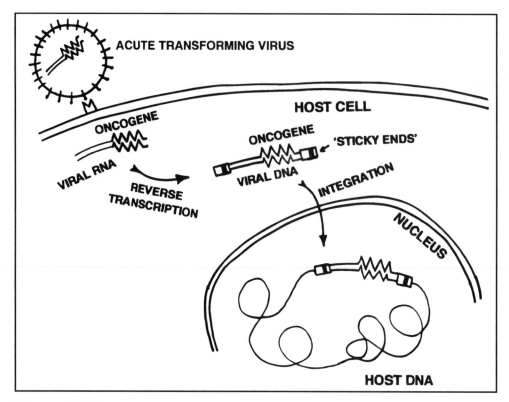

Figure 26-3

leukemia has also been described in the Caribbean and Japan.

A second human retrovirus was isolated from T-cells of patients with a T-cell variant of hairy cell leukemia. This is called HTLV-II, but this virus has no known role in producing disease.

In early 1980 a new epidemic was first noted that we now call the **Acquired Immunodeficiency Syndrome** (**AIDS**). Several factors suggested that this disease was caused by a retrovirus:

1) The infectious agent was present in filtered blood products, such as concentrated factor VIII given to patients with hemophilia. This suggested a viral etiology, as something small would be necessary to pass the filters.

2) There was a delayed onset between exposure (sexual or blood products) and the development of disease. This delayed onset had been observed in the other known retroviral diseases.

3) Immunodeficiency occurs with other animal retroviruses, such as feline leukemia virus. Even HTLV-I can cause immunosuppression.

4) AIDS patients have destruction of the T-helper lymphocytes. The known human retroviruses HTLV-I and II were both T-cell tropic.

Investigators stimulated T-cell culture growth (T-cells from patients infected with AIDS) with interleukin-2 and were able to find RNA and DNA, suggesting a retroviral etiology. The virus, which was subsequently identified, was called the **human immunodeficiency virus (HIV)**.

A second retrovirus, called HIV-2, causes a disease similar to AIDS in western Africa. It is a distantly related virus with 40% sequence homology with HIV-1.

A virus that causes an AIDS-like disease in primates, simian immunodeficiency virus (SIV), shares a close sequence homology with HIV-2.

HIV STRUCTURE

The structure of the virion and genome is similar for all of these retroviruses. We will focus on HIV, the cause of the world's most feared current epidemic. HIV is at the center of intensive research aimed at halting its devastating disease, AIDS.

Fig. 26-5.   Under the electron microscope, HIV appears as a spherical enveloped virion with a central cylindrical nucleocapsid.

To examine this structure fully, we will start from the inside and work outward:

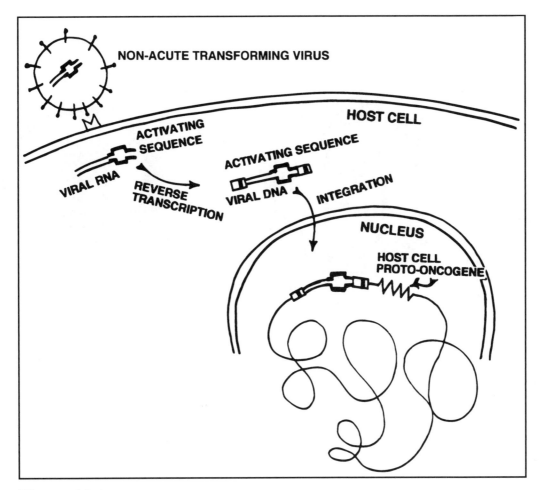

Figure 26-4

1) At the virion core lie 2 identical SS RNA pieces (a dimer). Associated with these are **nucleocapsid (NC) proteins** bound to the RNA and the 3 essential retroviral enzymes, **protease, reverse transcriptase** and **integrase**.

2) Surrounding the RNA dimer lies the capsid shell which has icosahedral symmetry. The proteins that constitute this shell are called **capsid proteins (CA)**. The major capsid protein is **p24**; this can be measured in the serum to detect early HIV infection.

3) The rest of the virus has the same structure described for the influenza virus (see Chapter 24). Proteins under the envelope are called **matrix proteins**. These proteins serve to hold the glycoprotein spikes that traverse the lipid bilayer membrane (envelope).

The surface **glycoproteins** are referred to as **gp** followed by a number: **gp 120**, and **gp 41**.

HIV Genome

Fig. 26-6. The HIV genome (simplified). All retroviruses possess, in their RNA genome, two ending **long terminal repeat** (LTR) sequences, as well as the **gag** gene, **pol** gene, and **env** gene.

1) **LTRs (long terminal repeat sequences)** flank the whole viral genome and serve 2 important functions.

   a) **Sticky ends**: These are the sequences, recognized by integrase, that are involved in insertion into the host DNA. Transposons, mobile genetic elements, have similar flanking DNA pieces.

   b) **Promotor/enhancer function**: Once incorporated into the host DNA, proteins bind to the LTRs that can modify viral DNA transcription.

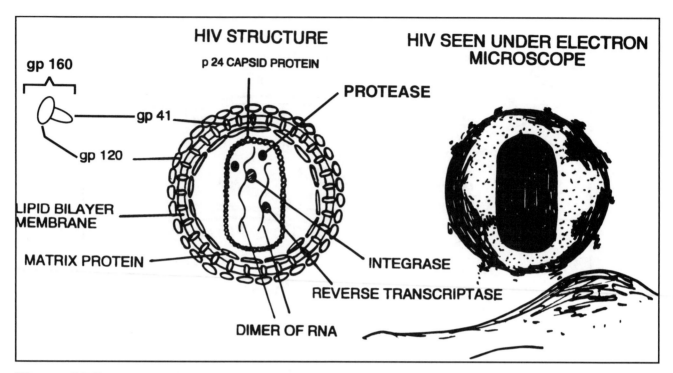

Figure 26-5

(Figure adapted from Fauci—In: Harrison's, 1994; and Haseltine, 1991)

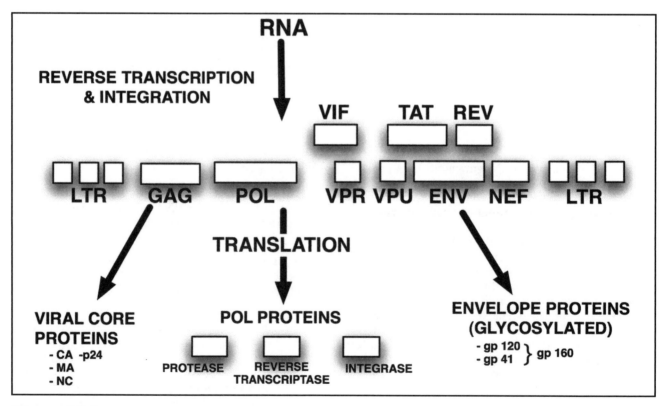

Figure 26-6

2) *gag* (**Group antigen**) sequences code for the proteins inside the envelope: Nucleocapsid (NC), capsid (CA)—called p24, and matrix (MA) proteins. Thus, *gag* codes for the virion's major structural proteins that are antigenic.

3) *pol* encodes the vital protease, integrase, and reverse transcriptase enzymes. The only way the retroviridae maintain their current **POL** position in the race to cause human disease is with these unique enzymes. Protease is a vital HIV enzyme that cleaves *gag* and *pol* proteins from their larger precursor molecules (post-translational modification).

Protease deficient HIV virions can not form their viral core and are non-infectious. New drugs have been developed that block the action of the HIV enzyme, protease. Therapy with these **protease** inhibitors reduces HIV levels and increases CD4 T-lymphocyte cell counts. Similar benefits occur with drugs that inhibit the reverse transcriptase enzyme.

4) *env* codes for the **ENV**elope proteins that, once glycosylated, form the glycoprotein spikes gp 120 and gp 41. Gp 120 forms the head and gp 41 the stalk. Together they are called gp 160 and bind to CD4 receptors on T-cells.

Regulatory and accessory proteins are encoded by the genes: *tat, rev, nef, vif, vpr,* and *vpu*. *Tat, rev, and nef* are termed early proteins, due to their activities in the early stages of the HIV life cycle, while *vif, vpr, and vpu* are termed late proteins, related to their activities in the late stages of HIV replication.

1) *tat* encodes the viral **TrAnsacTivator** protein. This protein binds to the viral genome and activates transcription (thus transactivates). This is a potent promoter of viral activity.

2) *rev* is another promoter that **REV**s up viral activity. It achieves this by a unique mechanism:

The HIV virus has multiple reading frames, producing different mRNAs, depending on where splicing occurs. It can be spliced into many pieces, producing the regulatory proteins such as *tat, rev, nef, vif, vpr,* and *vpu*. Alternatively, it can be spliced only a few times to produce the major *gag, pol,* and *env* products that form the virion. The *rev* protein binds to the *env* gene to decrease splicing. So it **REV**s up the reading of *gag, pol,* and *env* to produce virions!

3) *nef* is crucial to HIV virulence and evasion of the host immune system. It downregulates expression of both CD4 and MHC class I on the infected CD4 cell, thereby helping it escape host cytotoxic T cell mediated killing. (Note major histocompatibility complex I is used for recognition and binding by CD8 T cells who then can kill the infected cells.)

4) *vif* is required for double-stranded DNA to be produced from the HIV RNA genome. *Vif* has also been found to block the effects of **APOBEC3** enzyme. **APOBEC** is an exciting family of enzymes that can break down newly synthesized viral DNA. **APOBEC** appears to represent an innate antiviral defense!

5) *vpr* stands for "Viral Protein R". *Vpr* plays an important role in regulating nuclear import of HIV-1, and is required for virus replication in non-dividing cells, such as macrophages.

6) *vpu* is able to downregulate CD4 and MHC I expression on cell surfaces and it facilitates HIV virion release from infected cells.

Genome Heterogeneity

One of the reasons we are having so much trouble developing a vaccine against HIV is that it possesses the ability to change its genome in a critical area. Within the *env* gene, particularly the area encoding the gp 120 glycoprotein, lie hypervariable regions, where point mutations occur. In fact, duplications and deletions also occur here, and they occur in multiples of 3 to preserve the codon reading frame! Even the gene coding for reverse transcriptase undergoes frequent mutations. In fact, it has one of the highest error rates (mutation rates) described, leading to HIV strains resistant to zidovudine and other reverse transcriptase inhibitor medications. This heterogeneity protects the virus from the human immune system and vaccine induced antibodies.

As a result of viral variation by mutation, high rates of HIV virus turnover have led to the selection of viral variants or subgroups among populations. HIV can be subdivided into major (M) subgroups, classified A-K. Group M is responsible for 99% of the world's HIV/AIDS. M subtypes are analyzed based on variations in the **GAG** and **ENV** proteins. The predominant subtypes for HIV-1 vary by geographic region with subtype **B** predominant in North America and Europe and a mixture of subtypes throughout Africa and Asia. Variation between subtypes, which can be up to 30% between subtypes, is another obstacle in creating a universal HIV-1 vaccine.

HIV INFECTION

Epidemiology and Transmission

As everyone now knows, we are in the midst of a global pandemic of HIV infection. It is estimated by the World Health Organization that there were 34 million persons worldwide living with HIV/AIDS at the end of 2010. Over 40 million persons are estimated to have died due to AIDS since 1981. There were an estimated 2.7 million new infections and 1.8 million deaths due to HIV in 2010 alone. Ninety percent of infected persons today live in the developing world and two-thirds of these are in Africa. In fact, in several countries in Sub-Saharan Africa, over 10% of adults are infected! The HIV epidemic is rapidly spreading in South and Southeast Asia. In contrast, the total number of AIDS cases has changed little in Western Europe, Australia, and the United States. The CDC estimates there are over 1.1 million persons in the United States living with HIV/AIDS, of whom over 18% are undiagnosed or unaware of their HIV status. While African Americans make up only a small minority of the U.S. population, they now account for almost half of the new diagnoses of HIV/AIDS.

Patterns of acquisition have been changing in the last decade. Prior to 1990, in the U.S., Europe, and Australia, the main mode of transmission was men having sex with men (MSM) and via intravenous drug use. Currently in the U.S., the main mode of acquisition of HIV remains MSM, but heterosexual transmission has supplanted IV drug use as the second most common mode of transmission. In developing areas, namely Sub-Saharan Africa, spread has been primarily via heterosexual activity all along.

The HIV virus is spread by the parenteral route, much like hepatitis B virus. This occurs with:

1) Sexual activity: Heterosexual and homosexual activity is the most common mechanism of transmission of HIV. HIV is present in seminal fluid as well as vaginal and cervical secretions. During or following intercourse, the viral particles penetrate tiny ulcerations in the vaginal, rectal, penile, or urethral mucosa. Women are 20x more likely than men to get HIV with vaginal intercourse, likely because of the prolonged exposure of the vagina, cervix, and uterus, to seminal fluid. Receptive anal intercourse appears to increase the risk of transmission, likely secondary to mucosal trauma of the thin rectal wall. Sexually transmitted diseases also increase the risk of transmission. Organisms such as *Treponema pallidum*, herpes simplex virus, *Chlamydia trachomatis*, and *Neisseria gonorrhoeae* cause mucosal erosions and may even increase the concentration of HIV in semen and vaginal fluids. (Inflammation of the epididymis, urethra, and vaginal mucosa results in an increase in HIV-laden macrophages and lymphocytes.) Oral sex is much less likely to result in transmission.

2) Blood product transfusion: HIV can be transmitted in whole blood, concentrated red blood cells, platelets, white blood cells, concentrated clotting factors, and plasma. Gamma-globulin has not been associated with transmission. To reduce the risk of transmission via blood products, blood donors are screened for self reported risk factors and serologic markers of HIV infection. The latter includes screening for antibodies to HIV-1 and HIV-2 (by ELISA) and

for p24 antigen. Beginning in 1999, the U.S. started performing nucleic acid amplification testing of the blood for HIV-1 RNA. As a result, the estimated risk of HIV acquisition via a blood transfusion has decreased from 1 in 500,000 in 1995 to 1 in 1.6 million in 2003.

3) Intravenous drug use with needle sharing: This has led to growing numbers of infected persons in U.S. urban centers.

4) Transplacental viral spread from mother to fetus: The rate of transmission is about 30%, and infection occurs transplacentally, during delivery, and perinatally.

5) Note for students and health care providers: The risk of contracting HIV from a stick with a needle, contaminated with HIV infected blood, is 3 out of a thousand (0.3%). The risk is much lower for accidental body fluid contact with broken skin. There is virtually no risk in touching an HIV infected patient, unless there is contact with blood or body fluid. The risk goes up if the injury is deep, the needle was in a patient's artery or vein, or had blood visible on it, or if the patient has a high viral load (MMWR, 1995). To put the risk of transmission of HIV by needle stick (0.3% transmission risk) into perspective, the risk of transmission of Hepatitis B virus after a needle stick from a patient who is Hepatitis B_e antigen positive is about 30%, and for Hepatitis C virus is about 3%.

6) Epidemiologic evidence indicates that the virus is NOT spread by mosquito bites or casual contact (kissing, sharing food). There is NO evidence that saliva, urine, tears, or sweat, can transmit the virus.

Cell Infection

Once the HIV virion is in the bloodstream, its gp 160 (composed at gp 120 and gp 41) glycoproteins bind to the CD4 receptor on target cells. This CD4 receptor is present in high concentration on T-helper lymphocytes. These cells are actually referred to as CD4+ T-helper cells. Other cells that possess CD4 receptors in lower concentrations and which can become infected are macrophages, monocytes, and central nervous system dendritic cells. Following HIV binding to the CD4 receptor, the viral envelope fuses with the infected host cell, allowing capsid entry.

Part of the mystery of how HIV binds to the CD4 receptor is as follows. For fusion and translocation of HIV across the cell membrane both a CD4 receptor and a co-receptor, either CCR5 or CXCR4, must be present. These are found on helper T lymphocytes and macrophages.

Patients who fail to produce normal levels of CCR5 proteins appear to be resistant to HIV infection, and certain lymphocyte derived proteins (RANTES, MIP1-α, and MIP1-Beta) that bind to CCR5 appear to inhibit HIV infection. These findings have led to the recent development of a new class of HIV medications: CCR5 inhibitors. (Cohen, 1996; Deng, 1996; Dragic, 1996; Alkhatib, 1996).

The viral RNA is reverse transcribed into DNA in the cytoplasm. Double-stranded DNA is formed and transported into the nucleus, where integration into the host DNA occurs. The integrated DNA may lie latent or may activate to orchestrate viral replication. There is some evidence that certain infections, such as with tuberculosis, PCP, cytomegalovirus, herpes, Mycoplasma or immunizations, will activate T-cells and may promote viral replication within the T-cells. Stimulation of T-cells results in production of proteins that bind to the HIV LTR, promoting viral transcription.

Following viral replication the new capsids form around the new RNA dimers. The virion buds through the host cell membrane, stealing portions of the membrane to use as an envelope, leaving the T-cell dead.

Immunology and Pathogenesis

Following initial infection, HIV can begin replication immediately, resulting in rapid progression to AIDS, or there can be a chronic latent course. The former, most common pattern occurs in 3 stages starting with initial infection, marked by an acute mononucleosis-like viral illness. This progresses for a variable number of years (median 8 but range of less than 1 to greater than 20) of disease-free latency. After AIDS develops most patients die within 2 years if they do not receive effective anti-retroviral therapy (**Fig. 26-7**).

1) An **acute viral illness** like mononucleosis (fever, malaise, lymphadenopathy, pharyngitis, etc.) develops in 80% about 1 month after initial exposure. There are high levels of blood-borne HIV (**viremia**) at this stage, and the viruses spread to infect lymph nodes and macrophages. An HIV-specific immune response arises, resulting in decreased viremia and resolution of the above symptoms. However, HIV replication continues in lymph nodes and peripheral blood.

2) A **clinical latency** follows for a median of 8 years during which there are no symptoms of AIDS, although some patients develop a dramatic generalized lymphadenopathy (possibly secondary to an aggressive immune attack against HIV harbored in the lymph nodes). This is not a true viral latency without viral replication; HIV continues to replicate in the lymphoid tissue and there is a **steady gradual destruction of CD4 T-lymphocytes (helper) cells**. CD4+ T-helper cells are the number one target of HIV. The virus reproduces in these cells and **destroys** them.

Toward the end of the 8 years, patients are more susceptible to bacterial and skin infections, and can develop constitutional (systemic) symptoms such as fever, weight loss, night sweats, and adenopathy.

3) **AIDS** develops for a median of 2 years followed by death. AIDS is now defined as having a CD4 T-lymphocyte count of less than 200 (with serologic evidence of HIV infection such as a positive ELISA or western blot test) and/or one of many AIDS-defining opportunistic infections, which are infections that usually only patients with AIDS develop. These include *Candida* esophagitis, *Pneumocystis carinii* pneumonia, the malignancy Kaposi's sarcoma, and many others.

Fig. 26-7. The **clinical course of HIV infection** (acute viral illness, clinical latency, and AIDS) as CD4+ T-cells decline over time, and the opportunistic infections that develop at specific CD4 T-cell counts.

1) Normal CD4+ T-cell counts are 1000 cells/μl blood. In HIV-infected persons the count declines by about 60 cells/μl blood/year.

2) CD4+ T-cell count of 400–200 (about 7 years): Constitutional symptoms (weight loss, fever, night sweats, adenopathy) develop as well as annoying skin infections, such as severe athlete's foot, oral thrush (*Candida albicans*), and herpes zoster. Bacterial infections, especially *Mycobacterium tuberculosis*, become more common as CD4 counts drop below 400!!

3) CD4+ T-cell count less than 200 (about 8 years): As the immune system fails, the serious opportunistic killers set in, such as *Pneumocystis carinii* pneumonia, *Cryptococcus neoformans*, and *Toxoplasma gondii*.

4) CD4+ T-cell count less than 50: At this level the immune system is almost completely down. *Mycobacterium avium-intracellulare*, normally only causing infection in birds, causes disseminated disease in the AIDS patients. Cytomegalovirus infections also rise as the count moves from 50 to zero.

Viral Load

CD4 counts are used to determine severity of HIV infection, risk of opportunistic infection, prognosis, and response to anti-viral therapy. We can now measure plasma HIV RNA by the polymerase chain reaction (PCR) or branched chain DNA assay. There is mounting evidence that higher plasma HIV RNA levels (**viral load**) correlate with a greater risk of opportunistic infection, progression to AIDS, and risk of death (Mellors, 1996; Galetto-Lacour, 1996). CD4 counts are still the best predictor of a patient's current risk for particular opportunistic infections. David Ho has popularized the train analogy to explain the predictive values of HIV viral load and CD4 counts. Viral load tells you the speed at which the train is heading for the cliff (low CD4, development of opportunistic infections and death) while the CD4 count tells you where the train currently is! For example, if a patient has a CD4 count of 450 cells/μl and a viral load of >10^6 copies/ml that patient is at

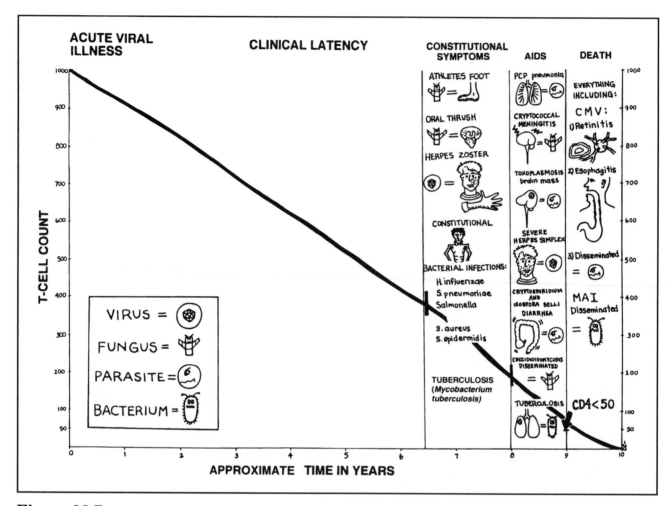

Figure 26-7

low risk for developing *Pneumocystis carinii* pneumonia today (CD4 > 200 cells/μl) but is at great risk in the future for rapid CD4 count decline, opportunistic infection and death if not treated.

Mechanism of T-Cell Death

The hallmark of HIV is the gradual depletion of CD4+ T cells, leading to a weakened immune system and secondary opportunistic infections. The pathogenesis of CD4+ T cell decline remains controversial. The major mechanism of CD4+ T cell depletion is programmed cell death (apoptosis). CD4+ T cells can be killed in multiple ways: 1) infected T cells may fuse together via their gp 160 envelope proteins and form short-lived multinucleated giant cells, 2) gp 160 on the T cell membrane may mark the cell as non-self, resulting in autoimmune T cell destruction by cytotoxic CD8 T lymphocytes, 3) uninfected

bystander CD4 T cells are destroyed by a Fas-mediated mechanism during activation induced cell death, or as a result of HIV proteins (Tat, gp120, Nef, Vpu) being released from infected cells stimulating apoptosis of these bystander cells. These multiple mechanisms of CD4+ T cell death are all possible targets for therapy.

What Is the Clinical Significance of These Events?

T-Cell death: A healthy person has about 1000 CD4+ T-helper cells per μl of blood. HIV-induced T-cell death results in a decline of about 60 CD4+ T-cells/μl a year. The T-helper cell plays a vital role in the orchestration of the cell-mediated immune system, activating and recruiting macrophages, neutrophils, B-lymphocytes, and other immune system effectors. A severe immunodeficiency state follows the loss of CD4+ T-cells.

Multinucleated giant cells: This T-cell to T-cell fusion allows the virus to pass from an infected cell to an uninfected cell **without contacting the blood**. This may protect the virus from circulating antibodies.

B-Lymphocytes: HIV does not actually infect the B-cells; however, B-cell dysfunction does occur with HIV infection. There is a polyclonal activation of B-cells, resulting in an outpouring of immunoglobulins. This **hypergammaglobulinemia** results in immune-complex formation and autoantibody production. The most important dysfunction that occurs is a diminished ability to produce antibodies in response to new antigens or immunization. This is very serious in infants with AIDS because they cannot develop humoral immunity to the vast number of new antigens they are exposed to.

Monocytes and macrophages: HIV infects these cells and actively divides within them. However, these cells are not destroyed by HIV. This is clinically significant in 2 ways:

1) Monocytes and macrophages serve as reservoirs for HIV as it replicates, protected within these cells from the immune system.
2) These cells migrate across the blood-brain barrier, carrying HIV to the central nervous system. HIV causes brain disease, and the predominant cell type harboring HIV in the central nervous system is the monocyte-macrophage line.

BIG PICTURE: HIV infection diminishes CD4 T-lymphocyte (helper) cell numbers and function. The CD4 T-helper cells are involved in all immune responses. So all immune cells have some kind of altered function. As T-cell numbers decline, the host becomes susceptible to unusual infections and malignancies that normally are easily controlled by an intact immune system.

ACQUIRED IMMUNODEFICIENCY SYNDROME (AIDS)

Fig. 26-8. The acquired immunodeficiency syndrome (AIDS) is an extremely complex disease. To better understand this complexity, consider 2 processes that occur. The HIV virus causes 1) direct viral disease and 2) disease secondary to the immunodeficiency state.

1) Direct viral disease
   Constitutional (widespread body) symptoms
   Neurologic damage
2) Disease secondary to the immunodeficiency state
   Failure of the immune surveillance system that prevents malignancies
   Secondary infections by pathogens and normal flora (opportunistic infections)

Constitutional Illness

AIDS patients suffer from night sweats, fevers, enlarged lymph nodes, and severe weight loss. The weight loss is often referred to as the wasting syndrome.

Neurologic Disease

The HIV virus is carried to the central nervous system by the monocyte-macrophage cells. It is unclear at this time whether the neuronal damage is caused by the inhibition of neuronal growth by the HIV envelope proteins or an autoimmune damage caused by the infected monocyte-macrophages themselves.

Many patients with HIV infection suffer from some form of neurological dysfunction. The brain can suffer diffuse damage (encephalopathy) resulting in a progressive decline in cognitive function referred to as the AIDS dementia complex. Meningeal infection results in aseptic meningitis. The spinal cord can become infected, resulting in myelopathy, and peripheral nerve involvement results in a neuropathy.

Malignancies

AIDS patients suffer from a high incidence of **B-cell lymphoma**, often presenting as a brain mass. Half of B-cell lymphomas in AIDS patients are found to contain Epstein-Barr virus DNA.

Another common AIDS associated malignancy is **Kaposi's sarcoma**. Most cases of Kaposi's sarcoma (96%) occur in homosexual men, which suggests that there may be a co-factor, which appears to be a herpes virus called **HHV-8. HHV-8** DNA sequences have been found in Kaposi's sarcoma, and antibodies to **HHV-8** are found in high concentrations in most patients (80%) with Kaposi's sarcoma and in 35% of homosexual HIV positive men (Moore, 1995; Kedes, 1996; Gao, 1996). The lesions are red to purple, plaques or nodules, and arise on the skin all over the body. The course can range from nonaggressive disease, with limited spread and only skin involvement, to an aggressive process involving skin, lymph nodes, lungs, and GI tract.

Additionally, persons with HIV infection are at increased risk of non-Hodgkin's lymphoma (>100X risk of the general population), Hodgkin's lymphoma (~10X risk), cervical cancer (with related to HPV co-infection), and anal intraepithelial neoplasia (in homosexual males).

Opportunistic Infections

The most common manifestation of AIDS is the secondary infection by opportunists. These are bugs that are normally pushovers to the intact immune system but wreak havoc in the absence of T-helper defenses (see **Fig. 26-7**).

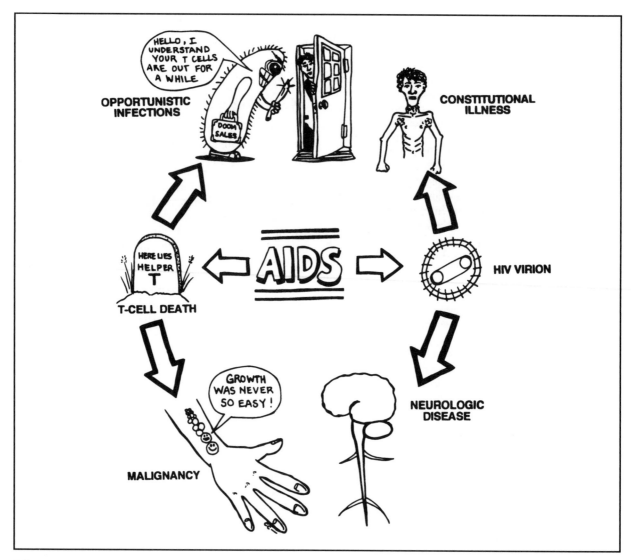

Figure 26-8

Bacterial and Mycobacterial Infections

Humoral immunity is impaired due to B-cell dysfunction related to HIV infection. As a result, HIV infected persons have a higher incidence of infections with encapsulated organisms, particularly with *Streptococcus pneumoniae*. In fact, the risk of invasive pneumococcal infection was found to be >20 fold in HIV-infected persons than in non-infected persons in a surveillance study performed in San Francisco.

Mycobacterium tuberculosis: Remember the pivotal role of cell-mediated immunity in defense against *Mycobacterium tuberculosis*. AIDS patients with latent infection have a higher chance of tuberculosis reactivation (about 10% chance per year) as compared to non-HIV patients (10% over their lifetime).

Mycobacterium avium-intracellulare (MAI): This nontuberculous mycobacterium, which is part of the *Mycobacterium avium-complex* (**MAC**), can be isolated from many sites (GI tract, liver, bone marrow, lymph nodes, lungs, blood) in infected patients. It causes a smoldering, wasting disease characterized by fever, night sweats, weight loss, and often diarrhea (GI tract infection) and elevated liver function tests. It is a common cause of **FUO** (fever of unknown origin) in persons with AIDS and CD4 counts <50 cells/mm³.

Fungal Infections

Candida albicans: This yeast is very common in HIV-infected patients. It causes oral thrush and esophagitis. Thrush looks like white plaques on the oral mucosa and when scraped off with a tongue blade leaves a red bleeding base. (More on page 208).

Cryptococcus neoformans: This fungus causes a meningitis in persons with AIDS and CD4 counts <100 cells/mm³. Fever, nausea, and vomiting may hint at cryptococcal meningitis. Important: AIDS patients are similar to the elderly and children: **Without a full immune system they often do not exhibit meningeal inflammation**. Only 25% of AIDS patients with cryptococcal meningitis will present with headache, mental status changes, or meningeal signs. For example: A normal host with meningitis would have meningeal inflammation with meningismus (positive Kernig's and Brudzinski's sign, stiff neck, headache). An AIDS patient can have a raging meningitis with only fever. You must have a high level of suspicion and always consider doing a lumbar puncture, for cerebrospinal fluid testing, on AIDS patients with fever. (More on page 207).

Histoplasma capsulatum and ***Coccidioides immitis***: These fungi produce disseminated disease in AIDS patients, infecting meninges, lungs, skin, and other areas. (More on page 205).

***Pneumocystis jiroveci* pneumonia (PCP)**: This is the most common opportunistic infection in the U.S. among patients with AIDS. Without prophylactic treatment there is a 15% chance each year of infection when the CD4 T-cell count is below 200. AIDS patients who develop PCP have cough and hypoxia. The chest X-ray can be normal or show interstitial infiltrates. Pneumothorax complicates 2% of PCP cases. About 80% of AIDS patients will get this at least once in their lifetime **unless** prophylactic antibiotics are taken. Why isn't it called **PJP**? Old habits die hard. In 2002 it was determined that pneumocystis carinii is the species that infects rats and pneumocytis jiroveci infects humans, but we got used to calling it **PCP** in humans and it stuck!! (More in Chapter 31/Protozoa).

Viral Infections

Herpes zoster: Painful vesicles develop in dermatomal distribution as the varicella-zoster virus ventures forth from its latency in the dorsal sensory ganglia. AIDS patients can also develop disseminated non-dermatomal zoster.

Epstein-Barr virus, another herpes family virus, is thought to cause **oral hairy leukoplakia (OHL)**. OHL usually develops when CD4 counts are <400 and is characterized by white hairlike projections arising from the side of the tongue. This is differentiated from Candidal thrush by the fact that OHL will not rub off with a tongue blade. Remember EBV is also associated with central nervous system B-cell lymphoma.

Herpes simplex: Viral infection results in severe genital and oral outbreaks.

Cytomegalovirus (CMV): This virus can cause chorioretinitis and blindness. Visual compromise must be looked into carefully in AIDS patients because it can represent the retinal lesions of CMV or brain masses of toxoplasmosis and lymphoma. CMV can also cause esophagitis (pain with swallowing) and diarrhea.

Protozoal Infections

Toxoplasma gondii: This parasite causes **mass** lesions in the **brain** in 15% of AIDS patients. Patients present with fever, headache, and focal neurologic deficits (seizure, weakness, aphasia). A CT scan will show contrast-enhancing masses in the brain. (More on page 329).

Cryptosporidium*, *Microsporidia*, and *Isospora belli: These parasites cause chronic diarrhea in patients with AIDS. (More on page 337).

DIAGNOSIS of HIV and AIDS

Following infection with HIV, viral RNA or antigens (such as p24) can be detected in the blood within weeks. Three to 6 weeks later antibodies against HIV antigens appear. The **enzyme-linked immunosorbent assay (ELISA)** test detects antibodies. This test is very sensitive at detecting HIV infection (sensitivity of 99.5%) but it often gives **false positive results**.

A **western blot test** is then routinely performed for confirmation of a positive ELISA. In this test, HIV antigens (*gag, pol*, and *env* proteins) are separated in bands on paper by molecular weight. The person's serum is then added to this paper. If the serum contains antibodies against HIV antigens they will stick to the antigens on the paper. Lastly, anti-human antibodies (labeled with enzymes) are added; these stick to the antibodies on the antigens, lighting up "bands" on the paper. A western blot is considered positive if it has bands to 2 HIV gene products (p24, gp41, gp120) (see **Figs. 26-5 and 26-6**).

Several rapid HIV tests for the presence of HIV antibodies are now available. Rapid testing leads to a result in 5 to 40 minutes. Specimens for rapid testing include blood, plasma, and saliva. These tests have excellent sensitivity and specificity (>99%), but must also be confirmed with a western blot test.

During the "window period", 3 to 6 weeks after HIV infection, persons may have HIV virus circulating in their bloodstream but have not yet formed antibodies to HIV. During this period clinicians can make a diagnosis by directly measuring HIV RNA with polymerase chain reaction (PCR) based studies. The major limitation of PCR for diagnosis is cost and an often prolonged time interval between sample collection and test results.

HIV infection should be suspected when an at-risk individual (homosexual, IV drug user, sexual partner of an at-risk individual, etc.) develops constitutional symptoms such as fevers, night sweats, and generalized adenopathy, or suffers from recurrent bacterial infections, tuberculosis, skin zoster or tinea infections, or oral thrush (*Candida*).

AIDS is diagnosed when the CD4 T-lymphocyte count is less than 200 (with serologic evidence of HIV infection such as a positive ELISA or western blot test) and/or the patient has one of many AIDS-defining opportunistic infections, which are infections that usually only patients with AIDS develop. These include *Candida* esophagitis, *Pneumocystis carinii* pneumonia, the malignancy Kaposi's sarcoma, and many others.

CONTROL, TREATMENT, CURE?

Efforts directed toward viral control are moving along 4 lines:

1) Prevention of HIV viral infection.
2) Vaccine development.
3) Limiting growth of HIV, once infection has occurred.
4) Treating the opportunistic infections that ultimately cause death.

Prevention

Education to avoid high-risk activities (needle sharing, multiple sexual partners, unprotected sex).
Screening blood products with ELISA, p24 antigen, and nucleic acid amplification for HIV-1 RNA.

Vaccine Development

The goal of vaccination is to stimulate an immune response that will counter a subsequent infection. Most vaccines stimulate an antibody response to a viral antigen, resulting in the neutralization of the virus. Persons infected with HIV develop antibodies against HIV determinants. Early in HIV infection antibodies arise that bind to a hypervariable portion of the envelope glycoprotein gp 120, called the **V₃ loop**. These V₃ specific antibodies will neutralize only the exact strain of HIV that elicited the antibody. Chimpanzees given V₃ neutralizing monoclonal antibody (passive immunization) and chimpanzees actively immunized with the V₃ envelope glycoprotein were protected from injected cell-free HIV virus of the same strain. These vaccines only protect against a virus with the exact same V₃ hypervariable region and only during peak immunity.

Another later developing antibody response occurs against the CD4 binding domain (gp 160, composed of gp 120 and gp 41); this domain is responsible for binding to T-lymphocyte CD4 receptors. The CD4 binding domain antigen is more conserved, meaning that it is similar in many HIV strains. Antibodies against this domain will prevent viral binding to the CD4 receptor, neutralizing HIV-1 in vitro.

Since the above antibody responses develop with HIV infection and with all the ongoing efforts to develop a vaccine, why are we told that successful vaccination is a distant reality?!?

There are many challenges to the development of a successful vaccine against HIV-1:

1) **Rapid mutation**: HIV envelope glycoproteins mutate rapidly, so there are many different strains. The rapidly mutating V₃ loop of gp 120 and the reverse transcriptase enzyme combined with rapid viral reproduction over a long disease course results in different "quasi-species," even in the same person. A vaccine would need to target a conserved region like the CD4 binding domain.

2) **HIV is transmitted from cell-to-cell**: With syncytial giant-cell formation, HIV is able to pass from one cell to another without contacting the bloodstream that carries antibodies. The virus can thus escape the antibody-mediated or **humoral** immune system. Protection against HIV infection also requires cell-mediated immunity: HIV proteins in an infected cell are expressed on the cell surface associated with class I molecules of the major histocompatibility complex. Circulating HIV-specific cytotoxic T-lymphocytes will recognize this complex and lyse the cell, destroying HIV inside. This cell-mediated response does occur in patients with HIV infection, and a successful vaccine would have to stimulate this arm of the immune system.

3) **Poor animal model**: One would like an animal model that is easily obtainable, cheap, and would develop an AIDS-like illness when infected with HIV. There is no such model. The only species that can be infected are the great apes; they are expensive, scarce, and do not get AIDS when infected with HIV. However, their antibody response to HIV can be followed, and they do get an AIDS-like illness when infected with SIV (which shares sequence homology with HIV-2).

Current Vaccine Research Efforts

1) **Live viruses** can be altered or attenuated so they lose their virulence while still stimulating immunity (e.g., polio and measles vaccine). This type of vaccine would stimulate cellular and humoral immunity, as

well as mucosal immunity. Studies with live viruses have used SIV and HIV-2, which have been made non-pathogenic secondary to gene deletions. The limitation of this approach involves the danger of new mutations occurring in the non-pathogenic live virus that might make it pathogenic. Also, the great apes do not develop an AIDS-like disease, so how can we be sure such a vaccine is non-pathogenic?

2) **A recombinant HIV-1 envelope glycoprotein vaccine** can be made by splicing the HIV gene, which codes for envelope glycoprotein antigens (V_3 loop, CD4 binding domain, or others), into the DNA of tumor-cell lines. The tumor cells will produce the HIV envelope glycoprotein in mass quantities that can be used as a vaccine. The hepatitis B vaccine is made in this manner, with cell lines producing the hepatitis B surface antigen (HBsAg). As mentioned above, however, these vaccines only protect against a virus with the exact same antigenic region used in the vaccine and only during peak immunity. They also fail to activate a cytotoxic T-lymphocyte response.

3) **Recombinant vector vaccines** use non-HIV viruses that are non-pathogenic in humans or highly attenuated (weakened) viruses to deliver copies of HIV genes into cells in the body. The vaccinia virus (live attenuated virus used for smallpox vaccine), canarypox virus (a live virus that causes disease in birds, but not in humans), bacille Calmette-Guerin (bacteria used for *Mycobacterium tuberculosis* vaccine), and adenovirus strains are some of the organisms used as vectors in vaccine experiments. Some studies have demonstrated an antibody (humoral) and T-lymphocyte (adaptive) response to proteins produced by these recombinant organisms.

4) **DNA vaccines** introduce a small number of HIV genes directly into the body via plasmids, hoping they will be picked up by cells and used to produce HIV proteins very similar to the ones from real HIV. This approach would be expected to elicit both humoral and an adaptive immune responses.

5) A **"Prime-boost strategy"** is the current most popular approach in vaccine development. This strategy uses any two vaccines, usually a combination of the above methods, to create a more vigorous immune response. A prime-boost strategy was used in the first moderately successful vaccine trial to date. Using a primer canary pox vaccine (**recombinant vector vaccine**), followed by a HIV-1 gp 120 vaccine (**a recombinant HIV-1 envelope glycoprotein vaccine**), researchers in Thailand were able to show a moderate reduction in HIV acquisition in vaccinated volunteers as compared to placebo controls. The results were modest, a 31% overall reduction in HIV infection, but encouraging never-the-less. This was the first even moderate success since HIV vaccine trials started in 1987!

Limiting Viral Growth

Triple drug therapies—Highly Active Antiretroviral Therapy (HAART) have been used to bolster the immune system in HIV-positive and AIDS patients, which has decreased the rate of development of opportunistic infections, including *Mycobacterium avium*, CMV retinitis, and oropharyngeal candidiasis. Please see Chapter 29, the anti-viral antibiotic chapter, which extensively discusses current anti-HIV drug strategies.

Treating the Opportunistic Infections

1) *Pneumocystis jiroveci* **pneumonia** (**PCP**): Trimethoprim and sulfamethoxazole are given prophylactically when CD4+ T-cell counts drop below 200–250. Greater than 90% of PCP infections are being prevented with this prophylactic intervention!

2) **Toxoplasmosis**: Brain lesions are treated with another tetrahydrofolate reductase inhibitor/sulfa combination called pyrimethamine/sulfadiazine. Patients improve rapidly. In fact, if there is no brain mass shrinkage (as seen by CT scan) by 2–3 weeks, then the diagnosis of toxoplasmosis is unlikely. Brain biopsy should then be done to determine whether the mass is a B-cell lymphoma.

The same medicine (trimethoprim and sulfamethoxazole), used for PCP prophylaxis, also prevents toxoplasmosis! It kills two birds with one stone.

3) *Mycobacterium tuberculosis* **and** *Mycobacterium avium-intracellulare*: The treatment of mycobacterial infections is covered in Chapter 15/ Mycobacterium.

Azithromycin or **clarithromycin** can be given daily for prophylaxis against future MAI infections.

4) **CMV**: Treatment with ganciclovir or foscarnet can prevent progression of visual loss.

5) **Herpes, Varicella-zoster**: Acyclovir.

6) *Candida albicans*: Oral clotrimazole, nystatin, or fluconazole preparations for thrush and esophagitis. Systemic fungal infections are treated with intravenous amphotericin B or fluconazole.

7) **Bacteria**: Appropriate antibiotics.

A Final Word

AIDS is a disease that has no dignity, a disease that cripples the immune system, allowing the scourge of all infestations. You are becoming a physician in the dawn of a new epidemic, and you will certainly play a role in the control of this epidemic.

References

Alimont JB, Ball TB, and Fowke KR. Mechanisms of CD4+ T lymphocyte cell death in human immunodeficiency virus infection and AIDS. J Gen Virol 2003; 84:1649–1661.

Alkhatib G, et al. CC-CKR5: a RANTES, MIP-1α, MIP-1 beta receptor as a fusion cofactor for macrophage-trophic HIV-1. Science 1996;272:1955–8.

Case-control study of HIV seroconversion in health-care workers after percutaneous exposure to HIV-infected blood—France, United Kingdom, and United States, January 1988–August 1994. MMWR 1995;44:929–33.

Cleghorn FR, et al. Human Immunodeficiency Viruses. In: Mandell GL, Bennett VF, Polin R, eds. Principles and Practice of Infectious Diseases. 6th edition. Philadelphia: Churchill Livingstone 2005;2119–2132.

Cohen J. HIV cofactor found. Science 1996;272:809–10.

Deng H, et al. Identification of a major co-receptor for primary isolates of HIV-1. Nature 1996;381:661–6.

Dragic T, et al. HIV-1 entry into CD4 T cells is mediated by the chemokine receptor CC-CKR5. Nature 1996;381: 667–73.

Fauci AS, Lane HC. Human immunodeficiency virus (HIV) disease: AIDS and related disorders. In: Isselbacher, Braunwald, et al., eds. Harrison's Principles of Internal Medicine. 13th ed. New York: McGraw-Hill, 1994; 1566–1618.

Galetto-Lacour A, et al. Prognostic value of viremia in patients with long-standing human immunodeficiency virus infection. J Infect Dis 1996;173:1388–93.

Gao SJ, et al. KSHV antibodies among Americans, Italians, and Ugandans with and without Kaposi's sarcoma. Nature Medicine 1996;2:925–8.

Goila-Gaur R and Strebel K. HIV-1 Vif, APOBEC, and Intrinsic Immunity. Retrovirology 2008;5:51:1–16.

Haseltine WA. Molecular biology of the human immunodeficiency virus type 1. The FASEB journal 1991:2349–2360.

Kedes DH, et al. The seroepidemiology of human herpes virus 8 (Kaposi's sarcoma-associated herpesvirus): distribution of infection in KS risk groups and evidence for sexual transmission. Nature Medicine 1996;2:925–8.

Mellors JW, et al. Prognosis in HIV-1 infection predicted by the quantity of virus in plasma. Science 1996;272:1167–70.

Moore PS, Chang Y. Detection of herpesvirus-like DNA sequences in Kaposi's sarcoma in patients with and those without HIV infection. N Engl J Med 1995;332:1181–5.

Nuorti JP et al. Epidemiologic relation between HIV and invasive pneumococcal disease in San Francisco County, California. Ann Intern Med. 2000;132(3):182–90.

Pantaleo G, Graziosi C, Fauci A. The immunopathogenesis of human immunodeficency virus infection. N Engl J Med 1993;328:327–335.

Recommended Review Articles:

DHHS Panel on Antiretroviral Guidelines for Adults and Adolescents. Guidelines for the use of antiretroviral agents in HIV-1 infected adults and adolescents. www.AIDSinfo.nih.gov. (Accessed on May 22, 2010)

Kaplan JE, et al. Guidelines for prevention and treatment of opportunistic infections in HIV-infected adults and adolescents: recommendations from CDC, the National Institutes of Health, and the HIV Medicine Association of the Infectious Diseases Society of America. MMWR Recomm Rep. 2009;58(RR-4):1–207.

Ross AL, et al. Progress towards development of an HIV vaccine: report of the AIDS Vaccine 2009 Conference. Lancet Infect Dis 2010;10:305–16.

Sterne JA, et al. Long-term effectiveness of potent antiretroviral therapy in preventing AIDS and death: a prospective cohort study. Lancet 2005;366(9483):378–84.

Acknowledgment:

A special thanks to Gabriela Perez, who contributed ideas and suggested changes to this chapter.

CHAPTER 27.   HERPESVIRIDAE

We all probably have at least 1 of the herpes family viruses living in a latent state in our bodies right now! Even before entering medicine we have seen people with herpesviridae infections. People with "fever blisters" of HSV-1, the child with multiple blisters covering the body with chickenpox (varicella-zoster), and the teenage friend who had to miss school with mononucleosis (Epstein-Barr virus).

There are some generalities that the herpesviridae share:

1) They can develop a **latent state**.

2) The members in the sub-family alpha have a cytopathic effect on cells, which become **multinucleated giant syncytial cells** with **intranuclear inclusion bodies**.

3) Herpesviridae are held at bay by the **cell-mediated immune response**.

Latency: During the primary infection the viruses migrate up the nerves to the sensory ganglia and reside there. The viruses rest there until reactivation occurs through some stress, such as menstruation, anxiety states, fever, sunlight exposure, and weakening of the cell-mediated immune system, as with AIDS or chronic disease. The viruses then migrate out to the peripheral skin via the nerves to cause local destruction.

Fig. 27-1. Captain Herpes hiding in latency in his dorsal sensory ganglia fortress.

Cytopathic effect: The herpesviridae that cause cell destruction are the alpha sub group viruses (herpes simplex virus 1 and 2, and varicella-zoster). This cell destruction results in the separation of the epithelium and causes blisters (vesicles). Microscopic study of skin biopsies or scrapings from blister bases in herpes simplex, chickenpox, and zoster all reveal multinucleated giant cells and intranuclear inclusion bodies. Viral proteins are inserted into the host cell plasma membranes, resulting in cell fusion to form multinucleated giant cells. Intranuclear inclusions are considered to be areas of viral assembly.

Both CMV (beta subgroup) and Epstein-Barr virus (gamma subgroup) have less cytopathic effects.

Patients with compromised cell-mediated immune status are more likely to suffer from severe herpesviridae infections such as disseminated HSV or multi-dermatomal zoster. CMV frequently causes disease in AIDS patients.

Herpes Simplex Virus 1 & 2 (HSV-1 & HSV-2)

By the fifth decade of life, more than 90% of adults will have antibodies to HSV-1 and more than 20% of adults will have antibodies to HSV-2. Those from lower socio-economic classes are more likely to have antibodies to these viruses. Transmission occurs by direct inoculation of muco-cutaneous surfaces (e.g., the oropharynx, cervix, conjunctivae, small cracks in the skin, etc). Clinical manifestations depend on the site of inoculation, and the immune status of the host, but may include:

1) **Gingivostomatitis**: Painful swollen gums and mucous membranes with multiple vesicles. Fever and systemic symptoms can accompany the infection, and the disease will resolve in about 2 weeks. Vesicles can also appear on areas of the skin where viral entry has occurred.

2) **Genital Herpes**: Despite common lore, genital herpes may be caused by HSV-1 or HSV-2, and are clinically indistinguishable. Fever, headache, vaginal and urethral discharge, and enlarged lymph nodes are early signs, followed by blisters and painful or painless ulcers.

3) **Herpetic keratitis**: the most common infectious cause of corneal blindness in the United States.

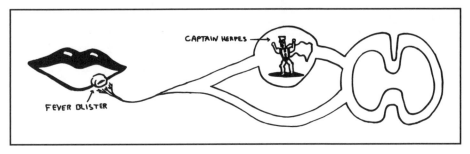

Figure 27-1

4) **Neonatal Herpes**: HSV infection during pregnancy can result in transplacental viral transfer. The infection of the fetus can cause congenital defects or intrauterine death. The neonate can also acquire the illness during delivery if the mother is having an active genital infection.

5) **Herpetic Whitlow**: Herpetic whitlow is an HSV infection of the finger. The finger becomes painful, bright red, hot and swollen. Before the common use of gloves, this infection occasionally occurred in health care workers in health care settings.

6) **Disseminated Herpes**: In immune compromised patients (organ transplant recipients, cancer chemotherapy, malnutrition, burns, etc) HSV may cause extensive mucocutaneous infections, or disseminate infection to organs such as liver, lung, and the gastrointestinal tract.

7) **Encephalitis**: HSV-1 is the most common cause of viral encephalitis in the U.S. Infection of the brain cells occurs, with cell death and brain tissue swelling. Patients present with sudden onset of **fever** and **focal neurological abnormalities**. HSV-1 must always be considered, because herpes is one of the few treatable causes of viral encephalitis!!

Reactivation: About one fourth of previously infected people have reactivation infection during stressed states. AIDS patients can present with severe reactivation of HSV.

Fig. 27-2. Death, carrying his **TORCH**, visits the pregnant mother and her baby. Remember that herpes is one of the **TORCHES** organisms that can cross the blood-placenta barrier:

Figure 27-2

TO: **TO**xoplasmosis
R: **R**ubella
C: **C**ytomegalovirus
HE: **HE**rpes, HIV
S: **S**yphilis

Varicella-Zoster Virus (VZV)

As the name implies, this virus causes 2 diseases: **varicella** (chickenpox) and **herpes zoster** (shingles). Chickenpox is not caused by the pox viridae!!! Varicella is usually a disease of children. After resolution the virus remains latent as described previously. Later in life, reactivation can cause the second disease, zoster. Once again, with stressors or depressed cell-mediated immunity (usually in the elderly), the virus will migrate out along sensory nerve paths and cause vesicles similar to those of chickenpox. However, with this reactivation infection, the vesicles appear in a dermatomal distribution, almost always unilaterally.

Varicella
(Chickenpox)

VZV is highly contagious, infecting up to 90% of those exposed: It occurs in epidemics, usually during winter and spring and involves children who have not previously been exposed. About 90% of the general adult population have contracted VZV in childhood.

The virus infects the respiratory tract and replicates for a 2-week incubation period, followed by viremia (viral dissemination in the bloodstream).

Fig. 27-3. In varicella (chickenpox), fever, malaise, and headache are followed by the characteristic rash. The rash of varicella starts on the face and trunk, spreading to the entire body, including mucous membranes (pharynx, vagina, etc.). The skin vesicles that form are described as dew on a rose petal: a red base with a fluid-filled vesicle on top. The fluid becomes cloudy, the vesicles rupture, and the lesions scab over. Multiple vesicles arise in patches (crops), and one crop will form as another crop scabs over. So there are lesions in different stages!

Figure 27-3

The vesicles will all scab in about 1 week, and the patient then ceases to be contagious.

It is important to distinguish chicken pox from small-pox. Remember:

Chicken Pox	Smallpox
Superficial lesions	Deep hard lesions
Lesions usually not umbilicated	Lesions often umbilicated (central depression)
Lesions at different stages of development	Lesions at the same stage of development
Lesions more common on the trunk	Lesions more common on the extremities

Zoster (Shingles)

Following a stressed state or lowered cell-mediated immunity, the latent varicella-zoster virus in the sensory ganglion begins to replicate and migrate to the peripheral nerves. Burning, painful skin lesions develop over the area supplied by the sensory nerves. The diagnosis of zoster is likely when a patient develops a painful skin rash that overlays a specific sensory dermatome.

Fig. 27-4. A) 55-year-old male with left, second trigeminal (V2) nerve involvement; B) 76-year-old female with left T5–6 involvement.

Since this is the same virus that causes chickenpox, children and adults who have never contracted varicella can get chickenpox from exposure to vesicles.

Control/Treatment

Vaccination with a single dose of a live attenuated varicella vaccine is now recommended for adults ages 60 and over without regard to whether they have had a prior episode of shingles. Large randomized controlled studies have shown the vaccine to reduce the incidence of zoster by about 50%.

In adults and immunocompromised patients (with leukemia or AIDS, for example), the infection can be more serious, leading to pneumonia and encephalitis. In these groups zoster immune globulin, which consists of antibodies against VZV isolated from patients with zoster, can be given. It will only help if given within days of exposure (not during rash development). Intravenous acyclovir, an antiviral drug, appears to decrease the severity and duration of the infection.

Cytomegalovirus (CMV)

CMV is so-named because infected cells become swollen (cytomegaly). As with the other herpesviridae,

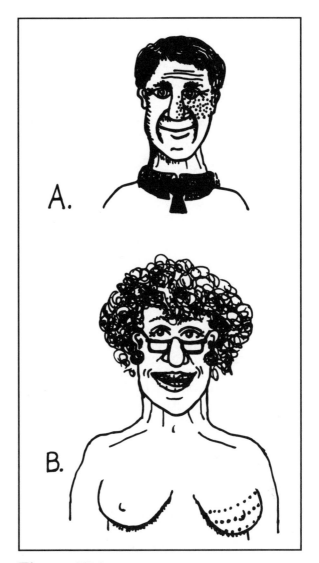

Figure 27-4

multinucleated giant cells and intranuclear inclusion bodies are present.

CMV causes 4 infectious states:

1) **Asymptomatic infection**: About 80% of adults in the world have antibodies against CMV. Most of these infections are asymptomatic.

2) **Congenital disease**: CMV is one of the TORCHES (see page 273) organisms that can cross the placenta and cause congenital disease. CMV is the most common viral cause of mental retardation. It also causes microcephaly, deafness, seizures, and multiple other birth defects. It is thought that this organism will reactivate during a latent state (as do all the herpesviridae). If reactivation occurs during pregnancy, the fetus may become infected.

3) **Cytomegalovirus mononucleosis**: CMV causes a mononucleosis syndrome in young adults,

Figure 27-5

similar to that caused by the Epstein-Barr virus. Patients with CMV as the cause of their mononucleosis syndrome will have **"Monospot negative mono"** (see Epstein-Barr virus for further explanation).

4) CMV can **reactivate** in the **immunocompromised patient** (as do all herpesviridae) to cause retinitis (blindness), pneumonia, disseminated infection, and even death.

Interestingly, CMV causes different diseases in 2 different immunocompromised populations: patients with AIDS versus patients who have undergone bone marrow transplantation. In AIDS patients, as the CD4 T-lymphocyte count drops below 50–100 cells per cc of blood, they frequently develop CMV viremia (CMV in the blood), CMV retinitis (leading to blindness unless treated), and CMV colitis (causing diarrhea). AIDS patients infected with CMV rarely develop pneumonia. In contrast, bone marrow transplant patients who are CMV antibody positive (representing prior infection and risk of reactivation) or receive a CMV positive donor bone marrow are at high risk of developing CMV pneumonitis. CMV pneumonitis is a severe pneumonia, often leading to death in this population. The transplant patients can also develop CMV viremia and colitis but do not develop retinitis.

Marrow Transplant = CMV pneumonitis
AIDS = CMV retinitis

Fig. 27-5. When you work in the hospital, you will frequently send special blood cultures for CMV from febrile organ transplant patients (on immunosuppressive drugs that prevent organ rejection), AIDS patients, and even children who have leukemia or lymphoma. The

CMV virus invades the white blood cells (see Epstein-Barr virus below) and so there is a higher yield if the buffy coat (layer of white cells in centrifuged blood) is cultured.

Diagnosis can be made several ways:

- Buffy Coat: Since CMV invades white blood cells, culturing the buffy coat (layer of white cells from centrifuged blood) increases the culture yield.
- Antigen: CMV antigen (protein recognized by specific antibody) can be detected in the blood, and is only present when the virus is replicating.
- Polymerase Chain Reaction (PCR): CMV DNA can be detected from the blood by PCR, and is only present in measurable levels when the virus is replicating.

Epstein-Barr Virus (EBV)

EBV, another of the herpesviridae, causes the famous disease **mononucleosis** and is involved in certain cancers such as **Burkitt's lymphoma** and nasopharyngeal cancers.

A general concept that helps us understand the way EBV is involved in these disease processes is **transformation**.

Transformation and Malignant Potential

In mononucleosis, EBV infects the human B-cells. EBV actually binds to the complement (C3d) receptor on cells. Once internalized, EBV will change the infected cell so that the cell does not follow normal growth controls. These changed or **transformed** cells proliferate and pass on copies of the EBV DNA to their progeny. The EBV DNA remains in the latent state as multiple copies of circular DNA. In some of the cells, the EBV activates and proliferates, and cell lysis with viral release occurs.

Interestingly, the transformed cells, which up to this point are acting as malignant (cancer) cells, suddenly disappear, with resolution of the mononucleosis illness. It is thought that the immune system destroys the infecting virus as well as the abnormal B-cells.

EBV has been found in the cancer cells of Burkitt's lymphoma, a B-cell lymphoma affecting children in central Africa. Since this cancer does not develop in other areas of the world where EBV infection occurs, it is thought that EBV may be just a co-factor in the malignant transformation. It has been discovered that all Burkitt's lymphoma cells carry a chromosomal arm translocation. This translocation activates a chromosomal oncogene. The infection of B-cells by EBV results in rapid uncontrolled B-cell growth, which may increase the frequency of this key and deadly chromosomal arm translocation (see more about oncogenes in Chapter 26).

Latent EBV infections in immunosuppressed patients can reactivate, resulting in transformation and uncontrolled growth of the B-cell line. Lymphoma and other lymphoproliferative diseases in these patients may be secondary to this EBV reactivation.

Mononucleosis

"Mono" is a disease of young adults. As with many viral infections, the lower the socioeconomic class, the earlier children are infected and the milder the disease (for example: chickenpox and polio). American teenagers, living in a higher socioeconomic class with better sanitation, handwashing, etc., are infected later in life through social contact, usually kissing. Thus the references to "kissing disease."

Patients with mononucleosis develop fever, chills, sweats, headache, and a very painful pharyngitis. Most will have enlarged lymph nodes (as the B-cells multiply) and even enlarged spleens. Blood work reveals a high white blood cell count with **atypical lymphocytes** seen on the blood smear. These are large activated T-lymphocytes. The blood also has **heterophile antibody**, which is an antibody against EBV that cross reacts with and agglutinates sheep red blood cells. This can be used as a rapid screening test for mononucleosis (**Monospot test**).

HHV 8

HHV 8 is a herpesvirus that appears to cause Kaposi's sarcoma! (See "Malignancy" subheading in Chapter 26 for further information.)

Fig. 27-6. Summary of the herpesviridae.

Recommended Review Articles:

Luzuriaga K, Sullivan JL. Infectious mononucleosis. N Engl J Med. 2010 May 27;362(21):1993–2000.

Oxman MN. Zoster vaccine: current status and future prospects. Clin Infect Dis. 2010;51(2):197–213.

Sen P, Barton SE. Genital herpes and its management. BMJ. 2007;334(7602):1048–52.

NAME	MORPHOLOGY	TRANSMISSION
Herpes simplex virus-1 (HSV-1)	1. Double-stranded linear DNA 2. Enveloped 3. Icosahedral symmetry	1. Direct contact of mucous membranes NOTE: viral shedding usually occurs in the presence of obvious herpetic lesions, but viral shedding can also occur when there are no visible lesions 2. Sexually transmitted 3. Herpes virus travels up sensory nerve fibers to the sensory nerve ganglia, where it replicates, then returns along the sensory nerve fibers to produce skin lesions
Herpes simplex virus-2 (HSV-2)	1. Double-stranded linear DNA 2. Enveloped 3. Icosahedral symmetry	1. Direct contact of mucous membranes NOTE: viral shedding usually occurs in the presence of obvious herpetic lesions, but viral shedding can also occur when there are no visible lesions 2. Sexually transmitted 3. Herpes virus travels up sensory nerve fibers to the sensory nerve ganglia, where it replicates, then returns along the sensory nerve fibers to produce skin lesions
Varicella-zoster virus	1. Double-stranded linear DNA 2. Enveloped 3. Icosahedral symmetry	1. Varicella is highly contagious! A. Aerosolized respiratory secretions B. Contact with ruptured vesicles 2. Zoster: reactivation from dorsal root ganglion

Figure 27-6   HERPESVIRIDAE

CLINICAL	TREATMENT & PREVENTION	DIAGNOSTICS
1. **Gingivostomatitis (cold sores)**: painful group of vesicles on the lips and mouth, which ulcerate, and heal usually without leaving a scar. Often accompanied by fever & "viral" symptoms 2. **Reactivation** of gingivostomatitis occurs in immuno-compromised individuals or when individuals are "stressed out." Similar eruption of vesicles as with primary gingivostomatitis, but the vesicles are less painful and last for fewer days 3. **Herpetic keratitis of the eye**: Recurrence is common. This is the most common cause of corneal blindness in the United States 4. **Encephalitis** (#1 cause of viral encephalitis in the United States): infection (most cases are reactivation of latent HSV-1) of the brain results in cell death and brain tissue swelling, manifested as fever, headache and neurologic abnormalities	1. Acyclovir 2. Valacyclovir 3. Famciclovir 4. Trifluridine eye drops: for corneal infection	1. Tzanck prep: reveals multinucleated giant cells & intranuclear inclusion bodies 2. Viral culture 3. Polymerase chain reaction 4. Serology 5. Direct Fluorescent Antibodies (DFA): Ulcer base scrapings can be tested with antibodies against HSV. The antibodies will latch onto HSV if present, & will fluoresce
5. **Genital herpes**: painful group of local vesicles on the cervix, or on the external genitalia of men and women. Often associated with fever and viral symptoms. These vesicles usually do not scar 6. **Reactivation** of genital herpes: similar eruption of vesicles, but less painful and vesicles last for fewer days 7. **Neonatal herpes**: acquired during the passage of a fetus through an infected birth canal. The risk of transmission is highest when a primary genital infection is present during delivery (One of the TORC*H*ES Organisms) A. Disseminated B. Central nervous system C. Skin D. Eye • Note: HSV-1 & HSV-2 can interchangeably cause any of the above diseases 8. **Herpetic whitlow** 9. Disseminated herpes infection of organs	1. Acyclovir 2. Valacyclovir 3. Famciclovir 4. Condom use	1. Tzanck prep: reveals multinucleated giant cells & intranuclear inclusion bodies 2. Viral culture 3. Polymerase chain reaction 4. Serology 5. Direct Fluorescent Antibodies (DFA): Ulcerbase scrapings can be tested with antibodies against HSV. The antibodies will latch onto HSV if present, & will fluoresce
1. **Varicella (Chicken Pox)** A. 2 week incubation period B. Fever and headache develop C. Rash: the vesicles first erupt on the trunk and face, and spread to involve the entire body (including mucous membranes). The vesicles rupture and scab over. Note that the vesicles erupt in crops, so one crop forms as another crop scabs over. Patients are infectious until all of their lesions scab over D. Pneumonia or encephalitis can occur in immunocompromised patients	1. Acyclovir 2. Valacyclovir 3. Famciclovir 4. Chickenpox vaccine 5. Zoster immune globulin 6. Zoster vaccine (recommended for > age 60)	1. Vesicles are described as dew drops on the top of a rose petal: a red base with fluid filled vesicle on top 2. Tzanck prep: reveals multinucleated giant cells

NAME	MORPHOLOGY	TRANSMISSION
Varicella-zoster virus (continued)		
Cytomegalovirus	1. Double-stranded linear DNA 2. Enveloped 3. Icosahedral symmetry	1. Virus present in milk, saliva, urine & tears 2. Transmission occurs with prolonged exposure, such as between children in households or day care centers 3. Sexual transmission
Epstein-Barr virus (EBV)	1. Double-stranded linear DNA 2. Enveloped 3. Icosahedral symmetry	1. Intimate contact from asymptomatic shedders of EBV 2. Infects human B-cells & transforms them
Human Herpesvirus 6 (HHV-6)	1. Double-stranded linear DNA 2. Enveloped 3. Icosahedral	• Transmitted by saliva
HHV-8	1. Double-stranded linear DNA 2. Enveloped 3. Icosahedral	• Sexual transmission, especially with homosexual men

Figure 27-6   (continued)

CLINICAL	TREATMENT & PREVENTION	DIAGNOSTICS
2. **Zoster (Shingles)**: painful eruption of vesicles isolated to a single dermatome distribution. The vesicles dry up and form crusts, which disappear in about 3 weeks. Pain in the dermatomal distribution can last for months in the elderly • **Herpes Zoster ophthalmicus**: Vesicles on one side of the forehead and on tip of the nose (the dermatomal distribution of the first division of cranial nerve V) may be associated with severe corneal involvement that (similar to HSV) can lead to blindness		

1. Asymptomatic infection (latent phase)
2. Congenital disease (TOR*C*HES)
3. CMV mononucleosis
4. Reactivation in immunocompromised patients
  A. Pneumonia
  B. Retinitis
  C. Esophagitis
  D. Disseminated disease
	1. Ganciclovir
	2. Foscarnet
	3. Cidofovir
	4. Fomivirsen
		1. CMV shell viral culture: Blood buffy coat (white cells) is cultured over night. The following morning, the cells are centrifuged. This breaks up the white blood cells, releasing CMV antigens, which are detected with monoclonal antibodies
		2. Serology
		3. Histology: reveals enlarged (Cytomegalic) cells with intranuclear & cytoplasmic inclusion bodies
		4. CMV early antigens can be detected in white blood cells. These antigens are an early marker for infection in bone marrow transplant patients.
		5. PCR testing for CMV DNA

1. **Infectious mononucleosis**
  A. Fever
  B. Sore throat
  C. Severe lethargy
  D. Enlarged lymph nodes and spleen
2. Associated with **Burkitt's B-cell lymphoma**
	• Supportive	1. Elevated heterophile antibodies
		2. Differential white blood cell count will show elevated "atypical lymphocytes"
		3. Serology: IgM against the viral capsid antigens (VCA)

• Roseola (exanthum subitum):
  1. High fever lasting 3-5 days, which resolves, and is followed by a. . . .
  2. Rash: located mostly on the trunk, which lasts just a day or two		

• Appears to be the cause of Kaposi's sarcoma		

M. Gladwin, W. Trattler, and S. Mahan, *Clinical Microbiology Made Ridiculously Simple* ©MedMaster

CHAPTER 28.   THE REST OF THE DNA VIRUSES

We have covered in detail the two H's of the HHAPPPy DNA viruses: herpesviridae and hepatitisviridae. We will now briefly cover the poxviridae, papovaviridae, adenoviridae, and parvoviridae.

POXVIRIDAE

Fig. 28-1. **Poxviridae** is structurally the most complex of all known viruses. It is a brick-shaped box, **POX in a box**, and has at its center a large complex DNA genome coding for hundreds of proteins. The DNA is organized into a dumbbell shape with structural proteins, surrounded by two envelopes. This virus carries many of its own enzymes and, unlike other DNA viruses, replicates in the **cytoplasm**.

Smallpox

Poxviridae do NOT cause chickenpox. Chickenpox is caused by the herpesviridae varicella-zoster. This big complex pox-box is no chicken! Poxviridae used to cause smallpox.

Why do we say *used* to cause smallpox? Answer: The last case of smallpox was in 1977 and it is thought that this virus has been eradicated from the planet earth!!! Pox has been placed in a box and buried.

For more than 3 thousand years this highly contagious virus spread via the respiratory tract, causing pox skin lesions and death. A concerted vaccination and surveillance program conducted by the World Health Organization brought this tyranny to an end.

The Vaccine and Why It Worked

1) A vaccine was developed that induced solid, lasting immunity. The vaccine contained vaccinia virus, an avirulent form of poxviridae, which induced immunity to virulent poxviridae.

2) Smallpox only infected humans. There are no animal reservoirs that can harbor this virus and protect it.

3) All poxviridae infections produced clinically overt smallpox. Every smallpox attack was obvious, so members of the World Health Organization could localize communities that needed vaccination. The virus could not hide as an asymptomatic infection or in a latent state.

Recently, there is concern that smallpox could be used as a weapon against a population, since many people in the world have no immunity against smallpox. This is an example of Bioterrorism. This concern has stimulated some groups (military, first responders, public health professionals) to start receiving the vaccine again.

Smallpox is one of the most serious biological threats encountered today because:

1. Routine vaccination ceased in the United States in 1972, so most people have either lost their immunity to this virus or have never been vaccinated. Smallpox is highly contagious, spreading directly from person to person primarily via large droplets (an aerosol in rare instances). **Droplet transmission** occurs when relatively large (< 5 micron) particles containing smallpox are propelled from an infected person over relatively short distances (3–6 feet) and deposited onto mucous membranes (usually mouth or nose) of another person or on an environmental surface. **Aerosol transmission**

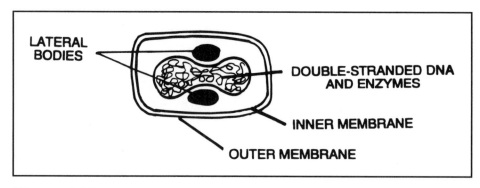

Figure 28-1

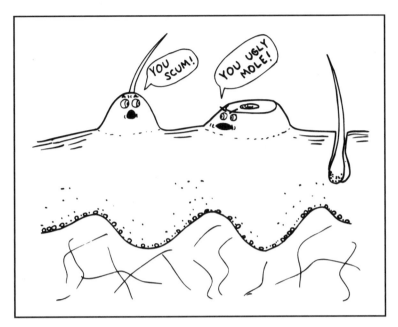

Figure 28-2

occurs when small (< 5 micron) particles containing smallpox, which may remain suspended in the air for long periods and may travel long distances, are propelled from an infected person and are inhaled by another person. Aerosols are typically generated by cough.

2. There is no known treatment for smallpox. Historically, approximately 30% of patients died, and those who survived became scarred for life (the pox lesions crust over, fall off, and leave pitted marks).

3. Although there are only 2 official repositories of smallpox virus in the world – in the CDC and in Novosibirsk, Russia, there are serious concerns that clandestine stocks remain in other nations.

The disease may be confused with chickenpox (described in **Chapter 27**), but there are several key differences summarized in the table below. The major differences are that smallpox pustules (which like chickenpox also look like dew on a rose petal) develop all at the same time (synchronous lesions), usually appear most dense on the face and palms (centrifugal spread) and ulcerate deeper into the skin. Chickenpox will appear in waves or crops of pustules and the lesions usually start on the face and torso, later spreading to the entire body, including the mucous membranes.

Smallpox	Chickenpox
Orthopoxviridae	Herpesviridae
Synchronous lesions	Asynchronous lesions
Deep lesions	Superficial lesions
Centrifugal spread	Centripetal spread
Prominent face, palms	Prominent in chest

Did you know?

Smallpox is the first disease for which a vaccine was developed. Edward Jenner developed the vaccine in the late 19th century after observing that milk maids did not acquire the disease. Milk maids were exposed to cowpox, a similar disease that affects cows. These milk maids developed antibodies against cowpox, which were cross protective with smallpox. Presently the vaccine is made with an unrelated virus, called vaccinia. Nobody is absolutely sure when vaccinia replaced cowpox for vaccination, but modern molecular microbiology has shown that cowpox and vaccinia are very different viruses. One more thing: Nobody believed Jenner at the time, so he had to finance his own work and publication.

Molluscum Contagiosum

Fig. 28-2. You **mole**! You **scum**! **molluscum**. A pox virus causes these small, 1–2 mm in diameter, white bumps that have a central dimple (seen to the right of the mole in this figure). They are similar to warts with benign hyperproliferation of epithelial cells. You will probably first see these lesions on AIDS patients, who frequently develop them.

PAPOVAVIRIDAE

There are 3 members of the **PA-PO-VA** VIRIDAE:
PA: PApilloma virus causes human warts and **cervi**cal cancer.

PO: POlyomavirus has 2 members: human BK and JC virus.

VA: Simian **VA**cuolating virus does not infect humans.

PAPOVAVIRIDAE:

With these viruses think **O**
O for circular double-stranded DNA (naked icosahedral capsid)
O for round warts
O for round cervix

Papilloma Virus

Different strains of the papilloma virus can cause warts and cervical cancer.

Warts

There are many strains of papilloma viruses. They have a tropism for squamous epithelial cells, and different strains like certain anatomic regions: common warts, genital warts, laryngeal warts. Warts are benign hyperproliferations of the keratinized squamous epithelium. Most will resolve spontaneously within 1–2 years. For unclear reasons many people do not develop warts despite the ubiquitous nature of the papilloma virus. Perhaps in these unaffected individuals the virus remains latent or is effectively controlled by the host immune system.

Cervical Cancer

Cervical dysplasia and carcinoma are associated with sexual activity and previous exposure to certain strains (type 16 and 18) of human papilloma virus.

Think of **PAP**illoma virus and the **PAP** smear, which is used to detect early dysplastic cellular changes. The Pap test has resulted in early detection of cervical dysplastic changes and has significantly reduced the progression to cervical cancer.

There are now two FDA–approved vaccines for human papilloma virus. There is a quadrivalent vaccine that is effective against HPV types 16 and 18, which cause approximately 70% of cervical cancers, and against HPV types 6 and 11, which cause approximately 90% of genital warts. This vaccine was **98% effective** in reducing cervical squamous cell cancer, adenocarcinomas, and their precursors. A new 9–valent vaccine was approved in 2015; it includes all 4 HPV types in the quadrivalent vaccine, as well as HPV types 31,33,45,52, and 58, which are the next most common types implicated in causing cervical cancer.

Polyomavirus

Two polyoma viruses infect humans. They were named after the initials of the patient from whom the virus was discovered. Both are ubiquitous and infect worldwide at an early age.

BK Polyomavirus

1) As ubiquitous as the **B**urger **K**ing at every highway exit.
2) Causes mild or asymptomatic infection in children.
3) BK virus primarily causes symptomatic disease in immunocompromised patients. The major diseases caused by BK virus are nephritis and ureteral stenosis in renal transplant patients and hemorrhagic cystitis in bone marrow transplant recipients.

JC Polyomavirus

Fig. 28-3. **JC P**olyomavirus is similar to BK but also causes an opportunistic infection in immunocompromised patients called **Progressive Multifocal Leukoencephalopathy (PML)**. In this disease, patients develop central nervous system, white matter damage. Visualize shoppers at **J.C. P**enney with **PML**, walking around the store with memory loss, poor speech, and incoordination.

ADENOVIRIDAE

The **ADEN**oviridae cause upper respiratory tract infections. Visualize **A DEN** full of coughing, sneezing children. Studies estimate more than 10% of childhood respiratory infections are caused by strains of adenoviridae, and virtually all adults have serologic evidence of prior exposure. Infection can result in rhinitis, conjunctivitis, sore throat, and cough. This can sometimes progress to lower respiratory tract pneumonia in children.

The most common respiratory illnesses in children: RSV, metapneumovirus, parainfluenza, rhinovirus, and adenovirus.

There are also **enteric adenoviruses**, which as the name suggests, can cause diarrheal illness. In fact these are the second most common viral cause of endemic diarrheal illness in infants and children across the world.

PARVOVIRIDAE

Fig. 28-4. **PAR**vovirus is the smallest icosahedral virus and has a single strand of DNA. Simple as a **Par-ONE** golf course!

It causes a childhood disease called **erythema infectiosum** (Fifth disease), characterized by fever and a "slapped face" rash on the cheeks.

Fig. 28-5. Summary of the Rest of the DNA Viruses.

Figure 28-3

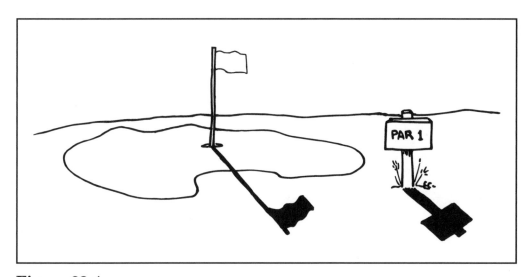

Figure 28-4

NAME	MORPHOLOGY	CLINICAL
POXviridae	1. Complex coat 2. Double-stranded linear DNA 3. The only DNA virus to replicate in cytoplasm	1. **Smallpox**: causes skin lesions and death. This disease has been eradicated from the earth! 2. **Molluscum contagiosum**: small white bumps with a central dimple (like a wart). Often found in the genital region

PAPOVAviridae	1. Naked icosahedral
	2. Double-stranded circular DNA
	3. Replicates in nucleus
		1. Human papilloma virus (HPV): cause **warts** (over 50 virals strains)
		  A. Common warts (types 1, 2, 4 & 7)
		  B. Genital warts (types 6, 11, 16, 18 and others)
		  C. Laryngeal warts (6 & 11)
		  D. Cervical cancer (types 16 & 18)
		2. BK Polyomavirus: causes **kidney disease** in renal transplant patients, **hemorrhagic cystitis** in bone marrow transplant patients, and mild respiratory illness in children.
		3. JC Polyomavirus: Progressive multifocal leukoencephalopathy, characterized by degenerative central nervous system white matter disease

ADENOviridae	1. Naked icosahedral
	2. Double-stranded linear DNA
	3. Replicates in nucleus
		1. Childhood upper respiratory tract infections:
		  A. Rhinitis
		  B. Sore throat
		  C. Fever
		  D. Conjunctivitis
		2. Epidemic keratoconjunctivitis (Pink Eye)
		3. Epidemic diarrheal illness in infants and children.

PARVOviridae strain B-19	1. Naked icosahedral
	2. The only single-stranded linear DNA virus (negative stranded)
	3. Replicates in nucleus
		1. **Erythema infectiosum** (Fifth disease): affects children between the ages of 4 to 12
		  A. Fever
		  B. "Slapped cheek" rash
		2. **Transient aplastic anemia crisis**: Occurs when the Parvo virus stops the production of red blood cells in the bone marrow

Figure 28-5   THE REST OF THE DNA VIRUSES

Recommended Review Articles:

Henderson DA, Inglesby TV, Bartlett JG, et al. Smallpox as a Biological Weapon. JAMA 1999;281:2127–2137.

Johnston O, Jaswal D, et al. Treatment of polyomavirus infection in kidney transplant recipients: a systematic review. Transplantation. 2010;89(9):1057–70.

Kahn JA. HPV vaccination for the prevention of cervical intraepithelial neoplasia. N Engl J Med. 2009;361(3):271–8.

TREATMENT & PREVENTION	MISCELLANEOUS
• Vaccine: an avirulent pox virus was developed that induced immunity to virulent pox virus	1. No animal reservoirs! 2. Codes for DNA and RNA polymerase
Methods of wart removal: 1. Liquid nitrogen (freeze them off): Best method 2. Surgical 3. Electrosurgery (Laser ablation) 4. Podophyllin: for genital warts • Many warts resolve spontaneously in 1–2 years • Relapses are common after treatment, because HPV DNA is found in **normal** appearing tissue around the wart	• Second smallest DNA virus
• Illness is self-limited	
• Illness is self-limited • I.V. immunoglobulin can be used with aplastic crisis	• Smallest DNA virus

M. Gladwin, W. Trattler, and S. Mahan, *Clinical Microbiology Made Ridiculously Simple* ©MedMaster

CHAPTER 29.   THE REST OF THE RNA VIRUSES
Pesky mosquitos, headaches, diarrhea, rabid dog bites, colds and bioweapons!

We have covered in previous chapters some of the RNA viruses: retroviridae, orthomyxoviridae, and paramyxoviridae. We will now briefly review:

1) The **ARthropod BOrne viruses**, which are called **arboviruses**: These RNA viruses include the **togaviridae, flaviviridae**, and **bunyaviridae**.

Although not transmitted by arthropods, rubivirus (which causes rubella) and hantavirus (hantavirus pulmonary syndrome) will be discussed here because they are members of the togaviridae and bunyaviridae, respectively.

The **West Nile Virus** is a flavivirus spread by mosquitos that will be highlighted because it has now become endemic in the United States.

2) The **picornaviridae**, which are a large group of enteroviruses (ENTERO = GI, fecal-oral transmission): poliovirus; coxsackie A and B; echovirus, and the new enteroviruses.

3) Viruses that cause the common cold: **rhinovirus** (really in the picornaviridae family) and **coronaviridae**.

A new terrifying member of the coronaviridae that causes one hell of a cold is the deadly **SARS virus** (Severe Acute Respiratory Syndrome associated coronavirus).

4) Viruses that cause diarrhea: **rotavirus**, and **caliciviridae** (which includes the Norwalk virus).

5) Rabies caused by the **rhabdoviridae**.

6) Viruses that cause hemorrhagic fever and are potential bioweapons. These include the deadly *Filoviridae*, namely **Ebola** and **Marburg Viruses**, and the *Arenaviridae*, such as the **Lassa Fever virus**.

THE ARBOVIRUSES

Fig. 29-1. The arboviruses include the **bunyaviridae, togaviridae**, and **flaviviridae**. All are transmitted by blood-sucking arthropods and cause fever and encephalitis. The legendary U.S. logger Paul **Bunyan**, wearing a **toga**, has a rich **flavor** that attracts mosquitos and other arthropods. He is about to chop down a tree (**ARBOL** in Spanish). You can well imagine Paul Bunyan has quite a headache (encephalitis) with that blood-sucker clinging to his toga!

Figure 29-1

Togaviridae

Two members of this family infect humans:

1) **Alpha viruses** in the Western hemisphere are mosquito-borne and cause encephalitis, an inflammation of the brain with fever, headache, altered levels of consciousness, and focal neurologic deficits.
2) **Rubivirus** causes rubella.

Alpha Viruses

The 3 main alpha viruses that cause encephalitis all infect horses, birds, and humans. They use the mosquito as a vector.

Fig. 29-2. The togaviridae alpha viruses. Picture Paul Bunyan riding on a roller-coaster wearing his

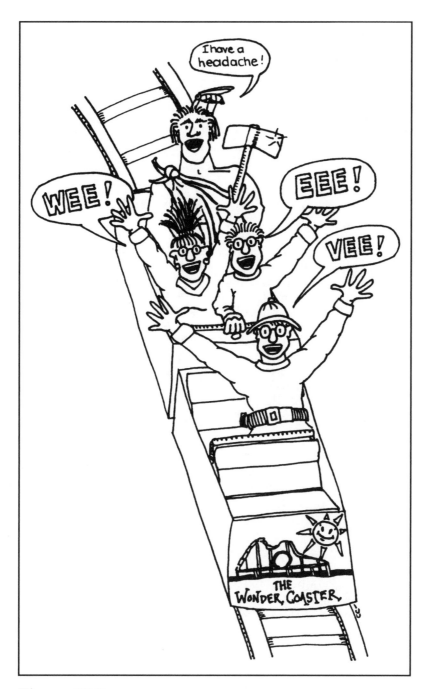

Figure 29-2

toga (**toga**viridae), with a **mosquito** (mosquito vector) on his head. The other passengers scream the names of the 3 main diseases caused by the togaviridae alpha viruses, which are named by geographic region:

WEE: **W**estern **e**quine **e**ncephalitis (western U.S. and Canada).

EEE: **E**astern **e**quine **e**ncephalitis (eastern U.S.).

VEE: **V**enezuelan **e**quine **e**ncephalitis (South and Central America, southern U.S.).

Chikungunya virus is a mosquito-borne alpha virus that **does not** typically cause encephalitis. Chikunguna virus is indigenous to tropical Africa and Asia where it is transmitted by *Aedes* mosquitoes. Initial symptoms include fever, rash, and **joint pain/swelling**. The fever and rash typically resolve after a few days, but the joint pains can last for a much longer period of time. The disease was first described in the 1950's. In late 2013 the first local transmission of chikungunya virus in the Americas was discovered in the Carribbean. Beginning in 2014, local transmission of chickungunya was reported in Florida and Puerto Rico. With no vaccine or treatment available it appears this nasty virus is here to stay.

Rubivirus

Rubivirus is a togavirus, but it is not an arbovirus because humans are the only infected creatures. Rubivirus causes **rubella**, which is a mild febrile illness with a rash. The importance of this virus lies in its ability to cross the placenta and cause terrible congenital defects, especially in the first trimester.

Rubella ("German measles") is a mild measles-like illness. Like measles, rubella is contracted by respiratory secretions and has a prodrome of fever, lymphadenopathy and flu-like symptoms. This is followed by a red maculopapular rash that spreads from forehead to face to torso to extremities. Unlike measles, patients are less "sick," complications such as encephalitis rarely occur, and the rash lasts only 3 days, not 6. Thus its other name: "3-day measles." Young women can develop self-limiting arthritis with the infection.

The **R** in TORCHES (see **Fig. 27-2**) stands for the feared **congenital rubella**. The risk of rubella-induced congenital defects is greatest early in fetal development when cell differentiation is at a peak. Rubivirus-infected human embryo cells demonstrate chromosomal breakage and inhibition of mitosis.

Body areas affected in congenital rubella include:

1) **Heart**: patent ductus, interventricular septal defects, pulmonary artery stenosis, others.

2) **Eye**: cataracts, chorioretinitis, others.

3) **CNS**: mental retardation, microcephaly, deafness.

A live attenuated rubella vaccine is given to all young children in the U.S. It is not recommended for pregnant women because of the theoretical risk of fetal infection. However, there is no evidence that this vaccine causes congenital defects.

Pregnant women are routinely screened for immunity to rubella. If they do not have antibody to rubivirus, they will receive immunization after delivery.

Flaviviridae

The flaviviridae share many similarities with the togaviridae:

The morphology is similar (see Fig. **29-10**).

Figure 29-3

They cause encephalitis, with names based on geographic location (Japanese encephalitis, Russian encephalitis, etc.).

St. Louis encephalitis is the second leading cause of epidemic viral encephalitis in the United States. Again, it is named after the geographic region it was first encountered. It accounts for tens to hundreds of cases in a typical year and up to 3000 cases in epidemic years.

Fig. 29-3.   The flaviviridae are spread by a mosquito vector, infecting humans and birds.

The flaviviridae also cause the febrile diseases **yellow fever** and **Dengue fever**.

1) **Yellow fever** was made famous by the Panama Canal project. This flavivirus (yellow fever virus) was transmitted to canal workers by mosquitoes. One week later they would develop hepatitis with jaundice (yellow appearance), fever, backache, nausea, and vomiting.

Once the vector was found to be a mosquito, insecticides were used to control the disease. Spraying continues in the southern U.S. and Latin American urban centers, virtually eliminating urban yellow fever.

Yellow fever remains a concern primarily in Africa where 90% of cases now occur. Vaccination is required for travelers going to and from most endemic areas.

2) **Dengue fever** is a mosquito-borne febrile disease that occurs primarily in the tropics, but has worldwide distribution, including periodic cases in the Southwestern U.S. along the border with Mexico (the only spared regions are Europe and Antarctica). As global warming trends continue you will probably hear more of this **painful** illness. It is also called break-bone fever because of the severe painful backache, muscle and joint pain, and severe headache. Painful fever!

There are 4 main dengue serotypes. Repeat infection with a second serotype, especially serotype 2, predisposes to a variant of disease called Dengue hemorrhagic fever, which causes hemorrhage or shock, especially in children, with mortality rates of almost 10%. (remember, **Dengue takes you down in the second – serotype 2 – round**).

West Nile Virus

West Nile Virus caused human disease in North America for the first time in 1999. An outbreak of meningitis and encephalitis in Queens, New York was determined to be caused by the West Nile virus, which had previously been found only in Africa, Europe, the Middle East and Asia. This virus is a member of a group of closely related flaviviruses that includes **St. Louis encephalitis, Japanese encephalitis** and **Kunjun** and **Murray Valley fever viruses**. The virus was most likely spread by either a bird or mosquito traveling on an airplane or by migrating birds. Its genetic sequence most closely matches a virus isolated from Israel. In subsequent years the virus marched across larger and larger geographic areas, causing epidemics throughout the United States.

Most experts believe the virus now has a permanent foothold in North America and will cause yearly outbreaks in animals and man during the mosquito season. The virus can infect many types of animals—among birds and horses it has proven most lethal. The vast majority of West Nile virus cases are transmitted by a mosquito bite, but since its arrival in the United States, it has been transmitted via blood transfusions and organ transplantation multiple times, transplacentally to a fetus, possibly through breast milk to a nursing infant, and in a laboratory accident that resulted in infected blood inoculation.

Clinical Manifestations of West Nile

Most patients are asymptomatic. Serologic studies have shown the symptomatic illness to "silent or occult" infection ratio to be approximately 1:150. Some infected people develop West Nile fever which often includes a headache and occasionally a maculopapular rash. While this form of the illness was previously thought to be self limited and to last about a week, recent reports indicate that in many people it may be followed by fatigue, weakness, and difficulty concentrating that can persist for months. Other infected people have neuroinvasive disease: either an aseptic meningitis, frank encephalitis with decreased level of consciousness, or dramatic motor paresis (muscle weakness) or paralysis. Most patients with these motor symptoms have a polio-like syndrome with spinal cord anterior horn pathology while in others the disease is more similar to Guillain-Barre syndrome with demyelination. This myelitis can occur independently or in the setting of encephalitis. In the New York outbreak, the mortality in patients with encephalitis was 19%; with encephalitis plus motor weakness the mortality was 30%. The elderly and immunocompromised patients (cancer patients who received chemotherapy, etc.) are the most likely to suffer severe clinical manifestations and have a poor outcome. Children seem to fare the best, developing neurologic manifestations less frequently, although reports of neuroinvasive disease in children are increasing for unclear reasons (likely increased recognition and diagnosis).

Diagnosis of West Nile

West Nile virus can be diagnosed one of 4 ways:

1) Detection of virus by isolation (culture) or nucleic acid amplification techniques like PCR
2) Serum IgM versus WNV with IgG neutralizing antibodies against WNV in the same specimen
3) Cerebrospinal fluid IgM against WNV
4) A fourfold rise in serum antibody titers

Serum or cerebrospinal fluid IgM (antibody versus the virus) is currently the most sensitive test. Polymerase chain reaction tests of cerebral spinal fluid can occasionally detect the viral genome and portends a worse prognosis, but most patients have already cleared the virus from blood and CSF and are developing antibody *at the time of presentation*. The diagnosis can easily be missed if only PCR is sent without antibody titers. There is some cross-reactivity among IgM assays for St. Louis encephalitis and Yellow fever and Dengue viruses. Furthermore, the IgM for West Nile can persist for greater than one year; thus a positive test is not a guarantee of the diagnosis. Thus, neutralizing IgG antibody against WNV must be present in the same sample to confirm a diagnosis.

Prevention and Treatment of West Nile

A vaccine is under development. Currently, the best prevention is to eliminate mosquito breeding areas (stagnant water) and for personal protection, to wear long sleeves and long pants and to use DEET-containing bug spray in outbreak areas, especially after dusk. **Multiple studies evaluating a variety of mosquito prevention strategies support the efficacy of DEET-containing sprays (increasing concentrations of DEET increase mosquito prevention).** There is currently no treatment for actual infection other than supportive care.

Bunyaviridae

Bunyaviridae also cause diseases characterized by fever and encephalitis, such as **California encephalitis** and **Rift Valley fever**. For comparison of the bunyaviridae with the other arboviruses (toga and flavi), see **Fig. 29-10**.

Hantavirus Pulmonary Syndrome

In May 1993 reports began to emerge from the Four Corners area of New Mexico, Arizona, Colorado, and Utah, of an influenza-like illness followed by sudden respiratory failure, frequently culminating in death. Many of these patients were previously healthy adults. The etiologic agent is a virus in the bunyaviridae family in the genus **hantavirus**. Hantavirus had previously only been associated with the disease **hemorrhagic fever with renal failure** seen in Asia and Europe. Four species of hantavirus are known to cause hantavirus pulmonary syndrome (HPS) in the United States, each with its own rodent vector. The deer mouse is the primary vector of Sin Nombre virus (the no name virus), which predominates in the Southwestern U.S. and is the hantavirus species that has been the primary cause of disease in the U.S.

As of February 2013, 617 cases of hantavirus pulmonary syndrome have been confirmed in the U.S. Patients typically present with high fevers, muscle aches, cough, nausea, and vomiting. Their heart and respiratory rate is rapid and blood work may reveal a high white blood cell count, low platelets, and an elevated red blood cell count. Diagnosis is confirmed by serologic identification of IgM and IgG antibodies to Sin Nombre Virus. The lung capillary per-meability is disrupted resulting in fluid leakage into the alveoli (pulmonary edema). The fluid-filled alveoli are unable to deliver oxygen to the bloodstream, and intubation with mechanical ventilation is required to enhance oxygenation in close to 90% of patients. The overall case fatality rate of HPS is approximately 40%

This illness should be considered in young adults with influenza-like symptoms who develop pulmonary edema. Therapy is supportive with no proven anti-viral therapy available, despite some early evidence that ribavirin might be beneficial.

PICORNAVIRIDAE

This family of viruses all have similar structure and replication.

There are 4 genera that contain important human pathogens: **enterovirus**, **rhinovirus**, **hepatovirus** (hepatitis A virus), and **parechovirus** (a recently defined genus, previously classified as echoviruses).

Enterovirus

1) **Enteroviruses** have 5 subgroups:

   a) Poliovirus
   b) Coxsackie A viruses
   c) Coxsackie B viruses
   d) Echovirus
   e) new enteroviruses (including Rhinoviruses)

These are all called enteroviruses because they infect intestinal epithelial and lymphoid (tonsils, Peyer's patches) cells. They are excreted in the feces and spread by the fecal-oral route. The replication in the tonsils also results in viral shedding from pharyngeal secretions.

Poliovirus will be discussed first as it causes the important paralytic disease, poliomyelitis. The remainder will be discussed together as there is significant overlap in the diseases they cause.

Rhinovirus causes the common cold and will be discussed last.

Poliovirus

Hopefully, learning about polio and poliomyelitis will soon be purely academic. We are in the midst of a worldwide polio eradication campaign with only a few remaining pockets of infection, primarily in India, Pakistan, and Nigeria. The goal for eradication was 2005, but unfortunately that goal remains stubbornly elusive.

Poliovirus has the ability to infect cells in the:

1) Peyer's patches of the intestine.
2) Motor neurons.

This tropism explains:

1) The fecal-oral mode of transmission.
2) The disease paralytic poliomyelitis.

Polio was one of the feared diseases of the 20th century. In the 1950's 6 thousand cases of paralytic polio occurred each year in the U.S.

This disease was in part due to **improvements** in sanitation. Children tend to have fewer paralytic complications with poliovirus infection than do adults. As sanitation improved and fecally contaminated substances were cleared, fewer people were exposed to poliovirus as children, and thus more adults were infected. The chances of developing paralytic poliomyelitis **increases** as one gets older, thus explaining the increase in paralytic poliomyelitis with improvements in sanitation.

Now that a vaccine has been developed, the incidence has markedly diminished.

The Disease Polio

The virus initially replicates in the tonsils and Peyer's patches, spreading to the blood, and across the blood-CNS barrier to the anterior horn of the spinal cord.

Because of the initial replication in the tonsils, the virus can be spread by respiratory secretions, as well as the usual fecal-oral route, early in the course of infection.

There are 3 disease manifestations:

1) **Mild illness**: An asymptomatic infection or a mild febrile viral illness is the most common form. This especially occurs in infants in less-developed nations, where the sanitation is poor.

2) **Aseptic meningitis**: Fever and meningismus can develop as the poliovirus infects the meninges. Recovery is complete in 1 week.

3) **Paralytic poliomyelitis**: This is the feared manifestation of poliovirus infection!!! The viral infection destroys presynaptic motor neurons in the anterior horn of the spinal cord as well as the postsynaptic neurons leaving the horn. The damage to the exiting motor neurons results in clinical manifestations of peripheral motor neuron deficits, while the presynaptic neuron damage causes central motor neuron deficits.

This disease is truly terrifying. A mild febrile illness resolves, 5 to 10 days later the fever recurs, followed by meningismus and then flaccid asymmetric paralysis. The paralytic disease can range from 1 leg or arm to paraplegia, quadriplegia, and even respiratory muscle dysfunction. The later more serious events usually occur in persons older than 15 years.

The affected extremities early in the course will have painful muscle spasms. Asymmetric muscle paralysis develops. Ultimately, atrophy and loss of reflexes occur (there are no sensory losses).

Vaccines

The **inactivated polio vaccine**, developed by Jonas Salk, contains **formalin-killed viruses** that are injected subcutaneously, provoking an IgG antibody response that will protect against future viremia. This is the primary vaccine used in the U.S. and developed countries worldwide.

The **oral polio vaccine** was developed by Albert B. Sabin. This vaccine contains **attenuated poliovirus** that has lost the ability to multiply in the CNS. It is taken orally and replicates and sheds in the feces in the normal fashion but does not cause paralytic poliomyelitis.

This vaccine works by supplanting the wild type (disease-causing) poliovirus with a docile attenuated counterpart. We are NOT eliminating poliovirus completely but are only trying to eliminate the virulent strain. In the case of the smallpox vaccine, immunization of the population depleted the viral reservoir (humans), and the virus was completely eliminated.

There are good and bad things about the oral Sabin vaccine:

Positives

1) It is an oral vaccine.

2) The oral route and full replication allow formation of both IgG in the blood and secretory IgA in the GI tract.

3) The attenuated virus is spread to contacts, resulting in a secondary infection and immunity in these individuals.

Negatives

Vaccine-associated paralytic poliomyelitis: The vaccine can pick up virulence and cause paralysis in the person taking the vaccine or in those exposed to shedding (parents changing a vaccinated infant's diapers). This is very rare (1/2.6 million doses), but out of the 138 cases of paralytic polio between 1973 and 1984, 105 cases were considered vaccine related. Persons with immunodeficiency states should not receive the oral attenuated vaccine. Since oral polio vaccine (Sabin) was totally replaced by the injectable polio vaccine (Salk) in the U.S. in 2000 there have been no cases of vaccine associated paralytic poliomyelitis.

Ultimately, both vaccines have been very effective, resulting in almost complete control of the virulent poliovirus in vaccinated geographic regions.

Coxsackie A and B, Echoviruses, New Enteroviruses

The remainder of the Enteroviruses in the Picornaviridae family are responsible for a variety of diseases. There is overlap since the different viruses can cause the same clinical symptoms.

The coxsackie viruses (A and B), the echoviruses, and the new enteroviruses **all** have multiple serotypes and **all** can cause:

1) Asymptomatic or mild febrile infections.
2) Respiratory symptoms ("cold").
3) Rashes.
4) Aseptic meningitis: The enteroviruses are the most common cause of non-bacterial (aseptic) meningitis in the U.S.

Coxsackie A

Coxsackie A can be differentiated from B by its effect on mice. Coxsackie A causes paralysis and death of the mouse with extensive skeletal muscle necrosis. Coxsackie A also causes:

Herpangina. A mild self-limiting illness characterized by fever, sore throat, and small red-based vesicles over the back of the throat.

Hand, foot, and mouth syndrome. A common acute illness, primarily in children, characterized by fever, oral vesicles, and small tender lesions on the hands, feet, and buttocks (I wonder how it got its name?!)

Coxsackie B

Coxsackie B causes less severe infection in mice but multiple organs can be damaged, such as heart, brain, liver, pancreas, and skeletal muscle. It also causes:

1) **Pleurodynia**. Fever, headache, and severe lower thoracic pain on breathing (pleuritic pain) mark the coxsackie B virus respiratory infection.

DISEASES	COXSACKIE		ECHOVIRUS &
	A	**B**	**NEW ENTEROVIRUS**
Asymptomatic infections	+	+	+
Respiratory infections	+	+	+
Rashes ("exanthems")	+	+	+
Aseptic meningitis	+	+	+
Herpangina	+	−	−
Pleurodynia	−	+	−
Myocarditis	−	+	−
Pericarditis	−	+	+

Figure 29-4   ILLNESSES CAUSED BY ENTEROVIRUSES

2) **Myocarditis/Pericarditis**. Infection and inflammation of the heart muscle and pericardial membrane can result in self-limited chest pain or more serious arrhythmias, cardiomyopathy, and heart failure. Many viruses can cause this, but coxsackie B is associated with 50% of the cases!

Enterovirus D68

A mix of enteroviruses circulates every year, with certain types being more prevalent in different years. In 2014, a large number of persons, especially children with a history of asthma, developed severe respiratory illness due to **enterovirus D68**. Over 1000 persons had confirmed infections in the U.S., but it is estimated that millions were infected. Additionally, there were several cases of a poliolike illness among persons in whom enterovirus D68 was detected in respiratory samples. Was enterovirus D68 the cause? Other enteroviruses, such as E71, have been associated with flaccid paralysis in Asia. Investigations into the link between the cases of polio-like illness and enteroviral infection in the U.S. are still ongoing.

Fig. 29-4.   Comparisons of the enteroviruses.

VIRUSES THAT CAUSE THE "COLD"

Fig. 29-5.   Rhino with the common cold, drinking a **Corona** beer. This will help you remember that the **rhinovirus** and the **coronaviridae** both cause the common cold.

More than 100 different serotypes of rhinovirus are responsible for the "common cold." Transmission occurs by hand-to-hand spread of mucous membrane secretions.

Figure 29-5

The coronaviridae cause a cold indistinguishable from the rhinovirus common cold. About 15% of adult common colds are caused by the coronaviridae.

THE DEADLY SARS VIRUS (SARS-LIKE CORONAVIRUSES OR SARS-COV)

In November 2002, there was an outbreak of severe atypical pneumonias in China. The cause of this outbreak was not known, and was initially confined to the mainland. In early 2003, this disease had spread to Hong Kong, Singapore, and Toronto, Canada, and the syndrome was called **SARS (Severe Acute Respiratory Syndrome)**. Within a few weeks of the description of this syndrome, it was discovered that SARS is caused by a virus, specifically a novel **coronavirus (SARS-CoV)**. Coronaviruses are a family of enveloped, single stranded RNA viruses, previously only known to cause the common cold in humans. Before the outbreak was controlled, it had spread to 29 countries and territories, and infected over 8000 people.

SARS is an example of an **emerging infectious disease**. This virus was not known to exist previously, and since the initial outbreak has not caused any additional natural outbreaks. An identical virus has not been found to exist in any natural reservoir, though similar viruses have been found in bats. SARS-like coronaviruses have been isolated from Himalayan **palm civets** and from **raccoon dogs** in markets in China. How people were first infected with this virus remains unclear but one theory is that it was a mutated virus of animals that crossed the species barrier and caused the epidemic.

Clinical Features of SARS

The primary mode of transmission is thought to occur via direct or indirect contact of mucous membrane (eyes, nose, or mouth) with infectious respiratory droplets. While this virus is not as transmissible as was previously thought, it appears that there were a few persons who were responsible for more transmissions, this is called **super-spreading events**. A lot of the transmission of this virus from one person to the next occurred in hospitals and other health care settings. After a 2–10 day incubation period, people infected with SARS often presented with fevers, myalgias, and chills, and later developed a dry cough, chest pain (pleurisy) and shortness of breath (dyspnea). Surprisingly, few patients develop sore throat or rhinorrhea as one might expect from a coronavirus. Most patients present to the doctor with an abnormal chest radiograph or chest CT scan showing alveolar consolidation, which can progress to frank **ARDS** (the acute respiratory distress syndrome). About 20 to 30 percent of patients required admission to an intensive care unit, and most of them required mechanical ventilation. About 8% of people with SARS died, and this was primarily due to respiratory failure.

Diagnosis

During the outbreak, a case definition was developed by the World Health Organization (WHO), which had a high sensitivity but a very low specificity. This is because the clinical features are not unique to this virus or disease. More accurate diagnoses can be made by Reverse-Transcriptase (quantitative)-Polymerase Chain Reaction (PCR) testing for the viral RNA in respiratory secretions, feces, urine and from lung biopsy tissue, or seroconversion (detection of antibodies in the blood to the virus).

Treatment

The optimal therapy of SARS is not known. Patients with suspected SARS are generally treated empirically with broad-spectrum antibacterial drugs that are effective against other agents that cause community acquired pneumonia. Ribavirin was often used in the treatment of SARS, but it was later shown in animal models that this drug had no effect against this virus. Corticosteroids were also used frequently, though there is no solid clinical or animal data to support their use. Generally treatment is supportive with mechanical ventilation and intensive care.

Recommended Review Article

Peiris JM, Phil D, Yuen KY, et al. Current concepts: The severe acute respiratory syndrome. N Engl J Med 2003; 349:2431–41.

Enserink M. SARS: Chronology of the Epidemic. Science (March 15) 2013:1269–73.

VIRUSES THAT CAUSE DIARRHEA

Viruses that cause gastroenteritis are acquired by the fecal-oral route and usually prey on infants and young children, although outbreaks among adults do occur.

Fever, vomiting, abdominal pain, and diarrhea follow a 1–2 day incubation, and symptoms resolve within 4–7 days.

Infants die secondary to loss of fluids and electrolytes.

DIARRHEA = DEATH BY DEHYDRATION

Four groups of viruses have been implicated in diarrhea: **caliciviruses** (including the *Norwalk virus* and noroviruses), **rotaviruses**, **adenoviruses** (discussed in Chapter 28), and **astroviruses**.

Fig. 29-6. If your **calico** cat develops diarrhea, **rotate** the kitty litter frequently, or **rotate** the **calico** cat off to **Norway**. This picture will help you remember that viral gastroenteritis (diarrhea) is caused by **caliciviridae**, including the **Norwalk** virus, and the very common **rotavirus**.

1) **Caliciviridae** primarily infects young children and infants. The gastroenteritis is indistinguishable from that of rotavirus, including diarrhea, vomiting, and fever.

2) **Norwalk virus** can occur in adults, but the virus is named after an outbreak in a Norwalk, Ohio,

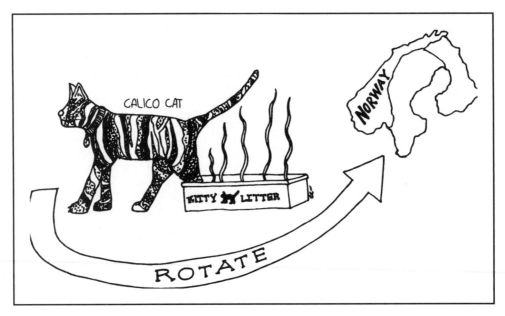

Figure 29-6

elementary school. In that episode, 50% of students developed diarrhea and severe vomiting.

3) **Norovirus** (or Norwalk-like virus) is a major cause of acute infectious diarrheal outbreaks such as on cruise ships or in the aftermath of Hurricane Katrina in 2005. Of 24,000 evacuees housed in the Louisana Superdome 1,619 reported symptoms of acute gastroenteritis. Norovirus was confirmed in 50% of specimens analyzed.

4) **Rotavirus** is a member of the **reovirus** family (Respiratory Enteric Orphan). It is one of the leading causes of acute infectious diarrhea and a major cause of infant mortality worldwide. A live attenuated rotavirus vaccine, **RotaShield**, was launched in the U.S. but withdrawn due to an association with intussusception. **Intussusception** is a condition where the small intestine collapses inward on itself like the closing of an old fashion telescope or a rolled up poster (the upper intestine) sliding into a poster tube (the lower intestine). Two new rotavirus vaccines have recently come to market, they are FDA approved, and are currently part of the CDC-recommended childhood vaccination schedule. They are **RotaTeq**, a human-bovine re-assortment vaccine, and **Rotarix**, a live attenuated vaccine. Both are given orally and neither has been associated with intussusception.

5) **Astroviruses** cause periodic outbreaks of diarrhea in infants, children and the elderly.

Treatment for all of these is supportive, with IV fluids and electrolytes. **Oral rehydration therapy** has revolutionized the treatment of diarrhea in underdeveloped nations, where IV fluids are a rare commodity.

RHABDOVIRIDAE AND RABIES

Fig. 29-7. Rhabdoviruses have bullet-shaped, enveloped, helical symmetry nucleocapsids.

Rabies virus is the only virus in this family that normally infects humans. The rabies virus can infect all warm-blooded animals, with dogs, cats, skunks, coyotes, foxes, raccoons, and bats serving as reservoirs. The infected creatures develop encephalitis and can become fearless, aggressive, and disoriented. The famous stories of the mad farm dog or the wild wolf that stumbles fearlessly into an Alaskan town have popularized this conception.

Fig. 29-8. When a human is bitten, the virus replicates locally at the wound site for a few days, then migrates (slowly over weeks to a year) up nerve axons to the central nervous system, causing a fatal encephalitis.

Fig. 29-9. Brain cells in rabies demonstrate neuropathic changes and pathognomonic collections of virions in the cytoplasm called **Negri bodies**.

Following the bite of a rabid animal there is an **incubation period** that has tremendous variability, ranging from a week to years! Once symptoms develop, there is rapid progression to death over 1–2 weeks:

1) **Prodrome**: Infected persons first develop nonspecific symptoms of fever, headache, sore throat, fatigue, nausea, and painfully sensitive nerves around the healed wound site. The muscles around the site may even fasciculate!

2) **Acute encephalitis**: Hyperactivity and agitation lead to confusion, meningismus, and even seizures. **Madness**!!

Figure 29-7

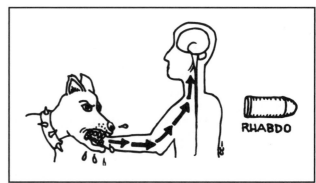

Figure 29-8

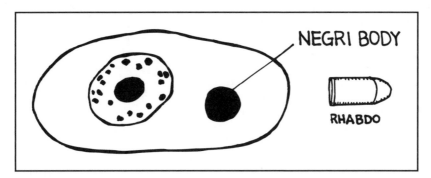

Figure 29-9

3) **Classic brainstem encephalitis**: The brainstem infection causes cranial nerve dysfunction and painful contractions of pharyngeal muscles with swallowing liquids ("hydrophobia"). This results in an inability to swallow saliva and "foaming of the mouth."

4) **Death** ultimately occurs secondary to respiratory center dysfunction. Rabies has the highest case fatality ratio of any infectious disease. The first confirmed recovery from an active rabies infection occurred in early 2005. The case involving a 15-year-old girl bitten by a bat received national attention (Willoughby, et al. NEJM 2005;352:2508). They treated her by inducing a coma with ketamine and midazolam while they waited for her native immune response to mature; rabies vaccine was not administered and she was also treated with the antiviral medications ribavirin and amantadine (see **Chapter 30**). She was discharged to her home after 76 days and remained alert and communicative, but with choreoathetosis, dysarthria, and an unsteady gait. In 2011, there was a confirmed second recovery from active rabies involving an 8-year-old girl in California. A similar treatment protocol was followed.

Through effective control and treatment strategies, the number of actual cases of rabies has been dramatically reduced.

1) Vaccination of dogs and cats has been very effective in the U.S., with only 44 cases reported from 1995 to 2009 (www.CDC.gov/rabies).

2) When a person has been bitten or an open wound licked by a possibly rabid animal, the wound should be aggressively cleaned with soap and water. Washing alone will significantly lower the risk of infection.

3) The animal should be captured or destroyed. Captured animals are confined, and if no symptoms develop in the animal within 10 days, the animal does not have rabies. If destroyed, the dead animal's brain can be examined for Negri bodies or tested for uptake of fluorescently labeled antibodies to rabies virus.

4) If the animal cannot be captured or the above tests are positive, the bitten individual should receive human rabies immune globulin (passive immunization), followed by 5 injections of the killed rabies virus vaccine (active immunization). The idea is to develop immunity while the virus is still in the prolonged (variable length) incubation period.

Notice the similarities in the treatment of rabies with that given to a person presenting with tetanus (see Chapter 6).

VIRUSES THAT CAUSE HEMORRHAGIC FEVER

1) *Filoviridae*: Ebola and Marburg Viruses

EBOLA VIRUS

Although all previous outbreaks had been confined to Central Africa, an epidemic struck the West African countries of Sierra Leone, Guinea, and Liberia in 2014. The initial case is believed to be a two-year old child who had fever, vomiting, and bloody stools in Guinea in late 2013. Although these three countries were hardest hit, isolated cases were also treated in Nigeria, Senegal, Mali, the United States, and Europe. The impact of Ebola on these West African nations, as well as the rest of the world was immense. As of May 2015, the number of cases attributed to Ebola was over 27,000 with more than 11,000 deaths. The epidemic appears to be subsiding, although new cases continue to be diagnosed in Sierra Leone.

The presentation of Ebola infection can be particularly gruesome. Initial symptoms include the abrupt onset of fever, chills and malaise. Nausea, vomiting, diarrhea and abdominal pain usually develop within the first few days. A rash may develop by day five to seven of the illness. Although not universal, about twenty percent of patients can develop hemorrhage, usually in the GI tract (Hemorrhage is a bad sign and portends a poor prognosis). Other symptoms include confusion, seizures, chest pain, and conjunctivitis. Patients who survive typically begin to improve during the second week of illness.

Diagnosis of Ebola is made by the detection of viral antigens or RNA in the blood or other body fluids. PCR for viral RNA is generally detectable within 3 days of symptom onset.

Filo, in filoviridae, means "filament" in Latin and describes the filamentous shape of the RNA viruses Ebola and Marburg that comprise the filovirus family. They are responsible for rare outbreaks of viral hemorrhagic fever that until recently had been confined to sub-Saharan Africa (Zaire, Sudan, Uganda, Kenya) or in the U.S. or Europe following contact with monkeys from these sub-Saharan African areas. Humans and monkeys are infected during outbreaks but it is not known what organism serves as reservoir between epidemics. Serologic studies have demonstrated a 17% seropositivity for Ebola in selected central African populations. This is highest in hunter-gatherers such as the Aka Pygmies (37.5% seropositivity) who handle freshly killed animals. (Johnson, 1993). It is still unclear what sets off the infrequent and deadly epidemics of viral hemorrhagic fever.

Both Ebola and Marburg are known to have been weaponized.

Transmission and Control

In various African outbreaks the most frequently infected groups were health care workers, home care givers, and family members (especially spouses). **Most were in direct contact with bodily fluids.** *However there are cases of nosocomial infections of health*

FAMILY	GENUS/SPECIES	MORPHOLOGY
ARBOVIRUSES (*AR*thropod-*BO*me)		
TOGAviridae	Alpha virus: 1. Western equine encephalitis (WEE) 2. Eastern equine encephalitis (EEE) 3. Venezuelan equine encephalitis (VEE) 4. Chikungunya virus	1. Positive (+) single-stranded RNA 2. Nonsegmented 3. Icosahedral symmetry 4. Replicates in the cytoplasm 5. Enveloped 6. Vector = mosquito for WEE, EEE, VEE, and Chikungunya 7. No insect vector for Rubivirus, so not an arbovirus. Spread by respiratory secretions
	Rubivirus	
*FLAVI*viridae	1. Yellow fever virus 2. Dengue virus 3. St. Louis encephalitis 4. Japanese encephalitis	1. Positive (+) single-stranded RNA 2. Nonsegmented RNA 3. Icosahedral symmetry 4. Replicates in the cytoplasm

Figure 29-10   THE REST OF THE RNA VIRUSES (chart continued)

care workers in which airborne transmission could not be ruled out.

Direct contact with blood, vomitus, urine, stool, or semen, from the living or dead patient, appears to be the most important route of transmission. This likely occurs via skin or mucous membrane contact with the virus-infected body fluids. Reuse of unsterile needles was significant in the Kikwit, Zaire epidemic.

Current CDC guidelines recommend strict contact and face shield protection, as well as *droplet* precautions (surgical mask when within 2 meters). Some African outbreaks have been controlled with these measures alone. But because of the possible cases of airborne spread, *airborne* precautions (N95 respirator masks and patient isolation in negative airflow rooms) may be more prudent. This is recommended by the Johns Hopkins Working Group on Civilian Biodefense.

Epidemics have been controlled by barrier precautions to avoid contact with infected body fluids, use of sterile needles, limiting laboratory blood work, and proper disposal of corpses (sealed in leakproof material and cremated or buried in sealed casket). Laundry and equipment must be incinerated, autoclaved, or washed with bleach.

There is no proven therapy for Ebola or Marburg infection. Therapy is largely supportive with an emphasis on fluid and electrolyte replacement and treatment of secondary infections. Several novel therapies were employed during the most recent West African Ebola epidemic, including convalescent plasma and whole blood donated from Ebola survivors, repurposed antiviral agents, monoclonal antibodies (ZMapp), and short interfering RNA molecules (TKM–Ebola). Studies are ongoing to evaluate their potential role and effectiveness.

2) *Arenaviridae*

This family of viruses includes **Lassa Fever virus** and the 4 South American Hemorrhagic Fever Viruses: Argentine (**Junin virus**), Bolivian (**Machupo virus**), Venezualan (**Guanarito virus**) and Brazilian (**Sabia virus**). This group produces a disease that is slower in onset, but has similar manifestations as the filoviruses once it is full blown. The means of transmission is primarily direct contact and droplet spread, but airborne spread cannot be ruled out. In contrast to the filoviruses, these viruses are treatable. High dose, intravenous ribavirin reduces mortality.

3) **Rift Valley Fever (family *bunyaviridae*)** and Yellow Fever virus (family *flaviviridae*) have been mentioned earlier, and also cause hemorrhagic fever. They are suspected to have been weaponized. Neither is transmissible person-to-person, although infections of laboratory workers have occurred from aerosolization of specimens. Rift Valley fever is treatable with high dose intravenous ribavirin, and yellow fever is not.

Fig. 29-10.   Summary of the "Rest of the RNA Viruses."

CLINICAL FINDINGS	TREATMENT	MISCELLANEOUS

1. WEE, EEE, VEE symptoms:
  A. Headache and fever
  B. Altered level of consciousness
  C. Focal neurologic deficits
2. Chikungunya symptoms:
  1. Initially: fever, rash, joint pain
  2. Chronic arthritis.		

• Rubella: (German measles/3 day measles)
  A. Fever, lymphadenopathy and mild flu-like symptoms
  B. Rash: from forehead to face to torso to extremities (lasts 3 days)
  C. Congenital defects: occurs when a women in her first trimester of pregnancy gets exposed. The fetus may develop defects of the heart, eyes, or central nervous system
	• Prevention: MMR vaccine (live attenuated)
	  A. Measles
	  B. Mumps
	  C. RUBELLA
		• The "R" in TORCHES

1. **Yellow fever**:
  A. Hepatitis (with jaundice)
  B. Fever
  C. Backache
	• Prevention: mosquito control	1. Diagnosis:
		  A. Viral culture
		  B. Serology

FAMILY	GENUS/SPECIES	MORPHOLOGY

*FLAVI*viridae (continued)	5. Hepatitis C virus
	6. West Nile virus
	7. Zika virus
		5. Enveloped
		6. Vector = mosquito
		  • *Aedes*: yellow fever, dengue, and Zika
		  • *Culex*: St. Louis, Japanese, and
		    West Nile encephalitis

*BUNYA*viridae	1. California encephalitis virus 2. Rift valley fever virus 3. Sandfly fever virus Hantavirus	1. Negative (−) single-stranded RNA 2. Segmented (3 segments) 3. Helical symmetry 4. Arthropod vector (except for Hantavirus)
	PICORNA VIRUSES	
*ENTERO*viridae	Poliovirus Coxsackie A Coxsackie B ECHOviruses Enteroviruses Rhinovirus 113 serotypes	1. Positive (+) single-stranded RNA 2. Naked icosahedral symmetry 3. Replication occurs in the cytoplasm

Figure 29-10   (continued)

CLINICAL FINDINGS	TREATMENT	MISCELLANEOUS

2. **Dengue fever**: "Break bone fever"
   A. "Painful fever": high fever along with
     1. Headaches
     2. Muscle aches
     3. Joint aches
     4. Backache
   B. Dengue hemorrhagic fever: hemorrhage, thrombocytopenia and septic shock
 3. St. Louis, Japanese, West Nile encephalitis: encephalitis and fever
 4. Hepatitis C virus
 5. West Nile: fever, meningitis, encephalitis or myelitis that produces flaccid paralysis.
 6. Zika: Fever, rash, joint pain, and conjunctivitis (red eyes)
	• Vaccination required when traveling to and from endemic countries	2. With repeat infections, individuals are at higher risk of developing the **hemorrhagic** form of dengue fever
		 3. West Nile: serology is much more sensitive than PCR, although because of cross-reactions with other flaviviruses, it is less specific.
		 4. Zika: Association with with microcephaly in infants born to infected mothers

• Most cause fever and encephalitis • Rift valley fever can also cause hemorrhagic manifestations.		• Rift Valley fever affected more than 1 million Egyptians in the 1970's. It killed more than 300 Kenyans in Winter 1998.

1. Hemorrhagic fever with renal syndrome: in Europe and Asia
 2. Hantavirus pulmonary syndrome: in New Mexico, Arizona, Colorado and Utah
   A. Fever, muscle aches, cough, nausea and vomiting
   B. Pulmonary edema, leading to respiratory failure
   C. 40% mortality
	• Investigational: extracorporeal membrane oxygenation	• Report suspected cases to the CDC

1. **Mild febrile illness**: often occurs in infants in less developed nations
 2. **Aseptic meningitis**: fever and stiff neck. Recovery in one week
 3. **Paralytic poliomyelitis**: get flaccid paralysis due to necrosis of the large motor neurons in the anterior horn of the spinal cord
	• Vaccine:
	 1. Salk vaccine: formalin-killed polio virus that is injected subcutaneously
	 2. Oral polio vaccine (developed by Sabin): attenuated (non-virulent) polio virus is ingested
		1. *Transmission*:
		   A. Fecal-oral
		   B. Respiratory secretions
		 2. The chance of developing paralytic poliomyelitis increases as one gets older

1. "Cold" rashes, viral meningitis 2. **Herpangina**: fever, sore throat and small red-based vesicles over the back of the patient's throat 3. **Hand Foot and Mouth Disease**: occurs in children less than 5, Vesicles erupt on hands, foot and mouth, which are highly contagious		• Note: The PICORNA viruses are the smallest RNA viruses
1. Viral meningitis 2. Myocarditis/pericarditis: arrhythmia, cardiomyopathy, heart failure 3. Pleurodynia: fever and sharp, pleuritic chest pain		
1. "Cold", rashes, viral meningitis 2. Pericarditis		
1. Respiratory infections (often severe with enterovirus D68) 2. "Polio-like" acute flaccid paralysis (associated with E71 and possibly D68)		
• Common cold		

FAMILY	GENUS/SPECIES	MORPHOLOGY
CALCI, REO, CORONA, RHABDO, FILO & ARENA viruses		
Astroviridae	Astrovirus	1. Positive (+) single-stranded RNA 2. Star-like morphology
CALICNiridae	Norwalk virus and other related Caliciviruses	1. Positive (+) single-stranded RNA 2. Naked icosahedral symmetry 3. Replication occurs in the cytoplasm 4. Fecal-oral transmission
REOviridae	Rota virus	1. Double stranded RNA 2. Segmented (11 segments) 3. Naked icosahedral symmetry 4. Fecal-oral transmission
CORONAviridae		1. Positive (+) single-stranded RNA 2. Nonsegmented 3. Helical symmetry. 4. Enveloped 5. Replication in the cytoplasm
RHABDOviridae	Rabies virus	1. Bullet shaped 2. Negative (−) single-stranded RNA 3. Nonsegmented 4. Helical nucleocapsid is coiled into a bullet shape 5. Replication in the cytoplasm 6. Zoonotic (all warm blooded animals): dogs, cats, skunks, coyotes, foxes, raccoons, and bats are reservoirs in the U.S. 7. Transmitted via an animal bite
Filoviridae	1. Marburg virus 2. Ebola virus	1. Negative (−) single-stranded RNA 2. Nonsegmented 3. Helical symmetry 4. Enveloped 5. Replication in the cytoplasm

Figure 29-10   (continued)

CLINICAL FINDINGS	TREATMENT	MISCELLANEOUS
Diarrheal illness	• Intravenous fluids and supportive care	
• **Viral gastroenteritis**: (explosive, but self-limited): A. Fever B. Abdominal pain C. Vomiting D. Diarrhea (no blood, no pus)	• Intravenous fluids	1. Infants die secondary to loss of fluids and electrolytes 2. Note: Hepatitis E is probably a species of Caliciviruses
• **Viral gastroenteritis**: causes profound dehydration, especially in infants. Fever, abdominal pain, vomiting and diarrhea • No blood, No pus in diarrhea • This is a major cause of infant death in under-developed countries and the most common cause of diarrhea in infants less than 3 years of age	1. Intravenous fluids 2. New oral rotavirus vaccine appears safe & effective in infants	• Responsible for almost 50% of infant diarrhea cases that require hospitalization in the United States
• Upper respiratory tract infection ("common cold")		
RABIES • Incubation can be from 2 weeks to a year 1. Prodrome: fever, headache, sore throat and very sensitive nerves around the healed wound site 2. Acute encephalitis: hyperactivity and agitation leading to confusion and seizures 3. Classic brain stem encephalitis: A. Cranial nerve dysfunction B. Painful contraction of pharyngeal muscles when swallowing liquids, resulting in hydrophobia and "foaming of the mouth" 4. Death: due to respiratory center dysfunction	1. Vaccination of animals 2. If bitten by a possibly rabid animal: 3 possibilities: A. Capture animal: observe for 10 days B. Destroy animal: examine brain for Negri bodies C. Treat immediately (if you can not capture the animal, or the animal is found to have rabies): 1. Clean wound 2. Passive immunization with rabies immune globulin 3. Active immunization with killed rabies virus vaccine	1. Diagnosis: Microscopic examination of the central nervous system reveals Negri bodies. These are collections of virions in the cytoplasm where replication occurs 2. Note: Spread of this virus though the peripheral nerves to the central nervous system is very slow 3. This is one of the only diseases where you can get vaccinated *after exposure*
• Acute Viral hemorrhagic fever: high mortality rate (50%–90%) 1. Transmission secondary to contact with infected body fluids: contaminated medical instruments, and close contact with sick or dead patients and their	1. Therapy is largely supportive 2. Experimental therapies: a. Convalescent plasma/blood b. ZMapp- monoclonal antibodies versus Ebola c. antivirals- Favipiravir, Brincidofovir	Diagnosis: PCR, ELISA

FAMILY	GENUS/SPECIES	MORPHOLOGY
*Filo*viridae (continued)		6. Humans and monkeys in Sub-Saharan Africa are infected in rare epidemics 7. Unknown Reservoir
*Arena*viridae	1. Lymphocytic choriomeningitis virus (LCM) 2. Lassa virus 3. South American hemorrhagic fever viruses	1. Negative (−) single-stranded RNA 2. Segmented (2 segments) 3. Helical symmetry 4. Enveloped 5. Replication in the cytoplasm 6. Zoonotic: responsible for asymptomatic infections in rodents 7. Spread of infection: Contact with rodent urine

Figure 29-10   (continued)

References

CDC. Outbreak of Ebola viral hemorrhagic fever – Zaire 1995. MMWR 1995;44(19)381–382.

CDC. Update: outbreak of Ebola viral hemorrhagic fever – Zaire, 1995. MMWR 1995;44(25):468–470.

CDC. Update: management of patients with suspected viral hemorrhagic fever – United States. MMWR 1995; 44(25):475–479.

Johnson ED, Gonzalez JP, Georges A. Filovirus activity among selected ethnic groups inhabiting the tropical forest of equatorial Africa. Transactions of the Royal Society of Tropical Medicine and Hygiene 1993;87:536–538.

Norovirus outbreak among evacuees from Hurricane Katrina—Houston, Texas, September 2005. MMWR. 2005;54:1016–1018.

Peters CJ. Marburg and Ebola virus hemorrhagic fevers. In: Mandell GL, Bennett JE, Dolin R, eds. Principles and Practice of Infectious Diseases; 4th edition. New York: Churchill Livingstone, 1995;1543–1546.

Willoughby, RE, Tieves KS, Hoffman GM, et al. Survival after treatment of rabies with induction of coma. NEJM. 2005; 352:2508–2514.

Recommended Review Articles:

Gould EA, Solomon T. Pathogenic flaviviruses. Lancet. 2008; 371(9611):500–9.

Howard RS. Poliomyelitis and the postpolio syndrome. BMJ. 2005;330:1318–1319.

Rossi SL, Ross TM, Evans JD. West Nile virus. Clin Lab Med. 2010;30(1):47–65.

Simpson SQ, Spikes L, et al. Hantavirus pulmonary syndrome. Infect Dis Clin North Am. 2010;24(1):159–73.

Wilder-Smith A, Schwartz E. Dengue in travelers. N Engl J Med. 2005 Sep 1;353(9):924–32.

CLINICAL FINDINGS	TREATMENT	MISCELLANEOUS
body fluids. Most likely mechanism is via skin or mucous membrane contact with virus-infected body fluids (blood, vomit, diarrhea, semen). Rare airborne transmission cannot be ruled out. 2. A 2 to 21 day incubation period followed by *abrupt* onset of fever, headache, and myalgia, abdominal pain, diarrhea, pharyngitis, hiccups, cough, and somnolence may develop. Progression to bleeding from needle stick sites and all mucous membranes follows. Death results from multi-organ failure.		
1. Lymphocytic choriomeningitis: Influenza-like illness, sometimes associated with a viral meningitis. Occasionally fatal 2. Lassa fever: fever, sore throat, abdominal pain, with intractable vomiting and hypotension. Fatal in up to half of cases 3. Gradual onset of fever, myalgias, nausea, abdominal pain, conjunctivitis, sometimes generalized lymphadenopathy. Later, bleeding/ petechiae. Sometimes tremor, seizures, diarrhea.	1. LCM: supportive treatment 2. Lassa fever: ribavirin 3. South American hemorrhagic fever viruses—high dose intravenous ribavirin. Locally, in South America, immune plasma is used successfully.	• Diagnosis: by examining the blood for a rise in titer of virus-specific antibodies

M. Gladwin, W. Trattler, and S. Mahan, *Clinical Microbiology Made Ridiculously Simple* ©MedMaster

CHAPTER 30.   ANTI-VIRAL MEDICATIONS

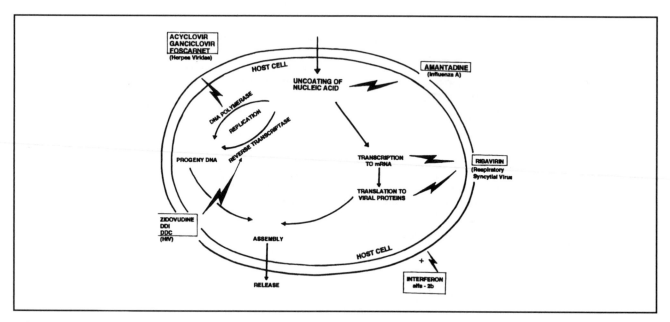

Figure 30-1

The virus is a tough creature to kill. It has NO peptidoglycan wall, NO ribosomes, and NO cell membrane. All it has is a protein coat, nucleic acid strand, and a few simple enzymes. The only thing these critters do is replicate and then hang out in a latent state. The current antiviral agents attack steps in viral replication much like the chemotherapeutic agents attack replicating tumor cells.

Fig. 30-1.   Site of action of antiviral drugs.
Two important concepts:

1) These drugs attack steps in actively replicating viruses and have no effect on latent viruses. For this reason these drugs are only **virustatic** (not **virucidal**).

2) Most of these drugs are nucleotide analogues. They look just like the viral nucleotides but do not function appropriately. As such, they are taken up and used by viral DNA polymerase or reverse transcriptase and are like monkey wrenches thrown into the gears of replication. They inhibit the DNA polymerase or reverse transcriptase and are also incorporated into the growing DNA strand, resulting in chain termination.

ANTI-HERPESVIRIDAE DRUGS

Both **acyclovir** and **ganciclovir** are guanine analogues that act against the herpes family. There is a key difference between them. To become active, acyclovir must first be phosphorylated by a virus-specific thymidine kinase. Most of the herpesviridae have this enzyme while human cells do not. For this reason acyclovir is active only against herpesviridae and has limited toxicity to our cells. One of the herpesviridae, cytomegalovirus (CMV), lacks thymidine kinase and so acyclovir is less active for CMV infections. On the other hand, ganciclovir is not dependent on a virus-specific thymidine kinase for phosphorylation. It kills ALL the herpesviridae including CMV. It is also toxic to some rapidly replicating human cells such as neutrophils and platelets (causes neutropenia and thrombocytopenia).

Acyclovir
("A cycle")

Fig. 30-2.   To remember that **ACYCLovir** is used to treat infections caused by the herpes family, visualize **A CYCLE** traversing the heights of a huge herpes cold sore.

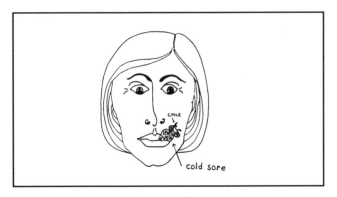

Figure 30-2

Clinical Uses

Studies demonstrate that if acyclovir is given very early, it reduces the severity and duration of all herpes simplex and varicella-zoster (V-Z) infections, such as cold sores (mucocutaneous herpes simplex infections), varicella (chickenpox), and zoster (shingles). However, because these infections are self-limiting and mild, acyclovir is currently not recommended for these diseases.

In the immunocompetent host, it is reserved for more serious infections such as herpes simplex encephalitis and herpes simplex and zoster infections of the eye. It is also approved for herpes simplex genital infections.

In the immunocompromised host, it is used for most herpes infections (mucocutaneous, varicella, and zoster).

Acyclovir is not used for CMV or Epstein-Barr virus infections.

Fig. 30-3. Adverse effects are minimal. With high intravenous doses acyclovir (A CYCLE) may crystallize in the renal tubules, resulting in reversible renal toxicity. About 1% of patients have CNS side effects, such as confusion or seizures.

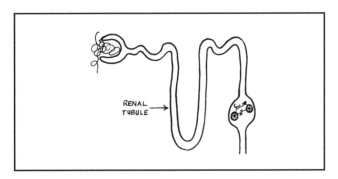

Figure 30-3

Famciclovir and Valacyclovir

These 2 drugs have the same mechanism of action as acyclovir but have the added punch of increased drug levels after oral absorption. One study (Tyring, 1995) compared famciclovir with placebo in the treatment of adults with herpes zoster and found that it reduced the time to lesion healing, viral shedding, and, more importantly, the duration of post-herpetic neuralgia by almost 2 months!

These drugs are currently indicated for herpes zoster and recurrent genital herpes in immunocompetent adults. Adverse effects are mild and include headache, nausea, diarrhea, and dizziness.

Ganciclovir and Valganciclovir
("Gang of cycles")

Ganciclovir, an analogue of acyclovir, has similar activity to acyclovir, except that it is much more active against CMV. Valganciclovir is the L-valyl ester of ganciclovir and has much better oral bioavailablity.

Fig. 30-4. A **GANG** OF **CYCLES** running over HSV, V-Z-V, CMV, EBV, and Mr. Neutrophil and Mrs. Platelet. **GANCICL**ovir has broader coverage of the herpesviridae than acyclovir, but it is more toxic.

Clinical Uses

Because of their toxicities ganciclovir and valganciclovir are used primarily for CMV infections in immunocompromised hosts:

1) **AIDS patients**: CMV retinitis, pneumonitis, esophagitis.
2) **Bone marrow transplant patients**: CMV pneumonitis and prophylaxis against CMV infection.

Adverse effects include reversible neutropenia and thrombocytopenia.

Foscarnet

This pyrophosphate analogue inhibits DNA polymerase **and reverse transcriptase**. It has extended anti-viral activity, covering the herpesviridae and HIV. It is important to stress that this anti-viral activity against HIV is very minimal and not adequate for treatment or viral suppression.

Foscarnet is used for AIDS patients with:

1) **CMV retinitis**.
2) Acyclovir-resistant strains of herpesviridae.

A big side effect, especially in AIDS patients, is reversible nephrotoxicity. Increased seizure potential is possible in patients with prior history of seizure, head

Figure 30-4

trauma, renal impairment or taking concomitant medications that increase seizure potential.

Cidofovir

Cidofovir is a phosphonate nucleoside analogue of cytosine. It is one of the new kids on the block. Its major use is in CMV infections, primarily retinitis. It has broad activity against all herpes viruses as well as several other DNA viruses such as the poxviruses, including smallpox. It is used primarily in patients who have failed treatment of CMV infections with ganciclovir and foscarnet. Its main toxicity is kidney damage.

TOPICAL AGENTS FOR COLD SORES

There are two FDA-approved topical agents for cold sores, penciclovir cream and docosanol. Both are underwhelming in their benefits. **Penciclovir**, which has a similar mechanism of action as acyclovir, has been shown in at least one trial to decrease time to healing by about 1 day and to decrease time to pain relief by ½ day. **Docosanol**, which blocks viral entry, is available over-the-counter and makes similar claims, although not well substantiated.

THE HUMAN IMMUNODEFICIENCY VIRUS (HIV)

There has been an explosive development of new antiretroviral medications. "Have a **HAART** (**H**ighly **A**ctive **A**nti**R**etroviral **T**herapy)" is the foremost theme for those physicians caring for HIV positive patients. HAART refers to the use of several very potent anti-HIV (aka antiretroviral) agents in combination to suppress viral replication and stop the spread of resistant viruses. There are at least 28 different anti-HIV medications in the United States: seven nucleoside reverse transcriptase inhibitors [zidovudine (AZT, ZDV), didanosine (ddI), zalcitabine (ddC), stavudine (d4T), lamivudine (3TC), emtricitabine

(FTC) and abacavir (ABC)]; one nucleotide reverse transcriptase inhibitor (tenofovir); five non-nucleoside reverse transcriptase inhibitors (nevirapine, delaviradine, efavirenz, rilpivirine and etravirine); ten protease inhibitors (saquinavir, indinavir, ritonavir, nelfinavir, amprenavir, fosamprenavir, atazanavir, tipranavir, lopinavir/ritonavir and darunavir); one entry inhibitor (enfuvirtide), one CCR5 inhibitor (maraviroc); and three integrase inhibitors (raltegravir, elvitegravir, and dolutegravir).

Before representing each of these drugs, let's focus on the big picture of how to use them.

1. Antiretroviral therapy should be started in all patients with HIV infection regardless of CD4 count or HIV viral burden.

2. Three or four drugs should be used because the data shows these combinations are more effective and prevent emergence of resistance. Choice of agents can be tailored to avoid side effects.

The classic principle behind HAART is the use of several different agents with varying mechanisms of antiviral activity and patterns of resistance. A three drug combination of two nucleoside reverse transcriptase inhibitors combined with either a protease inhibitor, a non-nucleoside analog, or an integrase inhibitor. Tailoring anti-HIV medications requires patience while finding the right mix of efficacy and tolerability. Several of the above agents have been co-formulated to lower overall pill burden and thereby increase compliance. Specific examples of these combinations are shown below.

Preferred initial drug regimens:
abacavir + lamivudine + dolutegravir
(tenofovir + emtricitabine)* + doltegravir
(tenofovir + emtricitabine)* + elvitegravir/cobicistat
(tenofovir + emtricitabine)* + raltegravir
(tenofovir + emtricitabine)* + darunavir/ritonavir
*Truvada

3. Physicians must follow CD4 T-lymphocyte counts, viral load assays, and the patient's clinical status to

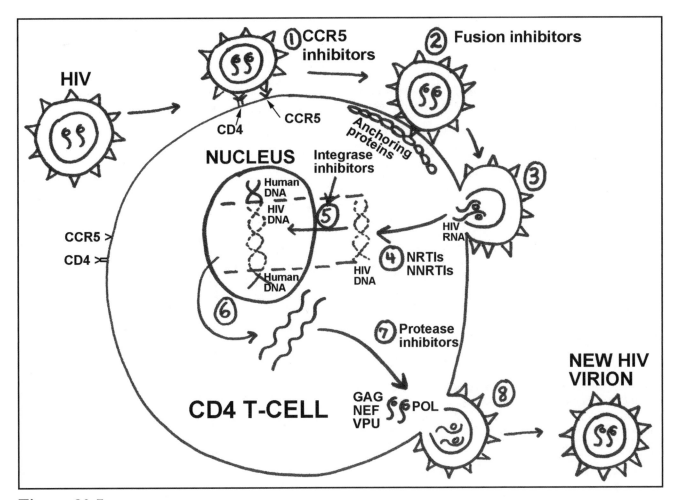

Figure 30-5

determine if treatments are effective. If CD4 counts drop, viral load increases, or opportunistic diseases develop, therapy should be changed. If side effects develop, drugs should likewise be changed.

There will be many more trials in the following years with the goal of completely suppressing viral load while minimizing pill burden. HIV may become a virus we harbor in a suppressed state and AIDS may be prevented indefinitely. A major hurdle in the battle against HIV has been the provision of affordable HIV medications, especially in resource poor settings. Fortunately, large donor programs and government assistance have brought the price of medications down and increased access to treatment across the globe.

The **HIV Life Cycle** and the steps at which the HIV anti-virals attack: 1) **HIV binds** to a CD4 receptor and one of two key co-receptors (either CCR5 or CXCR4). *CCR5 inhibitors* block HIV binding to the CCR5 co-receptor. 2) **Fusion**: HIV fuses to anchor proteins in the host cell

membrane. *Fusion inhibitors* block this process. 3) **Infection** occurs as the HIV virion empties its contents into the cell. 4) **Reverse transcription**: HIV uses its reverse transcriptase enzyme to convert single stranded RNA into double-stranded DNA. *Nucleoside/nucleotide reverse transcriptase inhibitors and non-nucleoside reverse transcriptase inhibitors* block this process. 5) **Integration**: HIV then uses an integrase enzyme to insert the viral DNA into the host DNA. Here is where *integrase inhibitors* work. 6) **Transcription**: a host enzyme called the RNA polymerase creates copies of the HIV genes as well as shorter strands of messenger RNA (mRNA). The mRNA is then used as the blueprint for long chains of HIV proteins. 7) **Assembly**: Proteases then cleave these long chains of HIV proteins into the smaller HIV proteins and they are packaged together to form a new HIV virus. You guessed it...*protease inhibitors* block this step! 8) **Budding**: In this final step, the new virus "buds" out of the host cell and takes part of the host cell's outer membrane, which is now covered with HIV glycoproteins.

NUCLEOSIDE/NUCLEOTIDE REVERSE TRANSCRIPTASE INHIBITORS (NRTIs)

Zidovudine (ZDV or AZT)

This was the first drug approved for HIV treatment, in 1987. Large studies have shown that zidovudine:

1) **Reduces mortality and opportunistic infections** in symptomatic HIV-infected patients with CD4 T-lymphocyte counts less than 200/mm³ (Fischl, 1987).

2) **Delays progression to AIDS** in HIV infected patients with CD4 T-lymphocyte counts less than 500/mm³ (Volberding, 1990; Fischl, 1990).

The problem with zidovudine is that HIV can rapidly develop resistance to zidovudine when it is used alone. This is the rationale for always starting with 2 agents.

3) **Reduces maternal-to-infant transmission of HIV** when given to the mother orally prior to birth, intravenously during delivery, and then to the baby orally for 6 weeks. In one study, this regimen reduced the transmission rate from 25% to 8% (Conner, 1994)!

Fig. 30-6. AIDS knocks out CD4 T-lymphocytes, and AZT (ZDV) knocks out red blood cells (anemia) and neutrophils (neutropenia).

It also causes other pesky adverse effects including headache, insomnia, myalgias, nausea, and CNS disturbances (confusion, seizures). If a patient develops these problems, the dose can be decreased or another anti-HIV drug can be used.

Lamivudine (3TC) and Emtricitabine

Lamivudine and its newer generation formulation, emtricitabine, are both well tolerated with no dose lim-

Figure 30-6

iting side-effects. Lamivudine is felt to be the backbone of most combination therapy regimens. When mutations to these agents do arise the resultant HIV strain may actually have enhanced susceptibility to AZT. In addition to anti-HIV activity these agents both suppress Hepatitis B viral DNA replication. This additional activity may be a plus when treating patients with HIV/Hepatitis B co-infection.

Didanosine (ddl)

This is a synthetic purine nucleoside analogue that is unstable in acid conditions, such as the gastric environment. Therefore, it is formulated with a buffer or antacids, and should be taken on an empty stomach.

Didanosine can cause pancreatitis which may be life threatening, in which case the drug should be discontinued. Other risk factors for pancreatitis, such as history of pancreatitis, alcoholism, and hypertriglyceridemia, may increase the likelihood of developing pancreatitis. Due to its multitude of side-effects ddI has largely fallen out of favor in first-line treatment regimens.

Zalcitabine (ddC)

Unlike didanosine, zalcitabine is well absorbed from the gastrointestinal tract. Pancreatitis occurs less commonly than with didanosine. Severe oral ulcers have been reported in up to 3% of zalcitabine-treated patients. Zalcitabine is no longer commonly used due to alternative agents that are felt to have superior efficacy.

Stavudine (d4T)

Mild increases of hepatic transaminases have also been noted during treatment with stavudine. Although a commonly used second-line agent, stavudine has the highest associated incidence of lipoatrophy, hyperlipidemia, and lactic acidosis of all NRTIs.

Abacavir

Hypersensitivity reactions have been the most concerning adverse effect, reported in approximately 5% of patients receiving abacavir. A rash is accompanied by systemic signs and symptoms such as fever, fatigue, nausea, vomiting, diarrhea or abdominal pain. These symptoms occur early and usually appear within the first 6 six weeks of treatment. Symptoms usually resolve rapidly after discontinuation of the drug. It is important to remember that once abacavir has been discontinued because of a hypersensitivity reaction, it should not be reintroduced. More severe outcomes, including death, have been reported to occur when abacavir was reinstituted. More recently, an association between the presence of the MHC class I allele HLA-B*5701 and abacavir hypersensitivity has led to recommendations to screen for this allele prior to beginning treatment with abacavir.

Tenofovir

Tenofovir is a nucleotide analog with similar mechanism of action as the nucleoside analogue. It was first approved for use in 2001, making it one of the newer kids on the block. Tenofovir has been shown to be well tolerated and highly effective, leading to its inclusion in one of the initial preferred combination drug regimens. It is primarily eliminated by the kidneys and renal impairment with hypophosphatemia can occur. It is contraindicated in patients with renal impairment. Tenofovir also has activity against hepatitis B.

Non-specific side effects: All of these agents can cause rash, fatigue, headaches, nausea, vomiting, diarrhea, abdominal pain, and insomnia. Physicians may need to juggle these medications to find the best agent with the least side effects.

Common Side Effects of the NRTI Class

Peripheral neuropathy, lactic acidosis, and fat redistribution (lipodystrophy) are all potential side-effects of the NRTI class of drugs. Peripheral neuropathy is most commonly associated with the "D" drugs: didanosine (**d**dI), zalcitabine (**d**dC), and stavudine (**d**4T). Peripheral neuropathy usually manifests with numbness or tingling of the feet or hands and seems to be dose related. It is generally reversible with discontinuation of these agents.

Remember that the D's (ddI, ddC, d4T) cause Neuropathy!!!

Lactic acidosis, a potentially life-threatening condition, can occur with any of these agents, but is most commonly associated with didanosine and stavudine, particularly if used together. Stavudine is most likely to cause lipodystrophy. One must be aware of these potential side-effects and take them into account when choosing combination therapy options. As you may have noted, the "D" drugs must be used with caution and rarely in combination with one another.

Non-Nucleoside Reverse Transcriptase Inhibitors (NNRTIs)

NNRTIs bind directly and noncompetitively to the enzyme reverse transcriptase. They block DNA polymerase activity by causing conformational change and disrupting the catalytic site of the enzyme. Unlike nucleoside analogues, NNRTIs do not need phosphorylation to become active, and they are not incorporated into viral DNA. When NNRTIs are administered as a single agent or as part of an inadequately suppressive treatment regimen, resistance emerges rapidly. Mutations conferring resistance to one drug in this class generally confer cross-resistance to most other NNRTIs. Cross-resistance to nucleoside analogues or protease inhibitors has not been observed.

Nevirapine

Nevirapine was the first NNRTI to be approved for treating HIV infection. This drug is frequently given to pregnant women in developing countries as a single dose at the time of delivery to prevent mother to infant transmission of HIV. This low cost intervention effectively reduces transmission to the infant by about 50%. Nevirapine is an inducer of the cytochrome P450 CYP3A system, including autoinduction of its own metabolism. Because of this induction potential, nevirapine is avoided in combination with other medications that are metabolized through the CYP450 system, especially antiviral medications.

Nevirapine commonly causes a rash which usually develops within the first six weeks of therapy. Severe rashes including ulcerations and Stevens-Johnson syndrome have been reported. In addition, elevation in liver function tests and occasionally liver failure has been documented (the greatest risk is in women placed on this agent with CD4 counts >250 cells/microliter).

Delavirdine

Delavirdine is metabolized by the cytochrome P450 system and also inhibits CYP450 activity, including its own metabolism. This inhibition may lead to increased plasma levels of concurrent medications metabolized through CYP450.

Delavirdine can cause similar side effects as Nevirapine, including rash and elevation of liver enzymes. This agent is seldom used due to its decreased potency relative to the other NNRTIs.

Efavirenz

Efavirenz, when combined with tenofovir and emtricitabine in the coformulation (Atripla) as a once a day combination pill has been one of the workhorse antivirals in HIV care. Central nervous symptoms have been reported in approximately 50% of patients treated with efavirenz. These symptoms include abnormal dreams that are often dysphoric in nature as well as insomnia, dizziness, impaired concentration, and somnolence. Symptoms tend to diminish with continued therapy but may increase with concomitant alcohol and psychoactive drug use. You may have to stop this drug if your patient starts dreaming about **axe-wielding Elves** (**Efav**irenz). Efavirenz is an inducer of the CYP450 system. This may lead to drug interactions similar to those with delaviridine.

Etravirine

Etravirine is one of the more recently approved NNR-TIs. It remains effective in HIV-1 infected persons with resistance to the other NNRTIs. It too is a potent inducer of CYP450. This agent retains activity against most HIV isolates despite their resistance to other NNRTIs. Rash is a frequent side effect, but rarely severe. Drug interactions are common, and it is not recommended to be used with several PI based regimens.

Rilpivirine

Rilpivirine is an NNRTI approved in 2011. Much like etravirine, rilpivirine retains effectiveness in many HIV infected persons with prior NNRTI resistance. It is also co-formulated with the emtricitabine/tenofovir into a single once daily tablet to increase compliance. Similar to the other drugs in its class (except delavirdine) it is a potent inducer of CYP450. The most common side effects are depression, insomnia, headache, and rash.

PROTEASE INHIBITORS (PIs)

HIV protease is required for production of infectious HIV particles. Protease inhibitors inhibit this vital enzyme. There are currently 10 protease inhibitors: saquinavir, ritonavir, indinavir, nelfinavir, amprenavir, fosamprenavir, lopinavir/ritonavir, atazanavir, tipranavir and the newest agent darunavir.

All PIs are associated with metabolic abnormalities including dyslipidemia, insulin resistance, and lipodystrophy. Lipodystrophy with PIs is associated with a central accumulation of fat (in the abdominal and visceral area), accumulation of fat in the dorso-cervical region creating a "buffalo hump" and loss of extremity fat.

All the PIs have **navir** at the end of their names. Think of the PIs causing a viral **nadir** or **No vir**us.

Saquinavir

Saquinavir was the first PI to be approved by the FDA. The biggest problem with saquinavir has been that a minimal amount of the drug gets absorbed when taken orally.

Fortovase is a soft gel formulation of saquinavir with enhanced bioavailability that has replaced hard gel saquinavir, Invirase. Fortovase should be taken with a meal to increase oral absorption. The main side effects are as you might have guessed, gastrointestinal. These effects include diarrhea, nausea, abdominal discomfort and pain, dyspepsia and vomiting.

Indinavir

Indinavir is the opposite of saquinavir because it is rapidly absorbed in a fasting state. If indinavir is given with a high-fat/high-protein meal, absorption is substantially reduced. Consumption of a light meal (e.g. dry toast and coffee) has minimal effect on absorption. The main side effects again are gastrointestinal including abdominal pain, nausea, vomiting and diarrhea. Nephrolithiasis or "kidney stones" occasionally occur with renal insufficiency or acute renal failure. Therefore caution is needed in patients with compromised kidney function. All patients receiving indinavir should drink at least 1.5L (48 oz.) of water daily to ensure adequate hydration and prevent development of kidney stones.

Ritonavir

Ritonavir is no longer used as a sole PI agent due to its high rate of gastrointestinal side effects at standard doses. Ritonavir is a very potent inhibitor of the cytochrome P450 system. As a result, ritonavir is used primarily to "boost" other PIs. Adding ritonavir to other PIs slows their metabolism, thereby extending their half life, and allows for less frequent dosing and improved drug levels. **"Boosting"** with ritonavir is currently very hot in the treatment of HIV! As you might guess, one must be wary of drug interactions due to ritonavir's inhibition of P450.

Nelfinavir

The most frequently reported side effect associated with nelfinavir is diarrhea, noted in up to 32% of patients. Nelfinavir-associated diarrhea is generally mild to moderate. Other reported side effects include nausea, vomiting, abdominal pain, and rash.

Amprenavir and Fosamprenavir

Amprenavir and its more recently approved prodrug, fosamprenavir, are newer PIs. The most frequently reported side effects are gastrointestinal (nausea, vomiting, diarrhea and abdominal pain); most are graded as mild to moderate. Other reported side effects include rash, parasthesias, and depressive or mood disorders.

Fosamprenavir is generally the preferred agent of these two due to less gastrointestinal side effects and a lower pill burden.

Lopinavir/ritonavir

The co-formulation of lopinavir/ritonavir capitalizes on ritonavir's ability to boost. This agent combined with combivir (zidovudine/lamivudine) is the preferred drug regimen in pregnancy. It is generally well tolerated except for occasional gastrointestinal side effects such as mild diarrhea, nausea, and vomiting.

Atazanavir

Atazanavir is one of the more frequently prescribed PI agents. It appears to have **less impact on lipids** than the other PIs. In addition it can be given **once daily**. One curious side effect is a rather consistent rise in **indirect bilirubin**, but rarely clinical jaundice. This asymptomatic rise in bilirubin can be used by the clinician as way to monitor whether the patient is taking his medication. Atazanavir requires an acidic environment for absorption, and, therefore, proton pump inhibitors, such as omeprazole, are contraindicated in patients on atazanavir.

Tipranavir

A recent addition to the PI class, approved in late 2005. It must be administered with ritonavir for "boosting." It has similar side effects and drug interactions as the other PIs. It is intended to be used as part of combination therapy by patients with resistance to multiple other protease inhibitors.

Darunavir

This is the newest PI on the block. Darunavir has been listed in the preferred drug regimens for anti-retroviral "naïve" patients. "Boosted" darunavir may be taken once daily in treatment naïve patients and should be taken twice daily in treatment experienced patients. It must be taken with food. Side-effects are similar to the other PI's, but due to a slightly different resistance profile from the other PI's, it is often still an option in persons who have developed resistance to the other PI's.

Fusion Inhibitors-Enfuvirtide (T-20)

This is the first approved agent in its class. Enfuvirtide binds HIV along the gp41 envelope protein and prevents fusion of HIV with the target cell membrane. This agent is given as a twice daily subcutaneous injection. Due to its requirement for injection, frequent injection site reactions, and occasional neutropenia it is used primarily as part of a "salvage" regimen. (A **"salvage"** regimen is the last ditch regimen when all the usual first and second line agents have failed).

Interleukin-2 infusion

Interleukin-2 is a cytokine released by T-lymphocytes that regulates the proliferation of CD4 (helper-T) T-lymphocytes. In two recent clinical trials of HIV-infected persons treated with HAART with or without infusions of interleukin-2 (IL-2), persons in the combined treatment arm had significant and sustained increases in their CD4 counts. Unfortunately, the additional of IL-2 yielded no clinical benefit (INSIGHT-ESPRIT Study Group, 2009).

CCR5 INHIBITORS

Maraviroc

Maraviroc is the first approved agent in this exciting new class of anti-retrovirals. It works by binding to the CCR-5 coreceptor which is vital for HIV attachment and entry into the CD4 T cells. Vicriviroc is a second drug in this class which is still undergoing study and has not yet been approved. Think of these agents as **"rocks"** being thrown at HIV and the CD4 cells to prevent them from hooking up! Maraviroc is generally well-tolerated and is being investigated for its ability to boost immune recovery (increase in CD4 counts) independent of its anti-viral activity. One caveat ... often in advanced HIV disease CD4 cells will switch to using CCR-4 as the primary co-receptor for HIV viral entry. In these cases, maraviroc and vicriviroc don't work. Therefore, it is standard practice to check a "tropism assay" to determine the dominant co-receptor on CD4 cells prior to prescribing this agent.

INTEGRASE INHIBITORS

Raltegravir

This recently approved agent has muscled its way on to the list of "preferred" options for treatment naïve persons beginning HAART. It blocks integration of reverse-transcribed HIV DNA into the chromosomes of host cells. Adverse effects of raltegravir include diarrhea, nausea, fatigue, and muscle aches.

Elvitegravir

Approved in 2012, this integrase inhibitor is only available as part of a combination pill, Stribild ©, (elvitegravir, cobicistat, emtricitabine, and tenofovir). Cobicistat is an inhibitor of cytochrome P450 allowing decreased metabolism of elvitegravir and hence once daily dosing (similar to the use of ritonavir for "boosting"). Stribild © is also approved as a first line agent for newly diagnosed persons with HIV infection. Most common side effects are nausea, diarrhea, and proteinuria.

Dolutegravir

Dolutegravir is an integrase inhibitor which is a preferred agent when combined with either tenofovir/emtricitabine or abacavir/lamivudine in HIV infected treatment naïve individuals. It is well tolerated. The main side effects are hyperglycemia and rise in serum liver enzymes (ALT).

Post-Exposure (i.e., Needle Stick) HIV Prophylaxis

After a needle-stick or other percutaneous exposure with HIV-infected blood the risk of seroconversion is

Figure 30-7

0.3%. This risk goes up if the injury is deep, the needle was in the patient's vein or artery, the needle had visible blood on it, or the patient died within 60 days of the stick (suggesting late-stage AIDS with high levels of viremia).

A four week course of 3 or more drugs is recommended for post-exposure prophylaxis (PEP). As of this writing emtricitabine/tenofovir combined with raltegravir is the recommended first line regimen. Always consult with an HIV specialist when determining the best PEP regimen for a patient. Recommended PEP regimens may vary based on the clinical status and treatment history of the source patient. This is a field that changes rapidly. For current recommendations for post-exposure prophylaxis go to *http://aidsinfo.nih.gov*.

INFLUENZA DRUGS

Adamantanes

Warning: Beginning with the 2005–2006 flu season, widespread resistance to this class of flu agents has been observed. Use of amantadine and rimantadine should only be considered based on the current circulating strain of influenza virus and after consulting current CDC seasonal guidelines (www.cdc.gov/flu). The adamantanes had little effect during the recent H1N1 "swine flu" scare of 2009–2010.

Amantadine
("A Man to Dine")

Amantadine has the narrowest spectrum, only inhibiting Influenza **A**, NOT B; the "**A**" is for "**A**mantadine." It is thought to do this by inhibiting viral genome **uncoating** in the host cell. It has minimal side effects.

Fig. 30-7. Amantadine. You have a hot dinner date with this stud of **A MAN**. You meet him **TO DINE** at a fancy restaurant in town; he sits at the table and takes off his coat (**uncoats**). To your intense chagrin, he then

begins to blow his nose loudly and drip strings of snot on his plate, explaining that he has a **terrible flu**.

If given early during an influenza A infection, amantadine will decrease the duration of flu symptoms. It also helps prevent influenza A if given prophylactically. For example, it can be given to nursing home residents if there is an outbreak of influenza A.

Rimantadine appears to be as effective as amantadine for the prevention of influenza A. It has less CNS side effects (anxiety and confusion) and does not require dose adjustments in renal failure, making it a safer agent for the elderly.

Neuraminidase Inhibitors

More recently a new class of agents, neuraminidase inhibitors, with clinical activity against both influenza A and B types have been introduced. (This includes activity against both avian influenza A, otherwise named "**bird flu**", and the 2009–2010 H1N1 strain of "**swine flu**".) These agents target neuraminidase (see page 235), which is responsible for cleaving the bonds between emerging virus and the cell and therefore freeing the virus to penetrate respiratory secretions and replicate. Resistance to neuraminidase inhibitors appears to be slow developing. These agents are indicated for the treatment of uncomplicated acute illness due to influenza and will decrease flu-symptoms by 1 to 2 days.

Oseltamivir

Oseltamivir is available as an oral tablet, which must be started within 2 days of onset of influenza symptoms. Oseltamivir is the only neuraminidase inhibitor used for flu prophylaxis as well. This agent is relatively well tolerated with dizziness, headache, fatigue, insomnia and vertigo being the most frequent side effects occurring in less than 2% of patients.

Zanamivir

Zanamivir is available as an intranasal spray and oral inhaler and should be initiated within 2 days of onset of influenza symptoms. Of course, because this drug is delivered through the upper respiratory tract, it is not recommended in patients with severe chronic obstructive pulmonary disease (COPD), or asthma. Nose bleed is the most characteristic side effect occurring with the nasal spray in up to 4% of patients.

Peramivir

Peramivir is a newer neuraminidase inhibitor, that was FDA approved in December 2014. It is given as a single IV dose. It is the only IV anti-influenza agent available and is indicated for persons unable to tolerate oral medication or critically ill persons who are not responding to oral regimens.

HEPATITIS C DRUGS

News alert...we can cure a virus! That is right. Hepatitis C has a cure! Up until now we could keep several viral infections at bay, but to cure a chronic viral infection is just about unheard of. Since late 2014/early 2015 there has been an explosion of new and very effective treatment options for hepatitis C. Several drug companies are jockeying for market share in this very lucrative field. A full treatment course can **cost over $100,000!** Hopefully, competition will bring down the cost. Preferred treatment regimens are determined based on disease characteristics, virus genotype, and prior treatment history. Interferon-based regimens, as well as the more recently approved drugs **boceprevir** and **telapravir** (protease NS3-4A inhibitors) are largely a thing of the past. They have been superseded by various direct-acting antiviral (DAA) containing regimens. As this goes to press the antiviral combinations of **ledipasvir-sofosbuvir, ombitasvir-paritaprevir-ritonavir-dasabuvir**, and **simprevir-sofosbuvir** are favored. To keep up with this rapidly evolving field get updates at: http://www.hcvguidelines.org/.

MISCELLANEOUS ANTI-VIRAL AGENTS

Ribavirin

Ribavirin has a wide spectrum of activity against many DNA and RNA viruses. However it is teratogenic in small mammals. Due to concerns about safety, it is only used for severe respiratory syncytial virus (RSV) infections in infants, for the rare case of Lassa fever (a severe influenza-like illness in Africa caused by the Lassa fever virus, in the family Arenaviridae), and for the hantavirus pulmonary syndrome (investigational use).

Ribavirin is used in its oral form in combination with interferon to treat hepatitis C virus (HCV).

Interferon
(alpha, beta, gamma)

Human interferons are cytokines that promote a cellular anti-viral state. They have been produced in large quantities by using recombinant DNA technology. Many studies are looking at these agents for the treatment of viral infections as well as cancers.

In the past, interferon has been used in combination with ribavirin for the treatment of hepatitis C. It is still used as a treatment option for young patients with well-compensated liver disease due to hepatitis B. Its biggest drawback, aside from questionable efficacy, is its high frequency of side-effects. The majority of patients receiving interferon develop flu-like symptoms shortly after drug administration. This is only one of a whole host of possible side-effects that limit its popularity among patients.

Fig. 30-7. Summary of anti-viral drugs. All of these medications have different side effects and have to be juggled around to find the combination with the least adverse efects and greatest sustained depression of viral load.

References

Balfour HH. Antiviral Drugs. N Engl J Med 340:1255–68, 1999.

Conner EM, et al. Reduction of maternal-infant transmission of human immunodeficiency virus type 1 with zidovudine treatment. N Engl J Med 331:1173–80, 1994.

Fischl MA, et al. AIDS Clinical Trials Group: The safety and efficacy of zidovudine (AZT) in the treatment of subjects with mildly symptomatic human immunodeficiency virus type 1 (HIV) infection: a double-blind, placebo-controlled trial. Ann Intern Med 112:727–737, 1990.

Keating MR. Antiviral Agents for Non-Human Immunodeficiency Virus Infections. Symposium on Antimicrobial Agents-Part XV. Mayo Clin Proc 74:1266–83, 1999.

Skowron G, et al. Alternating and Intermittent Regimens of Zidovudine and Dideoxycytidine in Patients with AIDS or AIDS-Related Complex. Ann Intern Med 118:321–329, 1993.

Treanor JJ, Hayden FG, Vrooman PS, et al. Efficacy and Safety of the Oral Neuraminidase Inhibitor Oseltamivir in treating Acute Influenza: a randomized controlled trial. JAMA 283:1016–1024, 2000.

Tyring S, et al. Famciclovir for the treatment of acute herpes zoster: effects on acute disease and postherpetic neuralgia. Ann Int Med 1995;123:89–96.

Volberding PA, et al. AIDS Clinical Trials Group of the National Institute of Allergy and Infectious Diseases: Zidovudine in asymptomatic human immunodeficiency virus infection: a controlled trial in persons with fewer than 500 CD4-positive cells per cubic millimeter. N Engl J Med 322:941–9, 1990.

Wong DKH, et al. Effect of Alpha-Interferon Treatment in Patients with Hepatitis B e Antigen-Positive Chronic Hepatitis B: A Meta-Analysis. Ann Int Med 119: 312–323;1993.

Zelalem T, Wright AJ. Antiretrovirals. Symposium on Antimicrobial Agents. Mayo Clin Proc 74:1284–1301, 1999.

Recommended Reading:

Huldrych FG, et al. Antiretroviral Treatment of Adult HIV Infection 2014. Recommendations of the International Antiviral Society–USA Panel. JAMA 2014;312(4):410–425.

Panel on Antiretroviral Guidelines for Adults and Adolescents. Guidelines for the Use of Antiretroviral Agents in HIV-1 Infected Adults and Adolescents. Department of Health and Human Services. April 8, 2015. Available at https://aidsinfo.nih.gov/guidelines

On the web: *www.hivinsight.org*

NAME	MECHANISM OF ACTION	PHARMOKINETICS
ANTI-HERPES: HSV & CMV		
Acyclovir	1. Guanosine analogue: looks like guanosine triphosphate (GTP) 2. Acyclovir requires the herpes encoded enzyme Thymidine Kinase for phosphorylation into its active form - acycloGTP 3. The abnormal GTP is inserted into the growing viral DNA chain, which terminates the growth of the DNA strand. This drug also inhibits viral DNA polymerase	1. Intravenous, oral or topical 2. Renal excretion
Famciclovir and Valacyclovir	• Mechanism of action similar to acyclovir	• Enhanced oral absorption results in higher drug levels
Ganciclovir	1. Guanosine analogue: looks like guanosine triphosphate (GTP) 2. Ganciclovir is activated (phosphorylated) by human thymidine kinase 3. The active triphosphate form inhibits viral DNA synthesis	1. Intravenous & oral 2. Renal excretion
Valganciclovir	Valganciclovir is a prodrug of ganciclovir	1. Oral 2. Converted to ganciclovir by intestine & liver 3. High fat diet increases absorption
Cidofovir	1. Cidofovir is a cytosine analogue, which is phosphorylated by host (not viral) cellular kinases to its active diphosphate form. 2. Activation of cidofovir occurs in both infected and uninfected cells. 3. Cidofovir diphosphate inhibits dCTP from being incorporated into viral DNA strands. Instead, cidofovir is incorporated into viral DNA chains, and serves to block further viral DNA chain elongation	1. Intravenous 2. Renal excretion
Foscarnet	1. Foscarnet is a pyrophosphate analogue 2. Inhibits viral DNA polymerase and reverse transcriptase 3. Does not require phosphorylation	1. Intravenous 2. Renal excretion

Figure 30-8   ANTI-VIRAL DRUGS

ADVERSE EFFECTS	THERAPEUTIC USES	MISCELLANEOUS

1. Nephrotoxic at high doses (because acyclovir precipitates in renal tubules)
2. Neurotoxic at high doses: confusion, lethargy or seizures
	1. Herpes virus 1 and 2
	  A. Herpes: treatment for primary and recurrent genital, oral, and ocular herpes
	  B. Herpes encephalitis
	2. Varicella-zoster virus
	  A. Varicella (chickenpox)
	    – Varicella pneumonia
	  B. Zoster (shingles), including:
	    – Herpes zoster ophthalmicus
		• Acyclovir is phosphorylated (activated) primarily by viral thymidine kinase (produced only in herpes infected cells). There is little toxicity to uninfected cells
		• Not as effective against CMV or EBV but is used for prophylaxis against CMV after bone marrow transplantation for low risk cases.

1. CNS: Headache, dizziness 2. G.I.: Occasional nausea and diarrhea	1. Herpes zoster: reduces the duration of postherpetic neuralgia (nerve pain) 2. HSV	• Also effective for reducing risk of HSV recurrence

• More toxic than acyclovir because it is activated by the human thymidine kinase
 	
1. Neutropenia
2. Thrombocytopenia
	1. Cytomegalovirus (CMV)
	  A. Retinitis, Pneumonia, Esophagitis
	  B. CMV prophylaxis: organ transplant patients
	2. HSV, HZV, EBV: but not commonly used due to high toxicity
		1. Ganciclovir is activated by a CMV protein kinase enzyme, which is a product of the UI.97 gene
		2. Viral resistance: altered viral DNA polymerase
		3. Ganciclovir implant: Intraocular implant filled with ganciclovir used for CMV retinitis

• Same as Ganciclovir	• Same as Ganciclovir	1. Valganciclovir requires fewer pills and gains higher levels as compared to oral ganciclovir 2. Ganciclovir has a much higher affinity to viral DNA polymerase than human DNA polymerase

1. Nephrotoxic
  A. Reduce kidney toxicity by hydration & oral probenicid
  B. Check renal function frequently while using cidovir, as damage can be reversible if caught early.
2. Neutropenia
	1. Cytomegalovirus (CMV)
	2. HSV (especially acyclovir-resistant strains)
	3. Other viruses: EBV, Adenovirus, HPV HHV-6
		1. Unlike ganciclovir, cidofovir activation occurs in both uninfected & infected cells. This helps uninfected cells from becoming infected
		2. Since cidofovir is phosphophorylated by human enzymes, it can be effective in strains with altered viral kinases

1. Reversible nephrotoxicity 2. Anemia 3. Alteration in serum electrolytes (calcium, phosphate)	1. Cytomegalovirus infections 2. HSV (especially acyclovir-resistant strains) 3. Other viruses: EBV, Adenovirus, HHV-6	1. Effective against Herpes viruses with altered thymidine kinase, since phosphorylation is not required 2. Viral resistance: altered viral DNA polymerase

NAME	MECHANISM OF ACTION	PHARMOKINETICS
Fomivirsen	1. Inhibits CMV replication through an anti-sense mechanism. 2. The nucleotide sequence of Fomiversen is complementary to a sequence of CMV mRNA that encode critical viral proteins. 3. Inhibits synthesis target proteins & therefore inhibits viral replication	• Injected into the eye
Anti-HIV Nucleoside reverse transcriptase inhibitors (NRTI) Zidovudine (ZDV) formerly Azidodideoxythymidine (AZT)	1. Nucleoside analogue: all look like thymidine triphosphate (TTP), adenosine triphosphate (ATP), and cytosine triphosphate (CTP) 2. Nucleoside must be phosphorylated by cellular enzymes into its triphosphate form to be activated 3. Potent inhibitors of reverse transcrip-tase (reverse transcriptase uses a viral RNA template to transcribe a DNA strand) 4. Activated nucleosides are preferentially incorporated into the growing DNA strand (instead of TTP), which terminates the growth of the strand • Thymidine analog: looks like thymidine triphosphate (TTP)	1. All absorbed orally 2. Most metabolized by cytochrome p450 system 3. Most are renally cleared • Renal and hepatic excretion
Didanosine (ddI) formerly dideoxyinosine (ddI)	• Adenosine analogue: looks like adenosine triphosphate (ATP)	• Food decreases absorption so take without food
Zalcitabine (ddC) formerly dideoxycytidine (ddC)	1. Cytosine analogue: looks like cytosine triphosphate (CTP) 2. Inhibits viral DNA synthesis	
Lamivudine (3TC)	• A cytosine analog	
Stavudine (d4T)	• Thymidine analogue: looks like thymidine triphosphate (TTP)	

Figure 30-8   (continued)

ADVERSE EFFECTS	THERAPEUTIC USES	MISCELLANEOUS
1. Ocular inflammation 2. Elevated eye pressure	• CMV retinitis	1. Fomiversen is 21 nucleotides long 2. First oligonucleotide drug approved in US
All can cause: 1. Nausea and vomiting 2. Hepatic steatosis (fatty liver) 3. lactic acidosis 4. lipodystrophy "D" drugs (ddl, ddC, and D4T) can all cause peripheral neuropathy and pancreatitis 1. Macrocytosis (enlarged red blood cells) 2. Neutropenia • Both anemia & neutropenia are dose related 3. Headache (60%), insomnia or confusion	• Human immunodeficiency virus (HIV) Common points for initiating "HAART": 1. CD4 count , 200 (regardless of viral load) 2. CD4 count 200–350 (influenced by high HIV viral load) 3. CD4 count .350 (HIV viral load .100,000 copies/ml) 4. An AIDS defining illness • Human immunodeficiency virus (HIV) • Reduces HIV transmission from mother to infant if given to mother before birth and during breast feeding, and to the infant.	Combination therapy: 1. Use Highly Active Anti-Retroviral Therapy-"HAART" 2. Always include drugs from two classes 3. Two nucleoside reverse transcriptase (NRTI) inhibitors + Protease inhibitor or nonnucleoside RT inhibitor 4. Consider cost, side-effects, potency, ease of use • Resistance to AZT develops via mutations of the reverse transcriptase enzyme
1. Peripheral neuropathy: numbness, tingling or pain in the feet or hands 2. Pancreatitis 3. Diarrhea, abdominal pain, increased liver enzymes (SGOT) 4. Retinal changes and optic neuritis	• Human immunodeficiency virus (HIV)	• Resistance to ddl develops via mutations of the reverse transcriptase enzyme
1. Painful peripheral neuropathy 2. Transient mouth ulcers and skin rashes 3. Pancreatitis	• Human immunodeficiency virus (HIV)	1. Not as effective as AZT or ddl alone 2. Similar resistance & toxicity as ddl 3. 3x's/day dosing (more than the other NRTI's) • NOTE: due to above – ddC is not used often

1. Well tolerated. Most adverse effects are secondary to ZDV used in combination
2. Occasionally:
  A. Pancreatitis
  B. Peripheral neuropathy
	1. Human immunodeficiency virus (HIV)
	2. Chronic Hepatitis B infection (relapses occur when drug is discontinued)
		1. Resistance to Lamivudine develops via mutations of the reverse transcriptase enzyme
		2. Combivir: Lamivudine co-formulated with zidovudine.

1. Most likely NRTI to cause lipodystrophy.
2. Associated with peripheral neuropathy and pancreatitis.
3. Macrocyosis (enlarged red blood cells)
	• Human immunodeficiency virus (HIV)	Drug interactions
		  A. Do not mix with Zidovudine – due to drop off in effects against HIV
		  B. Do not mix with ddl, due to increased risk of pancreatitis, hepatitis & neuropathy.
		2. Avoid alcohol – which can increase risk of pancreatitis & hepatitis

NAME	MECHANISM OF ACTION	PHARMOKINETICS
Abacavir (ABC)	• dGTP analogue: Intracellularly, abacavir is converted by cellular enzymes to the active metabolite carbovir triphosphate. Carbovir triphosphate is an analogue of deoxyguanosine-5′-triphosphate (dGTP). Carbovir triphosphate inhibits the activity of HIV-1 reverse transcriptase both by competing with the natural substrate dGTP & by its incorporation into viral DNA. The lack of a 3′-OH group in the incorporated nucleoside analogue prevents the formation of the 5′ to 3′ phosphodiester linkage essential for DNA chain elongation. This results in viral DNA growth termination.	
Emtricitabine	• Cytosine analogue: looks like cytosine triphosphate (CTP)	• Oral absorption
Tenofovir DF (TDF)	• Nucleotide reverse transcriptase inhibitor: similar to nucleoside analogues (AZT, d4T, 3TC). The difference is that nucleotide analogues are already phosphorylated, unlike nucleoside analogues. They therefore require less processing in the body to become active.	• Food enhances absorption
Non-nucleoside reverse transcriptase inhibitor	• Non-nucleoside reverse transcriptase inhibitors bind directly to the HIV reverse transcriptase and block the RNA-dependent and DNA-dependent DNA polymerase activities by causing a disruption of the enzyme's catalytic site. The activity of Non-nucleoside reverse transcriptase inhibitors does not compete with nucleoside triphosphates. Eukaryotic DNA polymerases (such as human DNA polymerases) are not inhibited. • Many more agents are being developed in this class.	1. All absorbed orally 2. Most metabolized by cytochrome p450 system
Nevirapine	• Non-nucleoside reverse transcriptase inhibitor	
Delaviridine mesylate	• Non-nucleoside reverse transcriptase inhibitor	• Take before antacids
Efavirenz (EFV)	• Non-nucleoside reverse transcriptase inhibitor	• Oral absorption: avoid high fat meals, which increases absorption

Figure 30-8   (continued)

ADVERSE EFFECTS	THERAPEUTIC USES	MISCELLANEOUS
1. Severe hypersensitivity: fever, rash, sore throat, & fatigue. Stop abacavir immediately 2. If abacavir is stopped, restarting abacavir can lead to severe & fatal hypersensitivity reactions within hours – even in patients with no history of hypersensitivity to abacavir. If abacavir is stopped for any reason, extreme caution should be exercised before reintroducing abacavir.	• Human immunodeficiency virus (HIV) Greater potency than other nucleoside reverse transcriptase inhibitors	1. Trizivir: Combination of abacavir, zidovudine & lamivudine 2. Test for HLA-B*5701 prior to starting. If present=increased risk for abacavir hypersensitivity.
1. Headache 2. GI: Diarrhea, Nausea	1. Human immunodeficiency virus (HIV) 2. Also active against Hepatitis B	1. Emtricitabine and Tenofovir (Truvada)-part of preferred 1st line drug regimen. 2. Similar resistance profile as 3TC.
1. Mild gastrointestinal disturbances 2. Acute renal insufficiency/ Fanconi syndrome	1. Human immunodeficiency virus 2. Hepatitis B virus	• First nucleotide reverse transcriptase inhibitor
All cause: 1. Rash (including Steven's Johnson syndrome) 2. Increased liver enzymes • Efavirenz causes CNS side effects: insomnia, abnormal thinking, dizziness, & bad dreams. These decrease after a few weeks.	• Antiretroviral therapy should be initiated in all patients with a history of an AIDS-defining illness or with a CD4 count <350 cells/mm³	Combination therapy: 1. Use Highly Active Anti-Retroviral Therapy-"HAART" 2. Always include drugs from two classes 3. Two nucleoside reverse transcriptase (RT) inhibitors + Protease inhibitor or non-nucleoside RT inhibitor 4. Consider cost, side-effects, potency, ease of administration
1. Rash (17%) 2. Increased liver enzymes (GGT is increased in 20% of patients) 3. Inhibits P450 system	• Human immunodeficiency virus (HIV)	• Rapid emergence of resistance
1. Rash (18%) 2. Inhibits p450 system: inhibition leads to increased levels of itself, plus other medications metabolized by p450 system	• Human immunodeficiency virus (HIV)	• Rapid emergence of resistance
1. Rash (27%) 2. CNS symptoms (Bad dreams, insomnia, abnormal thinking, dizziness) 3. Inhibits P450 system 4. Pruritis 5. Hepatitis	• Human immunodeficiency virus (HIV)	

NAME	MECHANISM OF ACTION	PHARMOKINETICS
Etravirine	• Non-nucleoside reverse transcriptase inhibitor	• Take with food
Rilpivirine	• Non-nucleoside reverse transcriptase inhibitor	• Take with food
Protease Inhibitors	• Inhibit HIV protease. Protease is a vital HIV enzyme that cleaves gag and pol proteins from their precursor molecules. Protease deficient virions can not form their viral core, and thus are non-infectious • Many more being developed	1. All are absorbed orally 2. All are metabolized by cytochrome p450 system
Saquinavir	• HIV-1 protease inhibitor	• Poorest absorption of the protease inhibitors • New gelatin formulation called fortovase allows for better GI absorption
Ritonavir	• HIV-1 protease inhibitor	• Binds avidly to p450 & increases plasma levels of drugs (indinavir, nelfinavir & saquinavir)
Indinavir	• HIV-1 protease inhibitor	• Excellent oral absorption but high fat/high protein meals reduce absorption
Loponivir/ritonavir	• HIV-1 protease inhibitor	• Food enhances absorption
Nelfinavir	• HIV-1 protease inhibitor	• Food enhances absorption

Figure 30-8   (continued)

ADVERSE EFFECTS	THERAPEUTIC USES	MISCELLANEOUS
1. Rash, nausea, diarrhea 2. May cause increase in hepatic transaminases, cholesterol, triglycerides. 3. Inhibits and induces different pathways of the cytochrome P450 systems.	• Human immunodeficiency virus (HIV)	Activity against strains of HIV resistant to other NNRTIs
1. Depression 2. Headache 3. Insomnia 4. Rash	• HIV	1. A newer NNRTI 2. Activity against strains of HIV resistant to other NNRTIs
All cause: 1. GI intolerance 2. Increased liver enzymes 3. Increased triglycerides 4. Fat redistribution syndromes with peripheral fat wasting, truncal obesity, and facial thinning 5. Insulin resistance: diabetes	• Antiretroviral therapy should be initiated in all patients with a history of an AIDS-defining illness or with a CD4 count <350 cells/mm^3	Combination therapy: 1. Use Highly Active Anti-Retroviral Therapy-"HAART" 2. Always include drugs from two classes 3. Two nucleoside reverse transcriptase (RT) inhibitors + Protease inhibitor or non-nucleoside RT inhibitor 4. Consider cost, side-effects, potency, ease of administration
	• Human immunodeficiency virus (HIV)	
• Many: 85–100% have GI side effects	• Human immunodeficiency virus (HIV) • Used only as a "booster" in combination with other protease inhibitors. Not used as a sole agent in its class.	Poorly tolerated when used at doses higher than the typical low "boosting dose"
1. Kidney stones (2–3%); drug precipitates in renal collecting system, forming indinavir crystals; patients have to drink buckets of water a day 2. Interstitial nephritis	• Human immunodeficiency virus (HIV)	• Alternative medications can interact: St. John's Wort stimulates p450 activity & decreases indinavir by 80%
1. Diarrhea (25%) 2. Increased cholesterol & triglyceride levels 3. Liver toxicity Note: People co-infected with hepatitis B and/or C may be at greater risk of developing liver toxicity 4. Increases blood levels of Viagra	• Human immunodeficiency virus (HIV)	• Randomized controlled trial showed it to be superior to nelfinavir
1. Diarrhea	• Human immunodeficiency virus (HIV)	• Decrease rifabutin dose when co-administering

NAME	MECHANISM OF ACTION	PHARMOKINETICS
Atazanavir	• HIV-1 protease inhibitor	First protease-inhibitor approved for once/day dosing
Amprenivir and Fosamprenivir	• HIV-1 protease inhibitor	• Food decreases absorption
Tipranavir	• HIV-1 protease inhibitor	1. Must "boost" with ritonavir 2. Take with food
Darunavir	• HIV-1 protease inhibitor	1. Must "boost" with ritonavir. 2. Take with food.
Raltegravir	1. HIV-1 integrase inhibitor 2. Inhibits integration of reverse-transcribed HIV DNA into the chromosomes of host cells.	• Take with or without food
Elvitegravir	• HIV-1 integrase inhibitor	• Take with food
Dolutegravir	• HIV-1 integrase inhibitor	• no food restrictions
Maraviroc	1. CCR5 Inhibitor 2. It binds to CCR5, one of two possible coreceptors used by HIV to enter CD4+ cells and thereby blocks its entry.	1. No food restrictions 2. Dose adjustments often required based on co-administered medications
Enfuvirtide (Fuzeon)	• Fusion Inhibitor: Blocks HIV from fusing with and delivering its contents into a healthy CD T-Cell	• Administered twice a day as subdermal injection
	ANTI-INFLUENZA	
Amantadine Rimantadine	• Inhibits uncoating of Influenza A ("A for Amantadine") after it enters the cell (So blocks the release of viral RNA into the cell)	1. Oral absorption 2. Rimantidine metabolized by liver, & mostly metabolites excreted in urine. 3. Renal excretion: amantadine levels increase with renal failure
Oseltamivir	• Neuraminidase inhibitor	• Oral absorption

Figure 30-8   (continued)

ADVERSE EFFECTS	THERAPEUTIC USES	MISCELLANEOUS
1. Elevated bilirubin levels, which can lead to jaundice 2. Diarrhea	• Human immunodeficiency virus (HIV)	Less negative impact on total cholesterol levels and trigyclerides than other protease inhibitors
1. G.I. side effects 2. Rash	• HIV: triple combination therapy	
1. Gastrointestinal upset 2. Elevated liver function tests	HIV-1 infection (used when other regimens have failed)	
1. Inhibits p450 system. 2. Diarrhea, nausea, headache, rash. 3. Contains sulfa moiety-use with caution in person with sulfa allergy. 4. Hepatitis.	HIV immunodeficiency virus (HIV)	Active against many HIV strains with extensive protease inhibitor resistance
1. Diarrhea, fatigue, myalgias. 2. May cause rise in hepatic transaminases or pancreatic amylase.	Human immunodeficiency virus (HIV)	1. Approved for treatment as part of elvitegravir/cobicistat + tenofovir/ emtricitab (Stribild©)
1. Nausea 2. Diarrhea 3. Proteinuria	• HIV	Only available as part of co-formulation: elvitegravir/cobicistat + tenofovir/ emtricitab (Stribild©)
1. Hyperglycemia 2. Elevated ALT	• HIV	
1. Cough, upper respiratory tract infections, myalgias, diarrhea, and sleep disturbances	Human immunodeficiency virus (HIV)	Must check "tropism assay" prior to use.
1. Injection site reactions: itching, swelling, redness, pain 2. Increased risk of bacterial pneumonia in clinical trials 3. Pain and numbness in feet or legs 4. Loss of sleep, appetite & strength	• Human immunodeficiency virus (HIV)	1. Mechanism: HIV surface glycoprotein gp120 attaches to T-cell. Viral gp41 then undergoes a conformational change, which enables fusion of virus and T-cell membranes. Enfuvirtide binds to gp41, which prevents the conformational change necessary for membrane fusion. 2. Use in combo with other anti-HIV drugs
1. Central nervous system effects, such as anxiety and confusion (more common in the elderly) 2. Anticholinergic effects: dry mouth and urinary retention 3. Teratogenic: pregnant women should not use this	• Influenza A (not effective against influenza type B)	• Also used to treat Parkinson's disease, as it potentiates anticholinergic drugs • Rimantadine: has less CNS side effects, and does not need dose adjustment in renal failure; may be a better choice for the elderly
• Less than 2% of patients have side effects • Reported side effects include dizziness, headache, fatigue, insomnia, & vertigo	1. Influenza A & B treatment 2. Can be used for prophylaxis of influenza A & B	

NAME	MECHANISM OF ACTION	PHARMOKINETICS
Zanamivir	• Neuraminidase inhibitor	1. Nasal spray 2. Oral inhaler
Peramivir	• Neuraminidase inhibitor	• Intravenous
MISCELLANEOUS		
Ribavirin	1. Guanosine analogue: looks like GTP 2. Requires phosphorylation by human kinases to be activated 3. Inhibits viral replication	1. Aerosol 2. Oral absorption

Interferon alpha 2b	1. Induces production of proteins that inhibit RNA synthesis
	2. Induces production of enzymes that chop up viral DNA (as well as cellular DNA)
	3. Inhibits messenger RNA
		• Administration:
		  1. Intravenous
		  2. Intramuscular
		  3. Subcutaneous

Boceprevir	• NS3-4A protease inhibitor	Take with food
Telaprevir	• NS3-4A protease inhibitor	Take with food
Vidarabine	1. Adenosine analogue 2. Activated by phosphorylation 3. Competitive inhibitor of viral DNA polymerase	1. Intravenous 2. Renal excretion
Trifluridine	1. Nucleoside analogue of thymidine (deoxyuridine) 2. Phosphorylated by cellular enzymes into its active triphosphate form in all cells (regardless of viral infection) 3. Incorporated into viral DNA (instead of thymidine), which inhibits viral replication	• Eye drop

Figure 30-8   (continued)

ADVERSE EFFECTS	THERAPEUTIC USES	MISCELLANEOUS
• Epistaxis (nose bleed) – (4% with nasal spray)	• Influenza A & B treatment	• Should not be used in patients with chronic obstructive pulmonary disease (COPD), or asthma
	1. H1N1 "swine flu" infection: A. Not responding to oral agents. B. Concern that patient is unable to absorb oral agents. 2. Effective against both influenza A and B.	
1. Little toxicity when given via aerosol 2. I.V. toxicity: hemolysis can occur	1. Respiratory syncytial virus 2. Lassa fever 3. Hantavirus pulmonary syndrome 4. Hepatitis C (HCV) (Ribavirin in combo with interferon)	• Extremely expensive
1. Flu like syndrome: fever, headache, nausea and myalgias 2. Neutropenia, anemia 3. Thrombocytopenia 4. Neurotoxic: confusion, dizziness	1. Chronic hepatitis C 2. Chronic hepatitis B 3. Condyloma acuminatum (genital warts): caused by human papilloma virus (inject into warts)	
1. CNS-Fatigue, chills, insomnia, dizziness 2. Alopecia, rash 3. Nausea, anorexia 4. Anemia, neutropenia 5. Arthralgia	Hepatitis C (genotype 1)	Used in combination with IFN and ribavirin
1. Fatigue 2. Rash 3. Pruritis 4. Nausea, Diarrhea 5. Anemia	Hepatitis C (genotype 1)	Used in combination with IFN and ribavirin
1. GI toxicity (anorexia, nausea/vomiting) 2. Bone marrow suppression	1. Herpes virus (but not as good as acyclovir) 2. HZV infections (but not as good as acyclovir!)	
• Only used topically: extremely toxic when used intravenously	• HSV infections of the cornea	

M. Gladwin, W. Trattler, and S. Mahan, *Clinical Microbiology Made Ridiculously Simple* ©MedMaster

PART 4. PARASITES

CHAPTER 31. PROTOZOA

Protozoa are free-living, single celled, eucaryotic cells with a cytoplasmic membrane and cellular organelles, including 1 or 2 nuclei, mitochondria, food vacuoles, and endoplasmic reticulum. They come in many sizes, from 5 micrometers to 2 millimeters. They have an outer layer of cytoplasm (ectoplasm) and an inner layer (endoplasm), which appear different from each other under the microscope.

The protozoa ingest solid pieces of food through a small mouth called the **cytostome**. For example, amoebas (*Entamoeba histolitica*) can ingest human red blood cells into their cytoplasm. The protozoa reproduce asexually, undergoing DNA replication followed by division into 2 cells. They also reproduce sexually by the fusion of 2 cells, followed by the exchange of DNA and separation into 2 cells again.

When exposed to new environments (such as temperature changes, transit down the intestinal tract, or chemical agents), the protozoa can secrete a protective coat and shrink into a round armored form, called the **cyst**. It is this cyst form that is infective when ingested by humans. Following ingestion it converts back into the motile form, called the **trophozoite**.

THE INTESTINAL PROTOZOA

There are 5 intestinal protozoa that cause diarrhea. *Entamoeba histolytica* causes a bloody diarrhea, and *Giardia lamblia* and *Cyclospora cayetanensis* cause a non-bloody diarrhea. Both occur in normal individuals. *Cryptosporidium* and *Isospora belli* cause severe diarrhea in individuals with defective immune systems (such as patients with AIDS).

Entamoeba histolytica

This organism is the classic amoeba we have all heard about. It moves by extending creeping projections of cytoplasm, called **pseudopodia** (false feet). The pseudopodia pull it along or surround food particles.

The exact incidence of infection is unknown, but it is highest in the developing world and among immigrants to the U.S.. Old prevalence data is flawed because stool microscopy, which was previously used for screening and epidemiologic surveillance, is unable to differentiate between *E. histolytica* and the non-pathogenic intestinal protozoa *Entamoeba dispar*. Most infections

with *E. histolytica* are asymptomatic, as the amoebas live in peace inside their host carriers. These carriers pass the infective form, the cyst, to other individuals by way of the fecal-oral route. It is noteworthy that homosexual men commonly are asymptomatic carriers.

Fig. 31-1. The motile feeding form of the amoeba is the trophozoite, which cruises along the intestinal wall eating bacteria, other protozoa, and even human intestinal and red blood cells. This trophozoite can convert to a precyst form, with two nuclei, that matures into a tetranucleated cyst as it travels down and out the colon. The precyst contains aggregates of ribosomes, called **chromotoid bodies**, as well as food vacuoles that are extruded as the cell shrinks to the mature cyst; it is the mature cyst that is eaten, infecting others.

Sometimes (10% of infected individuals) the trophozoites invade the intestinal mucosa causing erosions. This results in abdominal pain, a couple of loose stools a day, and flecks of blood and mucus in the stool. The infection may become severe, with bloody, voluminous diarrhea.

The trophozoites may penetrate the portal blood circulation, forming abscesses in the liver, followed by spread through the diaphragm into the lung. Here the trophozoite infection causes pulmonary abscesses and often death (worldwide: 100,000 deaths annually).

The stool is examined for the presence of cysts or trophozoites. Trophozoites with red blood cells in the cytoplasm suggest active disease, while cysts or trophozoites without internalized red cells suggest asymptomatic carriage. Antigen detection assays, on stool or serum, are currently the most sensitive and specific tests available for diagnosing infection with *E. histolytica* and in differentiating this infection from the microscopically identical yet non-pathogenic protozoa *E. dispar*. CAT scan or ultrasound imaging of the liver will reveal abscesses if present.

Prevention rests on good sanitation: proper disposal of sewage and purification (boiling) of water. Treatment generally is with metronidazole, but tinidazole is a newer FDA approved alternative. Generally, these agents are followed by an agent specifically for intraluminal killing such as paromomycin or iodoquinol.

Fig. 31-2. The **Metro** (**metronidazole**) runs over *Entamoeba histolytica, Giardia lamblia, Trichomonas*

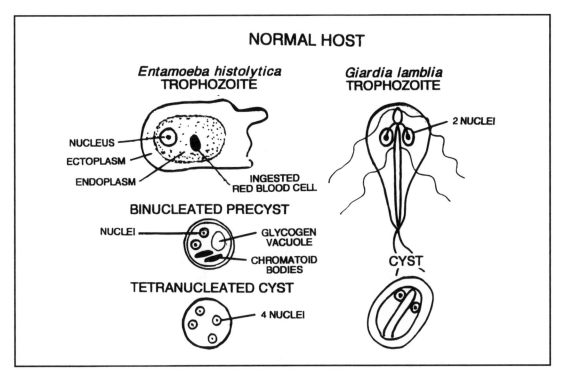

Figure 31-1

vaginalis, and the anaerobic cocci and bacilli including *Bacteroides fragilis, Clostridium difficile* and *Gardnerella vaginalis*. This drug is also called Flagyl (its trade name) because it kills the flagellated bugs, *Giardia* and *Trichomonas*.

Adverse effects of metronidazole

There is no drinking allowed on the train because it travels rapidly and jarringly, causing stomach upset to passengers that consume alcohol (Antabuse-disulfiram effect). If you eat the train, as King Kong once attempted, you would end up with a metallic taste in your mouth.

Giardia lamblia

Fig. 31-1. *Giardia lamblia* exists in 2 forms: as a cyst and as a mature, motile trophozoite that looks like a kite.

It is estimated that 5% of U.S. adults harbor this organism, mostly asymptomatically. Outbreaks occur when sewage contaminates drinking water. The organism is also harbored by many rodents and beavers; campers frequently develop *Giardia lamblia* infection after drinking from "clear" mountain streams.

After ingestion of the cyst, *Giardia lamblia* converts to the trophozoite form and cruises down and **adheres** to the small intestinal wall. The organism coats the small intestine, interfering with intestinal fat absorption. The stools are therefore packed with fat, which has a **horrific odor**! The patient will have a greasy, frothy diarrhea, along with abdominal gassy distension and cramps. Since *Giardia* do NOT invade the intestinal wall, there is NO blood in the stool!!!

For diagnosis and control of *Giardia*:

1) Examination of stool for cysts or trophozoites.
2) Commercial immunoassay kit to detect *Giardia lamblia* antigens in aqueous extracts of stool specimens.
3) Sanitation measures.

Treat these patients with **metronidazole** (see **Fig. 31-2**).

Cryptosporidium

It is now apparent that this critter is everywhere! Animals and humans are equally infected and about 25% of Americans show serologic evidence of previous infection. It can cause outbreaks of diarrhea from contaminated municipal water sources and in infants in day care centers. Sporadic cases can occur in travelers.

Cryptosporidium is ingested as a round oocyst that contains 4 motile sporozoites. Its life cycle occurs within the intestinal epithelial cells, and it causes diarrhea and

Figure 31-2

abdominal pain. These symptoms are self-limiting in immunocompetent individuals. However, in immuno-compromised patients (AIDS patients, cancer patients, or organ transplant recipients who are receiving immunosuppressive therapy), this organism causes a severe, protracted diarrhea that is life-threatening. These patients may have 3–17 liters of stool per day.

Treatment is based on the immune status of the patient. For persistent diarrhea in an otherwise normal host, nitazoxanide has been shown to be effective. Therapy in the immunocompromised has been discouraging. In these patients the main goal is restoration of their

immune function, often through highly active anti-retroviral therapy in AIDS patients. Either nitazoxanide or paromomycin with or without azithromycin can also be tried in these difficult to treat patients.

Isospora and Cyclospora

These organisms can cause severe diarrhea in immunocompromised individuals. They are transmitted by the fecal-oral route. Cyclospora has also been implicated in food borne outbreaks of diarrhea associated with contaminated raspberries. Both of these organisms

Figure 31-3

are diagnosed primarily by stool microscopy. Ask the pathologist to do an acid-fast stain . . . as *both of these organisms are acid-fast*!! Fluorescent microscopy is an alternative.

Both of these agents are treated with trimethoprim-sulfamethoxazole

THE SEXUALLY TRANSMITTED PROTOZOAN

Trichomonas vaginalis

Fig. 31-3. *Trichomonas vaginalis* is transmitted sexually and hangs out in the female vagina and male urethra. The trophozoite of *Trichomonas vaginalis* is a flagellated protozoon (as is *Giardia lamblia*).

A female patient with this infection may complain of itching (pruritus), burning on urination, and copious vaginal secretions. On speculum examination you will find a thin, watery, frothy, malodorous discharge in the vaginal vault. Males are usually asymptomatic.

Diagnosis of *Trichomonas*:

1) Microscopic examination of vaginal discharge on a wet mount preparation will reveal this highly motile parasite.
2) Examination of urine may also reveal *Trichomonas vaginalis*.

Treat your patient with **metronidazole** (see **Fig. 31-2**). Provide enough for sexual partners. Even though males are usually asymptomatic, they must be treated

or the female partner will be reinfected (since this organism is not invasive, no immunity is acquired).

THE FREE-LIVING MENINGITIS-CAUSING AMOEBAS

Naegleria fowleri, *Acanthamoeba*, and *Balamuthia mandrillaris* are free-living amoeba that live in fresh water and moist soils. Infection often occurs during the summer months when people swim in freshwater lakes and swimming pools that harbor these organisms. Although large numbers of persons are exposed, actual infection rarely occurs. When infection does occur it usually has rapid course and is ultimately fatal. In the case of *Naegleria fowleri* the organisms penetrate the nasal mucosa, pass through the cribiform plate, into the brain and spinal fluid. *Naegleria fowleri* will cause sudden deadly infection in immunocompetent persons. *Acanthamoeba* and *Balamuthia mandrillaris* are thought to be acquired via the respiratory route and via breaks in the skin. *Acanthamoeba* and *Balamuthia* tend to cause a more insidious granulomatous encephalitis with greatest incidence in those who are immunocompromised.

Naegleria fowleri

Fig. 31-4. *Naegleria* **fowl**eri is known for **FOWL PLAY**, since 95% of patients will die within 1 week. Infected persons will present with a fever, headache, stiff neck, nausea, and vomiting, which is very similar to a bacterial meningitis. If asked, they will give a history of swimming a week earlier. Examination of cere-

339

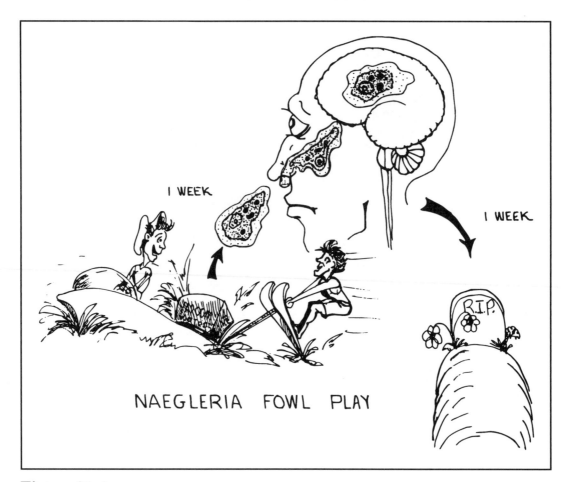

Figure 31-4

NAEGLERIA	ACANTHAMOEBA	BALAMUTHIA MANDRILLARIS
• ACUTE meningoencephalitis in normal hosts	• CHRONIC meningoencephalitis in immunocompromised hosts	• CHRONIC meningoencephalitis in normal and immunocompromised hosts
• Mature amoeba **only** in brain tissue (**NO** cysts)	• Cysts and mature amoeba found in brain tissue	• Cysts and mature amoeba found in brain tissue
• No corneal infection	• Corneal infection: diagnose by corneal scraping	• Granulomatous skin lesions- diagnose by skin biopsy

Figure 31-5   THE AMOEBAS

brospinal fluid (CSF) reveals a high neutrophil count, low glucose, and high protein, exactly like a bacterial meningitis!!! The Gram stain and culture will reveal NO bacteria, and microscopic examination may show the motile amoeba.

Two patients who survived were treated with intrathecal **amphotericin B**, an antifungal agent.

Acanthamoeba

Acanthamoeba is responsible for a chronic, granulomatous, brain infection in immunocompromised patients, such as those with AIDS. Over a period of weeks, they will develop headache, fever, seizures, and focal neurologic signs. Examination of the CSF and brain tissue will reveal *Acanthamoeba* in both the cyst stage and trophozoite stage. Treatment is difficult and involves multiple antifungal drugs with pentamidine.

This organism may also infect the cornea (in immunocompetent persons), often when contact lenses are not properly cleaned. This corneal infection (keratitis) can lead to blindness. Treatment is with antimicrobial eye drops.

Balamuthia mandrillaris

Balamuthia amebic encephalitis has been reported worldwide. It presents pathologically (with granulomas) and is clinically similar to *Acanthamoeba*. It differs in that it infects both immunocompetent and immunocompromised hosts. In addition, *Balamuthia mandrillaris* can produce chronic granulomatous skin lesions either alone or in conjunction with amebic encephalitis. Studies suggest that exposure to *B. mandrillaris* is common, yet most are asymptomatic. As with encephalitis caused by the other free-living amoebas, most cases are fatal. In vitro testing has shown fluconazole, azithromycin, and pentamidine to have some activity.

Fig. 31-5. Comparision of *Naegleria*, *Acanthamoeba*, and *Balamuthia mandrillaris*.

THE MAJOR PROTOZOAN INFECTIONS IN *AIDS* PATIENTS

The ineffective immune system in AIDS patients sets them up for certain infections that seldom affect the immunocompetent host. We have already discussed 2 parasites that can establish a severe, chronic diarrhea in AIDS patients: *Cryptosporidium* and *Isospora*. Two other parasites found more commonly in AIDS patients than in the general population are *Toxoplasma gondii* and *Pneumocystis carinii*. These organisms are harbored by most persons without problems. In AIDS, when the T-helper cell count drops below 200, these bugs flourish and cause disease.

Toxoplasma gondii

Many animals are infected with *Toxoplasma*, and humans are infected by the ingestion of cysts in undercooked meats (raw pork) or food contaminated with household cat feces. Kitty litter boxes are the most common source of exposure for humans, as up to 80% of cats are infected in the United States. *Toxoplasma gondii* undergoes sexual division in the cat and is excreted in the feces as the infectious cyst.

The protozoan causes disease by reactivation of a latent infection in an immunocompromised person or as a primary infection in a pregnant woman (leading to transplacental infection of the fetus).

1) Immunocompromised patients with AIDS or those who are taking immunosuppressive drugs (for cancer or organ transplantation) are susceptible to growth of the latent *Toxoplasma gondii*. The infection can present in many ways—with fever; lymph node, liver, and spleen enlargement; pneumonia; or frequently with infection of the meninges or brain. In fact, *Toxoplasma* encephalitis is the most common central nervous system infection in AIDS patients. The brain infection can involve a growing mass, much like a tumor, with symptoms of headache and focal neurologic signs (seizures, gait instability, weakness, or sensory losses). Infection of the retina, chorioretinitis, is also common, resulting in visual loss. Examination of the retina reveals yellow-white, fluffy (like cotton) patches that stand out from the surrounding red retina.

2) *Toxoplasma* is one of the transplacentally acquired TORCHES organisms that can cross the blood-placenta barrier (see **Fig. 27-2.**). Transplacental fetal infection can occur if a pregnant woman who has never been previously exposed to *Toxoplasma gondii* is infected. Congenital toxoplasmosis does not occur in pregnant women who have serologic evidence of previous exposure, most likely because of a protective immune response. **Pregnant women should avoid cats**!!

Like rubella (see Chapter 29), congenital toxoplasmosis can cause many problems, including chorioretinitis, blindness, seizures, mental retardation, microcephaly, encephalitis, and other defects. If the infection is acquired early during gestation, the disease is severe, often resulting in stillbirth. Interestingly, infants that appear normal can develop disease later in life. Clinical reactivation results most commonly in retinal inflammation (chorioretinitis, which can result in blindness) that flares late in life (peak incidence in second or third decade).

Note that immunocompetent adults (such as the pregnant women described above) who are infected with *Toxoplasma gondii* often develop generalized lymph node enlargement.

BIG PICTURE: In AIDS patients and fetuses *Toxoplasma gondii* is *TOXIC* to the *BRAIN* and *EYES*!!!

Diagnosis of toxoplasmosis can be made by:

1) CAT scan of brain will show a contrast-**enhancing** mass.

2) Examination of the retina of the eye will reveal retinal inflammation.

3) Serology: Elevated immunoglobulin titers suggest that the patient has at some time been exposed to this organism.

Sulfadiazine plus pyrimethamine can be used to treat patients with acute toxoplasmosis.

Pneumocystis carinii (P. jiroveci)

Pneumocystis carinii is a flying-saucer appearing FUNGUS that has previously been classified as a protozoan but now has been shown to be more closely related to fungi. Officially *Pneumocystis carinii* has been renamed *Pneumocystis jiroveci*, but this new nomenclature has been slow to stick . . . thus, for simplicity we will stick with the old name. This organism appears to invade the lungs of all individuals at an early age and persists in a latent state. In fact, based on IgM and IgG levels, it appears that about 85% of children have had a mild or asymptomatic respiratory illness with *Pneumocystis carinii* by age 4. In persons with a functioning immune system, this organism will live comfortably within the lung without causing symptoms. However, in immunocompromised patients (AIDS patients, cancer patients, and organ transplant recipients), this organism can multiply in the lungs and cause a severe interstitial pneumonia, called ***Pneumocystis carinii* pneumonia (PCP)**.

PCP is the most common opportunistic infection of AIDS patients. Without prophylactic treatment there is a 15% chance each year of infection, if the CD4+ T-helper cell count is below 200. Clinically, this illness presents with fever, shortness of breath, a nonproductive cough, and eventually death if not properly treated. Chest X-ray may show diffuse bilateral interstitial infiltrates, or it can be normal.

The Case of the Breathless Woman Who Had No Helpers

A 22 year-old female comes to the hospital with fever and a feeling of chest tightness. She says she has no medical problems but has lost weight. On physical examination you find large lymph nodes everywhere and numerous genital warts. You note that she is tachypneic, breathing 30 breaths per minute.

You look over her past record and find that she had a child that was born with AIDS.

Her chest film shows diffuse perihilar interstitial streaking bilaterally, sparing the outer lung margins. You order an absolute T4-cell count and find that she has 150 T-helper cells (Normal is greater than 1000).

Diagnosis of *Pneumocystis carinii* can be made by silver-staining or immunofluorescent staining alveolar lung secretions, revealing the flying saucer-appearing fungi. These secretions can be obtained as follows, in order of increasing yield:

1) Induce a sputum sample by spraying saline into the bronchioles and collecting the coughed material (60% sensitivity).

2) Insert a fiberoptic camera (bronchoscope) deep into the patient's bronchial tree, inject saline, and then wash it out (bronchoalveolar lavage) (98% sensitivity).

3) Biopsy the lung by bronchoscopy (100% sensitivity).

About 80% of AIDS patients will get PCP at least once in their lifetime unless prophylactic trimethoprim/

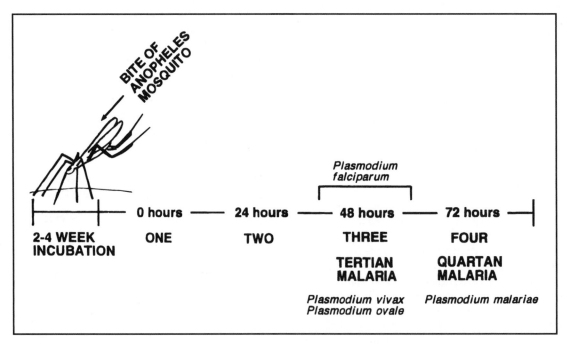

Figure 31-6

sulfamethoxazole is given when CD4+ T-cell counts drop below 200–250. More than 60% of PCP infections are being prevented with this prophylactic intervention! Symptomatic patients can be treated with high dose **trimethoprim/sulfamethoxazole** or intravenous **pentamidine**.

MALARIA
Plasmodium falciparum, Plasmodium vivax, Plasmodium ovale, Plasmodium malariae, and Plasmodium knowlesi

Malaria is a febrile disease caused by 5 different protozoa: *Plasmodium falciparum, Plasmodium vivax, Plasmodium ovale, Plasmodium malariae, and the more recently recognized Plasmodium knowlesi*. They infect about 300–500 million persons worldwide each year, resulting in approximately 1–3 million deaths. The anopheles mosquito carries the organisms within its salivary glands and injects them into humans while it feeds. The organisms then grow in the liver and spread to the human red blood cells, where they reproduce. The red cells fill with protozoa and burst. The red cells all burst at the same time, releasing the protozoa into the bloodstream and exposing them to the immune system, which results in fever.

Fig. 31-6. The different species of *Plasmodium* burst the red cells at different time intervals. *Plasmodium vivax* and *Plasmodium ovale* burst loose every 48 hours. The people who discovered all this started numbering things with one (they didn't have a zero) and so zero hours was called **one**, 24 hours called **two**, and 48 hours called **three**. So, *P. vivax* and *P. ovale* produce chills and fever followed by drenching sweats every 48 hours, which is called **tertian malaria**. *P. malariae* bursts loose every 72 hours, causing a regular 3-day cycle of fevers and chills, followed by sweats. This is called **quartan malaria**. *P. falciparum*, the most common and deadly of the *Plasmodia*, bursts red cells more irregularly, between 36–48 hours. Thus the chills and fevers tend to either fall within this period or be continuous, with less pronounced chills and sweats. *P. knowlesi* periodicity in humans has not yet been established.

Plasmodia Life Cycle

Fig. 31-7. The life cycle of the *Plasmodia* is complex, since the protozoa divide into many different forms with different names. This is clinically important because the diagnosis of this disease rests on being able to identify these forms on a slide of a patient's blood.

Imagine yourself in Kenya with a patient who has intermittent shaking chills, fevers, and soaking sweats. You pull out your little microscope with a reflecting mirror light source, tilt it to catch some of the deep red African sun, and focus your attention on the smear of red cells in front of you. What are those life cycle stages and what do they look like?

Plasmodia undergo sexual division in the anopheles mosquito and asexual division in the human liver and red cells. Let's start with the human:

Thin, motile, spindle-shaped forms of the *Plasmodia*, called **sporozoites**, swim out of the mosquito's sucker and into the human bloodstream. They wiggle their way to the liver and burrow into a liver cell. This marks the beginning of the **pre-erythrocytic cycle** in the liver, so-named because this cycle occurs before the red blood cells (erythrocytes) are invaded. The sporozoite rounds up to form a ball within the liver cell. This ball, now called a **trophozoite**, undergoes nuclear division, forming thousands of new nuclei. This big mass is now a cell with thousands of nuclei, called a **schizont**. A cytoplasmic membrane then forms around each nucleus, creating thousands of small bodies called **merozoites**. The new overloaded liver cell bursts open, releasing the merozoites into the liver and bloodstream. Some will reinfect other liver cells as the sporozoite did initially, repeating the same cycle discussed above, which is now called the **exo-erythrocytic cycle**.

Other merozoites will enter the bloodstream and enter red blood cells, starting the **erythrocytic cycle**. This cycle is similar to the exo-erythrocytic cycle, except that it occurs in the erythrocytes rather than the liver cells. The merozoite rounds up to form a trophozoite. In the red cells the trophozoite is shaped like a ring with the nuclear material looking like the diamond on the ring. Nuclear division then occurs with formation of a large multinucleated schizont. Cytoplasm surrounds each nucleus to form new merozoites within the late schizont. Red cell lysis occurs with release of merozoites. The released merozoites stimulate an immune response, manifested as fevers, chills, and sweats.

The merozoites can continue to invade other red cells and then grow for another 2–3 day cycle followed by rupture and release again. Some merozoites will change into male and female **gametocytes**. These cells circulate and will be taken up by a biting anopheles mosquito. If they are not, they will shortly die.

Two of the species, *P. vivax* and *P. ovale*, produce dormant forms in the liver (**hypnozoites**) which can grow years later, causing relapsing malaria. This is why you are asked if you have ever had malaria when you donate blood (an effort to screen out infected blood).

In the mosquito, the gametocytes are sucked into the stomach where the male and female gametocytes fuse. DNA is mixed and the fused gametocytes become an **oocyst**. The oocyst divides into many spindle-shaped wiggling sporozoites, which disseminate within the mosquito. They may find their way into the mosquito salivary gland and will be injected into the human for asexual reproduction.

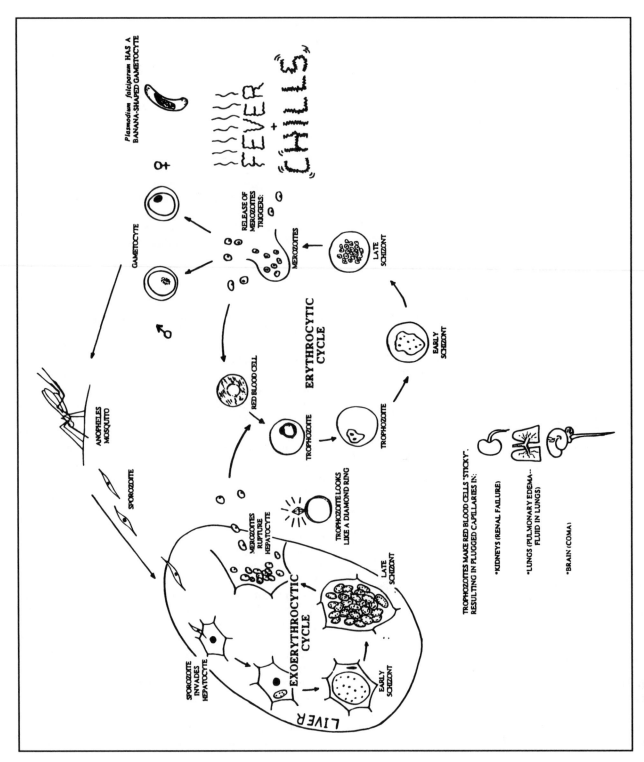

Figure 31-7

	PLASMODIUM FALCIPARUM	PLASMODIUM VIVAX & OVALE	PLASMODIUM MALARIAE	PLASMODIUM KNOWLESI
• Episodes of fever and chills	• Usually continuous	• 48 hour intervals (tertian form)	• 72 hour intervals (quartan form)	• Not known
• Continual reproduction in the liver (hypnozoites, latent liver stage)	• NO!	• **YES!!**	• NO!	• NO!
• Anemia	• **YES**	• **YES**	• **YES**	• **YES**
• Severe clinical manifestations	• **YES** (brain and kidney)	• NO	• NO	• **YES**
• Chloroquine sensitivity	• No- except in parts of Central America	• **YES**	• **YES**	• **YES**

Figure 31-8   MALARIA

The Disease Malaria

Malaria is well known for causing periodic episodes of severe chills and high fevers along with profuse sweating at 48–72 hour intervals. These episodes commonly last about 6 hours and are associated with the rupture of red blood cells.

You can imagine that all these cycles of red blood cell lysis must take their toll! In fact, *P. falciparum*, the most aggressive species, will invade up to 30% of erythrocytes, which results in anemia and sticky red blood cells. These sticky cells plug up post-capillary venules in organs such as the kidney, lung, and even brain, resulting in hemorrhages and blocked blood delivery to those tissues. Renal failure, lung edema, and coma may ensue, leading to death. Most deaths occur in children less than 5 years old in sub-Saharan Africa. These children often develop **cerebral malaria** characterized by seizures and impaired consciousness, leading to coma. Even with treatment, 20% of children with cerebral malaria will die.

Infected individuals also get hepatomegaly and splenomegaly. The spleen and liver enlarge as the fixed phagocytic cells (of the reticuloendothelial system) pick up large amounts of debris from the destroyed red blood cells. The enlarged spleen occasionally ruptures.

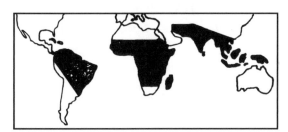

Figure 31-9

Many African-American and African blacks are resistant to *P. vivax* and *P. falciparum* infection. The resistance to *P. vivax* is based on the absence of red cell membrane antigens **Duffy a** and **b** that the *P. vivax* uses for binding. The sickle cell anemia trait (hemoglobin AS) appears to help protect the red cells from *P. falciparum* invasion. Endemic infection with malaria in the African continent is thought to have led to a Darwinian selection process, resulting in high levels of sickle trait and absence of Duffy a and b in many African and African-American blacks.

Fig. 31-8.   Comparison of the *Plasmodia* species.

Diagnosis

1) Examination of thin and thick smears (1000×) of blood, under oil-immersion magnification, reveals the trophozoites and schizonts within the erythrocytes. Sometimes the gametocytes can be visualized.

2) Several rapid antigen based diagnostic tests have been developed that are highly sensitive and specific. Their expense limits widespread use in developing countries.

Control of Malaria

1) Prevent mosquito bite:
   a) Eliminate vector with pesticides (pyrethins) at dusk in living and sleeping areas.
   b) Use insect repellants (containing DEET) and mosquito nets, and wear long-sleeved shirts and long pants.
2) Chemical prophylaxis for travelers: Atovaquone-proguanil, mefloquine, or doxycycline are the most common prophylactic regimens when traveling to

malaria endemic regions. Only in Central America north of the Panama Canal is chloroquine an option for *P. falciparum* prophylaxis. Chloroquine is still an effective drug to prevent the other forms of malaria.

3) Vaccination: This is the ultimate goal, but unfortunately not yet a reality. Phase 3 trial results of vaccine RTS,S/AS01 performed in Africa showed the vaccine induced partial protection against clinical malaria in children ages 5 to 17 months of age. Three doses of the vaccine led to a 36% reduction in clinical malaria. Further studies of this vaccine are ongoing.

Fig. 31-9. Chloroquine-resistant *P. falciparum* areas. Malaria is a disease of the tropics, cutting a swath across the equator. Chloroquine-resistant *Plasmodium falciparum* areas include most of Africa, Central America south of the Panama Canal, South America, India, and South East Asia (see map). **Chloroquine-sensitive** areas are limited to Central America North of the Panama Canal and Hispaniola.

Treatment of Malaria

To treat malaria you must understand two concepts: 1) the geographic pattern of susceptibility of *P. falciparum* to antimalarial drugs (**Fig. 31-9.**) and 2) the type of *Plasmodium* species causing the infection.

1) *P. malariae*, *P. vivax*, and *P. ovale* are all susceptible to **chloroquine** (except in New Guinea and rarely in Asia). Pushovers! But don't forget that *P. vivax* and *P. ovale* have exo-erythrocytic cycles in the liver and will be protected there from chloroquine. The acute infection will subside, but relapses will occur. Treatment with *primaquine* will kill the liver holdouts.

> **Prim**aquine is the **Prime** drug to kill
> *P. vivax* and *P. ovale* in the liver!

2) *Plasmodium falciparum*: This guy is nasty, causing the most hemolysis, organ damage, and death.

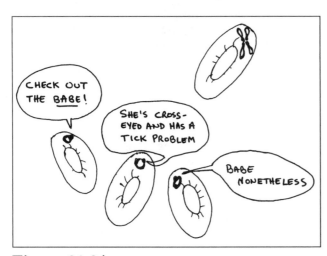

Figure 31-9A

a) Chloroquine-sensitive areas: Huh, I wonder which drug to use?? **Chloroquine** alone is enough as *P. falciparum* does not have an exo-erythrocytic cycle.

b) Chloroquine-resistant areas: Treatment options include **quinine** (**quinidine**, the antiarrhythmic drug, is more expensive but readily available in the United States and just as effective), **artemether** (see below), **atavaquone-proguanil, mefloquine**, and **pyrimethamine/sulfadoxine**. Severe infection (cerebral malaria) is treated with IV or IM quinine, quinidine, or artemether. Therapy with a combination of two anti-malarial agents has become the rule in the face of rapidly growing resistance by *Plasmodium falciparum*.

Newsflash!!! Artemether is new therapy for severe falciparum malaria in children and adults!

Chloroquine-resistant *P. falciparum* causes severe malaria in Africa, killing about 500,000 children a year (1–2 million world-wide). Quinine is the preferred therapy in these areas because it can be injected intramuscularly (IM). A newer drug named artemether (or its brother artesunate) is derived from a traditional Chinese malaria remedy (qinghaosu or wormwood!). Artemether is effective against chloroquine-resistant *P. falciparum* and has proven to work as well as quinine in the treatment of severe malaria. Unfortunately, even with therapy, 15–20% of children with cerebral malaria still die.

There are a number of common features of these drugs (also see **Fig. 31-13**).

1) All of the anti-malarial drugs can be taken **orally**.

2) All of the anti-malarial drugs cause **GI upset** as a primary adverse effect.

3) Chloroquine, primaquine, and quinine all cause hemolysis in patients with glucose-6-phosphate dehydrogenase deficiency (G-6-P-D deficiency is present in some Africans, Mediterraneans, and Southeast Asians).

4) Chloroquine, quinine, quinidine, and sulfadoxine/pyrimethamine are safe in all trimesters of pregnancy. Not enough data is available about the others.

The optimal management of *P. knowlesi* malaria is not known. Clinical experience suggests uncomplicated knowlesi malaria will respond to chloroquine, quinine, and artemether.

BABESIOSIS
(*Babesia microti, Babesia divergens*)

Babesiosis is an infection very much like malaria. It is transmitted by the bite of a blood sucker (tick in this case) and it invades and can be seen inside red blood cells. It also causes fever and hemolysis (anemia), as in malaria. It is *different* in that:

1) There are more than 100 species of *Babesia*, mostly causing disease in cattle and other domestic or wild animals.

2) *Babesia* are spread by tick bites, *not* mosquito bites.

3) They do not affect liver cells (so there is no exo-erythrocytic phase).

In the northeastern coastal United States (e.g. Nantucket Island) *Babesia Microti* is spread by the bite of the same tick that spreads lyme disease, *Ixodes scapularis*. After biting the white-footed mouse, the reservoir for *B. microti*, the tick will leap to the next carefree golfer who walks into the rough.

Like *Plasmodium, Babesia* sporozoites slither out of tick salivary glands into the blood of the hapless golfer. The sporozoites invade erythrocytes and differentiate into pear or ring-shaped trophozoites. Trophozoites asexually bud and divide into 4 merozoites that stick together, forming a cross or x-shaped tetrad ("Maltese cross"). Red cell infection results in only mild hemolysis, so infection is usually asymptomatic and sub-clinical. Asplenic patients are unable to clear the organisms as

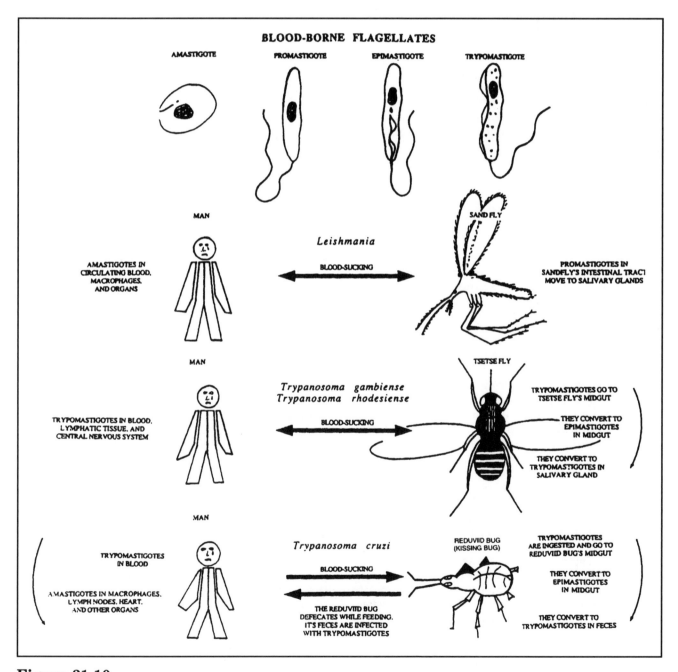

Figure 31-10

well and may have severe infection similar to *falciparum* malaria. (Gelfand, 1995; Persing, 1995). Treat infected patients with a combination of quinine and clindamycin or atovaquone and azithromycin.

Fig. 31-9A. Babesia sporozoites in the "hood". Giemsa or wright-stained thin and thick blood smears reveal ring-shaped trophozoites that look like *Plasmodium* and the distinctive **cross** or x-shaped tetrad of merozoites (Maltese cross). Transmitted by tick vector.

THE BLOOD-BORNE FLAGELLATES
(Leishmania and Trypanosoma)

Fig. 31-10. *Leishmania* and *Trypanosoma* are transmitted by the bite of a blood-sucking insect. Although they cause different diseases, they pass through similar morphologic states in the human and insect host. They can exist as rounded cells without flagella, called **amastigotes**, or as flagellated motile forms called **promastigotes, epimastigotes**, and **trypomastigotes**. These are named according to the insertion site of their single flagellum. All of these organisms cause an initial skin ulcer at the site of the insect bite, followed by systemic invasion.

These parasites can also be transmitted via blood product transfusion.

Leishmaniasis
(Leishmania tropica, Leishmania chagasi, Leishmania major, Leishmania braziliensis, Leishmania donovani)

Leishmania is zoonotic, carried by rodents, dogs, and foxes, and is transmitted to humans by the bite of the **sandfly**. The disease leishmaniasis is found in South and Central America, Africa, and the Middle East.

Following transmission from the sandfly, the promastigote invades phagocytic cells (macrophages) and transforms into the nonmotile amastigote. The amastigote multiplies within the phagocytic cells in the lymph nodes, spleen, liver, and bone marrow (the reticuloendothelial system) (see **Fig. 31-10**).

The different diseases caused by *Leishmania* depend on the invasiveness of the species as well as the host's immune response. Host immunity depends on a cell-mediated defense. It appears that some patients have genetically deficient defenses against *Leishmania* and will be afflicted with more severe disease. Leishmaniasis presents in a spectrum of disease severity: from a single ulcer that will heal without treatment; to widely disseminated ulcerations of the skin and mucous membranes; to the very severe infection striking deep into the reticuloendothelial organs, the spleen and liver. Note the similarity here to leprosy (see Chapter 15) in which differences in host cell-mediated defenses result in varied severity of disease.

There are 3 clinical forms of this disease:

1) Cutaneous leishmaniasis
   a) Simple cutaneous lesions
   b) Diffuse cutaneous lesions
2) Mucocutaneous leishmaniasis
3) Visceral leishmaniasis

Cutaneous Leishmaniasis

A sandfly injects *Leishmania* into the skin, where they migrate to reticuloendothelial cells (fixed phagocytic cells in lymph nodes). At the site of the sandfly bite, a skin ulcer develops, called an "oriental sore." This ulcer heals in about a year, leaving a depigmented (pale) scar. Diagnosis is made by observing *Leishmania* in stained skin-scrapings from the ulcer base.

Cell-mediated immunity is intact, so the immune system attacks the organisms resulting in skin destruction (ulcer formation) and clearance of infection (similar to the situation with tuberculoid leprosy). Because of the intact cell-mediated immunity, this organism invokes a delayed hypersensitivity reaction. Diagnosis can be made by injecting killed *Leishmania* intradermally (Leishmanin skin test). Just like the PPD test of tuberculosis, a raised indurated papule 48 hours later supports the presence of a *Leishmania* infection.

This disease is also seen in Latin America and Texas, where it is called American cutaneous leishmaniasis.

Diffuse Cutaneous Leishmaniasis

In Venezuela and Ethiopia, a chronic form of cutaneous leishmaniasis occurs in patients with deficient immune systems. A nodular skin lesion arises but does not ulcerate. With time, numerous nodular lesions arise diffusely across the body. There is often a concentration of lesions near the nose. The untreated infection can last more than 20 years.

The disease is diffuse because the host's immune system does not respond to the invasion by *Leishmania*, due to a defect in cell-mediated immunity. Therefore, the promastigotes are able to spread and infect the skin, causing the diffuse nodular lesions. Due to the defect in cell-mediated immunity, the Leishmanin skin test is negative (similar to the situation in lepromatous leprosy).

Mucocutaneous Leishmaniasis

Initially, a dermal ulcer, similar to cutaneous leishmaniasis, arises at the site of the sandfly bite and soon heals. However, months to years later, ulcers in the mucous membranes of the nose and mouth arise. If untreated, the infection is chronic, with erosion of the nasal septum, soft palate, and lips, over a course of 20–40 years. Death by bacterial secondary infection may occur.

Diagnosis is made via skin scrapings.

Visceral Leishmaniasis (Kala-azar)

The sandfly transmits *Leishmania donovani* or *Leishmania chagasi* to an individual (most commonly young malnourished children), who months later will complain of abdominal discomfort and distension, low-grade fevers, anorexia, and weight loss. This abdominal enlargement is due to *Leishmania donovani*'s invasion of the reticuloendothelial cells (fixed phagocytic cells) of the spleen and liver, causing hepatomegaly and **massive splenomegaly**. Patients also develop a severe anemia and the white blood-cell count can also be very low. Most cases (over 90%) are fatal if untreated.

Diagnosis is made by liver and spleen biopsies demonstrating these protozoa. Alternatively, the diagnosis may be made by demonstrating high titer anti-leishmanial IgG levels. The Leishmanin skin test is negative during active disease as cell-mediated immunity is deficient.

All forms of leishmaniasis can be treated with the pentavalent antimonial **stibogluconate**.

African Sleeping Sickness
(*Trypanosoma brucei rhodesiense* and *Trypanosoma brucei gambiense*)

Trypanosoma rhodesiense and *Trypanosoma gambiense* are responsible for African sleeping sickness, which is transmitted by the blood-sucking bite of a **tsetse fly**. Following this bite, the motile flagellated form of these 2 organisms, called a **trypomastigote**, spreads via the person's bloodstream to the lymph nodes and central nervous system (CNS) (see **Fig. 31-10**).

The first manifestation is a hard, red, painful skin ulcer that heals within 2 weeks. With systemic spread, the patient then experiences fever, headache, dizziness, and lymph node swelling. These symptoms can last a week, and then the fever subsides for a few weeks followed by renewed fevers. This pattern of fevers with fever-free intervals can occur for months. Finally, CNS symptoms develop, with drowsiness in the daytime (thus sleeping sickness), behavioral changes, difficulty with walking, slurred speech, and finally coma and death.

There are 2 forms of African sleeping sickness. **West African sleeping sickness**, caused by *Trypanosoma brucei* **gambiense**, is notable for slowly progressing fevers, wasting, and late neurologic symptoms. **East African sleeping sickness**, caused by *Trypanosoma brucei rhodesiense*, is similar to the West African variety but more severe, with death occurring within weeks to months. There is rapid progression from recurrent fevers to early neurologic disease (drowsiness, mental deterioration, coma, and death).

Q: Why the intermittent fevers???

A: Variable surface glycoproteins (VSG). The trypanosomes are covered with about 10 million molecules of a repeating single glycoprotein called the VSG. The trypanosomes possess genes that can make thousands of different VSGs. They will make and express, on their surface, a new VSG in a cyclical nature. Every time the human host develops antibodies directed against the VSG (and fever with this immune recognition), the trypanosomes produce progeny with a new VSG coat. Thus, there are waves of new antigens, producing recurrent fevers and protection from our immune defenses. This is similar to the antigenic variation of the spirochete *Borrelia recurrentis*, which causes relapsing fever (see **Fig. 14-12**).

Diagnosis consists of visualization of trypomastigotes in peripheral blood, lymph nodes, or spinal fluid. More recently a card agglutination test which assays for anti-*T. brucei gambiense* antibodies in the serum is available, but microscopy should still be performed.

Patients are treated with **suramin** or **pentamidine** if the central nervous system (CNS) is not involved (suramin does not penetrate into the CNS). With CNS involvement and *T. brucei gambiense* infection, **eflornithine** plus **nifurtimox** is preferred treatment, with the arsenical compound **melarsoprol** as an alternative. In *T. brucei rhodesiense* infections of the CNS, **melarsoprol** is the first-line agent.

Chagas' Disease
(*Trypanosoma cruzi, the American Trypanosome*)

Chagas' disease is caused by a trypanosome, but the pathogenesis and epidemiology differ greatly from the African trypanosomes.

This is truly a disease of the Americas, ranging from the southern U. S. (Texas), Mexico, Central America, and down into South America. *T. cruzi* survives in wild animal reservoirs such as rodents, opossums, and armadillos. The vector is the **reduviid bug**, also called the kissing bug. The bug feeds on humans while they sleep and defecates while it eats. *T. cruzi* trypo-mastigotes, which are present in the bug's feces, tunnel into the human host. The trypomastigote loses its undulating membrane and flagellum and rounds up to form the amastigote, which rapidly multiplies. Organisms invade the local skin, macrophages, lymph nodes, and spread in the blood to distant organs (see **Fig. 31-10**).

Acute Chagas' Disease

At the skin site of parasite entry, a hardened, red area develops, called a **chagoma**. This is followed by systemic spread with fever, malaise, and swollen lymph nodes. Organs that can be infected include the

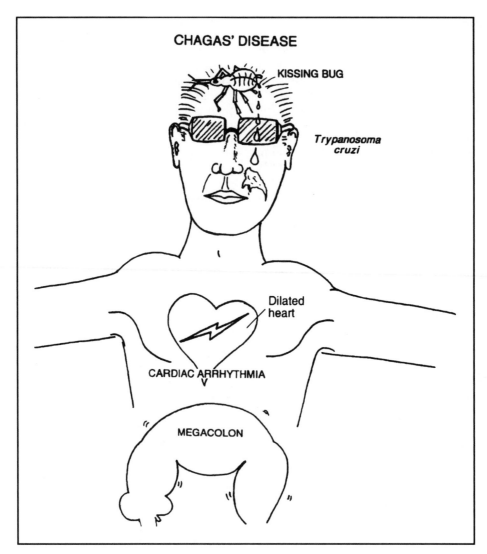

CHAGAS' DISEASE

Figure 31-11

heart and central nervous system (CNS). Heart inflammation results in tachycardia and electrocardiographic changes, while the CNS involvement can result in a severe meningoencephalitis (usually in young patients).

This acute illness resolves in about a month and patients then enter the **intermediate phase**. In this phase there are no symptoms, but there are persistently low levels of parasite in the blood as well as antibodies against *T. cruzi*. Most persons will remain in the intermediate phase for life.

For reasons that are poorly understood, some persons will develop chronic Chagas' disease years to decades later.

Chronic Chagas' Disease

The organs primarily affected are the heart and some hollow organs such as the colon and esophagus. Intracellular *T. cruzi* amastigotes cannot usually be found, and it is unclear why disease develops in these organs.

1) **Heart**: Arrhythmias are the earliest manifestation (heart block and ventricular tachycardia). Later there is an increase in heart size and heart failure (dilated cardiomyopathy).

2) **Megadisease of colon and esophagus**: A big, dilated, poorly functioning esophagus develops with symptoms of difficulty and pain in swallowing, and regurgitation of food. A dilated colon (megacolon) results

in constipation and abdominal pain. Patients can go weeks without bowel movements.

Fig. 31-11. *Trypanosoma cruzi* (the American Trypanosome). To remember that *T. cruzi* causes megacolon, electrical arrhythmias, and dilatation of the heart, and is transmitted by the feces of the kissing bug, picture Tom **Cruise** (the **American** actor).

Diagnosis and Treatment

Acute Chagas' disease:

1) Direct examination of the blood for the motile trypomastigotes.

2) **Xenodiagnosis**: This sensitive test is conducted as follows. Forty laboratory-grown reduviid bugs are allowed to feed on the patient, and one month later the bugs' intestinal contents are examined for the parasite.

Chronic Chagas' disease:

Classic clinical findings (cardiac and megadisease) along with serologic evidence of past *T. cruzi* infection allows for presumptive diagnosis.

Although **nifurtimox** and **benznidazole** can be used for acute cases, the response to therapy of chronic Chagas' disease has been less dramatic. Therefore, individuals should take precautions to prevent kisses by the kissing bug (insect repellent, bednets).

Balantidium Coli

If you do not want diarrhea, do not consume food or water contaminated by pig feces! This advice will prevent ingestion of *B. coli* cysts. These cysts mature into **ciliated** trophozoites, and travel to the intestinal tract. The trophozoites dig into the intestinal wall, where they exist happily consuming the native bacteria. Most infected individuals are asymptomatic, while some will develop diarrhea.

B. Coli trophozoites are notable for being the largest parasitic protozoans found in the intestine. Diagnosis is made by identifying the ciliated trophozoites or cysts in stool specimens. Tetracycline is effective at treating this infection.

Fig. 31-12. Summary of the protozoan diseases.
Fig. 31-13. Treatment of protozoan diseases.

References and Recommended Reviews

Baird JK. Effectiveness of antimalarial drugs. N Engl J Med 2005;352:1565–77.

Barrett MP, et al. The trypanosomiases. Lancet 2003;362: 1469–80.

Chen XM, Keithly JS, Paya CV, LaRusso NF. Cryptosporidiosis. N Engl J Med 2002;346:1723–31.

Freedman DO. Clinical practice. Malaria prevention in short-term travelers. N Engl J Med. 2008;359(6):603–12.

Gelfand JA. Babesia. In: Mandell GL, Bennett JF, Dolin R, eds. Principles and Practice of Infectious Disease. 4th ed. New York: Churchill Livingstone, 1995;2415–2427.

Greenwood BM, Bojang K, Whitty CJM, Targett GAT. Malaria. *www.thelancet.com* 2005;365:1487–98.

Haque R, Huston CD, Hughes M, Houpt E, Petri WA. Amebiasis. N Eng J Med 2003;348:1565–73.

Kirchhoff LV. American trypanosomiasis (Chaga's disease)-a tropical disease now in the United States. N Engl J Med 1993;329:639–44.

Murray HW, Berman JD, Davies CR, Saravia NG. Advances in leishmaniasis. *www.thelancet.com* 2005;366:1561–76.

Schuster FL, Visvesvara GS. Free-living amoebae as opportunistic and non-opportunistic pathogens of humans and animals. Int J for Parasitology 2004;34:1001–27.

The RTS, S Clinical Trials Partnership. A Phase 3 Trial of RTS,S/AS01 Malaria Vaccine in African Infants. N Engl J Med 2012;367:2284–95.

William T, et al. Severe *Plasmodium knowlesi* Malaria in a Tertiary Care Hospital, Sabah, Malaysia. Emerg Infect Dis 2011;17:1248–55.

PROTOZOA	TRANSMISSION	MORPHOLOGY	CLINICAL FINDINGS
Entamoeba histolytica	• Fecal-oral	*AMOEBA* 1. Oocyst 2. Trophozoite: motile • Bulls eye shaped nucleus, with red blood cells in the cytoplasm	1. Asymptomatic carriage 2. **Bloody** diarrhea: when trophozoites invade the intestinal mucosa, causing erosions 3. Liver abscess
Giardia lamblia	• Fecal-oral	1. Oocyst 2. **Flagellated** trophozoite	• Foul smelling, greasy diarrhea (with high fat content), and abdominal gassy distension
Cyclospora cayetanesis	• Oocysts from stool contaminate fruits & vegetables	• Oocysts	1. Watery diarrhea 2. Nausea & vomiting
Cryptosporidium	• Fecal-oral	1. Oocyst is infective agent (contains 4 sporozoites) 2. Life cycle occurs within infected epithelial cells	• Watery diarrhea, vomiting and abdominal pain (usually self-limiting, but can be life-threatening in immunocompromised patients)
Isospora species	• Fecal-oral	• Oocyst is infective agent (contains 8 sporozoites)	• Severe diarrhea and malabsorption in AIDS patients
Trichomonas vaginalis	• Sexually transmitted	1. No cyst stage 2. **Flagellated** trophozoite	1. Painful vaginal itching 2. Burning on urination 3. Yellow-green malodorous, frothy vaginal discharge
Naegleria fowleri	• Lives in freshwater lakes	• Amoeba	• Acute meningitis, which is usually fatal within a week
Acanthamoeba species	1. Lives in fresh water lakes 2. Eye infections from dirty contact lenses	1. Amoeba stage 2. Cyst stage in brain	1. Chronic, granulomatous brain infection, which is fatal in a year 2. Corneal infection: often associated with the use of contact lenses that are cleaned in non-sterile solutions
Balamuthia mandrillaris	Lives in fresh water and soil	1. Amoeba stage 2. Cyst stage in brain	1. Chronic, granulomatous brain infection- poor prognosis 2. Granulomatous skin infection

Figure 31-12   PROTOZOA

DIAGNOSIS	TREATMENT	MISCELLANEOUS
1. Fecal exam: look for cysts and trophozoites with red blood cells in their cytoplasm 2. Serology 3. Abdominal CT scan: look for liver abscesses	1. Metronidazole (first-line) 2. Tinidazole (alternative) Either of the above followed by an intraluminal agent such as: paromomycin or iodoquinol.	1. Not all species of *Entamoeba* are pathogenic 2. 90% of people infected with *Entamoeba histolytica* are **asymptomatic**
1. Fecal exam: look for cysts and trophozoites 2. Commercial immunoassay kit: Test stool	1. Metronidazole (first-line) 2. Tinidazole 3. Nitazoxanide	
• Stool exam reveals oocysts that fluoresce under UV light and are acid-fast	• Trimethoprim/sulfamethoxazole (Bactrim)	Associated with strawberries & raspberries
1. Fecal exam: look for oocyst 2. Biopsy of lining of the small intestine	1. Immunocompetent: no therapy needed 2. Persistent diarrhea or immunocompromised: nitazoxanide or paromomycin +/− azithromycin.	• Obligate intracellular parasite
1. Fecal exam: look for oocyst (acid-fast or fluorescent stain required) 2. Biopsy 3. **Eosinophilia**	• Trimethoprim/sulfamethoxazole (Bactrim)	• Obligate intracellular parasite
1. Examine vaginal discharge: identify these highly motile protozoa 2. Examination of urine for these protozoa	• Metronidazole	• Provide metronidazole to the patient's sexual partners
• CSF exam: look for motile amoeba, as well as white blood cells (suggesting an infection)	• Amphotericin B	
1. Examination of CSF and brain tissue: reveals both cysts and mature trophozoites 2. Examination of corneal scrapings	• Chronic meningoencephalitis: multiple antifungal drugs plus pentamidine, with questionable success	• Corneal infection: A. Topical antimicrobial agents B. Corneal transplant
1. Examination of brain tissue for cysts and trophozoites 2. Skin biopsy 3. Immunofluorescent staining	1. Fluconazole 2. Azithromycin 3. Pentamidine	Infects both immunocompetent immunocompromised hosts

PROTOZOA	TRANSMISSION	MORPHOLOGY	CLINICAL FINDINGS

Toxoplasma gondii	1. Ingestion of oocysts in raw pork
	2. Inhalation of oocysts from cat feces
	3. Congenital: if a pregnant women is exposed to *Toxoplasma* for the very first time, she can pass this infection to her fetus
		1. Oocyst is infectious
		2. Trophozoites
			1. Congenitally acquired (**TO**rches organism):
			  A. Still birth; chorioretinitis; blindness; seizures; mental retardation; microcephaly
			  B. Normal appearing infants may develop reactivation as adolescents or adults: chorioretinitis (which can lead to blindness)
			2. Immunocompromised patients: disseminated infection which may include:
			  A. Encephalitis presenting as a brain mass
			  B. Chorioretinitis
			  C. Lymph node, liver, and spleen enlargement
			  D. Pneumonia

Pneumocystis carinii	1. Acquired at an early age by respiratory route
	2. Remains latent in normal hosts
		• Flying-saucer appearing **Fungus** that was previously classified as a protozoan	• **Interstitial pneumonia**: with fever and a dry, nonproductive cough. Major pathogen in AIDS patients (when CD4 count is less than 200)
			  A. 15% chance of infection/year in AIDS patients
			  B. 80% lifetime risk without prophylactic trimethoprim/sulfa

Malaria:
1. *Plasmodium falciparum*
2. *Plasmodium vivax*
3. *Plasmodium ovale*
4. *Plasmodium malariae*
5. *Plasmodium knowlesi*
	• Female anopheles mosquito	1. See life cycle diagram (Fig 31-7)
		2. **Hypnozoite** stage: dormant form that hangs out in the liver
		  A. *P. vivax*
		  B. *P ovale*
			Malaria
			1. Periodic episodes of high fever and shaking chills, followed by period of profuse sweating (sweats occur when the red blood cells burst and release merozoites)
			  A. **Tertian** malaria: episodes occur every 48 hours (*P. vivax* and *P. ovale*)
			  B. **Quartan** malaria: episodes occur every 72 hours (*P. malariae*)
			  C. *P. falciparum* (most common and deadly): irregular episodes
			2. Anemia
			3. Hepatomegaly
			4. Splenomegaly (& occasionally splenic rupture)
			5. Brain, lung, and/or kidney damage with *P. falciparum*

Babesia microti	*Ixodes scapularis* tick	1. Sporozoites from tick cause infection 2. Mature into trophozoites in humans and infect red blood cells.	1. Immunocompetent: usually asymptomatic. 2. Immunocompromised: anemia due to hemolysis, fatigue and protracted course.
Leishmania 1. *Leishmania tropica*: 2. *Leishmania chagasi* 3. *Leishmania major*	1. **Sandfly** bite 2. Contaminated blood transfusion 3. Zoonotic: carried by rodents, dogs, and foxes	• *Promastigote*: **flagellated** (in sandfly) • *Amastigote*: intracellular & non-flagellated: in phagocytic cells of	1. **Cutaneous** leishmaniasis: single ulcer at site of the sandfly bite (oriental sore). Heals in a year, leaving a depigmented scar. 2. **Diffuse cutaneous**: nodules at bite site (which do not ulcerate) and over body (especially near the nose); can last 20 years without treatment

Figure 31-12   (continued)

DIAGNOSIS	TREATMENT	MISCELLANEOUS
1. Serology (high IgM and IgG titers) 2. Radiology: CT scan shows contrast-enhancing mass in the brain 3. Examination of the retina	1. Sulfadiazine & pyrimethamine 2. If allergic to sulfadiazine: clindamycin 3. Administer folinic acid when using pyrimethamine, since it decreases the risk of bone marrow depression without affecting pyrimethamine's anti-toxoplasmal effect 4. Pregnant women should avoid cats	• Obligate intracellular parasite
1. Silver stain: to see flying saucer appearing fungi in: A. Saline induced sputum B. Bronchoalveolar lavage with bronchoscope C. Bronchial wall biopsy with bronchoscope 2. X-ray: find an interstitial pneumonia, with diffuse infiltrates	*Agents used for treatment or prophylaxis:* 1. Trimethoprim/sulfamethoxazole (Bactrim) 2. Pentamidine 3. Atovaquone 4. Dapsone 5. Clindamycin	1. AIDS patients with CD4 positive T-cell counts less than 200 are susceptible 2. The most common opportunistic infection in AIDS
1. Examination of blood smear reveals: A. Trophozoites (diamond ring shaped) B. Schizonts C. Gametocytes 2. Rapid antigen diagnostics are available.	1. *P. ovale* and *P. vivax*: chloroquine plus primaquine 2. *P. malariae*: chloroquine 3. *P. falciparum*: Quinine (or quinidine), atovaquone, mefloquine, and artemesinin compounds combined with a second agent. 4. *P. knowlesi*: no firm guidelines, but clinically has responded to chloroquine, quinine and artemesinin compounds. 5. Prophylaxis: atavaquone-proguanil, mefloquine, doxycycline.	• Many African Americans are resistant to certain species of malaria: 1. Red blood cells that lack the cell membrane antigens **Duffy a** and **b** provides resistance to infection by *P. vivax* 2. **Sickle cell anemia trait** provides resistance to infection by *P. falciparum*
1. Blood smear 2. PCR (more sensitive than blood smear) 3. Serology	1. Quinine combined with clindamycin 2. Atovaquone combined with azithromycin.	1. Classic "maltese cross" on blood smear. 2. Patients may be co-infected with Lyme disease
1. Demonstration of protozoa A. Blood smear B. Biopsy of skin lesions, spleen or liver 2. Leishmanin skin test is negative in patients with low (defective) cell-mediated immunity	1. Stibogluconate 2. Amphotericin B 3. Miltefosine	• The different diseases caused by *Leishmania* depend on the species, as well as the patient's cell mediated immune response

PROTOZOA	TRANSMISSION	MORPHOLOGY	CLINICAL FINDINGS
4. *Leishmania braziliensis* 5. *Leishmania donovani*		the reticuloen-dothelial system (lymph nodes, spleen, liver and bone marrow)	3. **Mucocutaneous**: ulcers appear on mucous membranes after first ulcer at bite site heals. Ulcers erode the nasal septum, soft palate and lips. Can last 20–40 years 4. **Visceral** leishmaniasis (Kala-azar): common in young, malnourished children. Fever, anorexia, weight loss, & abdominal swelling (from hepatomegaly and massive splenomegaly). Often fatal

African Trypanosome
Trypanosoma rhodesiense
Trypanosoma gambiense
	1. **Tsetse fly** bite
	2. Contaminated blood transfusion
		1. *Trypomastigote* is the motile (**flagellated**) extracellular form living in blood, lymph nodes, and CNS
		2. *Trypomastigote* and *epimastigote*: in tsetse fly
			African Sleeping Sickness
			1. Hard, red painful skin ulcer at the site of the tsetse fly bite, which heals in 2 weeks
			2. Fever, headache, and lymph node swelling
			3. Fevers subside, then relapses can occur (and last for months)
			4. CNS symptoms develop: daytime drowsiness, behavioral changes, difficulty walking, slurred speech, coma and death
			  A. West African sleeping sickness (*T. gambiense*): slowly progressing fevers, wasting, and late neurologic symptoms
			  B. East African sleeping sickness (*T. rhodesiense*): more severe, with rapid cycling of fevers, leading to neurologic symptoms and death in weeks to months

American Trypanosome
Trypanosome cruzi
	1. **Kissing Bug** (reduviid bug): defecates on human skin while feeding. Trypomastigotes, present in the feces, tunnel into the skin
	2. Contaminated blood transfusion
		1. *Trypomastigote* is the motile (**flagellated**) extracellular form living in blood
		2. *Amastigote*: intracellular & non-motile: present in macrophages, lymph nodes, and organs (heart and brain)
		3. *Trypomastigote* and *epimastigote*: in kissing bug
			Chagas' disease:
			1. **Chagoma**: hardened red area at site of parasite entry
			2. **Acute Chagas' disease**: fever, malaise and swollen lymph nodes
			  A. Meningoencephalitis
			  B. Acute myocarditis with tachycardia and EKG changes
			3. **Intermediate** phase: low levels of parasites in blood and positive antibodies against *T. cruzi*, but **NO** symptoms. Most persons remain in this phase for life
			4. **Chronic** Chagas' disease (some people progress to this stage):
			  A. Cardiomyopathy: dilated heart, heart failure, and arrhythmias
			  B. **Megadisease**: large dilated poorly functioning hollow organs lead to:
			    1. Megacolon: constipation and abdominal pain
			    2. Megaesophagus: difficulty and pain with swallowing, and vomiting of food

Figure 31-12 (continued)

DIAGNOSIS	TREATMENT	MISCELLANEOUS
A. Diffuse cutaneous leishmaniasis B. Active visceral leishmaniasis 3. Elevated serum anti-leishmanial IgG Ab titers.		
1. Visualize trypomastigotes in blood, spinal fluid, or lymph nodes 2. Serology (high IgM titers) 3. Antibody agglutination test for *T. gamiense*	*T. gambiense-* 1. pentamidine (no CNS involvement) 2. eflornithine and nifurtimox (CNS involvement) *T. rhodesiense-* 1. suramin (no CNS involvement) 2. melarsoprol (CNS involvement)	• **ANTIGENIC VARIATION:** The trypanosomes express on their surface a new variable surface glycoprotein (VSG) in a cyclic nature. As antibodies form against a particular VSG, the trypanosome will produce progeny with a different VSG that can elude the immune response for a while. This is the mechanism for the recurrent fevers
Acute Chagas: 1. Visualize trypomastigotes in blood. 2. **Xenodiagnosis:** 40 laboratory grown reduviid bugs are allowed to feed on a patient. One month later the bugs' intestinal contents re examined for the parasite. Chronic Chagas: 1. Clinical diagnosis supported by serology and/or PCR.	1. Nifurtimox (for acute cases only) 2. Benznidazole	

M. Gladwin, W. Trattler, and S. Mahan, *Clinical Microbiology Made Ridiculously Simple* ©MedMaster

NAME	MECHANISM OF ACTION	PHARMOKINETICS
Chloroquine	1. Inhibits heme polymerase activity leading to a toxic build-up of heme. 2. Kills erythrocyte form only!! (so does not kill liver schizonts of *P. vivax* & *P. ovale*)	• Oral or IV
Primaquine	• Kills liver schizonts of *P. vivax* & *P. ovale*	• Oral
Quinine	1. Inhibits heme polymerase activity leading to a toxic build-up of heme. 2. Kills erythrocytic and gametocyte forms	• Oral, IV & IM
Mefloquine	• Kills erythrocyte forms only	• Oral
Doxycycline	Inhibits ribosome (see figure 17-14)	• Oral
Pyrimethamine/sulfadoxine (Fansidar)	• Inhibits synthesis of tetrahydro-**folate** (TH4), which is a crucial cofactor for the synthesis of purines (nucleic acids). Inhibition of TH4 production will therefore block DNA synthesis	• Oral
Artemether & Artesunate	• Blocks a P-falciparum-encoded endoplasmic reticulum ATPase	1. IM, IV or suppository 2. Fast acting

Figure 31-13   ANTI-PARASITIC DRUGS

ADVERSE EFFECTS	THERAPEUTIC USES	MISCELLANEOUS

1. Color vision changes, central visual loss & potentially permanent retinal damage. May reverse following discontinuation of therapy
2. GI disturbances
3. Pruritus, especially in dark skinned persons
4. Acute hemolytic anemia: in patients deficient in the enzyme glucose-6-phosphate dehydrogenase (G-6-P-D)
5. Safe in pregnancy
	1. Treatment & prophylaxis against malaria caused by non-resistant *P. falciparum* and *P. malariae*
	2. Used in combination with primaquine for *P. vivax* & *P. ovale*
		1. Only useful against *P. falciparum* in parts of Central America North of the Panama canal.
		2. *Also used for:*
		   A. Rheumatoid arthritis
		   B. Systemic lupus erythematosus

1. Acute hemolytic anemia: in G-6-P-D deficiency 2. Do not use in pregnancy	• For liver stage of *P. vivax* & *P. ovale*. Use in combination with chloroquine	

1. *CINCHONISM*:
   A. Ears: tinnitus, vertigo
   B. Eyes: visual disturbances
   C. GI: nausea, vomiting and diarrhea
   D. CNS: headache and fever
2. Acute hemolytic anemia: in G-6-P-D deficiency
3. Hypotension and heart block (when given IV)
4. Blackwater fever (rare): massive lysis of red blood cells, causing dark urine with hemoglobinuria, renal failure, intravascular coagulation, & possibly death
5. Safe in pregnancy
	• Use in combination with Fansidar (pyrimethamine/sulfadoxine) for chloroquine-resistant *P. falciparum*	• *Also used for:*
		   1. Nocturnal leg cramps
		   2. Local anesthesia

1. Transient nausea and vomiting 2. Do not use in pregnancy 3. Vivid dreams, dizziness, and reports of psychosis.	1. Acute malaria: used for treatment of chloroquine-resistant *P. falciparum* 2. Prophylaxis: Once weekly dosing and effectiveness make this a first line drug.	1. Mefloquine-resistant *P. falciparum* has been documented in Cambodia & Eastern Thailand
1. Photosensitivity, nausea 2. Stains permanent teeth & inhibits bone growth in children < 8 years 3. Do not use in children or pregnant women	• Prophylaxis: alternative to mefloquine	• Do not use in children or pregnant women
1. Bone marrow depression 2. Safe in pregnancy	• Used for treatment (with quinine) in areas of chloroquine-resistant *P. falciparum*	• Pyrimethamine has the same mechanism of action as trimethoprim (anti-bacterial), methotrexate (anti-cancer) & PAS (anti-tuberculosis)
• No serious toxicity reported	• Alternative for severe chloroquine-resistant malaria	• Both are Artemisinin (ginghaosu) derivatives: Old Chinese herbal remedy for malaria

NAME	MECHANISM OF ACTION	PHARMOKINETICS
Atovaquone-Proguanil	1. Atovaquone-inhibits parasite mitochondrial electron transport. 2. Proguanil-inhibits dihydrofolate reductase	Oral
ANTI-PROTOZOAL DRUGS		
Eflornithine	Inhibits the enzyme ornithine decarboxylase	Intravenous
Metronidazole		• Oral
Trimethoprim/Sulfamethoxazole (Bactrim)	• Together, these two drugs inhibit the synthesis of tetrahydro-**folate** (TH4) 2. Animal cells do not synthesize TH4. Therefore, this antibiotic does not block mammalian DNA synthesis	• Can be given orally or IV
Pentamidine	• Unknown	1. Intravenous 2. Aerosolized
Atovaquone		• Oral
Pyrimethamine plus sulfadiazine	Inhibits synthesis of TH4	
Nitazoxanide	Believed to interfere with electron transfer essential to anaerobic metabolism	• Oral
Stibogluconate	• Arsenical compound	• IV or IM
Suramin		1. IV 2. Does **NOT** penetrate into CNS
Melarsoprol	• Arsenical compound	1. IV 2. Penetrates into CNS
Nifurtimox		• Oral

Figure 31-13 (continued)

ADVERSE EFFECTS	THERAPEUTIC USES	MISCELLANEOUS
1. Occasional GI upset	Treatment and prophylaxis of *P.falciparum*	Most common agent used for prophylaxis in US travelers to *P.falciparum* endemic regions
1. Anemia 2. Leukopenia 3. Thrombocytopenia	*T. gambiense* (CNS involvement)	Given in combination with nifurtimox
1. Metallic taste in the mouth 2. Antabuse (disulfiram) effect: Severe hangover symptoms occur following consumption of alcohol	1. *Entamoeba histolytica* 2. *Giardia lamblia* 3. *Trichomonas vaginalis* 4. Anaerobic bacteria (including *Bacteroides* & *C. difficile*) 5. *Gardnerella* (bacterial vaginitis)	• Tindazole is an effective alternative for cases of metronidazole-resistant *Trichomonas vaginalis*. Similar side effects as metronidazole.
1. Gastrointestinal disturbances 2. Skin rashes 3. Patients with low folate levels can get macrocytic anemia. Coadministering folinic acid will prevent the anemia without affecting its antibacterial effect 4. Stevens-Johnson Syndrome (rare sulfa side effect)	1. *Pneumocystis carinii* (both prophylaxis & treatment): First line therapy 2. *Cyclospora cayetanensis* 3. *Isospora* 4. Many bacterial species	• See Miscellaneous antibiotic chart, Chapter 19, for more details
1. Hypoglycemia (6–9%) 2. Diabetes 3. Reversible renal failure 4. Many others: hypotension, nausea, rash, metallic taste, etc	1. *Pneumocystis carinii*: Second line therapy A. Treatment = IV administration B. Prophylaxis: inhalation 2. *T. gambiense* (no CNS involvement)	50% of treated patients experience adverse events
	• *Pneumocystis carinii*	MALARONE (atovaquone and proguanil hydrochloride) is a fixed-dose combination used for the prophylaxis of *P. falciparum* malaria
	• *Toxoplasma gondii*	• Provide folinic acid which decreases the amount of bone marrow suppression, without changing its effectiveness against toxoplasmosis
GI upset	1. *Cryptosporidium* 2. *Giardia lamblia* 3. *Clostridium difficile*	
	• Leishmania	
	1. *T. rhodesiense* (no CNS involvement) 2. Also used for *Onchocerca volvulus*	
	• *T. rhodesiense* (with CNS involvement)	
	• Drug of choice for American trypanosomiasis (*T. cruzi*)	

CHAPTER 32. HELMINTHS

Helminths is Greek for "worm." Worms are usually macroscopic, although diagnosis often requires the visualization of the eggs, which are microscopic, in the stool. We will discuss 16 types of worms that cause significant infections in humans. The first 10 are **roundworms** (**nematodes**), and the last 6 are the more primitive **flatworms (platyhelminthes)**. By understanding their life cycles, you can learn ways to prevent or eradicate helminthic infections.

Within the normal human host there is usually no immune reaction to living worms. However, there is often a marked response to dead worms or eggs. With many of the worm infections, our immune system is kind enough to raise a red flag—elevating the level of eosinophils in the blood—thereby assisting in diagnosis.

NEMATODES
(Roundworms)

Intestinal Nematodes

"Intestinal" nematodes all mature into adults within the human intestinal tract. The larval forms of many of these roundworms may be distributed widely throughout the body.

Three of the intestinal nematodes are acquired by the ingestion of eggs: *Ascaris lumbricoides*, *Trichuris trichiura* (whipworm), and *Enterobius vermicularis* (pinworm). Two worms, *Necator americanus* (hookworm) and *Strongyloides stercoralis*, are acquired when their larvae penetrate though the skin, usually of the foot. *Trichinella spiralis* is acquired by the ingestion of encysted larvae in muscle (pork meat).

With infection, most of these intestinal worms (except for *Enterobius* and *Trichuris*, which stay in the intestinal tract) invade other tissues at some stage of their life cycle. This stimulates our immune system to raise the number of eosinophils in the blood.

Fig. 32-1. The first 3 roundworms (*Ascaris lumbricoides, Necator americanus*, and *Strongyloides stercoralis*) all have a larval form that migrates through the tissue and into the lung at some stage of their life cycle. The larvae grow in the lung, are coughed up and swallowed into the intestine, where they grow into adult worms.

1) ***Ascaris lumbricoides***: Infection occurs in the tropics and mountainous areas of the southern U.S., when individuals **consume food** that is contaminated with eggs. Larvae emerge when the eggs reach the small intestine. The larvae penetrate through the intestinal wall and travel in the bloodstream to the lungs. The larvae grow in lung alveoli until they are coughed up and swallowed. These larvae again reach the small intestine and mature into adults. Here each adult worm produces over 200 thousand eggs per day, which are excreted in feces.

2) ***Necator americanus***: The larval form lives in the soil and eats bacteria and vegetation. After a week it transforms into a long, slender **filariform larva** that can penetrate human skin. The filariform larva penetrates between the toes of the hapless human who walks by shoeless. The larvae travel directly to the alveoli of the lungs, where they grow and are coughed up and swallowed. The adult worms develop as they arrive at the small intestine, where they attach by their mouths, and suck blood. At this point, the hookworms copulate and release fertilized eggs.

3) ***Strongyloides stercoralis***: The larval forms in the soil penetrate the human skin and travel to the lung. There they grow, are coughed up and swallowed into the small intestine, where they develop into adult worms that lay eggs. The **eggs are not passed in the stools**. Rather, filariform larvae hatch and can do 3 things:

   a) **Autoinfection**: The filariform larvae penetrate the intestine directly, without leaving, and go to the lung to continue the cycle.
   b) **Direct cycle**: The filariform larvae pass out in the feces, survive in the soil, penetrate the next passerby, and migrate to the lungs. This complete cycle is almost exactly the same as that of *Necator americanus* (hookworm).
   c) **Indirect cycle**: This is a sexual cycle where the filariform larvae are passed out in the stool and while in the soil develop into male and female adults. They mate in the soil and produce fertilized eggs. The filariform larvae then hatch and reinfect a human, moving to the lung.

Ascaris lumbricoides

Fig. 32-1. *Ascaris* infection may be mild or asymptomatic. With heavy infections the patient may develop abdominal cramping. Severe infections involve adult worm invasion into the bile ducts, gall bladder, appendix, and liver. Children with heavy worm loads may suffer

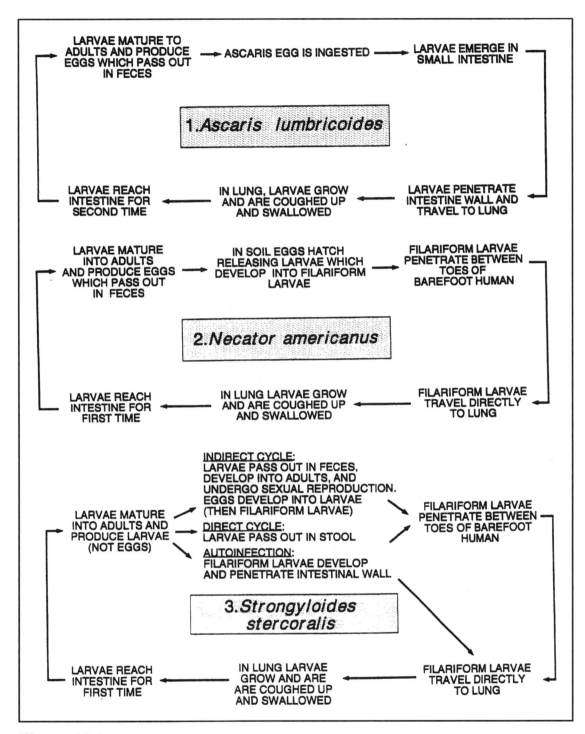

Figure 32-1

from malnutrition because the worms compete for the same food and sometimes a mass of worms can actually block the intestine. When the larvae migrate into the lung, the patient may develop cough, pulmonary infiltrate on chest x-ray, and a high eosinophil count in the blood and sputum.

Diagnosis is made by identification of eggs in feces. A sputum examination may reveal larvae. The peripheral blood smear may also reveal an increased number of eosinophils.

Treatment

The intestinal nematodes *Ascaris lumbricoides, Necator Americanus* (hookworm), *Strongyloides stercoralis, Enterobius vermicularis* (pinworm), *Trichuris trichiura* (whipworm), and *Trichinella spiralis* are all treated with the same drug:

Think of a worm,
Worms **BEND**,
BEND them a lot and you can kill them!

Me**bend**azole
Al**bend**azole
Thia**bend**azole (this one is the least well tolerated from a side effect standpoint and therefore the least used)

These drugs **paralyze** the roundworms so they are passed out in the stool. Other drugs will irritate the worms, causing them to migrate out of the small intestine to other organs, which can be fatal.

Pyrantel pamoate is an alternative agent for *Enterobius* (pinworm), *Necator* (hookworm), and *Ascaris*.

Necator americanus and Ancylostoma duodenale
(Hookworm)

Fig. 32-1. Notice that *Ascaris lumbricoides* and *Necator americanus* (hookworm) have very similar life cycles. They differ only in the path that each larvae form takes to reach the lung: *Necator*, foot to lung; *Ascaris*, intestine to lung.

The patient with hookworm can develop diarrhea, abdominal pain, and weight loss. Since each hookworm sucks blood from the wall of the intestine, hookworm infection may cause an iron deficiency anemia. There is also an intense itching and rash at the site of penetration through the skin (between the toes), and the local growth in the lung can result in a cough, infiltrate on chest x-ray, and eosinophilia.

Diagnosis is made by identifying eggs in a fresh fecal sample. This infection may be treated with **mebendazole** or **albendazole**. Also treat the iron deficiency anemia.

Strongyloides stercoralis

Fig. 32-1. Individuals infected with *Strongyloides* may complain of vomiting, abdominal bloating, diarrhea, anemia, and weight loss. Similar to hookworm infection, patients may develop a pruritic rash, lung symptoms (cough or wheezing), and/or eosinophilia. When patients infected with strongyloides are given immunosuppres-

sive medications, such as prednisone, they can develop a severe autoinfection. The filariform larvae will invade the intestine, lung and other organs, causing pneumonia, ARDS, and multi-organ failure. Patients with COPD and asthma living in areas endemic for strongyloides should have their stool and eosinophil count checked before steroid treatment!

Diagnosis is made by identifying larvae in feces (no eggs). The **enterotest**, where a long nylon string is swallowed and later pulled back out the mouth, may demonstrate larvae of *Strongyloides*. Sputum exam may also demonstrate larvae. Examination of the blood will reveal an elevated level of eosinophils. An enzyme-linked immunosorbent assay (ELISA) for antibodies to *Strongyloides* is a sensitive method of diagnosing uncomplicated infections.

Treat infected patients with **ivermectin, albendazole**, or thiabendazole.

Trichinella spiralis

You may have seen Gary Larson's "The Far Side" comic showing wolves at the edge of a pork farm, saying, "Let's rush them and *Trichinella* be damned." After reading about *Trichinella spiralis*, we can finally get the joke.

Infection occurs following the ingestion of the encysted larvae of *Trichinella spiralis*, which are often present in raw pork. After ingestion, the cysts travel to the small intestine, where the larvae leave the cysts and mature into mating adults. Following mating, the adult males are passed in the feces. The females penetrate into the intestinal mucosa, producing thousands of larvae. The larvae then enter the bloodstream and spread to organs and skeletal muscle. The larvae then become encysted in skeletal muscle, where they may last for decades.

Most patients are asymptomatic with the initial infection. Some patients will complain of abdominal pain, diarrhea, and fever as the worms mature in the small intestine and penetrate through the intestinal wall. About 1 week after the initial intestinal invasion, the larvae migrate into skeletal muscle, producing fevers and muscular aches. In severe (sometimes fatal) cases, larvae may invade heart muscle and brain tissue.

Diagnosis can be made with serologic tests or muscle biopsy, which will reveal the encysted larvae. Since there is significant muscle invasion, examination of the differential white blood cell count will reveal a marked increase in the percentage of eosinophils. The invasion of muscle also results in increased levels of serum muscle enzymes, such as creatinine phosphokinase (CPK).

Prevention is best carried out by killing cysts, either by cooking or freezing the pork meat prior to consumption.This is a helpful mnemonic to remember how we become infected with **Trichinella**: Raw bacon is delicious. It is **tric**ky not to eat it. (courtesy of Erik Madden from UCLA).

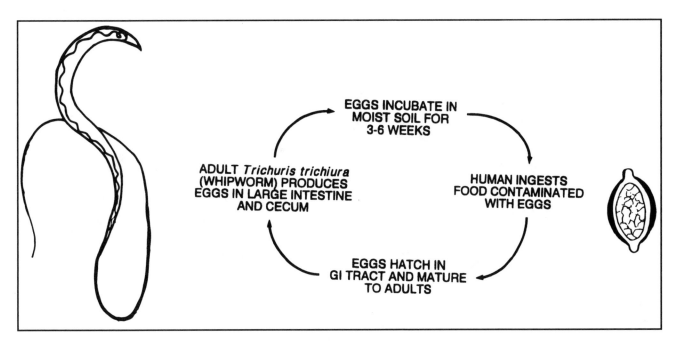

Figure 32-2

Albendazole, mebendazole and **thiabendazole** have little effect on muscle larvae but may be helpful against the enteric stages of the parasite.

Trichuris trichiura
(Whipworm)

The next 2 worms, *Trichuris trichiura* and *Enterobius vermicularis*, have very simple life cycles: there is no filariform larvae stage, no transit across through the intestinal wall, and no lung involvement. Since there is no transit outside the intestine, the eosinophil count is not elevated.

Fig. 32-2. *Trichuris trichiura* has a simple **slow** life cycle. Like the Congressional whip trying to move his party to a decision, the whipworm life cycle is **slow**. Transmission occurs by ingestion of food contaminated with infective eggs. The eggs hatch in the gastrointestinal tract and migrate to the cecum and ascending large intestine. The mature adult will then produce thousands of eggs per day for about one year. There is no autoinfection, since the eggs must incubate in moist soil for 3–6 weeks before they become infective.

Infected patients may have abdominal pain and diarrhea. Diagnosis is made by identifying eggs in fecal specimens. The eggs look like footballs with bumps on each end. The adults have the appearance of a bullwhip, thus the common name.

Treat patients with **albendazole** or **mebendazole**.

Enterobius vermicularis
(Pinworm)

The *Enterobius'* life cycle is very simple: The eggs are simply ingested, and the pinworms mature in the cecum and ascending large intestine. The female migrates to the perianal area (usually at night) to lay her eggs, which become infectious 4–6 hours later. This infection causes severe perianal itching. An infected individual will scratch the perianal area and then reinfect himself or others (hand-to-mouth) because his hands are now covered with the microscopic pinworm eggs.

This infection is more of a nuisance than it is dangerous. The infected individual has a terribly itchy behind, usually at night.

Diagnosis is made by placing scotch tape firmly on the perianal area. The scotch tape will pick up eggs, which can be viewed under a microscope. At night the larger adult females can sometimes be seen with the unaided eye, crawling across the perineal area. There is no eosinophilia, since there is no tissue invasion.

Treat with **albendazole, mebendazole**, or **pyrantel pamoate**, avoid scratching, and change the sheets daily.

Fig. 32-3. *Enterobius* vermicularis. The female worm is like an intestinal (**entero**) **bus**. She drives out of the anus every night and drops off her egg passengers on the perineum. If a **pin** (**pinworm**) pops her tire, seal the leak with some scotch tape. The **scotch tape test** is used to diagnose *Enterobius vermicularis* infection.

Blood and Tissue Nematodes

The blood and tissue nematodes are not spread by fecal contamination, but rather by the bite of an **arthropod**. These threadlike, round worms belong to the family Filarioidea and so are called **filariae**. Adult filariae live in the lymphatic tissue and give birth to prelarval

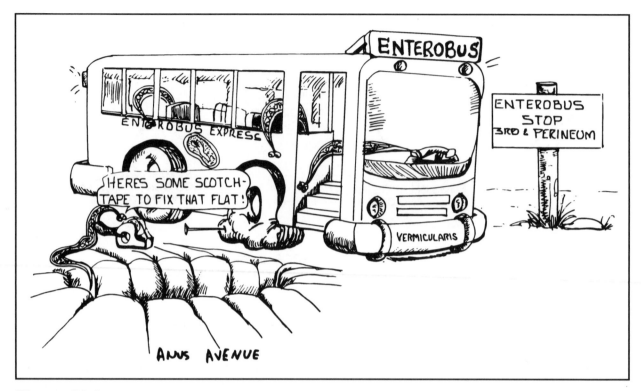

Figure 32-3

forms (they do not lay eggs) called **microfilariae**. The microfilariae burrow through tissue and circulate in the blood and lymphatic system. Microfilariae are picked up by bloodsucking arthropods, which transmit them to another human. Disease is caused by allergic reactions to both the microfilariae and dead adult worms in the lymphatic system as well as other organs (such as the eyes with *Onchocerca volvulus* infection).

Filarial parasites are themselves hosts to an endosymbiotic bacteria, *Wolbachia*, which they need to survive and thus represents an effective target for therapy. This rickettsial like bacteria was noted to live within filiarial worms (except loa loa) about 20 years ago, but just in past few years it was found that killing the *Wolbachia*, also led to sterility of the worms. Treatment with an antibiotic, doxycycline, significantly lowers microfilarial loads and even shows some activity against the adult worms.

Onchocerca volvulus

This filarial infection is found in Africa and Central and South America. The larvae are transmitted to humans by the bite of infected **black flies**. The microfilariae (larvae) mature into adults, which can be found coiled up in fibrous nodules in the skin and subcutaneous tissue. After mating, the adults produce microfilariae, which migrate through the dermis and connective tissue. Patients will develop a pruritic skin rash with darkened pigmentation. The skin may thicken with the formation of papular lesions that are actually **intraepithelial granulomas**. The thickened skin may appear dry, scaly, and thick ("lizard skin"). Microfilariae may also migrate to the eye, causing blindness (since the black fly vector breeds in rivers and streams, this is often referred to as "**river blindness**"). There are villages in endemic areas where most of the inhabitants are blind.

Diagnosis is made by demonstrating microfilariae in superficial skin biopsies, or adult worms in a nodule. Microfilariae can often be seen in the eye (cornea and anterior chamber) by slit lamp examination.

A new drug, **ivermectin**, has revolutionized the treatment of Onchocerciasis. It kills microfilariae and prevents the microfilariae from leaving the uteri of adult worms. The manufacturer (Merck) has donated the drug to the World Health Organization for a program to eradicate *Onchocerca* from the planet. As humans are the only reservoir, treating people in endemic areas for 10 years (as planned) will prevent the birth of new microfilariae while all the adult worms (which have long life spans) die of old age.

If you know Spanish, *ver* means "to see." So: **I VER with IVERmectin**!!

Wuchereria bancrofti and *Brugia malayi*
(Elephantiasis)

Wuchereria bancrofti and *Brugia malayi* both cause a lymphatic infection that can result in chronic leg swelling. *Wuchereria* infection is endemic to the Pacific Islands and much of Africa, while *Brugia* is endemic to the Malay Peninsula and is also seen in much of Southeast Asia. Transmission occurs by the bite of an infected mosquito. The transmitted microfilariae mature into adults within the lymphatic vessels and lymph nodes of the genitals and lower extremities. Mature adults mate, and their offspring (microfilariae) enter nearby blood vessels.

Small infections may only result in enlarged lymph nodes. Frequent infections in endemic areas result in acute febrile episodes, associated with headaches and swollen inguinal lymph nodes. Occasionally, following repeated exposures, fibrous tissue will form around dead filariae that remain within lymph nodes. The fibrous tissue plugs up the lymphatic system, which results in swelling of the legs and genitals. The swollen (edematous) areas become covered with thick, scaly skin. This chronic disfiguring manifestation is called **elephantiasis** because the extremities take on the appearance of elephant legs.

Diagnosis is made by the identification of microfilariae in blood drawn at nighttime. (For poorly understood reasons, very few organisms circulate during daylight hours. This is called **Nocturnal periodicity**.) Diagnosis can also be made by identification of positive antibody titers via immunofluorescence.

Diethylcarbamazine is the agent used to treat elephantiasis. Lymphatic damage may require surgical correction.

Fig. 32-4. There are two (**Di**) women named **Ethyl** in this **car**: **Di-ethyl-car**. You will notice that there is an elephant between Ethyl and Ethyl. You see, the drug **Diethylcar**bamazine is used to treat the filarial infections caused by *Wuchereria bancrofti*, *Brugia malayi*, *Loa loa*, and *Onchocerca volvulus*. These filariae chronically infect the lymphatics, causing lymph obstruction, giant swollen testicles, and elephant legs ("elephantiasis"). Thus the elephant between the Ethyls.

Dracunculus medinensis
(Guinea worm)

In 1986, there were an estimated 3.5 million cases of dracunculiasis reported in 20 countries. Eradication efforts led by the World Health Organization, combined with funding by the Gates Foundation, have led to a dramatic decrease in reported cases as of 2005. The remaining cases are all within Africa, primarily Ghana and the Sudan. Because there is no effective drug treatment, eradication efforts were via simple public health measures: identification of persons with worms that are actively migrating from their bodies and restricting these persons from bathing in the public water supply. Remember, when the worms get near water they release their progeny. Basically, education and simple sanitary measures do the trick. (MMWR 2005).

This very interesting tissue-invasive nematode lives as a larval form inside intermediate hosts: African and Asian freshwater copepods (tiny crustaceans). When a person drinks water containing the microscopic crustaceans, the larvae penetrate the intestine and move deep into subcutaneous tissue, where the adults develop and then mate. The male is thought to die, but the female grows over the course of a year to a size of 100 cm!!! She then migrates to the skin and a loop of her body pokes out and exposes her uterus. When the uterus comes into contact with water, thousands of motile larvae are released. Persons infected with *Dracunculus medinensis* will experience allergic symptoms, including nausea, vomiting, hives and breathlessness, during the larval release.

A common practice used to remove the worm involves driving a small stick under the part of the worm's body that is looped out of the skin. The stick is slowly twisted each day to pull out the 100 cm *Dracunculus*. Think of a wooden stake being used to get rid of Dracula! (mnemonic courtesy of Erik Madden from UCLA).

Cutaneous Larval Migrans

Also called **creeping eruption**, this intensely pruritic, migratory skin infection commonly occurs in the Southeastern U.S. The larvae of dog and cat hookworms penetrate the skin and migrate beneath the epidermis. As these larvae move (a few centimeters per day), an allergic response is mounted, resulting in a raised, red, itchy rash that moves with the advancing larvae.

The dog hookworm *Ancylostoma braziliense* and other species are responsible.

Human tissue-invasive nematodes such as the hookworm (*Necator americanus*) and *Strongyloides stercoralis* can produce similar creeping eruptions.

Diagnosis is made by biopsy of the advancing edge of the rash.

Treatment is with **ivermectin** or **albendazole.**

Figure 32-4

PLATYHELMINTHES
(Flatworms)

Flatworms do not have a digestive tract. There are 2 groups:

1) **Trematodes**, also known as flukes, include the freshwater-dwelling **schistosomes**. Both male and female members exist and mate within humans (not in the digestive tract, however). All flukes have a water snail species as an intermediate host.

2) **Cestodes**, also known as **tapeworms**, live and mate within the human digestive tract. Each tapeworm has both male and female sex organs (hermaphrodites), so individual tapeworms can produce offspring by themselves.

Trematodes
(Flukes)

Schistosoma
(Blood Flukes)

Schistosomal infections are extremely common worldwide, second only to malaria as a cause of sickness in the tropics. Schistosomes are found in freshwater. They penetrate through exposed skin and invade the venous system, where they mate and lay eggs. Since the eggs must reach freshwater to hatch, schistosomes cannot multiply in humans.

Fig. 32-5. The 3 major *Schistosoma* species worldwide.

The *Schistosoma* life cycle begins when their eggs hatch in freshwater. Larvae emerge and swim until they find a freshwater snail. After maturing within these snails, the larvae are released and are now infectious to humans. Mature schistosomal larvae (called **cercariae**, which look like little tadpoles with oral suckers on one end and a tail on the other), penetrate through exposed human skin, and travel to the intrahepatic portion of the portal venous system. At this location, the schistosomes mature and mate. Depending on the species, the pair of schistosomes will migrate to the veins surrounding either the intestine or the bladder, where they lay their eggs. These eggs may enter the lumen of the intestine or bladder, where they may be excreted via feces or urine into a nearby stream or lake so that they can continue their life cycle.

Interestingly, the adult worms in the venous system are able to survive and release eggs for years, since they are not killed by the immune system. It is thought that schistosomes practice **molecular mimicry** (incorporation of host antigens onto their surface, which fools the host's immune system into thinking that the schistosomes are NOT foreign).

However, cercariae (mature larvae) and eggs briskly stimulate the immune system, and are responsible for the systemic illness caused by this infection.

Clinical Manifestations

Schistosomiasis has 3 major disease syndromes that occur sequentially: 1) Dermatitis as the cercariae penetrate a swimmer's skin, 2) **Katayama fever** as the grown adults begin to lay eggs, and 3) Chronic fibrosis of organs and blood vessels from chronic inflammation around deposited eggs.

Upon penetration through the skin, patients transiently develop intensely itchy skin (swimmer's itch) and rash. **Katayama fever** follows 4–8 weeks later with symptoms that can include fever, hives, headache, weight loss, and cough. These symptoms may persist for 3 weeks. Lymph node, liver, and spleen enlargement with eosinophilia are common.

When the schistosomes set up their home in the veins surrounding the intestines or bladder, they begin releasing eggs, many of which do not reach the feces or urine. Instead, these eggs are whisked off by the bloodstream, where they are deposited in the liver, lung, or brain. The immune system reacts against these eggs, walling them off as granulomas. The deposition of eggs in the venous walls of the liver leads to fibrosis, which causes blockage of the portal venous system and a backup of venous pressure into the spleen and mesenteric veins (portal hypertension). Any blood vessels or organs that these eggs get stuck in can become inflamed, with formation of granulomas and ulcers. Patients may develop hematuria, chronic abdominal pain and diarrhea, brain or spinal cord injury, or pulmonary artery hypertension.

Diagnosis is made by demonstration of eggs in the stool or urine and serology for antibodies. Patients may also have high eosinophil counts, particularly during

SPECIES	GEOGRAPHIC DISTRIBUTION	RESIDES IN VEINS SURROUNDING	DEPOSITS EGGS IN:
Schistosoma japonicum	Eastern Asia	Intestinal tract	Feces
Schistosoma mansoni	South America and Africa	Intestinal tract	Feces
Schistosoma haematobium	Africa	Bladder	Urine

Figure 32-5 *SCHISTOSOMES*

acute Katayama fever. Control is directed at the proper disposal of human fecal waste and elimination of the snails that act as the intermediate host.

Treatment

A group of quantum physicists got together to eradicate this horrible disease. They wanted a drug that kills all the flukes and tapeworms also. They succeeded so **Praise the quantum physicists**! (Note: This is a lie).

Praziquantel: This is truly a broad spectrum anti-helminthic agent covering cestodes and trematodes alike. When treating patients with **praziquantel**, don't be surprised if there is an immediate exacerbation of symptoms. The death of these schistosomes evokes a vigorous immune response.

Cestodes
(Tapeworms)

Tapeworms are flatworms that live in the intestine of their host. Since they lack a true digestive tract, they suck up nutrients that have already been digested by their host. Tapeworms are hermaphrodites (both male and female organs in the same tapeworm), which enables a single tapeworm to produce offspring. Humans are usually the host of the adult tapeworm while other animals may serve as intermediate hosts for the larval stages.

Fig. 32-6. The tapeworm is long and flat (like a typewriter ribbon) and consists of a chain of boxlike segments called **proglottids**. Let us examine the tapeworm from its head down:

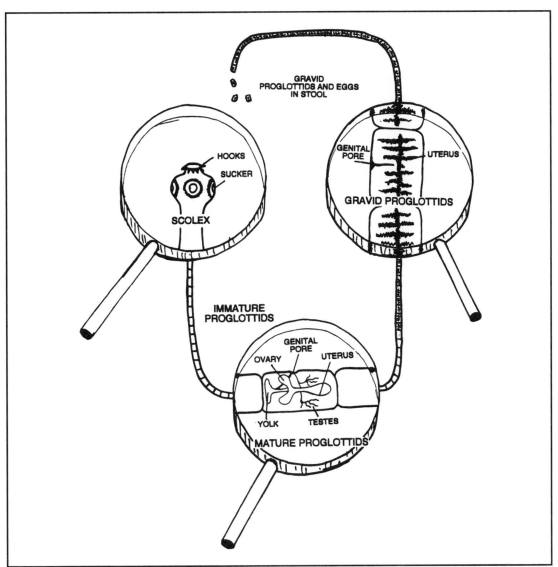

Figure 32-6

369

1) **Scolex**: The most anterior segment of the tapeworm (the head), which has suckers and sometimes hooks.

2) Immature proglottids.

3) Mature proglottids have both male and female sex organs.

4) Gravid proglottids contain fertilized eggs.

All of the tapeworm infections can be treated with **praziquantel** or **niclosamide**.

Fig. 32-7. Niclosamide is an alternative agent to praziquantel for treatment of tapeworm (cestode) infections. Picture a tapeworm wrapped around a **nickel** or a nickel under tape.

Albendazole and praziquantel are used for the treatment of *Taenia solium* larval cysticerci (see below).

Taenia solium
(Pork Tapeworm)

Humans acquire this infection by ingestion of undercooked pork infected with larvae. The pork tapeworm

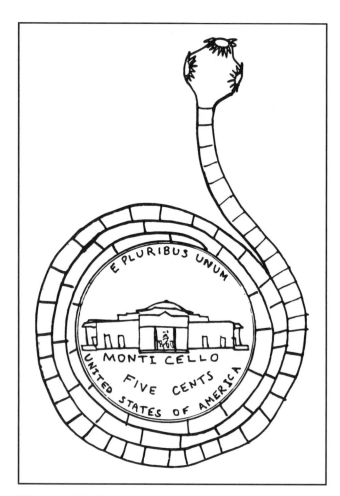

Figure 32-7

attaches to the mucosa of the intestine via hooks on its scolex and grows to a length of 2–8 meters. It releases eggs in the human feces. When pigs graze on fields contaminated with the egg-infested human feces, they become the intermediate host. The eggs develop into larvae that disseminate through the intestine and into muscle. In the animal's muscle tissue, the larvae develop into another larval form, the **cysticercus**. The **cysticercus** is a round, fluid-filled bladder with the larval form within. When a human eats insufficiently cooked pork muscle, the larval cysticercus converts to the adult tapeworm in the intestine, and the cycle continues (see **Fig. 32-8**).

Infected individuals usually have minimal symptoms. Diagnosis is made by demonstration of proglottids and/or eggs in a fecal sample.

Cysticercosis occurs when humans play the role of the pig and **ingest eggs rather than encysted larvae**. These eggs hatch within the small intestine, and the larvae migrate throughout the body, where they penetrate into tissue and encyst, forming cysticerci in the human. Most commonly, the larvae encyst in the brain and skeletal muscles (see **Fig. 32-8**).

Cysticerci in the brain tend to cause more symptoms, and this condition is called *Neurocysticercosis*. There are usually 7–10 cysts in the brain, causing seizures, obstructive hydrocephalus, or focal neurologic deficits. The cysts grow slowly and after 5–10 years begin to die and leak their fluid contents. The antigenic contents cause local inflammation and enhanced symptoms (more seizures, meningitis, hydrocephalus, and focal deficits).

In endemic areas such as Mexico, Central and South America, the Philippines, and SE Asia, cysticercosis is the most common cause of seizures.

The diagnosis of cysticercosis is made with the help of a CAT scan or biopsy of infected tissue (brain or muscle), both of which may reveal calcified cysticerci. Newer serologic tests are also proving helpful in the diagnosis of cysticercosis. Symptomatic disease, especially neurocysticercosis, can be treated medically with albendazole or praziquantel.

Note that our immune system raises a red flag to mark this invasion: the elevation of the eosinophil level in the blood.

Taenia saginata
(Beef Tapeworm)

Taenia saginata has the exact same life cycle as does *Taenia solium*, except that humans do not develop cysticerci when they ingest eggs, as do the intermediate hosts (cattle). For this reason, beef tapeworm infection is relatively benign.

The beef tapeworm is acquired by the ingestion of larval cysticerci in undercooked beef muscle. The adult beef tapeworm develops and adheres (via suckers on its scolex) to the intestinal mucosa, where it may reach a length of over 10 meters and contain more than 2 thousand proglottids.

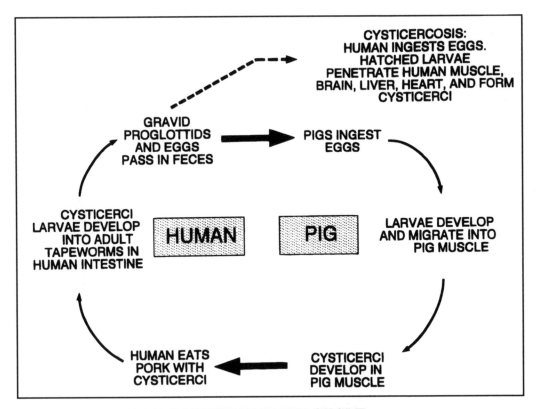

Figure 32-8   PORK TAPEWORM LIFE CYCLE.

Patients are usually asymptomatic, although they may develop malnutrition and weight loss. Diagnosis is made by identifying gravid proglottids and/or eggs in feces.

Diphyllobothrium latum
(Fish Tapeworm)

The fish tapeworm can grow to 45 meters in length. It is acquired by ingesting larvae in raw freshwater fish. The life cycle involves the human and 2 freshwater intermediate hosts, a crustacean and a fish. The adult tapeworms in the human intestine drop off their gravid proglottids loaded with eggs. When the eggs end up in water, they convert to a motile larval form, which is then ingested by a crustacean, which is then ingested by a freshwater fish (trout, salmon, pike, etc.), which is then ingested by a human—to end this long sentence!

The large intestinal *Diphyllobothrium latum* tapeworm provokes few abdominal symptoms, although it can absorb **vitamin B₁₂** to a significant degree. If vitamin B₁₂ deficiency develops, anemia will occur. Diagnosis of infection is made by identification of eggs in the feces.

Hymenolepis nana
(Dwarf Tapeworm)

This is the smallest tapeworm that infects humans (only 15–50 mm in length), and it has the simplest life cycle. There are NO **intermediate hosts**. Humans ingest eggs that grow into adult tapeworms, and the adult tapeworms pass more eggs that are again ingested by humans.

An infected patient will complain of abdominal discomfort and occasionally nausea and vomiting. Diagnosis is made by demonstration of eggs in a fecal sample.

Echinococcus granulosus and multilocularis
(Hydatid Disease, an Extra-intestinal Tapeworm Infection)

Fig. 32-9.   Dogs and sheep perpetuate the life cycle of *Echinococcus granulosus* and the human is only a dead-end in the cycle. *Echinococcus* shares many similarities with *Taenia solium*, with humans ingesting the **eggs**. These eggs hatch in the intestine and develop into larvae. After penetrating through the intestinal wall, the larvae disseminate throughout the body. Most larvae are concentrated in the liver, but larvae may also infect the lungs, kidney and brain.

Each larva forms a single, round fluid-filled "hydatid" cyst. These hydatid cysts can undergo asexual budding to form daughter cysts and protoscolices inside the original cyst. They can grow to 5–10 cm in size. Each cyst may cause symptoms because it compresses the organ around it (in the liver, lung, or brain). Humans are

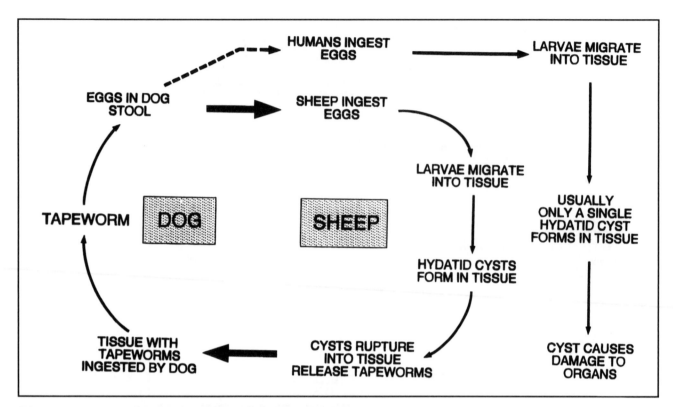

Figure 32-9   ECHINOCOCCUS LIFE CYCLE.

HELMINTHS	TRANSMISSION	EGGS	MORPHOLOGY
INTESTINAL NEMATODES (ROUNDWORMS)			
Ascaris lumbricoides	• Ingests eggs.		• Adult attains a length of 20–30 cm
Necator americanus (Hookworm)	• Larvae penetrate through skin		• Adults about 1 cm long
Strongyloides stercoralis	1. Larvae penetrate though skin 2. Autoinfection		• Adult females 2 mm long

Figure 32-10   HELMINTHS (Chart continued)

extremely allergic to the fluid within the cyst, and if the cyst bursts, the allergic reaction may be fatal. While the cysts of *Echinococcus granulosus* simply grow larger, only to spread if they rupture, *E. multilocularis* cysts undergo lateral budding and spread. These spreading cysts can be misdiagnosed as a slowly growing tumor.

Diagnosis of the hydatid cyst employs similar techniques as with *Taenia solium*'s cysticerci cysts, using CAT scanning and tissue biopsy. Only 10% of hydatid cysts cause symptoms, and treatment of these is difficult. The best thing to do is to cut them out of the liver, lung, or (yikes!) brain, but if the cyst fluid spills, out will pour daughter cysts, protoscolices, and highly allergenic fluid. Optional treatment usually starts with treatment for months with albendazole to kill the cysts (although this alone is rarely curative). The cyst is then exposed surgically and some of the cyst fluid is carefully aspirated. Saline, iodophore, or ethanol is next instilled into the cyst to make sure the contents are dead. After about 30 minutes the cyst is cut out.

When a hydatid cyst is inoperable due to a tricky location or a poor surgical candidate, therapy with albendazole is initiated and in some centers this is followed by CAT scan guided fine needle injection of ethanol into the cyst. This last method, called "PAIR" (**p**ercutaneous **a**spiration, **i**nfusion of scolicidal agents, and **r**easpiration), has also become the preferred treatment for uncomplicated cysts caused by *E. granulosus*.

Since dogs and sheep perpetuate the cycle, efforts toward control should target these animals. Dogs in grazing countries should not be fed uncooked animal meat and should be treated periodically with **niclosamide**.

Fig. 32-10. Summary of the helminths.

Fig. 32-11. Summary of the anti-helminths drugs.

REFERENCES

Garcia HH, et al. New concepts in the diagnosis and management of neurocysticercosis (*Taenia Solium*). Am J Trop Med Hyg 2005;72:3–9.

Gilbert DN, Moellering RC, Eliopoulos GM, Sande MA. The Sanford Guide to Antimicrobial Therapy 2006. Antimicrobial Therapy, Inc. Sperryville Virginia; 2006.

Taylor, MJ, Bandi C, Hoerauf A. *Wolbachia* bacterial endosymbionts of filarial nematodes. Advances in Parisitology 2005;60:245–84.

—PAIR puncture, aspiration, injection, re-aspiration: An option for the treatment of cystic echinococcosis. WHO/CDS/CSR/APH/2001.6 WHO, Geneva, 2001.

Recommended Review Articles

Hotez PJ, Molyneux DH, et al. N Engl J Med. Control of neglected tropical diseases. 2007;357(10):1018–27.

CLINICAL FINDINGS	DIAGNOSIS	TREATMENT	MISCELLANEOUS
1. Asymptomatic in many individuals 2. Abdominal cramping 3. Dry cough and fever while larvae are in the lungs 4. Children may develop malnutrition as worms compete for food	1. Fecal exam for eggs 2. Sputum exam may reveal larvae 3. Eosinophilia	1. Mebendazole: paralyzes worm and prevents it from migrating out of the small intestine to other organs 2. Albendazole (single dose) 3. Ivermectin 4. Nitazoxanide 5. Pyrantel pamoate (safe in pregnancy)	• If treated with the wrong antibiotic, *Ascaris* will migrate out of the GI tract
1. Diarrhea, abdominal pain, and weight loss 2. Iron deficiency anemia 3. Itching at site of skin penetration + rash 4. Occasional cough with bloody sputum	1. Fecal exam for eggs (examine quickly, as eggs hatch rapidly) 2. Sputum exam may reveal larvae 3. Eosinophilia	1. Mebendazole 2. Pyrantel pamoate 3. Albendazole	• *Ancyclostoma duodenale* is another species of hookworm
1. Vomiting, diarrhea, anemia and weight loss 2. Occasional fatal case caused by massive	1. Fecal exam for larvae (No eggs!) 2. Enterotest: swallow long nylon string and	1. Ivermectin 2. Albendazole 3. Thiabendazole	• Can undergo complete reproduction cycle in the soil

HELMINTHS	TRANSMISSION	EGGS	MORPHOLOGY
Strongyloides stercoralis (continued)			
Trichinella spiralis	• Ingestion of encysted larvae, often found in raw pork		• Cysts in skeletal muscle
Trichuris trichiura (Whipworm)	• Ingest eggs		• Egg looks like a football with polar bumps on each end. Adults whip-shaped, 3–5 cm long
Enterobius vermicularis (Pinworm)	• Ingest eggs		• Adult worms 1 cm long
BLOOD AND TISSUE NEMATODES (ROUNDWORMS)			
Onchocerca volvulus	1. Vector = black fly, which breeds in rivers and streams. Since cases cluster nearby, the disease is called "river blindness" 2. Found in Africa and Central & South America 3. Reservoir = humans		• **Filariae**: 1. Threadlike adult roundworms 2. Give birth to live off-spring called micro-filariae, which are transmitted via the black fly
Wuchereria bancrofti Pacific islands and Africa *Brugia malayi* Southeast Asia	• Vector = mosquito (transmits microfilariae)		• **Filariae**: 1. Threadlike adult roundworms 2. Give birth to live offspring called microfilariae

Figure 32-10   (continued)

374

CLINICAL FINDINGS	DIAGNOSIS	TREATMENT	MISCELLANEOUS
autoinfection (in immunocompromised host)	later pull out – may show larvae 3. Sputum exam may reveal larvae 4. Eosinophilia 5. Antibody and Antigen serum assay		
1. Fever, abdominal pain and diarrhea 2. Muscle aches, as larvae migrate to skeletal muscle 3. Severe cases: larvae migrate to heart and brain	1. Serologic tests 2. Muscle biopsy 3. Increased levels of muscle enzymes circulating in blood 4. Eosinophilia	1. Albendazole 2. Mebendazole 3. Thiabendazole 4. Cook or freeze pork prior to consumption	• Always cook pork products well
1. Diarrhea 2. Abdominal pain	1. Fecal exam for eggs 2. NO eosinophilia	1. Mebendazole 2. Albendazole	• Eggs must incubate in moist soil for 3–6 weeks before they become infective
• Severe perianal itching	1. Scotch-tape test 2. Examination of perineum at night may reveal adults seen with the unaided eye 3. NO eosinophilia	1. Mebendazole 2. Pyrantel pamoate 3. Albendazole	• Female migrates to perianal area at night to lay eggs
• Clinical findings: 1. Skin nodules: contain adult worms 2. Allergic reaction to micro-filariae migrating through the dermis leads to: A. Pruritic rash with dark-ened pigmentation B. Lizard skin: Intraepithelial granulomas, resulting in thick, dry, scaly skin 3. **River Blindness**: microfi-lariae migrate through the eye. A marked inflammatory response can occur upon their death, which can lead to blindness	• Skin biopsy reveals microfilariae	1. Ivermectin: kills micro-filarial stage only, and prevents them from leaving the uteri of adult worms 2. Doxycyline (kills endosymbiont *Wolbachia*) 3. Alternative: diethylcar-bamazine (but higher toxicity than ivermectin) 4. Excise adult worms in nodules	1. Disease is caused by allergic response to both microfilariae and dead adult worms 2. Hypersensitivity reac-tion may occur follow-ing administration of diethylcarbamazine
1. **Filarial Fever**: febrile episodes associated with headache and swollen lymph nodes 2. **Elephantiasis**: following repeat infections, fibrous tissue forms around the dead filariae that accumulate within the lymph nodes. This fibrous tissue plugs up the lymphatic system, resulting in swelling of the legs and genitals. Thick, scaly skin covers the edematous lower	1. Look for microfilaria in blood drawn at nighttime 2. Biopsy 3. Serology 4. Tropical pulmonary eosinophilia: A. Hypereosinophilia B. Elevated levels of IgE C. Granulomas within lymph nodes, spleen and lung	1. Diethylcarbamazine 2. Doxycycline (kills endosymbiont *Wolbachia*)	1. Disease is caused by allergic response to both microfilariae and dead adult worms located in the lym-phatic system 2. Tropical pulmonary eosinophilia: associ-ated with filarial infec-tion, and responsive to treatment with diethylcarbamazine, but exact etiology still unclear

HELMINTHS	TRANSMISSION	EGGS	MORPHOLOGY
Wuchereria bancrofti (continued)			
Dracunculus medinensis Guinea worm	1. Larvae within African, Middle Eastern, and Indian freshwater copepods (tiny crustaceans) are ingested when drinking freshwater 2. Larvae penetrate the intestine, and mature beneath the skin		• **Filariae**: 1. Threadlike adult roundworms: the female can grow to 100 cm in size 2. The adult female pokes a loop of her body through the skin, exposing her uterus. When her uterus is exposed to water, thousands of microfilariae are released
Cutaneous larva migrans (Commonly *Ancylostoma braziliense-* dog hookworm)	• Occurs in Southeastern U.S.		• Larvae of dog and cat tapeworms
Visceral larva migrans (most commonly *Toxocara canis-* dog roundworm)	• Ingestion of eggs		• Larve of dog roundworms, which can **NOT** mature in human
PLATYHELMINTHES (FLAT WORMS)			
Schistosomes Schistosoma japonicum Schistosoma mansoni Schistosoma haematobium	1. Penetrate through exposed skin 2. Since the eggs must reach freshwater to hatch, schistosomes cannot multiply in humans 3. Intermediate host = fresh water snail		**TREMATODES** 1. Eggs 2. Mature larvae (called cercariae) 3. Male and female adults

Figure 32-10   (continued)

CLINICAL FINDINGS	DIAGNOSIS	TREATMENT	MISCELLANEOUS
extremities, giving the appearance of elephant legs 3. **Tropical pulmonary eosinophilia**: hypersensitivity reaction that causes bouts of wheezing and coughing, associated with hypereosinophilia			
• Allergic symptoms occur during the release of microfilariae: nausea, vomiting, hives, and breathlessness		• Drive a small stick under the part of the worm's body that is looped out of the skin. The stick is slowly twisted each day to pull out the 100 cm *Dracunculus*	
• Creeping eruption: larvae of dog and cat hookworms penetrate the skin and migrate beneath the epidermis (a few centimeters per day). A raised, red itchy rash moves with the advancing larvae	• Biopsy of advancing edge of rash	1. Ivermectin 2. Albendazole	• Human tissue-invasive nematodes such as *Necator americanus* and *Strongyloides* can produce similar creeping eruption
Toxocariasis • Migration of larvae through the body results in fever, diarrhea, wheezing, hepatitis, and visual loss (from chorioretinitis)	1. Serology 2. Eosinophilia	1. Albendazole 2. Mebendazole	
• **Schistosomiasis** 1. Pruritic skin rash at site of penetration 2. Acute schistosomiasis (**Katayama fever**): 　A. Intense transient itching 　B. Weeks later: fever, hives, headache, weight loss, cough (lasts about 3 weeks) 3. Complications caused by immune reaction against eggs released by adults hanging out in the veins surrounding the intestine or bladder: 　A. Liver: fibrosis of portal venous system – leads to portal hypertension	1. Demonstration of eggs in stool or urine samples 2. Eosinophilia 3. Ultrasound of the liver will diagnose liver disease	• Praziquantel: results in immediate exacerbation of symptoms, followed later by improvement • Control: disposal of human fecal waste and destruction of intermediate host (snail)	1. **Molecular mimicry**: incorporation of host antigens onto their surface, which fools the host's immune system into thinking the schistosomes are **NOT** foreign 2. No person to person transmission

HELMINTHS	TRANSMISSION	EGGS	MORPHOLOGY
Schistosomes (continued)			
Taenia solium (Pork tapeworm)	1. Ingest undercooked pork containing larvae stage 2. Ingestion of eggs: results in **cysticercosis**		**CESTODES** 1. Scolex: head with **HOOKS** 2. Gravid proglottids: contains eggs & uterus has less than 15 pairs of lateral branches 3. Cysticercus: round, fluid-filled bladder with the larval form within 4. Adult can grow to 2–8 meters
Taenia saginata (Beef tapeworm)	• Ingest undercooked beef containing larvae stage		1. Scolex: head with **SUCKERS** (but **NO** hooks) 2. Gravid proglottids: contains eggs & uterus has more than 15 pairs of lateral branches 3. Can grow to 10 meters in length
Diphyllobothrium latum (Fish tapeworm)	• Ingest larvae in raw freshwater fish		• Can grow to 45 meters in length
Hymenolepis nana (Dwarf tapeworm)	1. Ingest fertilized eggs 2. **Auto-infection**		1. Adults grow to 15–50 mm 2. Eggs are infectious
Echinococcus (Hydatid Disease)	• Ingestion of fertilized eggs		

Figure 32-10   (continued)

CLINICAL FINDINGS	DIAGNOSIS	TREATMENT	MISCELLANEOUS
B. Lung: fibrosis of pulmonary arterioles can lead to pulmonary hypertension C. Intestine: deposits of eggs lead to inflammatory polyps			
1. Intestinal infection is usually asymptomatic 2. Cysticercosis: eggs hatch within the small intestine, and larvae travel to muscle, the central nervous system and/or the eye, where they eventually form calcified cysts that are inflammatory A. Blindness B. Neurologic manifestations: seizures, focal neurologic deficits, hydrocephalus blockage of CSF drainage	1. Fecal exam for eggs or gravid proglottids 2. CT scan or biopsy of brain or muscle may reveal calcified cysticerci 3. Eosinophilia occurs with cysticercosis	1. Praziquantel and albendazole 2. Second choice for all the tapeworms: niclosamide	Cysticerosis: occurs with ingestion of eggs Neurocyticercosis (cysts in brain)
1. Usually asymptomatic 2. Occasionally develop abdominal discomfort, weight loss and diarrhea	• Fecal exam for eggs or gravid proglottids	• Praziquantel	
1. Nonspecific abdominal symptoms 2. Vitamin B_{12} deficiency, leading to anemia	• Identify eggs or gravid proglottids	• Praziquantel	
1. Usually asymptomatic 2. Occasional abdominal discomfort, nausea and vomiting	1. Fecal exam for eggs 2. Proglottids are too small to see	• Praziquantel	• No intermediate host
1. **Hydatid cysts** form most often in the liver and lung: The cysts enlarge over 1 to 20 years, producing symptoms by mass effect. These cysts can calcify A. Liver: abdominal pain, palpable liver mass, biliary obstruction; can be fatal	1. CT scan or ultrasound reveals cysts in the liver or lung 2. Serology	1. Surgical removal of cysts: extreme caution is required, as leakage of cystic fluid can induce a severe anaphylactic reaction 2. **PAIR**- **P**ercutaneous **A**spiration, **I**nfusion of scolicidal agent, and **R**easpiration. 3. Albendazole or mebendazole are used adjunctively.	

HELMINTHS	TRANSMISSION	EGGS	MORPHOLOGY
Echinococcus (Hydatid disease) (continued)			

Figure 32-10 (continued)

NAME	MECHANISM OF ACTION	PHARMOKINETICS

1. Al**bend**azole
2. Me**bend**azole
3. Thia**bend**azole
	• Paralyzes worms!
	  1. These drugs bind to beta-tubulin, inhibiting microtubule synthesis
	  2. Microtubule-dependent uptake of glucose is blocked, depleting glycogen stores
		• Oral

Pyrantel pamoate	• Depolarizing neuromuscular junction blocker	• Oral
Diethylcarbamazine	• Increases susceptibility of microfilaria to phagocytosis	• Oral
Ivermectin	1. Kills helminths by opening chloride sensitive channels 2. Blocks GABA neurotransmitter in peripheral motor nerves 3. Kills microfilariae and impairs fertility of adult females; does not kill adult worms	• Oral
Praziquantel	• Increases calcium permeability, so calcium is lost, resulting in paralysis of worms	1. Oral & rapidly absorbed 2. CSF penetration
Niclosamide	• Inhibits oxidative phosphorylation	• Oral

Figure 32-11 ANTI-HELMINTHS DRUGS

CLINICAL FINDINGS	DIAGNOSIS	TREATMENT	MISCELLANEOUS
B. Lung: Cyst may rupture, causing cough or chest pain 2. Leakage of hydatid cyst fluid can cause a severe allergic reaction.			

M. Gladwin, W. Trattler, and S. Mahan, *Clinical Microbiology Made Ridiculously Simple* ©MedMaster

ADVERSE EFFECTS	THERAPEUTIC USES

• Transient abdominal pain
• Minimal side effects
  Thiabendazole is the least well tolerated (and therefore the least used)
	Intestinal nematodes
	  1. *Ascaris lumbricoides*
	  2. *Necator americanus* (hookworm)
	  3. *Strongyloides stercoralis*
	  4. *Trichinella spiralis*
	  5. *Enterobius vermicularis* (pinworms)
	  6. *Trichuris trichiura* (whipworm)
	  7. Cutaneous and visceral larva migrans
	  8. Adjunctive therapy for hydatid disease caused by *Echinococcus*
	  9. Albendazole now used to treat *Taenia Solium* (neurocysticerosis)
	 10. Microsporidia

• Transient nausea and vomiting, headache, and rash	• Alternative to mebendazole for *Ascaris, Necator americanus* (hookworm), and *Enterobius vermicularis* (pinworm)

• Severe reaction caused by death of parasites:
  1. Mazzotti reaction (with *Onchocerca*)
  2. *Wuchereria* and *Brugia*: fever, headache, rash and muscle aches
	• Used for the extraintestinal nematodes: the filaria
	  A. *Wuchereria bancrofti*
	  B. *Brugia malayi*
	  C. *Loa loa*
	  D. Second choice for *Onchocerca volvulus*:
	  1. Visceral larval migrans (toxocariasis)
	  2. Tropical pulmonary eosinophilia (probably caused by a species of filariae)

• Host reponse to dying microfilariae in tissue: pruritis, rash, dizziness, edema of face and limbs	1. Drug of choice for *Onchocerca volvulus* (which causes river blindness) 2. Also active against intestinal nematodes (*Ascaris, Trichuris, Enterobius*, and *Strongyloides*)
1. Abdominal pain, lethargy, diarrhea and/or fever 2. Exacerbation of symptoms of schistosomiasis can occur, as death of schistosomes evokes a vigorous immune response	1. Trematodes (flukes): *Schistosomes* 2. Cestodes (tapeworms) 3. *Taenia solium*: Neurocysticerosis
• Transient nausea and vomiting	• Second choice for tapeworm infections (after praziquantel)

M. Gladwin, W. Trattler, and S. Mahan, *Clinical Microbiology Made Ridiculously Simple* ©MedMaster

PART 5.   VERY STRANGE CRITTERS

CHAPTER 33.   PRIONS
(**Pr**oteinaceous **I**nfectious Particles):
Mad carnivorous cows, shaking cannibals, and a few good reasons to be a vegetarian . . .
by Hans Henrik Larsen, MD

Introduction

The fancy name for Prion diseases is the transmissible spongiform encephalopathies (TSE); so named because these diseases are transmissible and create spongiform pathological changes in the brain resulting in encephalopathy (i.e. causing brain damage). These diseases are fatal neurodegenerative disorders of humans and other animals. The best known animal diseases are scrapie in sheep and goats, and bovine spongiform encephalopathy (BSE or *mad cow disease*) in cattle. The first case of BSH in the US was diagnosed in 2003.

Four human prion diseases have been identified so far, all of which are VERY rare: Creutzfeldt-Jakob disease (CJD), kuru ("shivering"), Gerstmann-Sträussler-Scheinker disease (GSS), and fatal familial insomnia (FFI). Creutzfeldt-Jakob disease (CJD) is further characterized as sporadic (sCJD), new variant (vCJD), familial (fCJD) and as iatrogenic (iCJD). These distinctions will be described later in the chapter.

Prion diseases share the following characteristics:

- Long incubation time (months to years)
- Gradual increase in severity leading to death within months of onset
- No host immune response
- Non-inflammatory process in the brain
- Neuro-pathological findings may include:
  - Macroscopic examination is often normal.
  - Microscopic *spongiform* changes (small, apparently empty, microscopic vacuoles of varying sizes within the neural tissue), neuronal loss, and amyloid plaques (a pathological proteinaceous substance deposited between cells) with accumulation of the prion protein (PrP).

Prion diseases are unique in being both inherited and infectious. There is also a sporadic manifestation with no obvious inherited or infectious etiology. However, neural tissue from individuals affected by either inherited or sporadic (as well as infectious) form of prion diseases is infectious!

The Infectious Agent and Etiology

The nature of the infectious agent is still under investigation and debate. Briefly, two main theories have been debated: The *protein-only hypothesis* and the *viral hypothesis*. Advocates for the protein-only theory find that the infectious agent consists only of prion proteins with little, if any, nucleic acid, whereas advocates for the viral theory argue that it remains possible for prions to contain extremely small amounts, or protected, nucleic acid.

The *protein-only hypothesis* is by far the most widely accepted theory today.

Laboratory data indicate that the prion protein (PrP) is the major (if not exclusive) component of prions. In short, prions are resistant to agents that modify nucleic acids but susceptible to agents that modify proteins.

If prions do not contain nucleic acids, how can they be infectious? How do they replicate? What about the genetic code dogma? Isn't it heretical to suggest that an agent without nucleic acids can replicate?

Well, it turns out that PrP is encoded by an endogenous gene (PRNP, for humans, located on chromosome 20), but exists in two conformational isoforms; that is, PrP has two different (secondary and tertiary) protein structures:

- The cellular isoform of PrP is constitutively expressed by normal animals, primarily in the brain and is called PrP^C for *cellular*. The function of PrP^C remains to be clarified, but it may be involved in signaling processes or copper metabolism.
- The disease associated isoform of PrP is called PrP^{Sc} for *scrapie*.

What's the difference between the normal PrP^C and PrP^{Sc}? Again, the conformation (that is structure) of PrP^{Sc} is very different from PrP^C, and is thought to be responsible for the development of disease: Aberrant metabolism of PrP^{Sc} results in accumulation of the protein and brain injury.

The conformational change of PrP^C is thought to be *post-translational*, and can apparently be induced by the presence of PrP^{Sc}, maybe with the interaction of other proteins. Thereby a small amount of PrP^{Sc} will initiate a chain-reaction of conformational change of PrP^C into PrP^{Sc} (see **Fig. 33-1**).

This is the clue to the question raised above about replication of the prions if the infectious particles do not contain nucleic acids. The prions do not need the genetic software in the infectious particle—it's already present in the host as part of the genome as the PRNP gene! The amino acid sequence of the "sick" isoform of PrP (PrP^{Sc}) accumulating in a patient's brain is encoded by the PrP gene of that particular individual!

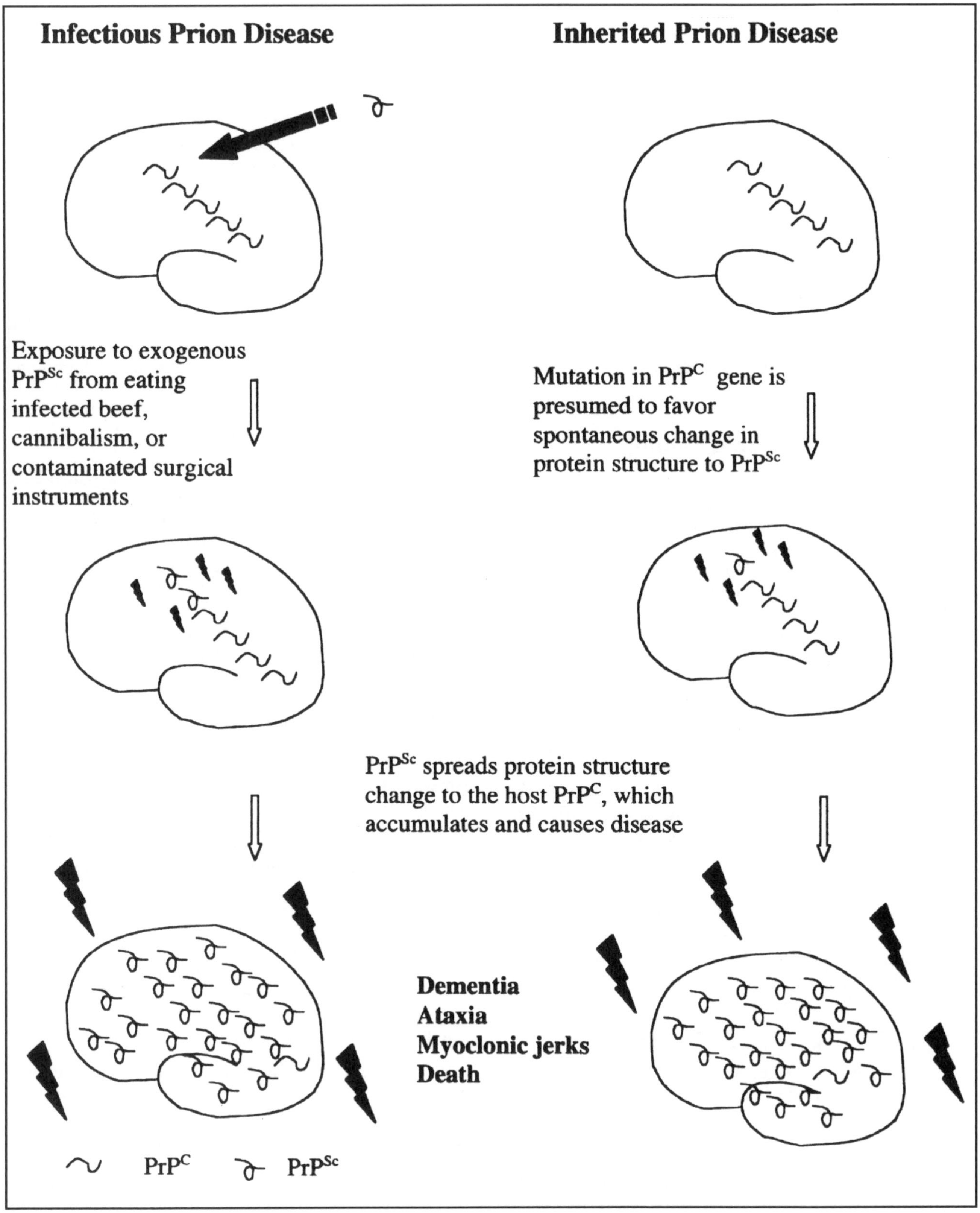
Infectious Prion Disease
Inherited Prion Disease
Exposure to exogenous PrP^{Sc} from eating infected beef, cannibalism, or contaminated surgical instruments
Mutation in PrP^{C} gene is presumed to favor spontaneous change in protein structure to PrP^{Sc}
PrP^{Sc} spreads protein structure change to the host PrP^{C}, which accumulates and causes disease
Dementia
Ataxia
Myoclonic jerks
Death
PrP^{C}
PrP^{Sc}

Figure 33-1

Dr. Stanley B Prusiner proposed the name *prion* for the agent causing transmissible spongiform encephalopathy to emphasize the nature of these *proteinaceous infectious* particles, and later concluded that the infectious agent is PrP^{Sc}. He was awarded the Nobel Prize in Physiology or Medicine in 1997 for the discovery of prions.

Basically three different etiologies are thought to be involved relating to the nature of the disease:

- Inherited: Mutations in PrP gene favoring spontaneous conformational change to PrP^{Sc}. The disease is following an autosomal dominant pattern.
- Infectious: Exogenous PrP^{Sc} inducing conformational change of host PrP^{C} into PrP^{Sc}.
- Sporadic: Unknown; probably spontaneous conversion of wild-type PrP^{C} or rare de novo mutations in PrP gene leading to the conformational change to PrP^{Sc}.

Transmission and Epidemiology

Contaminated neural tissue has been shown to transmit disease—even from one species to another (i.e. crossing the "species-barrier", e.g. sheep to cattle, cattle to human etc). The route of transmission can be inoculation but *also oral*. This has significant implications for the epidemiology of the disease. Ultimately, transmission of the disease from *mad cows* to humans causing a variant of Creutzfeldt-Jakob disease (new variant Creutzfeldt-Jakob disease, vCJD) has been established.

Infectivity of other tissues and body fluids than neural tissue (in particular blood and lymphatic tissue) are under investigation. There is now convincing experimental and clinical evidence for transfusion-associated vCJD. Furthermore, it turns out that the presence of PrP^{Sc} in lymphoreticular tissue is a defining feature of vCJD. The infectious particles are relatively resistant to heat and many commonly used chemical disinfectants as well as irradiation.

- The bovine spongiform encephalopathy (BSE) epidemic among cattle is attributed to the practice of feeding cattle (contaminated) sheep offal (the nice wording is *meat and bone meal*, MBM), which basically is the remainder of a butchered animal (bones, etc), actually turning cattle into carnivores. This practice has now been widely banned, initially with regard to cattle, and later in 2001 the European Union introduced a total MBM feed ban for all farm animals.
- Specified risk material (SRM) has to be removed before the products (e.g. meat) may enter the human food chain. SRM includes lymphatic tissue such as tonsils and neural tissue such as the brain, spinal cord and dorsal ganglia.

- The kuru epidemic among the Fore population of Papua New Guinea was probably transmitted through ritualistic cannibalism (as a rite of mourning the dead). Kuru gradually disappeared after the cessation of this practice. One can speculate whether that epidemic originated with an index case of sporadic Creutzfeldt-Jakob disease (sCJD).
- Iatrogenic cases of Creutzfeldt-Jakob disease (iCJD) have been established through contaminated (neuro) surgical instruments, dural and corneal grafts and administration of cadaveric pituitary hormones.

Clinical Presentation

Until the *mad cow disease* epidemic in the UK in the early 1990's causing vCJD in humans, prion diseases were widely unknown to many physicians. The human prion diseases are still VERY rare.

Between October 1996 and March 2011, there were 224 registered cases of vCJD, of which 175 (78%) have been in the United Kingdom.

Due to the large number of infected cattle which have entered the human food-chain and the possible long incubation time, it has been feared that new variant Creutzfeldt-Jakob disease (vCJD) would turn into an epidemic. However, recent surveillance data suggest that the human incidence has peaked. It remains possible however, that there will be a long "tail" of cases. Human prevalence studies on lymphoreticular tissue such as surgically removed tonsils and appendixes have not been conclusive, but it may be possible that a number of persons are still incubating the disease.

Furthermore, it turns out that ALL the human vCJD cases to date are homozygous for a particular polymorphism of the PRNP gene (methionine at codon 129), compared to about 40% of the background population. It remains to be determined if the homozygous are the only ones who are susceptible, or is the incubation time shorter in these individuals?

The prion diseases share many clinical features; ALL are characterized by neurological symptoms:

- Rapidly progressive dementia
- Psychiatric symptoms
- Cerebellar symptoms (e.g. ataxia)
- Involuntary movements (e.g. myoclonic jerks, choreoathetosis)
- Ultimately fatal disease

Please refer to **Fig. 33-2** for key features of individual diseases.

Diagnosis:

- The gold standard for diagnosis of prion diseases is histopathologic examination and immunostaining for PrP^{Sc} of brain tissue.

DISEASE	TYPICAL SYNDROMES	ETIOLOGY	DURATION OF ILLNESS BEFORE DEATH
Kuru	Ataxia, myoclonus followed by dementia	Infectious	Months
sCJD, fCJD and iCJD	Dementia, myoclonus followed by ataxia	Unknown (sporadic) Inherited (familial) Infectious (iatrogenic)	1 to few years
vCJD	Psychiatric changes, ataxia, dementia	Infectious	Months to few years
FFI	Sleep disturbances followed by dementia	Inherited	1 year
GSS	Ataxia followed by dementia	Inherited	Few years

sCJD: sporadic Creutzfeldt-Jakob disease

vCJD: new variant Creutzfeldt-Jakob disease

fCJD: familial Creutzfeldt-Jakob disease

iCJD: iatrogenic Creutzfeldt-Jakob disease

FFI: fatal familial insomnia

GSS: Gerstmann-Sträussler-Scheinker disease

Figure 33-2

- Cerebro-spinal fluid (CSF) is normal except that mildly elevated protein levels may be seen. Study of certain CSF proteins may be helpful (tau, 14–3–3 and S100 proteins). The tau protein has the highest sensitivity and specificity for vCJD, while 14–3–3 is a good marker for sCJD.
- Neuro-imaging tests: Several MRI findings have been described, such as "the hockey stick" sign (dorsomedial thalamic hyperintensity) and pulvinar sign in vCJD.
- Serial EEG can be very helpful in the diagnosis of Creutzfeldt-Jakob disease, in particular sCJD. An abnormal pattern (periodic sharp and slow wave complexes, PSWC) is ultimately seen in more than 2/3 of sCJD patients. However, this abnormality is NOT seen in vCJD.
- PrP^{Sc} is also, as mentioned previously, present in lymphoreticular tissue in vCJD. Tonsil biopsy has a high sensitivity and specificity for vCJD, and is an important tool allowing for diagnosis at an early clinical stage of vCJD.

Treatment

Unfortunately, this paragraph is going to be very short. No curative treatment is currently available.

Recommended Reviews

Belay ED. Transmissible spongiform encephalopathies in humans. Annu Rev Microbiol 1999; 53;283–314.

Johnson RT, Gibbs CJ. Creutzfeldt-Jakob disease and related transmissible spongiform encephalopathies. New Eng J Med 1998; 339;1994–2004.

Prusiner SB. Prions. PNAS 1998; 95;13363–13383.

Tyler KL. Prions and Prion Diseases of the Central Nervous System (Transmissible Neurodegenerative Diseases). In Mandell et al. Principles and Practice of Infectious Diseases 5th edition Churchill Livingstone (2000) pp. 1971–1980.

Additional Bibliography

Bruce ME et al. Transmissions to mice indicate that "New Variant" CJD is caused by the BSE agent. Nature 1997; 389:498–501.

Chesebro B. Prion protein and the transmissible spongiform encephalopathy diseases. Neuron 1999; 24:503–506.

Gajdusek DC, Zigas V. Degenerative disease of the central nervous system in New Guinea. New Eng J Med 1957; 257:974–978.

Gajdusek DC. Unconventional viruses and the origin and disappearance of Kuru. Science 1977; 197:943–960.

Harris DA. Prion diseases. Nutrition 2000; 16:554–556.

Hilton DA. Pathogenesis and prevalence of Variant Creutzfeldt-Jakob Disease. J Pathol 2006;208;134–141.

Head MW et al. Peripheral tissue involvement in sporadic, iatrogenic, and variant Creutzfeldt-Jakob disease: an immunohistochemical, quantitative, and biochemical study. Am J Pathol 2004;164;143–153.

Prusiner SB. Novel proteinaceous infectious particles cause scrapie. Science 1982; 216:136–144.

Prusiner SB. Molecular biology and genetics of neurodegenerative diseases caused by prions. Adv Virus Res 1992; 41:241–280.

Scott MR et al. Compelling transgenic evidence for transmission of Bovine Spongiform Encephalopathy Prions to humans. PNAS 1999; 96:15137–15142.

Internet Resources:

WHO site: http://www.who.int/zoonoses/diseases/bse/en/
CDC site: http://www.cdc.gov/ncidod/dvrd/vcjd/index.htm
FDA site: http://www.fda.gov/oc/opacom/hottopics/bse.html
The National Prion Disease Pathology Surveillance Center (NPDPSC) (US): http://www.cjdsurveillance.com/
The National Creutzfeldt-Jakob Disease Surveillance Unit (UK): http://www.cjd.ed.ac.uk/

PART 6.   THE END

CHAPTER 34.   ONE STEP TOWARDS THE POST-ANTIBIOTIC ERA?

DEVELOPMENT AND SPREAD OF ANTIMICROBIAL RESISTANCE

With the current **widespread overuse of antibiotics**, we are forcing bacteria to genetically change to survive. This **antibiotic Darwinism** has led to numerous highly resistant strains of bacteria that now threaten to create a **Post-Antibiotic Era**. Several factors are associated with emergence of resistance among organisms. These factors include:

1) Widespread, inappropriate use of broad-spectrum antibiotics, especially in daycare centers and ICUs. (e.g. treatment of viral illnesses with antibiotics).

2) Use of antibiotics in animal husbandry and fisheries to prevent infection and increase animal growth.

3) Excessive use of antimicrobial preparations in soaps and cleaning solutions in non-healthcare facilities.

4) Increased numbers of immunocompromised patients requiring prolonged courses of antibiotics.

5) Prolonged survival of debilitated patients.

6) International travel promoting the movement of resistant bacteria (e.g. *Mycobacterium tuberculosis*).

7) Poverty leading to inadequate antibiotic usage because of the increasing expense of adequate antimicrobial therapy.

Fig. 34-1 describes in more detail setting-specific contributing factors and resistant strains produced.

From examining antimicrobial resistance trends and outbreaks, there are certain principles that continue to hold true. First, given sufficient time and drug use, antimicrobial resistance will emerge. There are no antimicrobials to which resistance has not eventually appeared. Second, the development of resistance is progressive, evolving from low levels through intermediate to high levels. The exception to the rule is the direct transfer of genetic information by plasmid or transposon, which can result in immediate high level resistance. Third, organisms that are resistant to one drug are likely to become resistant to others. Cross-resistance among a class of drugs or resistance to multiple classes of antibiotics are both possible (e.g. *Streptococcus pneumoniae*). Fourth, once resistance appears, it is likely to decline slowly, if at all. Fifth, the use of antimicrobials by any one person affects others in the extended as well as the immediate environment.

MECHANISM OF BACTERIAL GENETIC VARIABILITY

Genetic variability is essential in order for microbial evolution to occur. There are 3 basic mechanisms of genetic variability leading to resistance among bacteria.

1) **Point mutations** may occur in a nucleotide base pair, which is referred to as **microevolutionary change**. These mutations may alter the target site of an antimicrobial agent, interfering with its activity.

2) **Macroevolutionary change** results in rearrangements of large segments of DNA as a single event. These rearrangements may include inversions, duplications, insertions, deletions or transposition of large sequences of DNA from one location of a bacterial chromosome to another.

3) **Acquisition of foreign DNA** carried by plasmids, bacteriophages or transposable genetic elements (see **Chapter 3**). These foreign elements give the organism the ability to adapt to antimicrobial activity.

MECHANISMS OF ANTIMICROBIAL RESISTANCE

This genetic variability can be further separated into more specific resistance mechanisms. These mechanisms of resistance are as follows:

1) **Enzymatic inhibition** of antibiotics leading to antibiotic inactivity.

2) **Alterations of bacterial membranes** to prevent entry of antibiotics into bacteria.

3) **Promotion of antibiotic efflux** which actively pumps the antibiotics out of the bacteria.

4) **Alterations of bacterial protein targets** which make these targets unrecognizable to antibiotics. Specific examples include:

- **Alterations of ribosomal target sites**
- **Alterations of cell wall precursor targets**
- **Alterations of critical enzymes**

5) **Bypass of antibiotic inhibition** allowing bacteria to find alternate pathways to survive when one pathway is blocked by an antibiotic.

Enzymatic Inhibition

Enzymatic inhibition is one of the most common modes of antimicrobial resistance. For example,

SETTING	CONTRIBUTING FACTORS	RESISTANT STRAIN PRODUCED
Day-care centers	• Crowding • Frequent respiratory infections • Lack of adherence to antibiotic regimens • Urinary/fecal incontinence • Inadequate handwashing by children • Lack of infection control by staff	*Streptococcus pneumoniae,* *Haemophilus influenzae,* *Moraxella catarrhalis,* *Neisseria meningitidis,* *Salmonella, Shigella,* *Escherichia coli*
Hospitals	• Immunocompromised patients • Patients with wounds, I.V. catheters, surgery, and hemodialysis • Clusters of patients on antibiotics • Common use of broad-spectrum antibiotics • Use of prophylactic antibiotics • Lack of infection control by staff	Coagulase-negative *Staphylococci,* *Staphylococcus aureus,* *Enterococci, Candida,* *Escherichia coli,* *Pseudomonas aeruginosa,* Multidrug-resistant *Tuberculosis,*
Long-Term Care Facilities	• Immunocompromised patients • Importation of resistant microbes from patients transferred from hospitals • Patients with infections, pressure ulcers, wounds, and urinary/fecal incontinence • Use of prophylactic and topical antibiotics • Inadequate hand washing by residents • Lack of infection control by staff	*Streptococcus pneumoniae,* *Haemophilus influenzae,* *Pseudomonas aeruginosa,* Coagulase-negative *Staphylococci,* *Staphylococcus aureus,* *Enterococci, Escherichia coli*
Animal Feed Lots	• Antibiotic use in healthy animals to promote growth • Subtherapeutic doses • Poor sanitation • Transfer of resistant microbes to humans through food chain	*Salmonella, Campylobacter,* *Enterococci, Escherichia coli*

Figure 34-1

Staphylococcus aureus' resistance to beta-lactam antibiotics (e.g. penicillin) is due mainly to the production of beta-lactamases, enzymes that inactivate these antibiotics by splitting the beta-lactam ring. Enterococci and gram-negative bacilli resistance to aminoglycosides is also commonly due to modifying enzymes that are coded by genes on plasmids or the chromosome.

There have been more than 340 different types of beta-lactamases described in the infectious disease literature. The most problematic of these enzymes are known as extended spectrum beta-lactamases (**ESBL**s). The ESBL producing organisms (e.g., *Klebsiella, Proteus, Pseudomonas, Citrobacter*, and *E. coli*) tend to be extremely resistant to beta-lactamases and the antibiotic options are limited to the carbapenem class (i.e., imipenem or meropenem).

Beginning in the late 1990's beta-lactamases that confer resistance to the carbapenem class (called **carbapenemases**) began to emerge and have subsequently spread worldwide. Of particular concern is a carbapenemase called **KPC (Klebsiella pneumoniae carbapenemase),** which was initially described in a clinical isolate of *Klebsiella pneumoniae* and can be transmitted via a plasmid to multiple other genera. In late 2010, researchers first published the discovery of a new carbapenemase called **NDM-1 (New Delhi metallo-beta-lactamase 1)** that is found primarily in *E.coli* and *Klebsiella pneumoniae* and is highly resistant to all antibiotics except tigecylcine and colistin!!!! (Kumarasamy et al, Lancet-Infection, 2010). As of this writing (June, 2015) carbapenemase-resistant enterbacteriaceae (CRE) are becoming a huge health concern in the U.S. and worldwide. The greatest risk for infection with CRE are in patients with long stays in critical care units and if they have been treated with carbapenems. A study done in Paris showed rates of intestinal colonization with CRE of >50% in persons staying in the intensive care unit longer than 6 weeks!

Alterations of Bacterial Membranes

Outer Membrane Permeability: Many penicillins have activity against gram-positive bacteria but not against gram-negative bacteria because gram-negative bacteria have a lipid bilayer that acts as a barrier to the penetration of antibiotics into the cell. Only when these penicillins are able to get inside the cell are they able to work. Passage of hydrophilic (water-soluble) antibiotics through this outer membrane is facilitated by the presence of porins, proteins that form water-filled diffusion channels through which antibiotics can travel. Mutations resulting in the loss of specific porins can occur and may lead to increased resistance to penicillins. *Pseudomonas aeruginosa* resistance to imipenem is a perfect example of this mechanism.

Inner Membrane Permeability: Aminoglycosides require active electron transport ("proton motive force") which means that a positively charged aminoglycoside molecule is "pulled" across cytoplasmic membranes of the internal negatively charged cell. The energy generation or the proton motive force that is required for substrate transport into the cell may be altered in mutants resistant to aminoglycosides. *Staphylococcus* resistant to aminoglycosides is an example. These aminoglycoside-resistant organisms with altered proton motive force occur rarely, but develop in the course of long-term aminoglycoside therapy.

Promotion of Antibiotic Efflux

The primary mechanism for decreased accumulation of tetracycline is due mainly to active efflux of the antibiotic across the cell membrane. Decreased uptake of tetracycline from outside the cell also accounts for decreased accumulation of tetracycline inside resistant cells. Tetracycline resistance genes are generally inducible by subtherapeutic concentrations of tetracycline which emphasizes the importance of adequate dosing. *Pseudomonas aeruginosa* and *Staphylococcus aureus* are bugs that display this type of resistance to tetracycline. This system may also represent a potential mechanism of resistance to the newer quinolones, but has not been found to be common among quinolone-resistant clinical isolates.

Alterations of Bacterial Protein Targets

Alterations of Ribosomal Target Sites: Resistance to a wide variety of antiribosomal agents, including tetracyclines, macrolides, clindamycin, and the aminoglycosides, may result from alteration of ribosomal binding sites. Failure of the antibiotic to bind to its target sites on the ribosome disrupts its ability to inhibit protein synthesis and cell growth. Ribosomal resistance to streptomycin may be significant but is fairly uncommon with gentamicin, tobramycin and amikacin. Ribosomal resistance can also be associated with decreased intracellular accumulation of the drug. Examples include *Staphylococcus aureus* and Enterococci species resistance to macrolides.

Alterations of Cell Wall Precursor Targets: Resistance of Enterococci to vancomycin has been classified as A, B, or C based on levels of resistance. Class

A resistance is considered high level resistance and is associated with the vanA gene. The vanA gene is carried on a plasmid and encodes an inducible protein that is involved in cell wall synthesis in *E. coli*. These proteins are responsible for synthesizing peptidoglycan precursors that have a different amino acid sequence from the normal cell wall peptidoglycan. This newly modified peptidoglycan binds glycopeptide antibiotics with reduced affinity, thus leading to resistance to vancomycin and teicoplanin. Classes B (vanB) and C (vanC) resistance phenotypes are considered to have moderate and low-level resistance respectively. The recent detection of decreased susceptibility to vancomycin among *Staphyloccus aureus* is also quite scary. Improvements are being made in the ability of clinical laboratories to characterize these resistant isolates.

Alterations of Critical Enzymes: Beta-lactam antibiotics inhibit bacteria by binding covalently to penicillin binding proteins (PBPs) also called transpeptidases (see **page 154**) in the cytoplasmic membrane. These target proteins are necessary for the synthesis of the peptidoglycan that forms the cell wall of bacteria. Alterations of PBPs that prevent successful binding can lead to beta-lactam resistance. In gram-positive bacteria, resistance to beta-lactam antibiotics may be associated either with a decrease in the affinity of the PBP for the antibiotic or with a decrease in the number of PBPs produced by the bacterium.

Bypass of Antibiotic Inhibition

Another mechanism for acquiring resistance to specific antibiotics is by the development of **auxotrophs**, which have growth factor requirements different from those of the wild strain. These mutants require substrates that normally are synthesized by the target enzymes, and thus, if the enzyme is blocked and the substrates are present in the environment, the organisms are able to grow despite inhibition of the synthetic enzyme. This is particularly concerning because the bacteria is able to create additional pathways to meet growth requirements in response to a particular pathway being blocked by the antibiotic. For example, "thymidine dependent" bacteria like enterococci are able to utilize exogenous supplies of thymidine for enzyme activity and are thus highly resistant to trimethoprim which blocks endogenous production of thymidine by bacterial enzymes.

DECREASING ANTIMICROBIAL RESISTANCE

In order to minimize antibiotic resistance in your patients you must employ these resistance management approaches:

BACTERIA	ANTIBIOTIC CLASS	MECHANISM OF RESISTANCE	GENETICS
Methicillin resistant *Staphylococcus aureus* (MRSA) or Oxacillin resistant *S. aureus* (ORSA)	Beta-lactams (i.e. oxacillin, methicillin, amoxicillin)	Enzymatic inhibition (i.e. beta-lactamases) & altered PBPs	Acquired chromosomal DNA segment (mecA) encoding PBP2A
Pseudomonas aeruginosa	Beta-lactams, aminoglycosides and macrolides	Permeability-uptake	Chromosomal
Enterobacter, Klebsiella, Citrobacter	3rd Generation Cephalosporins	Enzymatic inhibition	Chromosomal
Enterococci, gram-negative bacilli	Aminoglycosides	Enzymatic inhibition	Plasmid mediated except *Enterococcus faecium* (chromosomal)
Pseudomonas aeruginosa, Staphylococcus aureus	Quinolones, tetracyclines, chloramphenicol, and beta-lactams	Drug efflux	4 major effux systems of *Pseudomonas aeruginosa* are Mex AB-OprM and MexXY-OprM (intrinsic resistance), and MexCD-Opr and MexEF-OprN (acquired resistance)
Vancomycin resistant *Enterococcus* (VRE)	Glycopeptides	Altered cell wall precursors	High-level = VanA gene on the plasmid encodes an inducible protein; Low-level = VanB gene is self-transferable by conjugation with other enterococcal strains; VanC = constitutive plasmid
Staphylococcus aureus, Streptococci, Enterococci	Macrolides	Altered ribosomal target	Plasmid mediated
E. Coli, Staphylococcus aureus, Neisseria species	Sulfonamides	Bypass of antibiotic inhibition	Plasmid mediated
Pseudomonas aeruginosa, Klebsiella	Ceftazidime and other 3rd generation cephalosporins	Enzymatic inhibition	Plasmid mediated

Figure 34-2   COMMONLY ENCOUNTERED RESISTANT BUGS

1) **Withhold antibiotics** in situations where they are not likely to benefit the patient for self-limited viral infections such as "the common cold". Symptomatic treatment and supportive measures are the most appropriate care and antibacterial agents are not indicated.

2) Use the **narrowest spectrum antimicrobial agent** possible to treat an infection. For example, a semisynthetic penicillin or even oral penicillin would be a much better choice for treatment of a staphylococcal infection than a broad spectrum fluoroquinolone or cephalosporin. This works well provided the organism is known or likely to be susceptible to the narrower spectrum antibiotic.

3) **Base decisions about broadness of empiric antibiotic coverage on the severity of illness.** For example, in the case of a patient who is clinically stable and not at risk for significant morbidity if a resistant pathogen is not treated immediately, it may be appropriate to begin a narrow spectrum agent while awaiting culture and susceptibility data.

4) Emphasize prevention of infection through **careful hygiene**, especially **handwashing** and other measures to control the spread of pathogens. It sounds really simple, but proper and adequate handwashing by healthcare professionals can prevent many cases of infection due to virulent and antibiotic-resistant pathogens.

5) Utilize **education** to achieve therapeutic and preventative goals. Patients and families should be counseled as to when antibiotics are needed, how to take them correctly and for the **proper duration**. Education can also be used to foster **earlier detection of therapeutic failure**, which may be critical when treating patients who may be infected with antibiotic-resistant pathogens. Our communities must be cautioned against buying cleaning products with antimicrobial properties as well as using feed lot antibiotics.

References

Armand-Lefevre L, et al. Emergence of imipenem-resistant gram-negative bacilli in intestinal flora of intensive care patients. Antimicrob Agents Chemother 2013;57:1488.

Elliott TSJ, Lambert PA. Antibacterial Resistance in the Intensive Care Unit: Mechanisms and Management. British Medical Bulletin 55:259–276, 1999.

Fridkin SK, Gaynes RP. Antimicrobial Resistance in Intensive Care Units. Clinics in Chest Medicine 20:303–16, 1999.

Jones, RN. The Impact of Antimicrobial Resistance: Changing Epidemiology of Community-acquired Respiratory-tract Infections. Am J Health-Syst Pharm 56:S4–S11, 1999.

Liu HH. Antibiotic Resistance in Bacteria: A Current and Future Problem. RheumaDerm 1999;59:387–396.

Mayer KH, Opal SM, Medeiros AA. Mechanisms of Antibiotic Resistance. Basic Principles in the Diagnosis and Management of Infectious Diseases 2000; 15:236–252.

Neely AN, Holder IA. Antimicrobial Resistance. Burns 25:17–24, 1999.

Reece SM. The Emerging Threat of Antimicrobial Resistance: Strategies for Change. The Nurse Practitioner 24:70–86, 1999.

Virk A, Steckelberg JM. Clinical Aspects of Antimicrobial Resistance. Symposium on antimicrobial agents. Mayo Clin Proc 2000; 75:200–214.

Recommended Review Articles

Devasahayam G, Scheld WM, Hoffman PS. Newer antibacterial drugs for a new century. Expert Opin Investig Drugs. 2010;19(2):215–34.

Gandhi TN, DePestel DD, et al. Managing antimicrobial resistance in intensive care units. Crit Care Med. 2010;38 (8 Suppl):S315–23.

Giamarellou H. Treatment options for multidrug-resistant bacteria. Expert Rev Anti Infect Ther. 2006;4(4):601–18.

Goldstein FW. Combating resistance in a challenging, changing environment. Clin Microbiol Infect. 2007;13 Suppl 2:2–6.

CHAPTER 35. THE AGENTS OF BIOTERRORISM

Luciana Borio, MD and Clarence Lam

Introduction

Although the subject of Bioterrorism has recently received intense media attention in the United States and abroad, a related subject matter, Biological Warfare, has existed for many centuries. In the 14th century, the Tatar forces catapulted corpses of troops who had died of plague into the besieged strongholds of their enemies. In the 18th century, Lord Jeffrey Amherst (a British commander), gave smallpox-laden blankets to Native American Indians. Many countries had extensive biological warfare programs in the 20th century, including the United States, England, South Africa, Japan and the former Soviet Union.

President Nixon ended the American offensive biological program between 1969 and 1972. In 1972, several nations signed a document entitled *"The Convention on the Prohibition of the Development, Production and Stockpiling of Bacteriological (Biological) and Toxin Weapons and on Their Destruction,"* also known as the Biological and Toxin Weapons Convention (BWC). The BWC prohibits the development, production, stockpiling and acquisition of biological weapons.

Today a number of countries are still known or suspected to have extensive biological weapons development programs, and there are concerns that terrorist groups with access to technology are more able and willing to use biological weapons, in lieu of conventional bombs, against their perceived enemies. The bioterrorist acts in the United States in 2001, utilizing the bacteria *Bacillus anthracis*, are stark examples of such danger, although the perpetrators have not yet been identified.

In order to better prepare for bioterrorist acts, the Centers for Disease Control and Prevention (CDC) has created a list which classifies bioterrorism agents or diseases according to the level of disruption that they may cause to the civilian population. Those classified as Category A (the most dangerous) agents or diseases are: anthrax, plague, tularemia, botulism and the "viral hemorrhagic fevers." (http://www.bt.cdc.gov/agents/agentlist-category. asp)

Anthrax

The bacterium *Bacillus anthracis*, a spore-forming, nonhemolytic, non-motile, gram-positive rod, causes three forms of disease, depending on its mode of entry:

1. **Inhalation anthrax** – after breathing in the spores
2. **Cutaneous anthrax** – after direct contact between broken skin and the bacteria
3. **Gastrointestinal anthrax** – after the ingestion of contaminated meat
4. **Injectional anthrax** – after injectional recreational drug use.

Anthracis is the Greek word for *coal*, and the bacterium is given this name after its ability to cause a coal-like cutaneous lesion which is impressive but painless.

Inhalational anthrax is the most dangerous form of the disease, and it follows inhalation of the spores, which are between 1-5 microns in diameter. The spores are the ideal size to reach the deep end of the lungs, where they are ingested by pulmonary macrophages. The surviving spores travel to the lymph nodes in the mediastinum and the hilum of the lungs. Once in these lymph nodes, the spores vegetate (or sprout from their dormant state) which leads to problems. The bacteria have an antiphagocytic capsule and elaborate two factors (**edema factor – EF** and **lethal factor – LF**) and a **protective antigen** (**PA**). When the factors combine with the protective antigen, they become *toxins* (edema toxin and lethal toxin, respectively). Patients affected by this disease develop hemorrhagic mediastinitis (actual bleeding into the space surrounding the heart and great blood vessels), pleural effusions (often bloody), respiratory failure from the Acute Respiratory Distress Syndrome (ARDS; leakiness of the lungs so they fill up with water and cannot absorb oxygen!) and circulatory failure (shock or low blood pressure). Approximately 50% of patients also develop meningitis. Mortality is very high if antimicrobial therapy is delayed.

Fig. 35-1. Anthony has Anthrax. This figure demonstrates how the *Bacillus anthracis* spores are contracted from contaminated products made of hides and goat hair, or from bioterrorist delivery in the mail or by aerosol (crop dusters, air conditioners, etc.). The spores can germinate on skin abrasions (cutaneous anthrax), be inspired into the lungs (respiratory anthrax), or ingested into the gastrointestinal tract (GI anthrax). The spores are often phagocytosed by macrophages in the skin, intestine, or lung and then germinate, becoming active (vegetative) gram-positive rods. The bacteria are released from the macrophage, reproduce in the

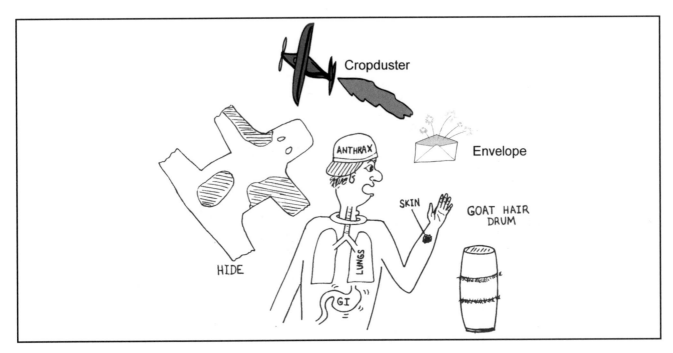

Figure 35-1

lymphatic system, and then invade the blood stream (up to 10–100 million bugs per milliliter of blood!!!)

The Office of Technology Assessment has estimated that if 100 kg of anthrax spores were aerosolized over Washington, D.C., an estimated 130,000 to 3 million deaths would follow. This lethality mirrors that of a hydrogen bomb! Disease may be averted by prompt administration of antibiotics (i.e. post-exposure prophylaxis). Unfortunately, antibiotics are recommended for at least 60 days following exposure to anthrax as spores can reside in the lungs for prolonged periods and germinate many days after exposure when antibiotics are discontinued. In 1979, there was an accidental release of anthrax spores from a biological weapons facility in Sverdlovsk (former Soviet Union) which killed 68 people. In this instance, some people developed the disease as late as 48 days following exposure to the spores.

Did you know?

Bacillus anthracis was isolated by Koch and used to develop his postulates in 1890?

1. The bacterium must be present in every case of the disease

2. The bacterium must be isolated from the diseased host and grown in pure culture

3. The specific disease must be reproduced when a pure culture of the bacterium is inoculated into a healthy susceptible host

4. The bacterium must be recoverable from the experimentally infected host

Smallpox

Smallpox is a disease caused by Variola, a DNA virus member of the family poxviridae and the genus orthopoxvirus. The viruses in this genus are very large and complex. Smallpox was declared eradicated by the World Health Organization in 1980 after a successful global vaccination campaign that lasted many years. It has been one of the most important accomplishments of modern medicine, because in the 20th century, smallpox killed more men than all the century's wars combined!

Smallpox is one of the most serious biological threats encountered today because:

1. Routine vaccination ceased in the United States in 1972, so most people have either lost their immunity to this virus or have never been vaccinated. Smallpox is highly contagious, spreading directly from person to person primarily via large droplets (an aerosol in rare instances. **Droplet transmission** occurs when relatively large (<5 micron) particles containing smallpox are propelled from an infected person over relatively short distances (3-6 feet) and deposited onto mucous membranes (usually mouth or nose) of another person or on an environmental surface. **Aerosol transmission** occurs when small (<5 micron) particles containing smallpox, which may remain suspended in the air for

long periods and may travel long distances, are propelled from an infected person and are inhaled by another person. Aerosols are typically generated by cough.

2. There is no known treatment for smallpox. Historically, approximately 30% of patients died, and those who survived became scarred for life (the pox lesions crust over, fall off, and leave pitted marks).

3. Although there are only 2 official repositories of smallpox virus in the world – in the CDC and in Novosibirsk, Russia, there are serious concerns that clandestine stocks remain in other nations.

The disease may be confused with chickenpox (described in **Chapter 27**), but there are several key differences summarized in the table below. The major differences are that smallpox pustules (which like chickenpox also look like dew on a rose petal) develop all at the same time (synchronous lesions), usually appear most dense on the face and palms (centrifugal spread) and ulcerate deeper into the skin. Chickenpox will appear in waves or crops of pustules and the lesions usually start on the face and torso, later spreading to the entire body, including the mucous membranes.

Fig. 35-2. Characteristic "dew on a rose petal" vesicles in chickenpox and smallpox.

Did you know?

Smallpox is the first disease for which a vaccine was developed. Edward Jenner developed the vaccine in the late 19th century after observing that milk maids did not acquire the disease. Milk maids were exposed to cowpox, a similar disease that affects cows. These milk maids developed antibodies against cowpox, which were cross protective with smallpox. Presently the vaccine is made with an unrelated virus, called vaccinia. Nobody is absolutely sure when vaccinia replaced cowpox for vaccination, but modern molecular microbiology has shown that cowpox and vaccinia are very different viruses. One more thing: Nobody believed Jenner at the time, so he had to finance his own work and publication.

SMALLPOX	CHICKENPOX
Orthopoxviridae	Herpesviridae
Synchronous lesions	Asynchronous lesions
Deep lesions	Superficial lesions
Centrifugal spread	Centripetal spread
Prominent face, palms	Prominent in chest

Figure 35-2

Plague

Plague is caused by the bacterium *Yersinia pestis*, a non-motile, gram-negative rod with a bipolar staining pattern which gives it the appearance of a safety pin. This bacterium is a member of the *Enterobacteriacae* family, and is a "lactose fermenter."

Fig. 35-3. *Yersinia pestis* is a gram-negative bacterium with a bipolar staining pattern. The ends of the rod-shaped bacterium take up more stain than the center. Three mammals fall prey to *Yersinia pestis*: wild rodents, domestic city rodents, and humans. The bacteria reside in the wild rodent population between epidemics and are carried from rodent to rodent by the flea. When wild rodents come into contact with domestic city rats (during droughts when wild rodents forage for food), fleas can then carry the bacteria to domestic rats. As the domestic rat population dies, the fleas become hungry and search out humans.

There are two major forms of the disease:

1. **Bubonic plague** – This is the most common and occurs after an infected flea bites a person, which causes the nearest lymph node to form a "bubo" (bubo means groin in Greek as it often spreads to the inguinal lymph nodes first and they can swell to the size of golf balls) and become tender, red and swollen. A small number of people develop "septicemic plague" after being bitten by an infected flea and never develop a "bubo." Another feature of this disease is dry gangrene of the extremities; this severe necrosis is why death due to plague has historically been called the "Black Death." The bacteria then spread to the blood and organs. Involvement of the lung can then lead to pneumonic plague.

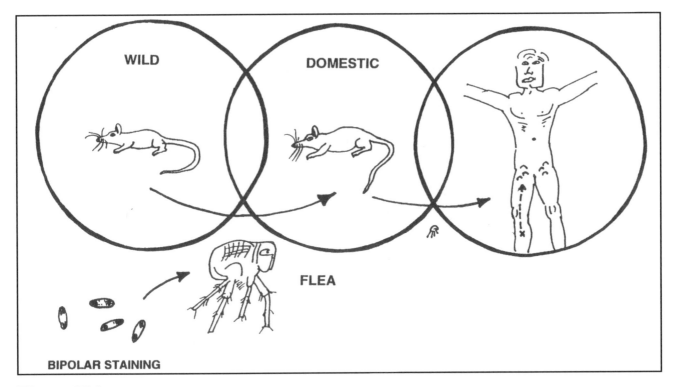

Figure 35-3

2. **Pulmonary or pneumonic plague** – occurs after the inhalation of the bacteria following an intentional aerosol release or via person-to-person transmission. This form occurs during a large epidemic or could occur with a bioterrorist attack, when large numbers of people are infected by flea bites and then develop lung involvement resulting in cough and then person to person spread of infected aerosol.

Plague can be spread from person to person and has killed millions during epidemics. Treatment involves an antibiotic active against gram-negative bacteria. Without treatment, mortality is very high (approximately 100%), but with treatment, most live. The most commonly used antibiotics are gentamicin or doxycycline. There is a widespread fear of the disease, and following a few cases in Surat (India) in 1994, approximately 500,000 people fled the city in panic.

Tularemia

Tularemia is caused by *Francisella tularensis*, a non-motile, gram-negative rod. Although the bacterium does not form spores, it does have a thin lipopolysaccharide-containing envelope making it very hardy. It is a facultative intracellular bacterium, which multiplies inside macrophages.

There are three major forms of the disease:

1. **Pneumonic tularemia** – As few as 10 organisms are all that is needed to cause disease, manifested by high fever, sore throat, pneumonia and pleuritis with occasional septic shock and meningitis.

2. **Ulceroglandular tularemia** – occurs following the bite of an infected tick or deerfly, or by handling a contaminated wild rabbit. The bacteria multiply at the site of inoculation, forming a papule. In a few days, it becomes pustular and ulcerates. In the meantime, the bacteria also travel to regional lymph nodes and to distant body sites via the blood stream.

3. **Oropharyngeal tularemia** – occurs after drinking contaminated water or ingesting contaminated food. Affected people develop a severe sore throat, tonsillitis and neck lymphadenopathy.

When inhaled, plague and tularemia share many similar features. Treatment also involves using an antibiotic effective against gram-negative bacteria, such as gentamicin or doxycycline.

Fig. 35-4. Francis (Francisella) the rabbit (rabbit vector) is playing in the **Tul**ips (**Tul**arensis). One ear has a tick, the other a deerfly.

Like *Yersinia pestis*, this organism is extremely virulent and can invade any area of contact, resulting in more than one disease presentation. The most

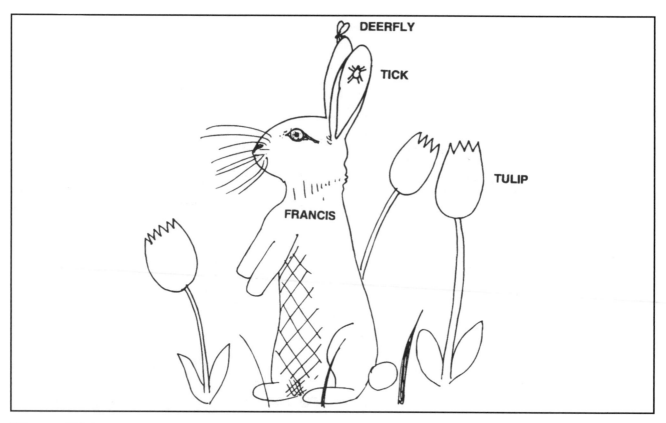

Figure 35-4

important diseases caused by *Francisella tularensis* are the ulceroglandular and pneumonic diseases:

1) **Ulceroglandular tularemia**: Following the bite of a tick or deerly, or contact with a wild rabbit, a well-demarcated hole in the skin with a black base develops. Fever and systemic symptoms develop, and the local lymph nodes become swollen, red, and painful (sometimes draining pus). The bacteria can then spread to the blood and other organs. Note that these symptoms are almost identical to bubonic plague, but the skin ulcer is usually absent in the plague and the mortality rate is not nearly as high as in bubonic plague, reaching 5% for ulceroglandular tularemia.

2) **Pneumonic tularemia**: Aerosolization of bacteria during skinning and evisceration of an infected rabbit or hematogenous spread from the skin (ulceroglandular tularemia) to the lungs can lead to a lung infection (pneumonia).

Francisella tularensis can also invade other areas of contact such as the eyes (**oculoglandular tularemia**) and the gastrointestinal tract (**typhoidal tularemia**).

Because this bacterium is so virulent (just 10 organisms can cause disease), most labs will not culture it from blood or pus. For the same reason it is not advisable to drain the infected lymph nodes. Diagnosis rests on the clinical picture, a skin test similar to the PPD for tuberculosis, and the measurement of the titers of antibodies to *Francisella tularensis*.

Botulinum Toxins

Botulinum toxins are the most poisonous substances known to man. These toxins bind irreversibly to peripheral cholinergic synapses in the neuromuscular junction. The toxins are all zinc proteinases that impede the release of acetylcholine by neuronal vesicles into the synaptic cleft of the neuromuscular junction. The end result is a neuromuscular paralysis. These supertoxins (there are seven distinct ones, designated as A through G) are produced by the bacterium *Clostridium botulinum*, a gram positive, spore-forming, obligate anaerobe, whose natural habitat is the soil.

In nature, botulism occurs after ingestion of contaminated food or water, or via direct contact into a wound. However, the toxins can also be absorbed through inhalation, occurring after an intentional aerosol release. Clinical manifestation includes a symmetric,

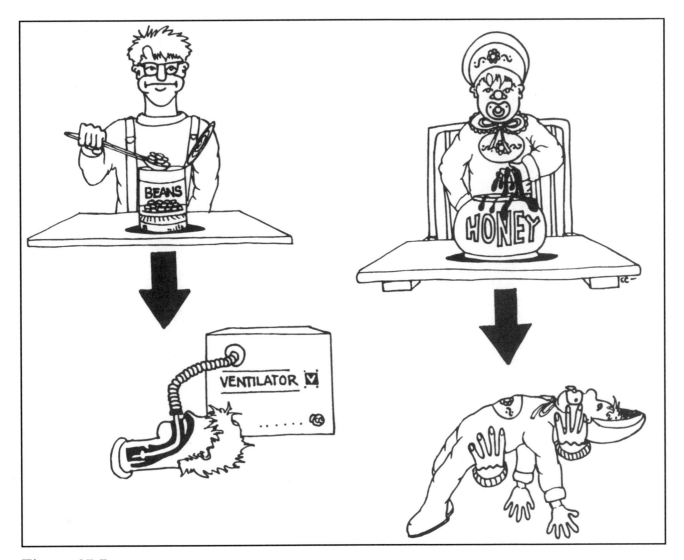

Figure 35-5

descending flaccid paralysis that always begins in the bulbar region (affecting the muscle supplied by the cranial nerves first). Patients have difficulty seeing, speaking and swallowing. Because the toxins bind irreversibly, recovery depends on regeneration of new motor neurons, which may take weeks to months. Danger arises when the diaphragm and accessory muscles of respiration become impaired, and such patients need to be placed on mechanical ventilation.

Fig. 35-5. Botulism. The adult is eating home-canned beans with neurotoxin while the infant is eating honey with spores. The adult often requires intubation and ventilatory support while the baby is merely "floppy."

Do you know the 4 Ds of botulism?

1. Diplopia (double vision)
2. Dysarthria (difficulty talking)
3. Dysphonia (abnormal voice quality, pitch or intensity)
4. Dysphagia (difficulty swallowing)

The treatment consists of passive immunization with botulinum antitoxins, which neutralizes the toxins and prevents further neurological deterioration. Because many of these antitoxins are made in horses, approximately 9% of patients develop urticaria, serum sickness or other hypersensitivity reactions to the antibody product; anaphylaxis occurs in approximately 2% of cases.

Viral Hemorrhagic Fevers

Many different viruses can cause hemorrhagic fevers, and they belong to a number of families. They are all difficult to make into biological weapons: It would be very difficult to work with them since vaccines or therapies are unavailable. Scientists who study them work in "space suits" with oxygen lines, in fully contained labs where nothing that enters (except for the scientists themselves) is ever allowed to leave.

All the agents are small, enveloped, RNA viruses that are very contagious from person-to-person. The most important ones are:

1. Filoviruses: Ebola (4 subtypes) and Marburg virus
2. Bunyaviruses: Crimean Congo Hemorrhagic fever virus
3. Arenaviruses: Lassa fever virus

These viruses may all also cause a hemorrhagic fever syndrome (conjunctival injection, mucosal bleeding with bleeding from gums, stomach and intestine, bloody diarrhea, petechiae (small pinpoint red hemorrhages under the skin), and ecchymoses-deep bruises under skin), but many patients only develop muscle aches, joint pains, rash or inflammation of the brain (encephalitis). These viruses are discussed in detail in **Chapter 29**.

Recommended Reviews

1. Inglesby TV, O'Toole T, Henderson DA, et al. Anthrax as a Biological Weapon JAMA 2002;287:2236–2252.
2. Henderson DA, Inglesby TV, Bartlett JG, et al. Smallpox as a Biological Weapon. JAMA 1999;281:2127–2137
3. Inglesby TV, Dennis DT, Henderson DA, et al. Plague as a Biological Weapon. JAMA 2000;283:2281–2290
4. Arnon SS, Schechter R, Inglesby TV, et al. Botulinum Toxin as a Biological Weapon. JAMA 2001;285: 1059–1070
5. Dennis DT, Inglesby TV, Henderson DA, et al. Tularemia as a Biological Weapon. JAMA 2001;285: 2763–2773
6. Borio L, Inglesby T, Peters CJ, et al. The Hemorrhagic Fever Viruses as Biological Weapons. JAMA 2002;287: 2391–2405.
7. Adelja AA, Toner E, Inglesby TV. Clinical Management of Potential Bioterrorism-Related Conditions. N Engl J Med 2015;372(10):954–62.

INDEX

NOTES

NOTES

NOTES

NOTES